Topics in Applied Physics

Volume 157

Topics in Applied Physics is a well-established series of review books, each of which presents a comprehensive survey of a selected topic within the domain of applied physics. Since 1973 it has served a broad readership across academia and industry, providing both newcomers and seasoned scholars easy but comprehensive access to the state of the art of a number of diverse research topics.

Edited and written by leading international scientists, each volume contains high-quality review contributions, extending from an introduction to the subject right up to the frontiers of contemporary research.

Topics in Applied Physics strives to provide its readership with a diverse and interdisciplinary collection of some of the most current topics across the full spectrum of applied physics research, including but not limited to:

- Quantum computation and information
- Photonics, optoelectronics and device physics
- Nanoscale science and technology
- Ultrafast physics
- Microscopy and advanced imaging
- Biomaterials and biophysics
- Liquids and soft matter
- Materials for energy
- Geophysics
- Computational physics and numerical methods
- Interdisciplinary physics and engineering

We welcome any suggestions for topics coming from the community of applied physicists, no matter what the field, and encourage prospective book editors to approach us with ideas. Potential authors who wish to submit a book proposal should contact Zach Evenson, Publishing Editor:

zachary.evenson@springer.com

Topics in Applied Physics is included in Web of Science and indexed by Scopus.

Vishnu Chauhan · Manish Kumar · Rajesh Kumar
Editors

Sustainable Lead-Free Perovskite Materials

Unique Potential of Applications in New Generation Devices

Springer

Editors
Vishnu Chauhan
Department of Physics
Ramjas College
University of Delhi
New Delhi, Delhi, India

Manish Kumar
Department of Physics
Atma Ram Sanatan Dharma College
University of Delhi
New Delhi, Delhi, India

Rajesh Kumar
University School of Basic and Applied Sciences
Guru Gobind Singh Indraprastha University
Dwarka, Gujarat, India

ISSN 0303-4216 ISSN 1437-0859 (electronic)
Topics in Applied Physics
ISBN 978-3-032-24872-5 ISBN 978-3-032-24873-2 (eBook)
https://doi.org/10.1007/978-3-032-24873-2

This Springer imprint is published by the registered company Springer Nature Switzerland AG
The registered company address is: Gewerbestrasse 11, 6330 Cham, Switzerland

This work is dedicated to our mentors and supervisors for their invaluable guidance, and to our colleagues and collaborators for their constant support. We also dedicate this book to our families, whose encouragement and patience have been a source of strength throughout this journey.

Preface

We are delighted to present this book, entitled "**Sustainable Lead Free Perovskite Materials: Unique Potential of Applications in New Generation Devices**", sum up the exceptional research and innovative advancements on lead free perovskite materials. This book provides an in-depth understanding of the synthesis, properties, and performance of emerging lead-free perovskite materials. It highlights recent advancements, innovative design strategies, and applications in areas such as solar cells, LEDs, sensors, catalysis, energy storage, and hydrogen production. Overall, the book presents a comprehensive overview of research and technological progress, emphasizing the potential of lead-free perovskites in modern sustainable applications. This book is divided into the 16 dedicated chapters:

Chapter 1 establishes the conceptual and scientific foundation of the book by introducing the structural chemistry, electrical behavior, chemical stability, and environmental advantages of lead-free perovskites. These materials are systematically classified based on both chemical composition (e.g., tin-, bismuth-, and antimony-based systems, including hybrid organic–inorganic structures) and structural dimensionality (0D, 1D, 2D, and 3D). A critical comparison with conventional lead-based perovskites highlights intrinsic material differences governing device performance. Key challenges, such as lower efficiency and stability, are also identified, thereby framing the scientific questions addressed in subsequent chapters.

Building on this foundation, Chapter 2 focuses on synthesis strategies, crystal structures, and optoelectronic properties of lead-free perovskites. Advanced fabrication techniques, including solution processing, hot-injection, and solid-state methods, are discussed for both thin films and nanocrystals. Particular attention is given to Sn-, Ge-, Bi-, and Sb-based systems, as well as double perovskites, emphasizing their environmental compatibility and structural stability. The chapter also critically examines persistent challenges such as oxidation, limited charge transport, and performance gaps relative to lead-based counterparts, offering directions for future research.

Chapter 3 addresses the development of lead-free perovskite solar cells as sustainable alternatives to conventional photovoltaic technologies. It highlights the growing interest driven by concerns over lead toxicity and the demand for eco-friendly energy solutions. The discussion is framed within circular economy principles, emphasizing safer recycling and reduced hazardous waste. Although current efficiencies remain lower than those of lead-based systems, rapid research progress is documented through a comprehensive literature review. A bibliometric analysis of 1894 Scopus-indexed publications further provides quantitative insights into research trends, emerging themes, and existing gaps.

Expanding into broader optoelectronic applications, Chapter 4 reviews advances in lead-free perovskite-based devices, including LEDs, photodetectors, and neuromorphic memristors. It evaluates alternative cations such as Sn, Ge, Bi, and Cu as substitutes for toxic Pb^{2+}, focusing on their band structures, optical properties, carrier transport, and device performance. Despite notable progress, challenges related to recombination losses, defects, and material instability are discussed, along with strategies such as compositional engineering, defect passivation, and interface optimization.

Chapter 5 explores environmental monitoring applications, particularly gas sensing using lead-free perovskites. The advantages of nanoparticle-based systems, including high surface area and enhanced sensitivity, are highlighted. Sensing mechanisms such as surface adsorption, ion exchange, and trap modulation are analyzed, alongside challenges in selectivity and long-term stability.

Chapter 6 examines photocatalysis and renewable energy conversion using lead-free perovskites. It presents a systematic discussion of synthesis routes, structural classifications, and structure–property relationships relevant to photocatalytic performance. Emphasis is placed on defect chemistry, charge separation, stability, and scalability.

Chapter 7 reviews energy storage applications, focusing on the use of lead-free perovskites in supercapacitors, batteries, and hybrid systems. The relationship between crystal structure, composition, and electrochemical performance (capacity, rate capability, and cycling stability) is thoroughly analyzed.

Chapter 8 discusses lead-free perovskite thin-film transistors (TFTs), covering device architectures, charge transport mechanisms, fabrication techniques, and performance metrics. Challenges such as mobility, stability, and large-area processing are critically examined.

Chapter 9 is dedicated to lead-free piezoelectric materials, including key systems such as KNN, BNT-BT, and BZT-BCT. Advanced processing methods, domain engineering, and phase boundary effects are discussed in relation to enhanced electromechanical properties and applications in sensors, actuators, and biomedical devices.

Chapter 10 explores hydrogen production using lead-free perovskites, covering photochemical, electrochemical, and photoelectrochemical water-splitting processes. Strategies such as doping, defect engineering, and surface modification are presented to improve efficiency and durability.

Chapter 11 focuses on emerging computing technologies, particularly memristors and neuromorphic devices based on lead-free perovskites. It discusses material selection, device design, fabrication, and operational principles for low-power and scalable computing systems.

Chapter 12 reviews thermoelectric applications, highlighting the potential of lead-free perovskites for waste heat recovery. Key parameters influencing the figure of merit (zT) and strategies for performance enhancement through material design and nanostructuring are discussed.

Chapter 13 addresses magnetic applications, emphasizing environmentally sustainable and multifunctional materials. Substitution with magnetic ions (e.g., Bi^{3+}, Mn^{2+}, Fe^{3+}) enables diverse magnetic phenomena, including ferro-, antiferro-, and multiferroicity, with applications in spintronics and data storage.

Chapter 14 investigates imaging and detection technologies, ranging from X-ray to near-infrared applications. The chapter highlights tunable optical properties, low-temperature processing, and environmental compatibility.

Chapter 15 examines antenna applications, focusing on high-dielectric lead-free perovskites such as $BaTiO_{3-}$ and BCZT-based systems for miniaturization and enhanced performance in next-generation wireless technologies.

Finally, Chapter 16 provides a forward-looking perspective on lead-free perovskite solar cells, covering material systems, synthesis, efficiency trends, commercialization challenges, and future opportunities. It integrates insights from earlier chapters, emphasizing that the advancement of lead-free perovskites is not only an environmental necessity but also a pathway toward robust, scalable, and high-performance technologies.

In summary, this book offers a comprehensive and interdisciplinary perspective on lead-free perovskites, bridging fundamental science, device engineering, and sustainability. Its systematic approach to material design, performance optimization, and application development makes it a valuable resource for researchers, engineers, and policymakers working toward environmentally sustainable perovskite technologies.

New Delhi, India Vishnu Chauhan
New Delhi, India Manish Kumar
New Delhi, India Rajesh Kumar

Acknowledgement

The authors extend the deepest gratitude to *Prof. Ajay Kumar Arora*, Principal, Ramjas College, University of Delhi, New Delhi, India and *Prof. Gyantosh Kumar Jha*, Principal, ARSD College, University of Delhi, New Delhi, India and for their unwavering support. Prof. (Dr.) Rajesh Kumar also extends his deepest gratitude to Padma Shri Prof. (Dr.) Mahesh Verma, Vice-Chancellor of Guru Gobind Singh Indraprastha University (Govt. of NCT of Delhi), New Delhi, India, for his constant support, motivation, and guidance.

Authors would also like to express the appreciation to whole Editorial team of Springer Nature, for their resolute initiate and hard work during the process of the publication of this book. Moreover, in continuation, it's our pleasure to acknowledge the support and assistance provided by the production team and their staff through their devotion, proficiency, and commitment to excellence.

We want to acknowledge the sincere thanks to our students specially, Mr. Abhishek Raj, Ms. Khushboo Gupta, Mr. Manish Kumar, Mr. Akshay, Mr. Uttkarsh and Mr. Manish Kumar, Dr. Deepika, Mr. Ashutosh Anand, Mr. Aniket, Mr. Sreejan Rayhan, Mr. Abhishek Singh Parihar, Mr. Utkarsh, Ms. Nivedita Dhiman and for their unwavering support throughout this work.

We would like to express our deepest gratitude to our family members, friends (Mr. Umesh Kumar, Mr. Devender, Mr. Hoshiyar, Mr. Shiv Kumar, Dr. Sachin Kumar, Dr. Tarun Kumar, Dr. Ganesh Gurjar, Dr. Divya Deep Yadav, Dr. Devi Puttar & special thanks to Dr. Deepika Dwivedi who inspired Dr. Vishnu Chauhan for this book), and colleagues who have provided us with unwavering support, inspiration, and understanding throughout this journey. Dr. Vishnu Chauhan is thankful to his grandparents (Late Shri Birbal & Smt. Kela Devi) and parents (Shri Ram Sharan & Smt. Maya Devi) and all family members. Dr. Manish Kumar is also thankful to his grandparents and parents (Late Shri K. L. Vats (Grandfather), Smt. Shima Devi (Grandmother), Shri Pravind Kumar (Father), Smt. Anita Sharma (Mother) and lovely wife Tanu and cute daughter Vanya and all family members. Prof. (Dr.) Rajesh Kumar expresses his heartfelt gratitude to his parents, Late Shri D.L. Sharma and Late Smt. Shakuntala Devi, as well as to his wife, Dr. Shweta Sharma, his beloved daughter

Palak Sharma, his son Reyansh Swaraj, and all family members for their constant support and encouragement.

We also want to acknowledge to one and all involved, whether directly or indirectly, in the foundation and publication of this book. Your assistance, support, and commitment have made this effort a really worthwhile and satisfying experience.

We believe that this book will be greatly useful for the undergraduate, postgraduate students and research scholars, working in the field of materials science, physics, chemistry and various interdisciplinary engineering branches along with the industry and academia.

Contents

Editors and Contributors

About the Editors

Dr. Vishnu Chauhan is an Assistant Professor at Ramjas College, University of Delhi, New Delhi, India Since 2023. He received his Ph.D. in Physics from Guru Gobind Singh Indraprastha University in 2021. Previously, he served as a Research Associate-II at IUAC, New Delhi, and held a DST Indo-German postdoctoral industrial fellowship at GSI Helmholtz Centre, Germany. His research focuses on materials science and nanotechnology, particularly the synthesis and ion beam modification of oxide and nanocomposite thin films. He specializes in characterizing structural, optical, electrical, and dielectric properties of low-dimensional oxide materials. Dr. Chauhan has published over 40 research papers and 17 book chapters in International Journals and is lead editor of Elsevier/Springer 02 books. With an h-index of 19 and over 1100 citations, he actively reviews leading journals published by Nature, Elsevier & Springer etc.. He has received several awards, including the Young Scientist Award and Best Paper recognitions.

Dr. Manish Kumar is an Associate Professor in the Department of Physics at ARSD College, University of Delhi, with over 13 years of teaching experience. He holds a Ph.D. in Physics from Banaras Hindu University and M.Tech. in Ceramic Engineering from IIT-BHU. His research spans multifunctional and green energy materials, multiferroics, and perovskite photovoltaics. He has published more than 152 research papers, 35 book chapters, one Indian patent, and authored/edited 8 Springer Nature books. Listed among the world's top 2% scientists by Stanford University (2021 and 2025), he has an h-index of 33 and over 3490 citations. Dr. Kumar has served as editor for high-impact journals like Scientific Reports and Materials Today Communications. He has supervised numerous students and completed research projects funded by the University of Delhi. He has organized several national and international conferences.

Prof. Dr. Rajesh Kumar is a professor of Physics at Guru Gobind Singh Indraprastha University, New Delhi. He earned his Ph.D. from Aligarh Muslim University and pursued postdoctoral research at Rensselaer Polytechnic Institute, USA. His research focuses on nanostructured materials for energy storage, gas sensing, and radiation applications. Prof. Kumar has published over 185 journal papers, contributed to 11 book chapters, and presented at 136+ conferences. He has received multiple awards, including the Young Scientist Award (DST) and Faculty Achievement Award (GGSIPU). Listed among the world's top 2% scientists by Stanford University, he has an h-index of 32 with over 3500 citations. He has led or participated in over 18 funded projects, supervised 11 Ph.D. Scholars and currently supervises 07 Ph.D. scholars. Prof. Kumar serves on editorial boards of international journals and has held leadership roles in scientific societies, including Vice President of the International Nuclear Track Society.

Contributors

R. Adithya Nath Photovoltaic Research Laboratory, Department of Physics, University of Kerala, Karyavattom, Thiruvananthapuram, India

Anamika Centre for Advanced Materials and Devices, School of Engineering and Technology, BML Munjal University (Hero Group), Gurugram, India

Ashutosh Anand University School of Basic and Applied Sciences, Guru Gobind Singh Indraprastha University, New Delhi, India

J. S. Anaswara Photovoltaic Research Laboratory, Department of Physics, University of Kerala, Karyavattom, Thiruvananthapuram, India

Manisha Bajpai Department of Physics, Siddharth University, Pachangawa, Uttar Pradesh, India

Taame Abraha Berhe Department of Chemical Engineering, National Taiwan University, Taipei, Taiwan;
Department of Chemistry, College of Natural and Computational Sciences, Adigrat University, Adigrat, Ethiopia

Ishika Bhullar Department of Physics, Indian Institute of Technology Roorkee, Roorkee, India

Abd. Rashid bin Mohd Yusoffe Department of Physics, Faculty of Science, University Teknologi Malaysia, Johor Bahru, Malaysia

Prachi Chaudhary Experimental Research Laboratory, Department of Physics, ARSD College, University of Delhi, New Delhi, India

Vishnu Chauhan Department of Physics, Ramjas College, University of Delhi, New Delhi, India

Chu-Chen Chueh Department of Chemical Engineering, National Taiwan University, Taipei, Taiwan

Srimanta Das Department of Physics, Indian Institute of Technology Roorkee, Roorkee, India

Amare Aregahegn Dubale Center for Material Science and Engineering, College of Natural and Computational Sciences, Addis Ababa University, Addis Ababa, Ethiopia

Deepika Dwivedi Dr. B.R. Ambedkar Center for Biomedical Research (ACBR), University of Delhi, New Delhi, India

Rohit Kumar Singh Gautam Department of Mechanical Engineering, Teerthanker Mahaveer University, Moradabad, India

Khushboo Gupta Department of Physics, School of Sciences, IILM University, Greater Noida, Uttar Pradesh, India

Rekha Gupta Department of Physics, Noida Institute of Engineering and Technology, Greater Noida, Affiliated to Dr. A. P. J. Abdul Kalam Technical University, Lucknow, Uttar Pradesh, India

Hardik Department of Applied Physics, Delhi Technological University, Delhi, India

R. Jayakrishnan Photovoltaic Research Laboratory, Department of Physics, University of Kerala, Karyavattom, Thiruvananthapuram, India

Manjunatha Kempasiddaiah Department of Physics, Indian Institute of Technology Roorkee, Roorkee, India

Saarthak Kharbanda Amity Institute of Nanotechnology, Amity University Haryana, Gurugram, India

Yogesh Khrolia Centre for Advanced Materials and Devices, School of Engineering and Technology, BML Munjal University (Hero Group), Gurugram, India

Abhishek Kumar Department of Applied Physics, Delhi Technological University, Delhi, India

Anirudh Kumar Biomaterials and Sensor Laboratory, Department of Physics, Ch. Charan Singh University, Meerut Campus, Meerut, Uttar Pradesh, India

Arvind Kumar School of Physical Sciences, Jawaharlal Nehru University, New Delhi, India

Deepak Kumar Biomaterials and Sensor Laboratory, Department of Physics, Ch. Charan Singh University, Meerut Campus, Meerut, Uttar Pradesh, India

Manish Kumar Experimental Research Laboratory, Department of Physics, Atma Ram Sanatan Dharma College, University of Delhi, New Delhi, India

Narendra Kumar Department of Chemical Engineering, Indian Institute of Technology Roorkee, Roorkee, Uttarakhand, India

Rajesh Kumar University School of Basic and Applied Sciences, Guru Gobind Singh Indraprastha University, New Delhi, India

Tarun Kumar Department of Physics, Ramjas College, University of Delhi, New Delhi, India

Maning Liu Centre for Analysis and Synthesis, Department of Chemistry, Lund University, Lund, Sweden;
Wallenberg Initiative Materials Science for Sustainability, Department of Chemistry, Lund University, Lund, Sweden;
NanoLund, Lund University, Lund, Sweden

Aman Mahajan Department of Physics, Guru Nanak Dev University, Amritsar, India

Parika Mahajan Department of Physics, Guru Nanak Dev University, Amritsar, India

Suyash Mishra Department of Applied Physics, Delhi Technological University, Delhi, India

Supratim Mitra Department of Physical Sciences, Banasthali Vidyapith, Banasthali, Rajasthan, India

Chia-Hsun Nieh Department of Chemical Engineering, National Taiwan University, Taipei, Taiwan

Niwesh Ojha Department of Chemical Engineering, Indian Institute of Technology Roorkee, Roorkee, Uttarakhand, India

C. K. Pandey Department of Physics, S.N.S. College, Motihari, BRABU, Muzaffarpur, Bihar, India

Kamal Kishore Pant Department of Chemical Engineering, Indian Institute of Technology Roorkee, Roorkee, Uttarakhand, India

Namrata Pant Department of Physics, Indian Institute of Technology Roorkee, Roorkee, India

Brijmohan Prajapati Department of Physics, Deshbandhu College, University of Delhi, New Delhi, India

Sandeep Kumar Pundir Department of Physics, University of Lucknow, Lucknow, Uttar Pradesh, India

Sunardi Rahman Centre for Analysis and Synthesis, Department of Chemistry, Lund University, Lund, Sweden;
Wallenberg Initiative Materials Science for Sustainability, Department of Chemistry, Lund University, Lund, Sweden;
NanoLund, Lund University, Lund, Sweden

Aruna Raj Photovoltaic Research Laboratory, Department of Physics, University of Kerala, Karyavattom, Thiruvananthapuram, India

Abhimanyu Singh Rana Centre for Advanced Materials and Devices, School of Engineering and Technology, BML Munjal University (Hero Group), Gurugram, India

Omwati Rana Department of Physics, Daulat Ram College, University of Delhi, Delhi, India

Soumitra Satapathi Department of Physics, Indian Institute of Technology Roorkee, Roorkee, India;
Center for Sustainable Energy, Indian Institute of Technology Roorkee, Roorkee, India

Manisha Sharma Biomaterials and Sensor Laboratory, Department of Physics, Ch. Charan Singh University, Meerut Campus, Meerut, Uttar Pradesh, India

Sanjeev Kumar Sharma Biomaterials and Sensor Laboratory, Department of Physics, Ch. Charan Singh University, Meerut Campus, Meerut, Uttar Pradesh, India

Savita Sharma Department of Physics, Kalindi College, University of Delhi, Delhi, India

Manvandra Kumar Singh Department of Mechanical Engineering, Graphic Era (Deemed to be University), Dehradun, Uttarakhand, India

Abhishek Kumar Srivastava Department of Physics, Siddharth University, Pachangawa, Uttar Pradesh, India

Shilpi Srivastava Department of Chemistry, Siddharth University, Pachangawa, Uttar Pradesh, India

Pratigya Tripathi Department of Chemistry, Siddharth University, Pachangawa, Uttar Pradesh, India

Vivek Mani Tripathi Department of Mechanical Engineering, Uttar Pradesh Textile Technology Institute, Kanpur, India

Anurag Tyagi Department of Physics, Noida Institute of Engineering and Technology, Greater Noida, Affiliated to Dr. A. P. J. Abdul Kalam Technical University, Lucknow, Uttar Pradesh, India

Komal Verma School of Physical Sciences, Jawaharlal Nehru University, New Delhi, India

Upkar Kumar Verma Department of Physics, B. S. N. V. P. G. College, Lucknow, Uttar Pradesh, India

Vinay Pratap Verma Department of Physics, University of Lucknow, Lucknow, Uttar Pradesh, India

Divya Deep Yadav Department of Physics, Ramjas College, University of Delhi, New Delhi, India

Chung-Yen Yang Department of Chemical Engineering, National Taiwan University, Taipei, Taiwan

Yukta Centre for Analysis and Synthesis, Department of Chemistry, Lund University, Lund, Sweden;
Wallenberg Initiative Materials Science for Sustainability, Department of Chemistry, Lund University, Lund, Sweden;
NanoLund, Lund University, Lund, Sweden

Qinqi Zhou Centre for Analysis and Synthesis, Department of Chemistry, Lund University, Lund, Sweden;

Wallenberg Initiative Materials Science for Sustainability, Department of Chemistry, Lund University, Lund, Sweden;
NanoLund, Lund University, Lund, Sweden

Chapter 1
Introduction and Classification of Lead-Free Perovskite Materials

Suyash Mishra, Abhishek Kumar, Hardik, and Savita Sharma

Abstract As worries about the environmental effects and health hazards of lead continue to develop, lead-free perovskite materials are becoming more and more popular as safer substitutes for conventional lead-based perovskites. This chapter study the fundamental structure, chemical stability, electrical behaviour, and ecological advantages of specific materials to determine what makes them useful. These materials will be arranged primarily according to their chemical composition (including tin, bismuth, antimony, and hybrid organic–inorganic) and structural dimensions (from 0 to 3D configurations). Understanding how lead-free perovskites differ from their conventional counterparts is crucial. These differences directly affect their performance in real-world applications, including solar cells, LED displays, and other sensors. Although recent studies have demonstrated encouraging advancements in lead-free perovskites, several obstacles still need to be overcome before they can be used as viable substitutes for lead-based materials. This chapter lays the groundwork for comprehending these remarkable materials and prepares the reader for the following chapter in depth examination of their particular applications. The aim is elucidating the clear picture of the current state of the discipline and the research challenges.

Keywords Lead-free perovskite materials · Lead-based perovskites · Environmental effects · Health hazards · Chemical stability · Electrical behaviour · Chemical composition · Tin · Bismuth · Antimony · Hybrid organic–inorganic · Structural dimensions · 0D–3D configurations · Solar cells · LED displays · Sensors · Applications · Challenges

S. Mishra · A. Kumar · Hardik
Department of Applied Physics, Delhi Technological University, Delhi, India

S. Sharma (✉)
Department of Physics, Kalindi College, University of Delhi, Delhi, India
e-mail: savitasharma@kalindi.du.ac.in

V. Chauhan et al. (eds.), *Sustainable Lead-Free Perovskite Materials*, Topics in Applied Physics 157, https://doi.org/10.1007/978-3-032-24873-2_1

Abbreviation

0D	Zero-dimensional
1D	One-dimensional
2D	Two-dimensional
3D	Three-dimensional
ABX_3	Archetypal perovskite crystal structure formula
ACN	Acetonitrile
AIMD	Ab initio molecular dynamics
ALD	Atomic layer deposition
BA	Butylammonium
BIPV	Building-integrated photovoltaics
CAGR	Compound annual growth rate
CBM	Conduction band minimum
CERCLA	Comprehensive environmental response, compensation, and liability act
CSTL	Charge state transition level
CVD	Chemical vapor deposition
DCB	1,2-Dichlorobenzene
DFE	Defect formation energy
DFT	Density functional theory
DJ	Dion-Jacobson phase structure
DMA	Dimethylammonium
DMF	Dimethylformamide
DMSO	Dimethyl sulfoxide
DOS	Density of states
EA	Ethyl acetate
EDTA	Ethylenediaminetetraacetic acid
EL	Electroluminescence
EQE	External quantum efficiency
ESG	Environmental, social, and governance
ETL	Electron transport layer
EVA	Ethylene vinyl acetate
FA	Formamidinium ($CH(NH_2)_2^+$)
FF	Fill factor
GBL	γ-Butyrolactone
GGA	Generalized gradient approximation
GRAS	Generally recognized as safe
HDP	Halide double perovskite
HEPA	High-efficiency particulate air
HCVD	Hybrid chemical vapor deposition
HSE06	Heyd-scuseria-ernzerhof hybrid functional
HTL	Hole transport layer
HTM	Hole transport material
ICSD	Inorganic crystal structure database

IoT	Internet of things
ITO	Indium tin oxide
JSC	Short-circuit current density
LCA	Life cycle assessment
LCOE	Levelized cost of electricity
LDA	Local density approximation
LED	Light-emitting diode
MA	Methylammonium ($CH_3NH_3^+$)
MOCVD	Metal-organic chemical vapor deposition
MSP	Module selling price
NMR	Nuclear magnetic resonance
OLED	Organic light-emitting diode
PCE	Power conversion efficiency
PDF	Powder diffraction file
PEA	Phenylethyl ammonium
PEDOT	Poly(3,4-ethylenedioxythiophene)
PL	Photoluminescence
PLQY	Photoluminescence quantum yield
PSC	Perovskite solar cell
PV	Photovoltaic
QD	Quantum dot
QR	Quantum rod
RCRA	Resource conservation and recovery act
RH	Relative humidity
RoHS	Restriction of hazardous substances
RP	Ruddlesden-popper phase structure
SCLC	Space charge limited current
SOC	Spin-orbit coupling
STE	Self-trapped exciton
TBAI	Tetrabutylammonium iodide
TCLP	Toxicity characteristic leaching procedure
TDDFT	Time-dependent density functional theory
THF	Tetrahydrofuran
THP	Tin halide perovskite
TWA	Time-weighted average
UV	Ultraviolet
VBM	Valence band maximum
VOC	Open-circuit voltage
XRD	X-ray diffraction

1.1 Introduction

Lead contamination poses a danger to human health and the environment through various forms of exposure. The World Health Organization does not define a safe dose of lead, but the long-term increase of lead will produce irreparable neurological effects, kidney failure, heart problems and developmental illnesses [115]. Exposure to lead sources during perovskite research lab studies is associated with inhalation of particulates, dermal exposure, and accidental ingestion; therefore, highly sensitive measures must be followed, such as special ventilation, personal protective gear, and decontamination of materials using chelating agents [43]. Halide perovskites of lead are highly soluble in water solutions, with studies of leaching indicating concentrations exceeding those allowed by regulators under simulated end-of-life conditions. According to the Toxicity Characteristic Leaching Procedure (TCLP) tests, lead perovskite modules have the capacity to leak 10–50 mg/L of lead, which is significantly higher than the EPA hazardous waste limit of 5 mg/L [48]. The Institute of Health Metrics and Evaluation estimates that exposure to lead resulted in 1 million deaths and 24.4 million disability-adjusted life years worldwide in 2017 [87]. Most recently, epidemiological studies have implicated even low-level exposure to lead as causing cognitive impairment and behavioural disorders in children [41].

To ensure the safety of lead-free materials, the laboratory safety measures created for perovskites focus on containment measures, such as the use of specific high-efficiency particulate air (HEPA) filters. Additionally, the purpose of decontamination procedures using a chelating agent, e.g., EDTA-2Na [65], should be mentioned. To address these issues, the global law establishes a framework for utilising lead-free electronic resources. The Restriction of Hazardous Substances (RoHS) Directive 2011/65/EU sets the limit at 0.1%. By weight of lead in electrical and electronic equipment, with exemptions only to fixed solar installations and to certain industrial use cases [63]. In the ongoing assessment by the directive, the procedure outlines additional restrictions on the use of lead. Apple, Samsung, and Sony, the electronics makers, have implemented lead-free policies in PVs that address consumer needs by prioritising environmental, social, and governance values. To initiate the process of removing lead-free materials, electronic manufacturers are allocated $500 million in research financing per year across the sector [118]. Concurrently with these financing programs are life cycle assessment (LCA) studies, which indicate that the perovskite module poses considerable environmental pressures during both its operational period and at the end of its operational duration [67]. Accelerated ageing studies indicate that lead perovskite modules are capable of discharging 15–30% of the lead in the modules under simulated hail damage, thermal image conditions, and UV exposure conditions [126]. In a cradle-to-grave comparison, lead-free systems have 60–80% reductions in human toxicity potential, 70–90% reductions in freshwater ecotoxicity, and 45–65% reductions in terrestrial ecotoxicity, compared to lead-based alternatives [49]. Such benefits, however, will have to be offset against the possible increases in processing energy needs and material usage, especially for materials that require special synthesis conditions or inert atmospheres, which emerging lead-free compositions may demand. Economic evaluation of end-of-life management reveals that the overall cost of ownership for lead-free systems is 15–25% lower than that of lead systems, considering the costs of disposal, regulation, and liability [90]. This economic benefit would be more pronounced because the cost of environmental controls could be stricter, and the costs of discarding the technology could be high throughout its life cycle.

In addition to the environmental and economic aspects, the shift towards lead-free perovskite materials is motivated by the more abstract concept of technological sustainability and long-term stability of the devices. Although perovskites have proven to exhibit superb optoelectronic performance using lead, a significant challenge remains their natural chemical instability when conditions of operation are considered more realistic. Devices exposed to moisture, oxygen, thermal, and ultraviolet radiation accelerate the degradation paths of phase separation and ion movement, leading to performance and structural decline. This chemical frailty and the toxicity issue have been the driving factors for extensive research into other compositions that can be more tolerant of moisture and have lower ion mobility [5, 7, 65]. Otherwise, most lead-free perovskite systems incorporate different cations of higher valence or ordered vacancy groups, which lower ion movement and destabilize the system in the presence of water. Examples of vacancy-ordered double perovskites, such as $Cs_2AgBiBr_6$, are more stable than standard lead halides over time [76]. As operational lifetime has become an important factor controlling commercial deployment, these stability advantages increasingly favor lead-free systems in material selection.

From a materials-design perspective, the lead-free perovskites thus constitute a step away from 'pure' efficiency for a more comprehensive balance of optoelectronic functionality, environmental compatibility and structural stability for manufacturing. Unlike their lead-based counterparts, lead-free compositions generally require higher control of stoichiometry, crystallization kinetics and microstructure for them to deliver competitive performance [7, 24, 76]. This has widened the range of research into perovskites, leading to development of new crystal chemistries, dimensional architectures and defect management. In parallel, lead-free perovskite research has spread beyond its traditional domains of application in photovoltaics to other domains, e.g. light emitting diodes, photodetectors, radiation sensors, optoelectronics in general, where often stability, low toxicity and controlled emission characteristics are more important than slight differences in power conversion efficiency [7, 38]. In these contexts, features such as strong excitonic effects and self-trapped emission, usually an unfavourable step with respect to the conversion of solar energy into electricity, can instead be used to increase device robustness and functionality.

Industrial scalability further makes the case for lead free perovskites. Commercial manufacturing requires compatibility with the existing processing infrastructure, reproducibility over large areas, and compliance with strict environmental and safety regulations, criteria that are more easily met by lead-free compositions and these compositions simplify waste management and reduce regulatory burdens associated with transport, storage and end-of-life disposal. These advantages result in lower capital and operational costs across the device lifecycle, which becomes increasingly important as production scales up. In addition, the global interest in the criticality of

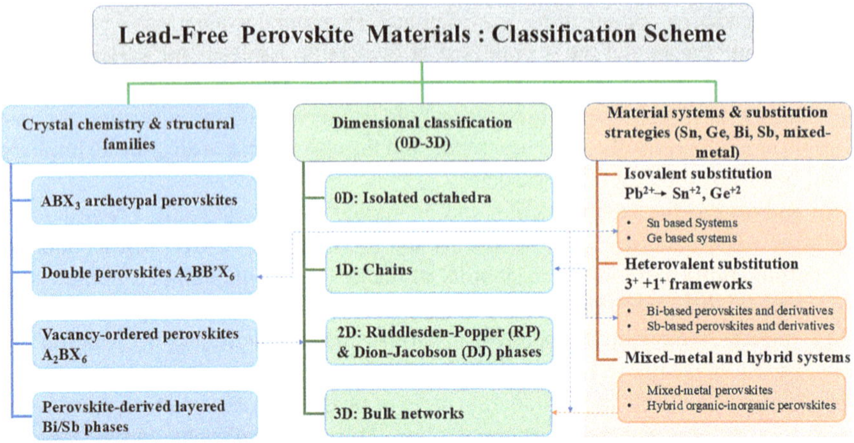

Fig. 1.1 Schematic classification scheme of lead-free perovskite materials, illustrating crystal-chemical families, dimensional architectures (0D–3D), and material substitution strategies including isovalent, heterovalent, and mixed-metal systems

materials and supply chain resiliency has also provided increased impetus to reduce dependence on toxic or highly regulated elements. Lead-free perovskites have the potential to diversify material inputs. While alternatives to lead-free perovskites have their drawbacks in terms of precursor availability or synthesis complexity, ongoing improvements in materials chemistry and processing are gradually eliminating these hurdles [7, 76]. Collectively, these considerations demonstrate that the development of lead-free perovskite materials is not merely a response to regulatory demands but a strategic advancement towards more environmentally responsible, longer-lasting, and more economically viable optoelectronic technologies. The literature has developed into numerous reports that have established lead-free perovskites as a materials class with its own distinct advantages and limitations, requiring specific design strategies, rather than merely substituting lead-based analogues [5, 7, 38]. A deep understanding of their material chemistry, defect physics, and structure–property relationships is therefore critical to guiding future innovation. A deep understanding of their material chemistry, defect physics and structure–property relationships is therefore critical to guide future innovation. To visually summarize this design framework, Fig. 1.1 presents a classification scheme for lead-free perovskite materials.

1.2 Synthesis and Processing

The successful development of lead-free perovskite materials requires sophisticated synthesis and processing methods that address the unique challenges associated with alternative metal cations while achieving the film quality and scalability necessary for commercial applications.

1.2.1 Solution-Based Processing Methods

The combination of low processing temperatures, the ability to use flexible substrates, and scalability makes solution-based methods the primary choice for synthesising lead-free perovskite.

One-step deposition strategies, which rely on the direct crystallisation of precursor solutions, are the most widely used pathways to laboratory-scale perovskite synthesis, with optimisation strategies tailored to each material system and processing needs. The approaches are the basis of solution-based processing of perovskite, systematic increase in the solvent selection, concentration of the precursor, atmosphere and thermal control of the solvent to obtain uniform, high-quality thin films [20]. Polar aprotic solvents, including dimethylformamide (DMF), dimethyl sulfoxide (DMSO), and γ-butyrolactone (GBL) are utilise because of their strong solubility to perovskite precursors, where DMF (donor number 26.6) and DMSO (donor number 29.8) show a high affinity to metal halide ions that allow a homogeneous solution, so that it can form controlled crystallisation [84, 108]. The standard precursor concentration is between 0.8 and 1.5 M to control film coverage and process stability. An inert gas environment is generally necessary to oxidise Botox tin-based systems during spin coating.

Additionally, heating the substrate to temperatures of 60–100 °C increases crystallisation rates by enhancing motion and reducing supersaturation, resulting in better film uniformity and lower defect density [69, 75]. Antisolvent engineering also enables the accurate control of crystallisation dynamics during spin coating, where antisolvents (e.g., chlorobenzene, toluene, and ethyl acetate) are selected based on their miscibility with host solvents and the low solubility of perovskite precursors [106, 128]. Time, the use of an antisolvent, and time are the critical factors in determining nucleation density and grain growth. The temperature of the substrates and the antisolvents used is also critical in determining the crystallisation pathways. It demonstrates that gas-pulse methods with a nitrogen flow can further improve the uniformity of the film, as they facilitate solvent extraction and homogeneous nucleation throughout the substrate surface [9, 72]. Interestingly, long application times of antisolvents, approximately 1.3 s, have been shown to have a higher solvent extraction rate than short application times of 0.18 s, in the case of type II antisolvents like ethyl acetate, which offers an optimal balance between the miscibility of the solvent and the solubility of its precursors [77].

Two-step sequential deposition systems offer improved crystallisation control by decoupling the template development of metal halides from the introduction of organic cations, thereby allowing for more precise control of film morphology and phase purity. In binary halide precursor methods, uniform films of PbI_2 or SnI_2 are spin-coated or deposited by thermal deposition, followed by an organised conversion reaction that involves the application of organic halide solutions, which penetrate the template and transform it into a perovskite phase [14, 60]. Optimization of reaction temperature and duration is essential to achieve a complete conversion reaction with minimal phase segregation. The resultant films display large grains and a low density of pinholes compared to films produced by single-step deposition [33, 77]. The Interdiffusion-based methods further generalise this methodology by proposing that sequential deposition of A-site and B-site precursor layers is employed, and low-temperature thermal activation enables the diffusion between layers to be controlled and stoichiometrically tuned by adjusting the layer thickness [15]. Solution-processing approaches to overcome unique fabrications have been advanced, particularly in the case of lead-free perovskite systems, where small processing dimensions require a higher degree of control over the crystallisation dynamics. Microwave-assisted synthesis can, in turn, take advantage of rapid and homogeneous heating, enhanced crystallinity due to limited thermal gradients, and the use of solvents such as DMF and DMSO, thus enabling efficient coupling to microwaves and scalability with continuous-flow processing ideas [92]. Ultrasonic processing advances solution chemistry by inducing nucleation through cavitation, enhancing precursor dissolution, and driving acoustic streaming that eliminates bubbles and ensures crystal homogeneity [1, 123]. Together, these high-order processing methods demonstrate outstanding potential for the development of lead-free perovskites that balance high-performance films with low-complexity processes for scalable manufacturing.

1.2.2 Vapor-Phase Deposition Methods

Vapour-phase synthesis and processing methods offer better compositional control and material purity than solution-based methods, albeit at the cost of higher processing temperatures and the need for specialised equipment. They are therefore significant for industrial production [47]. Among these techniques, thermal evaporation offers outstanding control, allowing for the sequential or simultaneous co-evaporation of constituent materials. Single-source evaporation utilises pre-synthesised perovskite powders as the evaporation sources, allowing for near-congruent sublimation and stoichiometric film growth at substrate temperatures that generally range from 100 to 200 °C. However, single-source evaporation has issues due to the thermal decomposition of the evaporation sources and non-congruent sublimation of complex compositions [18, 133].

In contrast, multi-source co-evaporation allows for the independent control of individual components using quartz crystal microbalances for real-time flux measurements, with separate heating systems for optimal stoichiometric correction, and with spatial composition gradients for systematic investigations of structure–property correlations. However, this approach introduces added complexity to the process and requires sophisticated control architectures [47, 89]. Chemical vapour deposition (CVD) further extends the vapour phase processing capabilities in terms of gas phase processing, which allows for highly uniform and conformal films as well. Metal–organic CVD (MOCVD) utilises volatile organometallic precursors with carefully matched decomposition temperatures, optimised gas flow mixing, and controlled carrier gas transport to achieve excellent uniformity over large areas, making it suitable for industrial deployment [31, 93]. Plasma-enhanced CVD involves introducing radio-frequency or microwave excitation to induce reactive radical species, thereby accomplishing crystallisation at lower temperatures. However, the effects of ion bombardment require careful tuning to achieve proper crystallisation without film degradation [80, 88]. New successes in sequential CVD have demonstrated the realisation of two-dimensional Sn–Pb mixed-halide perovskites by converting $SnCl_2/PbCl_2$ building blocks in phenylethylamine iodide atmospheres at temperatures ranging from 80 to 120 °C. At the ideal value of 100 °C, the resulting films were strongly crystalline and included grains with a size of up to around 5 μm [68].

Hybrid approaches to vapour solution processing offer further versatility in fabrication strategies by integrating the precision of vapour processes with the adaptability of solution processing. Hybrid chemical vapour deposition (HCVD) combines initial inorganic halide deposition and subsequent organic cation infiltration, using temperature cycling to optimise the formation of phases and provide independent compositional control over both organic and inorganic components [54, 105]. This strategy offers significant scalability benefits, including compatibility with roll-to-roll manufacturing methods. Recent developments in rapid HCVD (RHCVD) have cut processing times to <10 min and effectively suppress the hysteresis effects, achieving processing times comparable to those obtained with the solution coating technique, allowing for the fabrication of large-area modules with estimated module efficiencies of 12.3% for device areas of 22.4 cm^2 industrial implementations of vapour-phase techniques [89]. There has been significant progress (Von Ardenne adapted linear thermal evaporation techniques for perovskite manufacturing) with coating widths of >1000 mm and continuous campaign lengths of at least 10 days, showing that this approach can provide deposition with high-throughput, industrial-scale deposition capabilities [2]. These systems also enable the conformal coating of microtextured surfaces, a key requirement for silicon-based tandem architectures. Vapour-phase techniques also have unique advantages, including the elimination of toxic solvents, control over thickness, excellent batch-to-batch reproducibility, and easy integration with existing semiconductor manufacturing lines [2, 93]. Recent demonstrations include the single-source thermal evaporation of two-dimensional perovskites, such as $(BA)_2(MA)_3Pb_4I_3$, with power conversion efficiencies ranging from 2.54 to 4.67%, enhanced by solvent annealing and by showerhead-based CVD of lead-free methylammonium bismuth iodide, resulting in large-area films with crystallites up to 500 nm [2]. Moreover, co-evaporation processes have been brought to high-acceleration rates with sixfold increases in deposition speed whilst preserving film quality and high-power conversion efficiencies, without post-annealing requirements, highly simplifying the manufacturing workflows [85]. Collectively, these progresses make vapour-phase processing a leading pathway for industrial perovskite device fabrication that is effectively bridging the gap between laboratory-scale research and large-scale commercial production with proven scalability, improved material purity and compatibility with existing industrial infrastructure.

1.2.3 Scalable Manufacturing Approaches

The requirement for a transition from laboratory-scale synthesis to industrial manufacturing necessitates the design of scalable processing routes that can maintain the material's quality while meeting cost targets that align with the commercial context [22]. Slot-die coating has been a leading method among continuous large-area fabrication methods, including optimisation of process parameters to create a uniform meniscus using ink formulation and rheological regulation, coating speed of 1–10 m/min to balance throughput and film quality, dynamic substrate heating profiles to maximise crystallisation kinetics and the maintenance of inert atmospheres with continuous coating [107, 112]. Industrial slot-die processes combine high-speed optical inspection with surface morphology analysis. Real-time thickness uniformity coating can involve laser interferometry or beta-ray transmission analysis, while in situ X-ray crystallinity monitoring and structural integrity during coating require a high-quality in-line control system [124]. Simultaneously, screen printing provides an economical pathway to thick-film perovskite layers by attentively designing paste formulations that can result in thixotropic behaviour to define edges, incorporation of sub-micron-sized precursor particles to prevent nozzle or mesh clogging, organic binders to allow subsequent thermal processing and use of high-boiling-point solvents to slow drying [21, 86]. Systematic optimisation of printing parameters, such as screen mesh of 200–400, in order to fit resolution and ink transfer, squeegee pressure control to have full ink delivery, printing rate of 50–200 mm/s to balance productivity and film quality, as well as registration accuracy of <25 mm in multilayer device architectures, is required [86, 99]. Recent proclamations of mechanically durable, screen-printed, flexible perovskite solar cells highlight the practicality of this approach, achieving power transformation efficiencies of 18.6% with siloxane coupling agent interface adjustment. The devices retain 90% of their original performance after 6000 bending cycles at a radius of 5 mm [21].

Roll-to-roll (R2R) processing also allows high-throughput production of flexible perovskite gadgets through incorporation of improved substrate management systems that uphold constant web tension to avert atrophy, plasma or corona treatments to boost surface vitality, high-accuracy registration to line up multilayer assemblies and in-line defect detection with web-mark capabilities to manage yields [3]. R2R platforms integrate processes that include multi-station coating and sequential layer deposition, intermediate thermal or chemical treatment, a zone heating system, and tight local temperature control. They also feature closed-loop solvent management to ensure environmental compliance and statistical process control frameworks that facilitate product traceability and production consistency [44, 58]. Using perovskite solar cells, 100 m-long rolls of fully roll-to-roll processed material achieved direct efficiencies of 16.7% when R2R-coated with SnO_X layers, 13.8% when R2R-coated with P3HT hole transport layers, and fully R2R-processed [134]. Sophisticated ink formulation techniques also enable a higher level of industrial compatibility by solvent engineering techniques employing environmentally friendly solvents, like tert-butanol, due to its wide processing range, rheological regulatability to produce viscosity levels ideal in meniscus stabilisation, and thermal control measures allowing premature drying whilst maintaining total solvent removal [55]. It is also worth noting that slot-die coating of triple-halide perovskites, with or without nitrogen-knife drying methods, has reached efficiencies of up to 25.2% in tandem systems, no antisolvents are needed, and film uniformity is achievable over extensive layers [82]. Researchers use statistical process control methodologies in quality assurance to continuously monitor coating thickness, crystalline quality, and electrical performance across production batches. They detect deviations in the process and take corrective steps to maintain device efficiency (20). Advanced manufacturing scalability tests also reflect business viability, as well as cost–benefit calculations of possible production expenses of <$0.10/W, energy payback periods of <6 months, and material use efficiencies of over 90% based on optimised coating plans [83]. Combination with current manufacturing platforms, taking advantage of available roll-coating tools, quality systems, and photovoltaic chains of supply, renders capital investment requirements significantly lower, and scalable methods of printing and coating perovskite photovoltaics come into view as a realistic route to delivering perovskite PVs out of the laboratory to the commercial landscape [13, 110].

1.3 Properties

1.3.1 Structural Features

The versatility of the lead-free perovskite structure is the basis for customising its optoelectronic characteristics and its stability in the environment. These materials can be categorised systematically based on crystal symmetry, octahedral tilting behaviour, and dimensional architecture, as these structural features dictate electronic band dispersion, exciton dynamics, charge-carrier transport, and vulnerability to degradation processes [71]. Higher-symmetry ideal perovskites have lattices with high symmetry, typically cubic in type, composed of a network of pure BX_6 octahedra sharing corners, with a Pm3M structure. This form is typically limited to high temperatures or situations where the relationships between ionic sizes allow for near-ideal tolerance factors. With decreasing symmetry, structural distortions appear as octahedral tilting and A-site cation displacement, which result in tetragonal phases, like P4/mmm or I4/mcm and orthorhombic phases, like Pnma or Pbnm, which are usually preferred at a lower temperature or in compositions that have smaller A-site cations [29]. These symmetry changes are not merely crystallographic curiosities, they directly alter B-X-B bond angles, decrease electronic bandwidths, modify density-of-states distributions, and enhance electron–phonon interactions, which have a direct impact on exciton binding energies and charge-carrier mobilities, and therefore on device performance and stability.

The tilting in lead-free perovskites is explained by the Glazer notation, which expresses the magnitude of tilting as well as the phase of the tilting with respect to crystallographic axes. Representative tilt systems include $a^0a^0a^0$ for ideal cubic frameworks, $a^0a^0c^+$ for tetragonal structures exhibiting single-axis in-phase tilting, and $a^-a^-c^+$ for orthorhombic lattices combining out-of-phase and in-phase rotations [50]. The patterns of tilt space amplitude oscillation of orbital overlap between metal cation s/p and halide p orbital bands modulate carrier mass effective masses and the presence or absence of direct versus indirect optical transitions. Generally, this is evident in halide double perovskites, where the rules of parity selection and momentum-space extrema govern the absorption edges and excitonic recombination paths. Outside of local octahedrally distorted structures, a more general dimensional geometry of the perovskite lattice would be another lever to facilitate the design of electronic coupling and excitonic behaviour. Dimensional architecture has provided an additional and more substantial design approach in lead-free perovskites. Three-dimensional corner-sharing structures are fully three-dimensional, maximising both electronic coupling and carrier diffusion lengths, which is desirable for photovoltaic applications. On the contrary, reduced-dimensional derivatives-two, one and zero-dimensional forms-bring about the dielectric and quantum confinement effects, which substantially enhance the exciton binding energy and allow control over emission colour and spectral purity. Cation-ordered and vacancy-ordered perovskites, like $Cs_2AgBiBr_6$ and $Cs_2In_2X_6$, have order to stabilise the lattice, with cation arrangement manipulation and electronic dimensionality alteration in them to produce strong optoelectronic behaviour in extreme environmental models [11, 79]. A comparative summary of representative vacancy-ordered perovskites (A_2BX_6), highlighting their compositions, band gaps, electronic dimensionalities, and stability characteristics, is provided in Table 1.1.

1.3.2 Electronic Properties

The electronic properties of lead-free perovskites are intrinsically linked to their structural motifs and charge-compensation strategies, which broadly fall into isovalent and heterovalent substitution pathways. Isovalent substitution involves the replacement of Pb^{2+} with another divalent cation while preserving the three-dimensional ABX_3 framework. Among potential candidates, Sn^{2+} is the closest analogue to Pb^{2+} due to its comparable ionic radius ($Sn^{2+} \approx 1.35$ Å versus $Pb^{2+} \approx 1.49$ Å) and ns^2 lone-pair electronic configuration, which preserves favourable antibonding interactions at the valence and conduction band edges [16]. As a result, tin-based perovskites typically exhibit direct band gaps in the range of 1.2–1.4 eV, high absorption coefficients exceeding 10^5 cm^{-1}, and carrier mobilities above 100 cm^2 V^{-1} s^{-1} in optimized films and single crystals, while remaining compatible with established solution-processing routes [95]. However, the practical deployment of Sn-based perovskites is hindered by the propensity of Sn^{2+} to oxidize into Sn^{4+} upon

Table 1.1 Key properties of vacancy-ordered perovskites (A_2BX_6)

Composition	Band gap (eV)	Electronic dimensionality	Stability and notes
Cs_2SnI_6	1.3–1.5	Quasi-0D (isolated octahedra)	Excellent air stability (Sn^{4+}); weak absorption edge
Cs_2TiBr_6	~1.8 *DFT*	Quasi-0D	Superior chemical/thermal stability; high-T synthesis
Cs_2ZrCl_6	3.4 *DFT*	Quasi-0D	Predicted stability; experimental work ongoing
Cs_2HfBr_6	2.9 *DFT*	Quasi-0D	Theoretical direct gaps; synthesis under study
Cs_2PdCl_6	2.0 (Cl) *DFT*	Quasi-0D	Predicted strong absorption; experimental validation pending
Cs_2GeI_6	1.6 (I) *DFT*	Quasi-0D	Mixed-dimensional behavior; stability under investigation
Cs_2SnBr_6	2.2 (Br)	Quasi-0D	Moderate gap; high humidity stability
Cs_2TiCl_6	2.4 (Cl) *DFT*	Quasi-0D	Predicted high exciton binding; synthesis challenges
Cs_2ZrBr_6	2.8 (Br) *DFT*	Quasi-0D	Theoretical optical performance; requires inert processing
Cs_2HfCl_6	3.1 (Cl) *DFT*	Quasi-0D	High-energy UV applications predicted; stability high

exposure to air or moisture, leading to p-type self-doping, increased defect densities, and rapid performance degradation. Mitigation strategies, including reducing-atmosphere processing, incorporation of antioxidant additives, and robust encapsulation layers, have been developed to suppress oxidation, while in situ spectroscopic techniques enable real-time monitoring of oxidation states during film formation and device operation [8].

Germanium-based perovskites provide an alternative isovalent substitution pathway with improved oxidation resistance; however, their smaller ionic radius ($\approx$0.73 Å) induces stronger lattice distortions and less dispersive valence bands, resulting in wider band gaps typically ranging from 1.8 to 2.2 eV [114]. Figure 1.2 schematically illustrates the crystal structure of a germanium-based perovskite, highlighting the corner-sharing octahedral framework and associated lattice distortions. These wider band gaps, combined with indirect optical transitions and higher effective carrier masses (0.3–0.6 m_0), limit optical absorption and charge transport, thereby constraining photovoltaic efficiency [32]. Additional challenges include precursor instability, limited availability of GeI_2, and sensitivity to synthesis temperature and atmosphere, which collectively hinder large-scale processing.

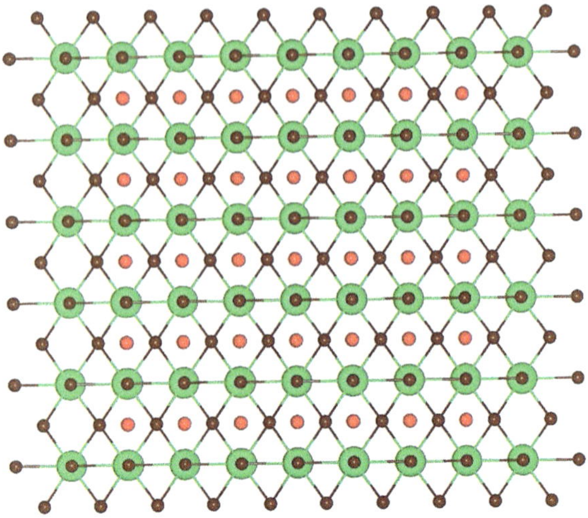

Fig. 1.2 Crystal structure of $CsGeBr_3$, showing the corner-sharing $GeBr_6$ octahedral framework

The compositional flexibility enabled by heterovalent substitution and cation ordering has given rise to a broad chemical design space for lead-free halide perovskites. Figure 1.3 summarizes the elemental combinations experimentally realized in halide double perovskites with general composition $A_2BB'X_6$, indicating site-specific cation occupancy within the cubic framework. Heterovalent substitution strategies maintain charge neutrality through mixed-valence cation combinations or ordered vacancy formation, often resulting in reduced electronic dimensionality and modified optical selection rules [28]. Bismuth- and antimony-based perovskites exemplify this approach, offering enhanced chemical stability at the expense of wider band gaps. Bismuth-based systems typically exhibit band gaps in the range of 2.0–2.8 eV and demonstrate air-stable operation exceeding 1000 h, rendering them particularly suitable for light-emitting diodes, radiation detectors, and ultraviolet optoelectronics [51]. Antimony-based compounds, such as $A_3Sb_2X_9$, provide intermediate band gaps (1.8–2.2 eV) and enhanced absorption coefficients relative to bismuth analogues, positioning them as promising bridge materials for tandem photovoltaic architectures and niche optoelectronic applications [28].

Double perovskites ($A_2B'B''X_6$) and vacancy-ordered perovskites (A_2BX_6) constitute structurally distinct, charge-neutral families in which cation ordering or systematic vacancy formation stabilizes the lattice while reshaping electronic band topology as depicting in Table 1.2. Silver–bismuth double perovskites, such as Cs_2AgBiX_6 (X = Cl, Br, I) as shown in the Fig. 1.4 which illustrates the crystal structure of $Cs_2AgBiBr_6$, emphasizing the ordered distribution of $AgBr_6$ and $BiBr_6$ octahedra, exhibit indirect, parity-forbidden band gaps in the range of 2.1–2.8 eV, combined with exceptional environmental stability and enhanced mechanical hardness relative to conventional ABX_3 perovskites, supporting durability-driven applications despite suppressed near-edge absorption [51]. A comparative overview of

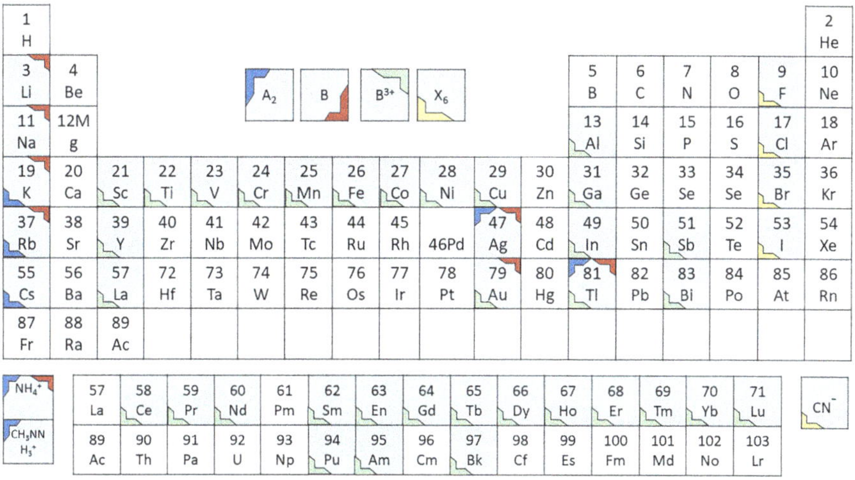

Fig. 1.3 Elements forming halide double perovskites with general composition $A_2BB'X_6$. Elements shown in light blue indicate experimentally synthesized compounds. The half-frame color denotes crystallographic site occupancy, as defined in the legend. Only compounds crystallizing in the Fm3m space group at room temperature are included

representative double perovskites ($A_2B'B''X_6$), summarizing their band gaps, transition types, and stability characteristics, is provided in Table 1.2. Vacancy-ordered systems such as Cs_2SnI_6 display quasi-zero-dimensional electronic behavior arising from isolated $[SnI_6]^{2-}$ octahedra, resulting in indirect band gaps of 1.3–1.5 eV, weak absorption edges, and enhanced air stability due to the tetravalent Sn^{4+} state, making them particularly attractive where long-term environmental durability is critical [25].

1.3.3 Optical Properties

Lead-free perovskites offer remarkable flexibility in bandgap engineering owing to diverse compositional and structural modification routes. The bandgap typically changes by 0.3–0.5 eV per step when halides are replaced (from Iodide to bromide to chloride) [125]. The identity of the cation substituting the A-site is adjusted, and typically the large organic cations (formamidinium > methylammonium > caesium) decrease the bandgap [4, 125]. It is also important that the metal used as the B-site be tin-based, as perovskites with tin have lower band gaps (1.2 eV) than those with lead (1.6 eV), germanium (1.82–2.2 eV), and bismuth (2.8 eV) [94, 125]. Moreover, by combining various compositions, it is possible to operate with bandgap tuning, which behaves according to the Vegard law [125]. Besides compositional and structural tuning, external inputs are also a viable path to alter the electronic structure further. Another general PHE strategy for bandgap control is the use of

Table 1.2 Key properties of double perovskites ($A_2B'B''X_6$)

Composition	Band gap (eV)	Transition type	Stability and notes
$Cs_2AgBiCl_6$	2.8 (Cl)	Indirect, parity-forbidden	>1000 h at 85% RH; high hardness vs. ABX_3
$Cs_2AgBiBr_6$	2.3 (Br)	Indirect, parity-forbidden	>1000 h at 85% RH; strong air stability
Cs_2AgBiI_6	2.1 (I)	Indirect, parity-forbidden	Lower absorption edge; robust mechanical properties
$Cs_2NaBiCl_6$	3.0 (Cl) *DFT*	Indirect, weak absorption	Predicted high stability; limited experimental data
Cs_2KBiCl_6	3.2 (Cl) *DFT*	Indirect, parity-forbidden	Theoretical stability; synthesis under development
$Cs_2CuSbCl_6$	1.0–1.5 (Cl)	Direct allowed	Near PV-grade gap; phase-control required; air-sensitive
$Cs_2AgSbCl_6$	2.5 (Cl) *DFT*	Indirect, parity-forbidden	Predicted stability; experimental validation pending
$Cs_2NaSbBr_6$	2.6 (Br) *DFT*	Indirect	Computational studies suggest tunable gaps
Cs_2KBiBr_6	2.4 (Br) *DFT*	Indirect	Potential for visible emission; requires inert synthesis
Cs_2CuBiI_6	1.2 (I) *DFT*	Direct allowed	Theoretical PV potential; synthesis challenges

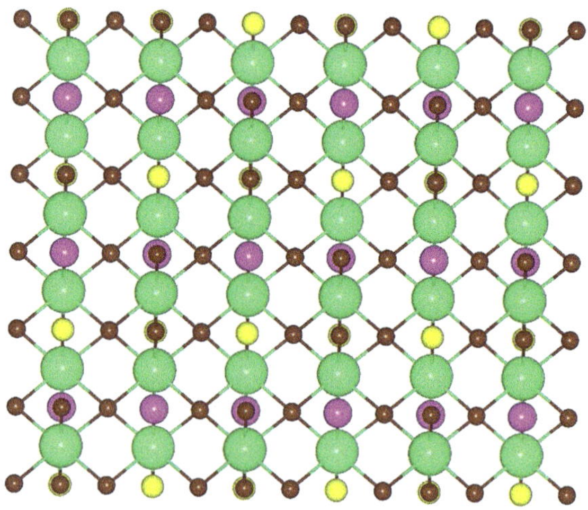

Fig. 1.4 Crystal structure of $Cs_2AgBiBr_6$, illustrating the ordered rock-salt arrangement of $AgBr_6$ and $BiBr_6$ octahedra within the three-dimensional double perovskite framework

hydrostatic pressure, with the majority of lead-free perovskites exhibiting a coefficient of pressure between −0.1 and −0.3 eV/GPa [17]. Such pressure not only alters the electronic bandwidths and optical properties but can also induce structural phase transitions, which can further modify the bandgap, making it suitable for use in flexible device applications. These are optical changes in the response of lead-free perovskites resulting from temperature and pressure-dependent transformations of the electronic structure. High-performance lead-free perovskites have a high optical constant, and the absolute value of the absorption coefficient exceeds 10^6 cm^{-1} at the band edge in direct-gap semiconductors [59]. The absorption exhibits notable spectral dependence, while sub-bandgap Urbach tails (10–50 meV) indicate disorder-induced absorption. Temperature variations influence both band broadening and the temperature dependence of the bandgap. Beyond absorption behaviour, refractive indices play a critical role in optical-device design. Refractive indices are critical for designing optical devices; visible-range values usually lie between 2.0 and 2.8 at 600 nm. Many perovskites show pronounced wavelength-dependent dispersion and, in anisotropic structures, polarization-dependent birefringent properties. In addition, their nonlinear optical behaviours (second and third-order susceptibilities) extend their scope of application. Beyond light-matter interaction, charge-transport characteristics play a decisive role in determining overall device performance [103]. Moreover, room-temperature carrier mobilities of more than 100 cm^2 V^{-1} s^{-1} have been achieved for tin perovskite lattices, governed by many scattering obstacles such as ionized imperfections, phonons, and grain boundaries. Anisotropic field-contingent characteristics such as mobility saturation and hot carriers have been encountered for reduced-dimensional perovskites [56]. Diffusion lengths of these materials, determined by photoluminescence quenching and device efficiency evaluation, span from 100 nm to more than 1 μm, depending on quality and have an inverse dependence on defect density [56, 59]. These transport mechanisms have a direct impact on the photoluminescent behaviour of lead-free perovskites, specifically their quantum yield and emission spectral characteristics, and are crucial to the application of such materials in optoelectronics. Quantum yield is studied through sphere analysis and comparative analysis, whereby temperature changes and excitation intensity changes to reveal the underlying recombination process, which influences the radiative efficiency through environmental stability. The emission may also be tuned from visual wavelengths to near-infrared wavelengths, allowing both the linewidth and Stokes shift analysis to provide useful information on material disorder and electron structures. These properties are related to LED specifications using colour coordinates within the CIE 1931 system [56, 103].

1.3.4 Chemical Stability

The long-term stability and environmental friendliness of lead-free perovskites necessitate oxidation resistance. In a tin system, the thermodynamic preference of Sn^{2+} to oxidise to Sn^{4+} is undesirable because its formation energy is negative and thus prefers to degrade instead of stabilising some solutions that are employed to extend the rate of the process and increase the working life of devices, including minimised passivation of surfaces through compositional factor optimisation. By contrast, germanium-containing perovskites have superior oxidation resistance by virtue of improved thermodynamic stability; likewise, compounds with bismuth sidestep further oxidation by a markedly different fundamental process. The long-term stability and environmental friendliness of lead-free perovskites necessitate oxidation resistance. In a tin system, the thermodynamic preference of Sn^{2+} to oxidise to Sn^{4+} is undesirable because its formation energy is negative and thus prefers to degrade instead of stabilising some solutions that are employed to extend the rate of the process and increase the working life of devices, including minimised passivation of surfaces through compositional factor optimization [10, 39]. Comparatively, germanium-based perovskites exhibit higher resistance to oxidation due to their higher thermodynamic stability, bismuth-based compounds also circumvent further oxidation through a significantly distinct fundamental reaction [36, 39, 46]. In addition to the thermodynamic factors inherent in it, kinetic considerations also contribute to the overall stability profile. Rates of oxidation are also affected by the partial pressure of oxygen available in the atmosphere, as well as working temperature and moisture, which act as an activator to facilitate a faster degradation process. Attack points of oxidative reactions are optimised areas and at grain boundaries; thus, following surface passivation and barrier layers with long durability is an important factor in device lifespan. Besides the issue of oxidation, moisture stability is another significant concern. Chemical breakdown and ion exchange occur when hydrolysis reactions, protonation, and structural deterioration take place under acidic and moist conditions. The primary protection measures include the use of a hydrophobic organic cation, a low gas transmission rate encapsulation barrier material, and the application of a hydrophobic coating. Engineers can achieve better interior coverage through dimensional engineering (e.g., incorporating 2D perovskite crystal structures with organic spaces) [46, 53]. Stability at high temperatures also determines the stability of devices during high-temperature performance. Sn-based perovskites self-decompose between 150 and 250 °C, depending on the conditions, whereas Bi-based compounds maintain their structure above 400 °C [70]. Mechanical reliability issues and crack propagation can occur due to the presence of volatile organic cations, phase change structures, and thermal expansion. The potential for stress caused by anisotropic thermal expansion, where failure occurs, necessitates matching the substrate coefficient in maintainable device design [36].

1.4 Application and Device Performance

Lead-free perovskite materials have tunability for various optoelectronic applications, for which specific material properties and device structures need to be optimized for performance and stability some of the applications which are shown in the Fig. 1.5.

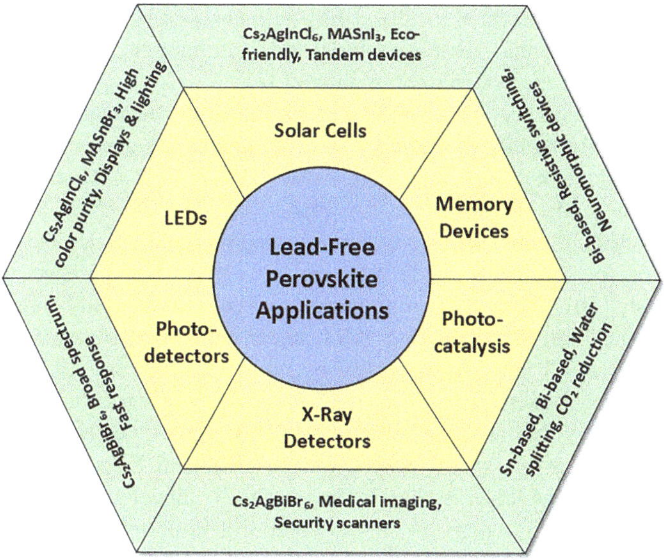

Fig. 1.5 Overview of major application domains of lead-free perovskite materials, highlighting their use in photovoltaics, light-emitting devices, photodetectors, X-ray detectors, memory devices, and photocatalysis, along with representative material systems

1.4.1 Solar Cell Applications

The evolution of device architecture has been instrumental in improving the performance of lead-free perovskite solar cells, whereby a systematic progression of both planar and mesoporous devices has enabled increasing efficiency without increasing the complexity of processing. Most planar architectures of devices employ either a traditional n-i-p architecture, which includes electron transport layers or perovskite absorbers, combined with hole transport layers, or can create an inverted p-i-n structure. This configuration enhances interface stability and is well-suited for low-temperature processing [35, 73]. Contact engineering has become a key factor in optimising performance in these architectures, as selective materials are used to reduce recombination loss by ensuring that functional and work recombination are carefully aligned. To accomplish interface passivation, buffer layers and surface treatments are utilised to decrease the density of defects at the interface [19, 52]. Conversely, mesoporous designs use scaffold designs and these scale to TiO_2 frameworks, which offer nanostructured pathways of electron collection, increased charge collection branching area, and mechanical stability of perovskite infiltration, which make improved charge collection and uniformity of films [120]. Other light-management approaches also enhance the performance of devices, such as antireflection finishes, surface texturing to enhance light trapping, optimised back reflectors, and plasmonic designs that take advantage of localized EM field enhancement [102], and charge-transport optimisation that aims at material choice, thickness, defect, and contact resistance optimisation to reduce series losses [37].

Such developments have been indicated by recent performance breakthroughs in lead-free (highly rapidly maturing) photovoltaic technologies, especially in the case of tin-based technologies, which are already taking the lead in terms of highest efficiency ratings. Mixed-cation (Cs, FA, PEA) $Sn(I, Br)_3$ absorbers based on 2D/3D heterojunction architectures have demonstrated certified efficiencies of 16.65% at silicon-type performance, while retaining cost benefits associated with solution processing [30]. These systems demonstrate significant stability enhancements, with operational lifetimes exceeding 500 h under encapsulated conditions and an area scaling of over 1 cm^2, accompanied by minimal efficiency degradation [26]. Improved atmospheric sensitivity, resulting from processing innovations that enhance precursor chemistry, has enabled commercialization routes, including the partnership between the University of Queensland and Halocell Energy, as part of the Australian Economic Accelerator Ignite program [27]. Tandem device architectures offer an alternative path to high efficiency, where silicon-perovskite tandem devices can operate with a certified efficiency of 25.5% and excellent operational stability. All-perovskite tandem devices, which also achieve a certified efficiency of 25.5%, demonstrate excellent stability.

In contrast, silicon-perovskite tandem devices show compatibility with established crystalline silicon technologies [109, 117]. Further engineering of devices, such as interface passivation, compositional grading, and hybrid mesoporous-planar geometries, is underway to improve performance. Hybrid architectures, such as optimised interfacial control, have achieved an efficiency of 10.7% [74]. Taken together, these breakthroughs make lead-free perovskite technologies viable contenders for commercial use, enabling a rapid transition from laboratory research to large-scale industrial manufacturing, with improved stability and processing capabilities [119].

1.4.2 Light-Emitting Devices (LEDs)

Figure 1.6 illustrates that the electroluminescence of lead-free perovskite light-emitting diodes (PeLEDs) results from a combination of charge injection, carrier recombination, and light extraction processes, which determine the device's efficiency and spectral performance. The alignment of energy levels between electrode–transport layer-perovskite interfaces is important in controlled injection efficiency, where work-function match reduces carrier injection barriers and carrier equilibrium in the emissive layer [113, 135]. Radiative recombination occurs due to band-to-band recombination, resulting in the emission of the desired band. Other recombination processes through defects result in spectral broadening and a loss of efficiency. The competing non-radiative mechanisms, such as the Auger recombination and defect-mediated trapping, become more apparent at high current densities and are the cause of efficiency droop, one of the most severe effects in PeLED operation [132]. Traditional devices tend to have a roll-off of efficiency of up to 50% at high current densities, but current developments of quantum-well width control and optimisation of cation ratios have greatly eased this problem with only 21% roll-off at high current densities of up to 500 mA cm^{-2} [78, 130]. Engineers can alter the composition through compositional engineering to fabricate a narrow spread of emission characteristics across the visible spectrum, as well as into the infrared. Halide substitution allows peak wavelengths to be fine-tuned through the compositional engineering of emission processes. In contrast, quantum confinement effects enable the emission linewidth to be reduced to values below 30 nm full width at half maximum energy, thereby fulfilling the colour purity demands as defined by the CIE 1931 colour space [45]. Depending on whether phosphor down-conversion or direct broadband emission happens, the generation of white light in mixed-dimensional perovskite formations has enabled hybrid devices to achieve external quantum efficiencies of up to 43% and luminance values of up to 176,000 cd/m^2 [45, 98].

Device architecture optimization further enhances performance by addressing charge balance, optical outcoupling, and operational stability. Balanced carrier injection is achieved through judicious selection of electron and hole transport layers, including metal oxides, PEDOT: PSS, and NiO_x, combined with interface dipole layers and graded energy-level structures that reduce injection barriers and suppress carrier accumulation [62, 121]. Transport layer thickness optimization simultaneously influences electrical transport and optical cavity effects, necessitating careful trade-offs between recombination efficiency and light extraction. Optical losses arising from refractive index mismatch between perovskite emitters ($n > 2$) and glass substrates ($n \approx 1.45$) are mitigated through advanced light-extraction strategies such

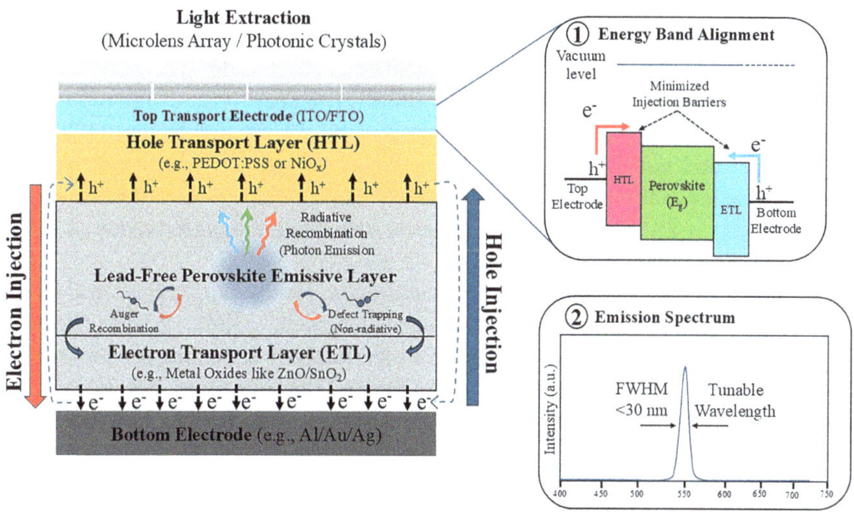

Fig. 1.6 Charge injection, recombination, and light-extraction mechanisms in lead-free perovskite light-emitting diodes (PeLEDs)

as microlens arrays, photonic crystals, and outcoupling layers, which have enabled extraction efficiencies exceeding 30% [91, 97]. As a result of these combined strategies, lead-free PeLEDs have achieved peak EQE values surpassing 20%, rivaling state-of-the-art lead-based devices. Notable breakthroughs include 26.7% EQE using ultrathin $CsPbBr_3$ emissive layers with NiO surface modification and 16.8% EQE in chiral perovskite quantum-dot spin LEDs exhibiting circularly polarized emission [31]. Stability improvements have been equally significant, with mixed-dimensional $FASnI_3$ devices reaching T_{50} lifetimes of 10.3 h at 100 mA cm^{-2} and all-inorganic $CsSnI_3$ systems achieving 39.5 h at similar current densities for near-infrared emission above 900 nm [129]. Accelerated lifetime testing reveals that encapsulated red PeLEDs can achieve T_{50} lifetimes exceeding 4800 h at 100 cd m^{-2}, while hybrid perovskite–organic tandem LEDs demonstrate operational lifetimes beyond 42,000 h through reduced Joule heating and improved thermal management [31, 97].

1.4.3 Photodetectors and Sensors

Lead-free perovskite photodetectors employ detection principles based on the fundamental principles of photoconductive and photovoltaic processes, enabling a sensitive response to light over a wide spectral range over a wide spectral range, as schematically illustrated in Fig. 1.7, specifically, the photoconductive mode. The manipulation of surplus charge carriers, generated through illumination and subsequently driven by an applied voltage, results in variations in current. Responsivity is defined as the amount of photocurrent generated for each unit of incident optical power. The lead-free system, primarily utilising metal-ion doping, has been found to demonstrate responsivities exceeding 1600 A/W, which highlights the significance of engineering defects and carrier lifetimes in enhancing performance [61, 111]. Spectral response generally lies between the visible and near-infrared parts of the spectrum, whereas attributes of spectral response are of the order of nanoseconds to microseconds, depending on the material quality, trap density, and device architecture [12, 122]. Dark current and noise eventually restrict the sensitivity of the detector, characterising noise-equivalent power and detectivity. A value of 1.56×10^{13} Jones has been reported for Zn^2-doped Cs_2SnI_6 devices, surpassing the performance of many lead-based photodetectors [96].

Photovoltaic sensing enables self-powered operation by utilising electric fields intrinsic to the device to isolate photogenerated carriers, eliminating the need for external bias. Consequently, such devices are particularly appealing for low-power sensing applications. Researchers have illuminated that open-circuit voltages of a few hundred millivolts can be achieved, and short-circuit current and fill factor dictate output power and conversion efficiency [6, 66]. The architecture of devices is a crucial factor in performance optimisation. To reduce recombination losses and improve carrier collection, PIN photodiodas consisting of layers of intrinsic

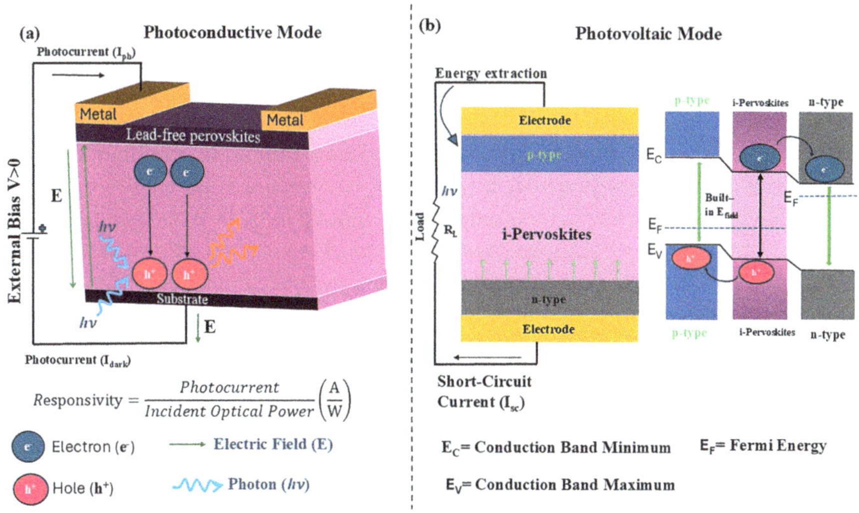

Fig. 1.7 Schematic illustration of charge-generation and carrier-transport mechanisms in lead-free perovskite photodetectors operating in (**a**) photoconductive and (**b**) photovoltaic modes

perovskite between charge-selective contacts have been developed [81]. The junctions of Schottky diodes are simplified to facilitate fabrication, whereas heterojunction designs permit band engineering through interface engineering [42]. Phototransistor designs also enhance photoresponse through intrinsic gain processes, with a photoconductive gain (ratio of carrier recombination to transit times) of up to 10^9 A/W, but at the expense of gain versus bandwidth [101, 122]. Detailed performance metrics such as detectivity, dynamic range that extends over many orders of magnitude, as well as frequency response, as characterised by RC time constants, show that even recent approaches to dielectric and interface engineering can significantly enhance bandwidth and noise levels without deteriorating thermal stability over a wide range of practice operating conditions [34, 131].

1.4.4 Other Optoelectronic Applications

More recently, lead-free perovskite materials have become useful as general substrates for new optoelectronic functionality, not necessarily limited to photovoltaics, with high absorption coefficients, desirable charge-transport characteristics, and material stability. High atomic number (Z) direct-conversion detector architectures with the lead-free perovskite Cs_2 SnI_6 and $Cs_2BiAgBr_6$ make contact with high-energy photons efficiently, enabling mass attenuation coefficients of up to 15 cm^2g^{-1} on photons in the 60 keV region [40]. These characteristics result in 10–20 mm of brilliant resolution, as confirmed by suppressed point-spread functions and modulation transfer functions values above 0.5 at 1 line pair per millimetre, and revealed by grandiose detection and collection quantum efficiencies exceeding 70% [104]. Notably, radiation hardness tests indicate that even the long-term performance of materials exposed to radiation doses of around 10^6 Gy is minimal, lending significance to the durability of these materials in severe ionising conditions [23]. Fast radiative recombination in lead-free perovskites is utilised in scintillation applications to efficiently produce visible photons by converting high-energy radiation into visible light. Cs_2SnBr_6 is reported to yield light up to 60,000 photons per MeV, which is comparable to traditional inorganic scintillators. Their scintillation decay kinetics feature ultrafast components down to 10 s and insignificant afterglow values of <0.1% at 1 megahertz, providing the high time resolution necessary in dynamic imaging and time-resolved detection [64, 127]. Additionally, temperature-responsive studies demonstrate a constant light output and a decay behaviour over a broad operating temperature range of −20 to 50 °C, which confirms the reliable operation of the deteriorators under changing environmental conditions.

In addition to radiation sensing, lead-free perovskites have demonstrated impressive prospects in coherent light generation, neuromorphic computing, and direct applications of their high optical gain, tunable emission properties, and ion migration behaviours. These materials are high optical-gain media, exhibiting an optical gain of 300–500 cm^{-1} when excited by pulsed methods and low population-inversion thresholds of approximately 10^{-2}, as well as saturation fluences of around 0.5 mJ cm^{-2}, which enables the generation of lasers at a lower threshold [100]. Wavelength tunability in the visible and near-infrared enables wavelength tunability over broad spectral gain bandwidths of 20–30 nm. At the device level, the Fabry–Perot microcavities made of $Cs_3Bi_2Cl_9$ films have demonstrated single-mode lasing at 520 nm with quality factors reaching over 2000 [116], and distributed feedback structures have demonstrated single-mode lasing at longer wavelengths and random lasing behaviour in polycrystalline films with no defined cavity mirrors [57]. Neuromorphic and memristive applications are built upon the inherent properties of ion migration and resistive switching of lead-free perovskites. Mixed-cation tin-based perovskite displays exhibit filamentary switching stability, with ON/OFF resistance ratios exceeding 10^4, and lasting longer than 10^5 switching cycles, meeting the primary criteria of non-usvolatile memory technologies [24, 104]. Those also enable analogue multilevel conductance modulation, which makes it possible to simulate synaptic plasticity and spike-timing-dependent plasticity learning rules with conductance change rates of about 1% per pulse. At the system level, with 16 × 16 arrays of memristors, a pattern recognition accuracy of >90% has been demonstrated, showcasing the potential of lead-free perovskite materials in scalable and energy-efficient neuromorphic hardware [104].

1.5 Conclusion and Outlook

Comprehensive environmental assessment demonstrates the sustainability advantages of lead-free perovskite materials while identifying opportunities for further environmental impact reduction. The comparative human health risk evaluation of lead-free systems with lead-based materials identifies strong quantitative benefits across several dimensions of toxicity, paths-of-exposure, and regulatory compliance. Measures of toxicity identify strong safety upgrades with acute toxicity data identifying lead-free alternates with 10–100 times less adverse effect relative to lead-based systems as shown by comparative LD_{50} findings that identify lead's oral lethal dose near 450 mg/kg relative to substantially larger threshold values for alternates such as tin and bismuth. Chronic exposure guidelines further distinguish these benefits with occupational regulations such as the US OSHA's permitted exposure level for lead near 50 $\mu g/m^3$ as an 8-h time-weighted average necessitating widespread monitoring, physician monitoring, and protective gear strategies largely not needed for lead-free systems. Lead's characteristic bioaccumulation profile that concentrates the element over decades in bone and soft tissues with resulting irreversible neurological impairment contrasts with alternates with faster elimination kinetics and lower long-term accumulation potential. Toxicity evaluations on the nervous system specifically identify lead's distinctive mechanisms for adverse effect including disruption of calcium-dependent neuron signalling, induction of oxidative stress, and disruption of key developmental processes. Exposure pathway analysis identifies overall risk mitigation throughout the supply chain, consumer use, end-of-life, and environmental contamination pathways. Manufacturing workers benefit from eliminating lead-specific hazards, reducing the complexity of personal protective equipment requirements, air monitoring protocols, and medical surveillance programs mandated under current OSHA lead standards. Consumer exposure during regular product operation is minimised by removing lead's propensity for surface migration and particulate generation, while end-of-life handling benefits from non-hazardous waste classification under the Resource Conservation and Recovery Act regulations, significantly reducing disposal costs and regulatory burden. Environmental exposure pathways are similarly improved, as lead-free systems eliminate the persistent contamination potential that characterises lead in soil, water, and air matrices. Regulatory compliance advantages provide additional compelling benefits for lead-free adoption. The European Union's RoHS Directive automatically accepts lead-free products without requiring application-specific exemptions that are increasingly time-limited and subject to periodic review, ensuring long-term market access certainty. US RCRA regulations classify lead-free materials as non-hazardous waste, streamlining disposal procedures and reducing associated costs. Emerging global regulations continue to tighten lead restrictions, with recent amendments to China RoHS adding additional phthalate restrictions and the EU progressively limiting lead exemptions, trends that favour lead-free systems' regulatory durability. The LCA methodology establishes consistent assessment frameworks using functional units such as per kWh electricity generation for energy applications, with system boundaries encompassing cradle-to-grave analysis, including raw materials, manufacturing, use, and disposal phases. Impact categories systematically evaluate climate change, acidification, eutrophication, human toxicity, and ecotoxicity effects, utilising primary data from manufacturing processes and secondary data from established databases to ensure data quality and reliability. The raw material effects suggest separate environmental advantages for lead-free substitutes on many fronts. Extraction processes for tin and other replacement components suggest substantially reduced environmental impact by comparison with lead extraction, involving energy-intensive processing and generating high acidification potential and ecotoxicity problems. The processing energies required in cleaning and prepping precursors for lead-free devices are primarily due to the lowered temperature processing and simpler chemical paths. Environmental aspects of fabrication demonstrate notable reductions in energy consumption and waste production for lead-free devices. Energy requirements of synthesis, deposition, and assembly of device processes, which make the process less so that lead-free perovskite processing takes place at 200–400 °C lower than lead-based equivalents. Non-toxic products utilise less energy in factories to clean rooms and the environment. Waste production trends exhibit significant improvements for chemical, solid, and liquid waste streams. Waste treatment for chemicals improves by avoiding special lead-containing solvent recovering and treating systems, and ordinary waste treatment procedures become adequate for lead-free substitutes. Solid waste production is reduced by enhancing material utilisation effectiveness and lowering process losses, and water usage requirements and effluent treatment become easier without lead pollution issues. Emission patterns of air improve significantly by avoiding lead-bearing volatile organic compounds and particulate matter, lowering workplace exposure hazards and environmental release possibilities. The use phase environmental benefits provide compelling long-term advantages through renewable energy production capabilities. Carbon displacement effects enable substantial fossil fuel substitution and greenhouse gas emission reductions, with typical solar installations offsetting manufacturing emissions within 1–3 years. Energy payback time calculations demonstrate lead-free systems achieving faster payback periods due to reduced manufacturing energy requirements, while net environmental benefits accumulate more rapidly throughout the operational lifetime. Grid integration advantages emerge from reduced regulatory barriers and simplified permitting processes, facilitating distributed energy generation deployment. Performance factors significantly influence environmental impact profiles through efficiency and durability considerations.

1.6 Future Outlook

The field of lead-free perovskite materials stands at a critical juncture where fundamental scientific advances are converging with market demands for sustainable semiconductor technologies. As demonstrated throughout this introductory chapter, the theoretical foundation for rational materials design is well-established, with clear pathways for achieving high-performance devices while eliminating toxicity concerns and meeting increasingly stringent environmental regulations.

References

1. K. Abdel-Latif, F. Bateni, S. Crouse, M. Abolhasani, Flow synthesis of metal halide perovskite quantum dots: from rapid parameter space mapping to AI-guided modular manufacturing. Matter **3**(4), 1053–1086 (2020). https://doi.org/10.1016/j.matt.2020.07.024
2. T. Abzieher, D.T. Moore, M. Roß, S. Albrecht, J. Silvia, H. Tan, Q. Jeangros, C. Ballif, M.T. Hoerantner, B.S. Kim, H.J. Bolink, P. Pistor, J.C. Goldschmidt, Y.H. Chiang, S.D. Stranks, J. Borchert, M.D. McGehee, M. Morales-Masis, J.B. Patel et al., Vapor phase deposition of perovskite photovoltaics: short track to commercialization? Energy Environ. Sci. **17**(5), 1645–1663 (2024). https://doi.org/10.1039/d3ee03273f
3. A.S.A. Ahmed, F.S.M. Hashem, A.B.A. El-Adasy, T.A. Seaf El-Nasr, Flexible photovoltaics based on perovskite materials. RSC Adv. **15**(29), 23266–23301 (2025). https://doi.org/10.1039/d5ra02563j
4. I. Ahmed, K. Prakash, S.M. Mobin, Lead-free perovskites for solar cell applications: recent progress, ongoing challenges, and strategic approaches. Chem. Commun. **61**(37), 6691–6721 (2025). https://doi.org/10.1039/d4cc06835a
5. S. Ahmed, M.A. Gondal, A.S. Alzahrani, M. Parvaz, A. Ahmed, S. Hussain, Recent trends and challenges in lead-free perovskite solar cells: a critical review. ACS Appl. Energy Mater. **7**(4), 1382–1397 (2024). https://doi.org/10.1021/acsaem.3c02327
6. M.P. Aleksandrova, Study of lead-free perovskite photoconverting structures by impedance spectroscopy. Energy **273**, 127141 (2023). https://doi.org/10.1016/j.energy.2023.127141
7. M.K.M. Ali, A.A. Mohsen, N.K. Allam, Lead-free perovskite materials for optoelectronic and solar energy applications. Sol. Energy Mater. Sol. Cells **295**, 114025 (2026). https://doi.org/10.1016/j.solmat.2025.114025
8. K.M. Anoop, D. Devadiga, M.S. Sunitha, T.N. Ahipa, Advancements in preventing Sn^{2+} oxidation in tin-based perovskite solar cells: a review, in *Physica Status Solidi (A) Applications and Materials Science*, vol. 222 (Wiley, Amsterdam, 2025). https://doi.org/10.1002/pssa.202400569
9. I.M. Asuo, A. Mahdavi Varposhti, E.D. Gomez, N.Y. Doumon, Anti-solvent engineering enables efficient ambient-processed halide perovskite solar cells. J. Mater. Chem. C **12**(21), 7562–7571 (2024). https://doi.org/10.1039/d4tc01305k
10. W. Ayaydah, E. Raddad, Z. Hawash, Sn-based perovskite solar cells towards high stability and performance. Micromachines **14**, 806 (2023). https://doi.org/10.3390/mi14040806
11. M. Baskurt, P. Erhart, J. Wiktor, Direct, indirect, and self-trapped excitons in $Cs_2AgBiBr_6$. J. Phys. Chem. Lett. **15**(33), 8549–8554 (2024). https://doi.org/10.1021/acs.jpclett.4c01604
12. S.A. Behera, P.G.R. Achary, Lead free perovskite based heterojunction photodetectors: a mini review. Appl. Surf. Sci. Adv. **14**, 100393 (2023). https://doi.org/10.1016/j.apsadv.2023.100393
13. S. Castro-Hermosa, G. Lucarelli, M. Top, M. Fahland, J. Fahlteich, T.M. Brown, Perovskite photovoltaics on roll-to-roll coated ultra-thin glass as flexible high-efficiency indoor power generators. Cell Rep. Phys. Sci. **1**(5), 100045 (2020). https://doi.org/10.1016/j.xcrp.2020.100045
14. M. Chen, Z. Qin, Z. Zhang, W. Xiang, Y. Liu, C. Tian, S. Chen, Y. Wang, L. Han, Low-temperature sequential deposition for efficient inverted perovskite solar cells. Nat. Commun. **16**(1), 61144 (2025). https://doi.org/10.1038/s41467-025-61144-y
15. N. Cheng, Z. Yu, W. Li, B. Lei, W. Zi, Z. Xiao, Z. Zhao, P.A. Zong, A modified two-step sequential spin-coating method for perovskite solar cells using CsI containing organic salts in

mixed ethanol/methanol solvent. Sol. Energy Mater. Sol. Cells **250**, 112107 (2023). https://doi.org/10.1016/j.solmat.2022.112107
16. T.H. Chowdhury, Y. Reo, A.R.B. Yusoff, Y.Y. Noh, Sn-based perovskite halides for electronic devices. Adv. Sci. **9**, 3749 (2022). https://doi.org/10.1002/advs.202203749
17. M. Coduri, T.B. Shiell, T.A. Strobel, A. Mahata, F. Cova, E. Mosconi, F. De Angelis, L. Malavasi, Origin of pressure-induced band gap tuning in tin halide perovskites. Mater. Adv. **1**(8), 2840–2845 (2020). https://doi.org/10.1039/d0ma00731e
18. H.A. Dewi, L.R.W. White, D. De Luca, R. Ahmad, A. Bruno, Thermally evaporated metal halide perovskites for optoelectronics. ACS Appl. Energy Mater. **8**(12), 7769–7779 (2025). https://doi.org/10.1021/acsaem.5c00706
19. D. Döhler, P. Büttner, F. Scheler, D. Thiel, B. Puscher, S. Bochmann, J. Mitrovic, P.P. Boix, D.M. Guldi, I. Mínguez-Bacho, J. Bachmann, A geometrically well-defined and systematically tunable experimental model to transition from planar to mesoporous perovskite solar cells. ACS Appl. Energy Mater. **5**(10), 11977–11986 (2022). https://doi.org/10.1021/acsaem.2c00870
20. A.J. Doolin, R.G. Charles, C.S.P. De Castro, R.G. Rodriguez, E.V. Péan, R. Patidar, T. Dunlop, C. Charbonneau, T. Watson, M.L. Davies, Sustainable solvent selection for the manufacture of methylammonium lead triiodide (MAPbI3) perovskite solar cells. Green Chem. **23**(6), 2471–2486 (2021). https://doi.org/10.1039/d1gc00079a
21. M. Duan, J. Yang, T. Li, J. Wen, B. Ren, K. Wang, Y. Xia, H. Zhang, Y. Chen, Mechanically stable screen-printed flexible perovskite solar cells via selective self-assembled siloxane coupling agents. NPJ Flexible Electr. **9**(1), 407 (2025). https://doi.org/10.1038/s41528-025-00407-6
22. V.C.M. Duarte, L. Andrade, Recent advancements on slot-die coating of perovskite solar cells: the lab-to-fab optimisation process. Energies **17**, 3896 (2024). https://doi.org/10.3390/en17163896
23. K.R. Dudipala, T.H. Le, W. Nie, R.L.Z. Hoye, Halide perovskites and their derivatives for efficient, high-resolution direct radiation detection: design strategies and applications. Adv. Mater. **36**, 4523 (2024). https://doi.org/10.1002/adma.202304523
24. S. Ghosh, H. Shankar, P. Kar, Recent developments of lead-free halide double perovskites: a new superstar in the optoelectronic field. Mater. Adv. **3**(9), 3742–3765 (2022). https://doi.org/10.1039/d2ma00071g
25. F. Giustino, H.J. Snaith, Toward lead-free perovskite solar cells. ACS Energy Lett. **1**(6), 1233–1240 (2016). https://doi.org/10.1021/acsenergylett.6b00499
26. M.A. Green, Silicon solar cells: evolution, high-efficiency design and efficiency enhancements. Semicond. Sci. Technol. **8** 11993 (1993), Printed in the UK Silicon solar cells: evolution, high-efficiency design and efficiency enhancements, *Semiconductor Sciences Technology*, vol. 8. http://iopscience.iop.org/0268-1242/8/1/001
27. M.A. Green, Australian photovoltaics research and development. ACS Energy Lett. **1**(3), 516–520 (2016). https://doi.org/10.1021/acsenergylett.6b00283
28. M.A. Hadi, M.N. Islam, J. Podder, Indirect to direct band gap transition through order to disorder transformation of $Cs_2AgBiBr_6$ via creating antisite defects for optoelectronic and photovoltaic applications. RSC Adv. **12**(24), 15461–15469 (2022). https://doi.org/10.1039/d1ra06308a
29. M.P. Hautzinger, W. Mihalyi-Koch, S. Jin, A-site cation chemistry in halide perovskites. Chem. Mater. **36**(21), 10408–10420 (2024). https://doi.org/10.1021/acs.chemmater.4c02043
30. D. He, P. Chen, J.A. Steele, Z. Wang, H. Xu, M. Zhang, S. Ding, C. Zhang, T. Lin, F. Kremer, H. Xu, M. Hao, L. Wang, Homogeneous 2D/3D heterostructured tin halide perovskite photovoltaics. Nat. Nanotechnol. **20**(6), 779–786 (2025). https://doi.org/10.1038/s41565-025-01905-4
31. S. He, W. Lin, D. Yu, J. Shi, Z. Yin, C. Sun, H. Liu, C. Zhang, J. Yuan, S. Bai, S. Xiao, G. Long, M. Yuan, Y. Jiang, Y. Chen, Q. Song, Perovskite spin light-emitting diodes with simultaneously high electroluminescence dissymmetry and high external quantum efficiency. Nat. Commun. **16**(1), 8 (2025). https://doi.org/10.1038/s41467-025-57472-8

32. S.F. Hoefler, G. Trimmel, T. Rath, Progress on lead-free metal halide perovskites for photovoltaic applications: a review. Monatshefte fur Chemie **148**(5), 795–826 (2017)
33. F. Hou, H. Yang, H. Guo, X. Ning, X. Ren, J. Guo, J. Wei, R. Liu, T. Li, *Efficient Two-Step Sequential Deposition Perovskite Solar Cells Via PbCl2 Enhanced PbI2 Precursor* (n.d.). https://ssrn.com/abstract=4564483
34. C.M. Hun, C.H. Tien, K.L. Lee, H.Y. Lai, L.C. Chen, The effects of temperature on the growth of a lead-free perovskite-like $(CH_3NH_3)_3Sb_2Br_9$ single crystal for an msm photodetector application. Sensors **21**(13), 4475 (2021). https://doi.org/10.3390/s21134475
35. I. Hussain, H.P. Tran, J. Jaksik, J. Moore, N. Islam, M.J. Uddin, Functional materials, device architecture, and flexibility of perovskite solar cell. Emerg. Mater. **1**(3–4), 133–154 (2018). https://doi.org/10.1007/s42247-018-0013-1
36. S.M. Jain, T. Edvinsson, J.R. Durrant, Green fabrication of stable lead-free bismuth based perovskite solar cells using a non-toxic solvent. Commun. Chem. **2**(1), 195 (2019). https://doi.org/10.1038/s42004-019-0195-3
37. Q. Jiang, Y. Zhao, X. Zhang, X. Yang, Y. Chen, Z. Chu, Q. Ye, X. Li, Z. Yin, J. You, Surface passivation of perovskite film for efficient solar cells. Nat. Photon. **13**(7), 460–466 (2019). https://doi.org/10.1038/s41566-019-0398-2
38. R. Jindal, A. Tripathi, C. Mohan, A. Garg, R.S. Varma, Lead-free halide double perovskites for sustainable environmental applications. Chem. Phys. Impact **9**, 770 (2024). https://doi.org/10.1016/j.chphi.2024.100770
39. W. Kaiser, D. Ricciarelli, E. Mosconi, A.A. Alothman, F. Ambrosio, F. De Angelis, Stability of tin- versus lead-halide perovskites: Ab initio molecular dynamics simulations of perovskite/water interfaces. J. Phys. Chem. Lett. **13**(10), 2321–2329 (2022). https://doi.org/10.1021/acs.jpclett.2c00273
40. G. Kakavelakis, M. Gedda, A. Panagiotopoulos, E. Kymakis, T.D. Anthopoulos, K. Petridis, Metal halide perovskites for high-energy radiation detection. Adv. Sci. **7**(22), 98 (2020). https://doi.org/10.1002/advs.202002098
41. M. Khan, M. Zahidur Rahaman, M. Lokman Ali, High-throughput screening of inorganic lead-free halide perovskites $CsCu_2X_3$ (X = Cl, Br, I) for optoelectronics applications. Mater. Sci. Eng. B **299**, 6928 (2024). https://doi.org/10.1016/j.mseb.2023.116928
42. H. Kim, Comparative review of field-effect transistors based on three-dimensional, two-dimensional, and double halide perovskites. Electron. Mater. Lett. **21**(5), 650–666 (2025). https://doi.org/10.1007/s13391-025-00588-2
43. W. Kim, B. Koo, M.J. Ko, H. Jung, Hot-injection synthesis of lead-free pseudo-alkali metal-based perovskite ($TlSnX_3$) nanoparticles with tunable optical properties. Front. Mater. **10**, 8188 (2023). https://doi.org/10.3389/fmats.2023.1298188
44. Y.Y. Kim, T.Y. Yang, R. Suhonen, A. Kemppainen, K. Hwang, N.J. Jeon, J. Seo, Roll-to-roll gravure-printed flexible perovskite solar cells using eco-friendly antisolvent bathing with wide processing window. Nat. Commun. **11**(1), 18940 (2020). https://doi.org/10.1038/s41467-020-18940-5
45. L. Kong, Y. Luo, Q. Wu, X. Xiao, Y. Wang, G. Chen, J. Zhang, K. Wang, W.C.H. Choy, Y.B. Zhao, H. Li, T. Chiba, J. Kido, X. Yang, Efficient and stable hybrid perovskite-organic light-emitting diodes with external quantum efficiency exceeding 40 per cent. Light Sci. Appl. **13**(1), 1500 (2024). https://doi.org/10.1038/s41377-024-01500-7
46. B.P. Kore, M. Jamshidi, J.M. Gardner, The impact of moisture on the stability and degradation of perovskites in solar cells. Mater. Adv. **5**(6), 2200–2217 (2024). https://doi.org/10.1039/d3ma00828b
47. F.U. Kosasih, E. Erdenebileg, N. Mathews, S.G. Mhaisalkar, A. Bruno, Thermal evaporation and hybrid deposition of perovskite solar cells and mini-modules. Joule **6**(12), 2692–2734 (2022). https://doi.org/10.1016/j.joule.2022.11.004
48. P. Kumar, A. Kumar, Device engineering of lead-free double perovskite ($Cs_4CuSb_2Cl_{12}$ and $Cs_2AgBiBr_6$)/crystalline silicon high-performance eco-friendly tandem solar cells. SILICON **16**(8), 3343–3357 (2024). https://doi.org/10.1007/s12633-024-02920-1

49. *Lead-free perovskite solar cell XX CAGR growth analysis 2025–2033* (n.d.). https://www.marketreportanalytics.com/reports/lead-free-perovskite-solar-cell-84188. Accessed 16 Aug 2025
50. J.H. Lee, N.C. Bristowe, J.H. Lee, S.H. Lee, P.D. Bristowe, A.K. Cheetham, H.M. Jang, Resolving the physical origin of octahedral tilting in halide perovskites. Chem. Mater. **28**(12), 4259–4266 (2016). https://doi.org/10.1021/acs.chemmater.6b00968
51. H. Lei, D. Hardy, F. Gao, Lead-free double perovskite $Cs_2AgBiBr_6$: fundamentals, applications, and perspectives. Adv. Funct. Mater. **31**(49), 5898 (2021). https://doi.org/10.1002/adfm.202105898
52. T. Leijtens, G.E. Eperon, S. Pathak, A. Abate, M.M. Lee, H.J. Snaith, Overcoming ultraviolet light instability of sensitized TiO_2 with meso-superstructured organometal tri-halide perovskite solar cells. Nat. Commun. **4**, 3885 (2013). https://doi.org/10.1038/ncomms3885
53. T.L. Leung, I. Ahmad, A.A. Syed, A.M.C. Ng, J. Popović, A.B. Djurišić, Stability of 2D and quasi-2D perovskite materials and devices. Commun. Mater. **3**(1), 9 (2022). https://doi.org/10.1038/s43246-022-00285-9
54. M.R. Leyden, L.K. Ono, S.R. Raga, Y. Kato, S. Wang, Y. Qi, High performance perovskite solar cells by hybrid chemical vapor deposition. J. Mater. Chem. A **2**(44), 18742–18745 (2014). https://doi.org/10.1039/c4ta04385e
55. C.F. Li, S.H. Huang, Y.R. Chen, H.C. Cha, S.Y. Chung, Y.H. Hsiao, F.Y. Tsai, Y.C. Huang, Sequential slot-die coating of perovskite solar cell modules under ambient conditions with precise phase-transition control. Chem. Eng. J. **517**, 164194 (2025). https://doi.org/10.1016/j.cej.2025.164194
56. C. Li, A. Wang, X. Deng, S. Wang, Y. Yuan, L. Ding, F. Hao, Insights into ultrafast carrier dynamics in perovskite thin films and solar cells. ACS Photon. **7**(8), 1893–1907 (2020). https://doi.org/10.1021/acsphotonics.0c00677
57. D.L. Li, J.Y. Chen, Y. Xiao, W.M. Zhong, Y.P. Jiang, Q.X. Liu, X.G. Tang, Perovskite photoelectric memristors with biological synaptic properties for neuromorphic computing. Adv. Sens. Energy Mater. **4**(4), 100159 (2025). https://doi.org/10.1016/j.asems.2025.100159
58. H. Li, C. Zuo, D. Angmo, H. Weerasinghe, M. Gao, J. Yang, Fully roll-to-roll processed efficient perovskite solar cells via precise control on the morphology of PbI2:CsI layer. Nano-Micro Lett. **14**(1), 815 (2022). https://doi.org/10.1007/s40820-022-00815-7
59. M. Li, W. Li, H. Wei, Unique optoelectronic properties and applications of lead-free perovskites and derivatives. Adv. Photon. Res. **6**(2), 95 (2025). https://doi.org/10.1002/adpr.202400095
60. T. Li, Q. Xiong, C. Hu, C. Wang, N. Zhang, S.Y. Lien, P. Gao, Improving crystallization and stability of perovskite solar cells using a low-temperature treated a-site cation solution in the sequential deposition. Molecules **28**(10), 4103 (2023). https://doi.org/10.3390/molecules28104103
61. X. Li, J. Shi, J. Chen, Z. Tan, H. Lei, Lead-free halide double perovskite for high-performance photodetectors: progress and perspective. Materials **16**(12), 90 (2023). https://doi.org/10.3390/ma16124490
62. Z. Li, Z. Chen, Z. Shi, G. Zou, L. Chu, X.K. Chen, C. Zhang, S.K. So, H.L. Yip, Charge injection engineering at organic/inorganic heterointerfaces for high-efficiency and fast-response perovskite light-emitting diodes. Nat. Commun. **14**(1), 9 (2023). https://doi.org/10.1038/s41467-023-41929-9
63. A.R. Lim, S.H. Park, Characterization on lead-free hybrid perovskite $[NH_3(CH_2)_5NH_3]CuCl_4$: thermodynamic properties and molecular dynamics. Molecules **27**(14), 546 (2022). https://doi.org/10.3390/molecules27144546
64. S. Liu, J. Zeng, Q. Chen, G. Liu, Recent advances in halide perovskite memristors: from materials to applications. Front. Phys. **19**(2), 1344 (2024). https://doi.org/10.1007/s11467-023-1344-9
65. I. López-Fernández, D. Valli, C.Y. Wang, S. Samanta, T. Okamoto, Y.T. Huang, K. Sun, Y. Liu, V.S. Chirvony, A. Patra, J. Zito, L. De Trizio, D. Gaur, H.T. Sun, Z. Xia, X. Li, H. Zeng, I. Mora-Seró, N. Pradhan et al., Lead-free halide perovskite materials and optoelectronic devices: progress and prospective. Adv. Func. Mater. **34**(6), 896 (2024). https://doi.org/10.1002/adfm.202307896

66. F. Lu, Y. Ye, X. Wang, N. Zhang, X. Xu, W. Lv, X. Cao, Towards high-performance photodiodes based on p-Si/perovskite heterojunction. Mater. Sci. Eng. B **301**, 169 (2024). https://doi.org/10.1016/j.mseb.2023.117169
67. A.E. Magdalin, P.D. Nixon, E. Jayaseelan, M. Sivakumar, S.K.N. Devi, M.S.P. Subathra, N.M. Kumar, N. Ananthi, Development of lead-free perovskite solar cells: opportunities, challenges, and future technologies. Results Eng. **20**, 101438 (2023). https://doi.org/10.1016/j.rineng.2023.101438
68. S.S. Magubane, R. Burns, S. Ngqoloda, C.J. Oliphant, P.F. Miceli, C.J. Arendse, Sequential chemical vapor deposition of two-dimensional Sn–Pb compound perovskite thin films and its exciton transport. ACS Appl. Electron. Mater. **5**(10), 5352–5361 (2023). https://doi.org/10.1021/acsaelm.3c00266
69. S.N. Manjunatha, Y.X. Chu, M.J. Jeng, L.B. Chang, The characteristics of perovskite solar cells fabricated using DMF and DMSO/GBL solvents. J. Electron. Mater. **49**(11), 6823–6828 (2020). https://doi.org/10.1007/s11664-020-08283-8
70. S.M. Masawa, C. Zhao, J. Liu, J. Xu, J. Yao, Fabrication and characterization of a lead-free cesium bismuth iodide perovskite through antisolvent-assisted crystallization. Nanomaterials **14**(7), 626 (2024). https://doi.org/10.3390/nano14070626
71. E. Meyer, D. Mutukwa, N. Zingwe, R. Taziwa, Lead-free halide double perovskites: a review of the structural, optical, and stability properties as well as their viability to replace lead halide perovskites. Metals **8**(9), 667 (2018). https://doi.org/10.3390/met8090667
72. G. Min, R.J.E. Westbrook, M. Jiang, M. Taddei, A. Li, T. Webb, S. Sathasivam, A. Azaden, R.G. Palgrave, D.S. Ginger, T.J. Macdonald, S.A. Haque, The effect of antisolvent treatment on the growth of 2D/3D tin perovskite films for solar cells. ACS Energy Lett. **10**(1), 254–262 (2025). https://doi.org/10.1021/acsenergylett.4c02745
73. Y. Ming, Y. Jiang, J. Li, J. Huang, P. Xiang, C. Qiu, Y. Rong, Design and upgrade of mesoporous perovskite solar cells. Chem. Inorg. Mater. **3**, 100050 (2024). https://doi.org/10.1016/j.cinorg.2024.100050
74. R.K. Misra, S. Aharon, M. Layani, S. Magdassi, L. Etgar, A mesoporous-planar hybrid architecture of methylammonium lead iodide perovskite based solar cells. J. Mater. Chem. A **4**(37), 14423–14429 (2016). https://doi.org/10.1039/c6ta06960f
75. M. Nur-E-Alam, M.A. Islam, Y.B. Kar, T.S. Kiong, H. Misran, M.U. Khandaker, Y. Fouad, M.E.M. Soudagar, E. Cuce, Anti-solvent materials enhanced structural and optical properties on ambiently fabricated perovskite thin films. Sci. Rep. **14**(1), 344 (2024). https://doi.org/10.1038/s41598-024-70344-3
76. D.O. Obada, S.B. Akinpelu, S.A. Abolade, E. Okafor, A.M. Ukpong, S.R. Kumar, A. Akande, Lead-free double perovskites: a review of the structural, optoelectronic, mechanical, and thermoelectric properties derived from first-principles calculations, and materials design applicable for pedagogical purposes. Crystals **14**(1), 86 (2024). https://doi.org/10.3390/cryst14010086
77. J. Oh, H. Lee, M.Y. Ryu, Effect of antisolvent application time on $CH_3NH_3PbI_3$ solar cells in ambient atmosphere. Curr. Appl. Phys. **56**, 119–125 (2023). https://doi.org/10.1016/j.cap.2023.10.009
78. Y. Oteki, Y. Okada, External-quantum-efficiency enhancement in quantum-dot solar cells with a Fabry-Perot light-trapping structure. Heliyon **9**(8), e19312 (2023). https://doi.org/10.1016/j.heliyon.2023.e19312
79. M. Palummo, E. Berrios, D. Varsano, G. Giorgi, Optical properties of lead-free double perovskites by Ab initio excited-state methods. ACS Energy Lett. **5**(2), 457–463 (2020). https://doi.org/10.1021/acsenergylett.9b02593
80. S.V.N. Pammi, R. Maddaka, V.D. Tran, J.H. Eom, V. Pecunia, S. Majumder, M.D. Kim, S.G. Yoon, CVD-deposited hybrid lead halide perovskite films for high-responsivity, self-powered photodetectors with enhanced photo stability under ambient conditions. Nano Energy **74**, 104872 (2020). https://doi.org/10.1016/j.nanoen.2020.104872
81. Y. Pan, X. Wang, Y. Xu, Y. Li, E.E. Elemike, A. Shuja, Q. Li, X. Zhang, J. Chen, Z. Zhao, W. Lei, Enhanced performance of perovskite single-crystal photodiodes by epitaxial hole blocking layer. Front. Chem. **8**, 791 (2020). https://doi.org/10.3389/fchem.2020.00791

82. R. Patidar, D. Burkitt, K. Hooper, D. Richards, T. Watson, Slot-die coating of perovskite solar cells: an overview. Mater. Today Commun. **22**, 808 (2020). https://doi.org/10.1016/j.mtcomm.2019.100808
83. N.K. Pendyala, A. Kolay, Y.N. Ponnada, A. Guerrero, L. Etgar, Evolving solar cell manufacturing: the promising outlook of open-air perovskite printing. Sustain. Energy Fuels **9**(7), 1633–1655 (2025). https://doi.org/10.1039/d5se00002e
84. A.A. Petrov, A.A. Ordinartsev, S.A. Fateev, E.A. Goodilin, A.B. Tarasov, Solubility of hybrid halide perovskites in dmf and dmso. Molecules **26**(24), 7541 (2021). https://doi.org/10.3390/molecules26247541
85. M. Piot, J.E.S. Alonso, K.P.S. Zanoni, N. Rodkey, F. Ventosinos, C. Roldán-Carmona, M. Sessolo, H. Bolink, Fast coevaporation of 1 μm thick perovskite solar cells. ACS Energy Lett. **8**(11), 4711–4713 (2023). https://doi.org/10.1021/acsenergylett.3c01724
86. D. Pourjafari, N.G. García-Peña, W.Y. Padrón-Hernández, D. Peralta-Domínguez, A.M. Castro-Chong, M. Nabil, R.C. Avilés-Betanzos, G. Oskam, Functional materials for fabrication of carbon-based perovskite solar cells: ink formulation and its effect on solar cell performance. Materials **16**(11), 113917 (2023). https://doi.org/10.3390/ma16113917
87. A.K. Pramod, M. Raj Subramaniam, S.A. Hevia, S.K. Batabyal, Synthesis of lead-free $Cs_3Sb_2Cl_3Br_6$ halide perovskite through solution processing method for self-powered photodetector applications. Mater. Lett. **306**, 130874 (2022). https://doi.org/10.1016/j.matlet.2021.130874
88. L. Qiu, S. He, Y. Jiang, Y. Qi, Metal halide perovskite solar cells by modified chemical vapor deposition. J. Mater. Chem. A **9**(40), 22759–22780 (2021). https://doi.org/10.1039/d1ta06459b
89. L. Qiu, S. He, Z. Liu, L.K. Ono, D.Y. Son, Y. Liu, G. Tong, Y. Qi, Rapid hybrid chemical vapor deposition for efficient and hysteresis-free perovskite solar modules with an operation lifetime exceeding 800 hours. J. Mater. Chem. A **8**(44), 23404–23412 (2020). https://doi.org/10.1039/d0ta09007g
90. S.N. Rahane, G.K. Rahane, A. Mandal, Y. Jadhav, A. Godha, A. Rokade, S. Shah, Y. Hase, A. Waghmare, N.G. Saykar, A. Roy, K.N. Salgaonkar, D. Dubal, S.K. Makineni, N.Y. Dzade, S.R. Jadkar, S.R. Rondiya, Lead-free $Cs_2AgBiCl_6$ double perovskite: experimental and theoretical insights into the self-trapping for optoelectronic applications. ACS Phys. Chem. Au **4**(5), 476–489 (2024). https://doi.org/10.1021/acsphyschemau.4c00008
91. S. Rahimi, M. Eskandari, D. Fathi, New nanostructure perovskite-based light-emitting diode with superior light extraction efficiency enhancement. Sci. Rep. **14**(1), 55951 (2024). https://doi.org/10.1038/s41598-024-55951-4
92. E. Reyes-Francis, C. Echeverría-Arrondo, D. Esparza, T. López-Luke, T. Soto-Montero, M. Morales-Masis, S.H. Turren-Cruz, I. Mora-Seró, B. Julián-López, Microwave-mediated synthesis of lead-free cesium titanium bromide double perovskite: a sustainable approach. Chem. Mater. **36**(3), 1728–1736 (2024). https://doi.org/10.1021/acs.chemmater.3c03108
93. S. Sanders, D. Stümmler, P. Pfeiffer, N. Ackermann, G. Simkus, M. Heuken, P.K. Baumann, A. Vescan, H. Kalisch, Chemical vapor deposition of organic-inorganic bismuth-based perovskite films for solar cell application. Sci. Rep. **9**(1), 46199 (2019). https://doi.org/10.1038/s41598-019-46199-4
94. M.A. Sarker, M.M. Hasan, M.A. Momin, A. Irfan, M.R. Islam, A. Sharif, Band gap engineering in lead free halide cubic perovskites $GaGeX_3$ (X = Cl, Br, and I) based on first-principles calculations. RSC Adv. **14**(12), 9805–9818 (2024). https://doi.org/10.1039/d4ra00224e
95. K.J. Savill, A.M. Ulatowski, L.M. Herz, Optoelectronic properties of tin-lead halide perovskites. ACS Energy Lett. **6**(7), 2413–2426 (2021). https://doi.org/10.1021/acsenergylett.1c00776
96. J. Shen, W. Zhu, Z. Lian, A. Lin, S.F. Shi, K. Yang, M. Li, D. Zhao, Y.Y. Sun, J. Lian, Metal ion-incorporated lead-free perovskites toward broadband photodetectors. ACS Appl. Electron. Mater. **5**(10), 5291–5302 (2023). https://doi.org/10.1021/acsaelm.2c01265

97. Y. Shen, L.P. Cheng, Y.Q. Li, W. Li, J.D. Chen, S.T. Lee, J.X. Tang, High-efficiency perovskite light-emitting diodes with synergetic outcoupling enhancement. Adv. Mater. **31**(24), 1517 (2019). https://doi.org/10.1002/adma.201901517
98. M.D. Smith, H.I. Karunadasa, White-light emission from layered halide perovskites. Acc. Chem. Res. **51**(3), 619–627 (2018). https://doi.org/10.1021/acs.accounts.7b00433
99. L. Song, Perovskite solar cells toward industrialization: Screen printed perovskite films. Mater. Rep. Energy **2**(4), 100171 (2022). https://doi.org/10.1016/j.matre.2022.100171
100. X. Song, T. Li, H. Li, S. Lin, J. Yin, K. Zhao, High hardness metal-free perovskite based on hexamethylenetetramine for efficient X-ray detection. Sci. China Mater. **67**(4), 1348–1355 (2024). https://doi.org/10.1007/s40843-023-2815-x
101. V.J. Sorger, R. Maiti, Roadmap for gain-bandwidth-product enhanced photodetectors: opinion. Opt. Mater. Exp. **10**(9), 2192 (2020). https://doi.org/10.1364/ome.400423
102. S.D. Stranks, H.J. Snaith, Metal-halide perovskites for photovoltaic and light-emitting devices. Nat. Nanotechnol. **10**(5), 391–402 (2015). https://doi.org/10.1038/nnano.2015.90
103. H. Tahara, T. Yamada, T. Handa, Y. Kanemitsu, *Nonlinear Optical Properties of Lead Halide Perovskites* (n.d.)
104. Y. Tang, G. Pu, Y. Tang, T. Sun, M. Wang, J. Wang, Recent advances in fast-decaying metal halide perovskites scintillators. J. Phys. Chem. Lett. **15**(27), 7036–7044 (2024). https://doi.org/10.1021/acs.jpclett.4c01310
105. M.M. Tavakoli, L. Gu, Y. Gao, C. Reckmeier, J. He, A.L. Rogach, Y. Yao, Z. Fan, Fabrication of efficient planar perovskite solar cells using a one-step chemical vapor deposition method. Sci. Rep. **5**, 14083 (2015). https://doi.org/10.1038/srep14083
106. A.D. Taylor, Q. Sun, K.P. Goetz, Q. An, T. Schramm, Y. Hofstetter, M. Litterst, F. Paulus, Y. Vaynzof, A general approach to high-efficiency perovskite solar cells by any antisolvent. Nat. Commun. **12**(1), 22049 (2021). https://doi.org/10.1038/s41467-021-22049-8
107. C. Teixeira, R. Fuentes-Pineda, L. Andrade, A. Mendes, D. Forgács, Fabrication of low-cost and flexible perovskite solar cells by slot-die coating for indoor applications. Mater. Adv. **4**(17), 3863–3873 (2023). https://doi.org/10.1039/d3ma00285c
108. H. Tokizawa, X. Zhao, M. Murayama, Influence of DMSO/DMF ratio on the crystal growth and optical properties of Sn-based perovskite films. Micro Nano Eng. **28**, 100302 (2025). https://doi.org/10.1016/j.mne.2025.100302
109. J. Tong, Q. Jiang, A.J. Ferguson, A.F. Palmstrom, X. Wang, J. Hao, S.P. Dunfield, A.E. Louks, S.P. Harvey, C. Li, H. Lu, R.M. France, S.A. Johnson, F. Zhang, M. Yang, J.F. Geisz, M.D. Mcgehee, M.C. Beard, Y. Yan, et al., *Carrier Control in Sn–Pb Perovskites Via 2D Cation Engineering for All-Perovskite Tandem Solar Cells with Improved Efficiency and Stability* (n.d.)
110. Y. Tu, J. Ye, G. Yang, Y. Zang, L. Zhang, Y. Wang, G. Li, L. Chu, W. Yan, Slot-die coating fabrication of perovskite solar cells toward commercialization. J. Alloys Compd. **942**, 169104 (2023). https://doi.org/10.1016/j.jallcom.2023.169104
111. E. Ünver, A. Toprak, A.A. Hussaini, M. Kuş, M. Yıldırım, Investigation and calculation of electrical performance of lead-free $AgBiI_4$ perovskite based Schottky photodiode using machine learning. J. Mater. Sci. Mater. Electron. **36**(11), 9 (2025). https://doi.org/10.1007/s10854-025-14728-9
112. L. Vesce, M. Stefanelli, F. Rossi, L.A. Castriotta, R. Basosi, M.L. Parisi, A. Sinicropi, A. Di Carlo, Perovskite solar cell technology scaling-up: eco-efficient and industrially compatible sub-module manufacturing by fully ambient air slot-die/blade meniscus coating. Prog. Photovolt. Res. Appl. **32**(2), 115–129 (2024). https://doi.org/10.1002/pip.3741
113. Q. Wan, W. Zheng, C. Zou, F. Carulli, C. Zhang, H. Song, M. Liu, Q. Zhang, L.Y. Lin, L. Kong, L. Li, S. Brovelli, Ultrathin light-emitting diodes with external efficiency over 26% based on resurfaced perovskite nanocrystals. ACS Energy Lett. **8**(2), 927–934 (2023). https://doi.org/10.1021/acsenergylett.2c02802
114. A. Wang, C. Zuo, X. Niu, L. Ding, J. Ding, F. Hao, Recent promise of lead-free halide perovskites in optoelectronic applications. Chem. Eng. J. **451**, 138926 (2023). https://doi.org/10.1016/j.cej.2022.138926

115. M. Wang, W. Wang, B. Ma, W. Shen, L. Liu, K. Cao, S. Chen, W. Huang, Lead-free perovskite materials for solar cells. Nano-Micro Lett. **13**(1), 578 (2021). https://doi.org/10.1007/s40820-020-00578-z
116. Q. Wang, X. Zhao, M. Luo, Y. Li, J. Liu, X. Wu, High-Q fabry-pérot cavity based on microlens array for refractive index sensing. Photon. Sens. **14**(4), 716 (2024). https://doi.org/10.1007/s13320-024-0716-2
117. R. Wang, T. Huang, J. Xue, J. Tong, K. Zhu, Y. Yang, Prospects for metal halide perovskite-based tandem solar cells. Nat. Photon. **15**(6), 411–425 (2021). https://doi.org/10.1038/s41566-021-00809-8
118. S. Wang, D. Han, C. Maheu, Z. Xu, A. Biewald, H. Illner, R. Hooijer, T. Mayer, A. Hartschuh, H. Ebert, T. Bein, Room-temperature synthesis of lead-free copper(I)-antimony(III)-based double perovskite nanocrystals. APL Mater. **11**(4), 4708 (2023). https://doi.org/10.1063/5.0144708
119. T. Wu, Z. Qin, Y. Wang, Y. Wu, W. Chen, S. Zhang, M. Cai, S. Dai, J. Zhang, J. Liu, Z. Zhou, X. Liu, H. Segawa, H. Tan, Q. Tang, J. Fang, Y. Li, L. Ding, Z. Ning et al., The main progress of perovskite solar cells in 2020–2021. Nano-Micro Lett. **13**(1), 313 (2021). https://doi.org/10.1007/s40820-021-00672-w
120. Z. Wu, S. Bai, J. Xiang, Z. Yuan, Y. Yang, W. Cui, X. Gao, Z. Liu, Y. Jin, B. Sun, Efficient planar heterojunction perovskite solar cells employing graphene oxide as hole conductor. Nanoscale **6**(18), 10505–10510 (2014). https://doi.org/10.1039/c4nr03181d
121. X. Xiao, K. Wang, T. Ye, R. Cai, Z. Ren, D. Wu, X. Qu, J. Sun, S. Ding, X.W. Sun, W.C.H. Choy, Enhanced hole injection assisted by electric dipoles for efficient perovskite light-emitting diodes. Commun. Mater. **1**(1), 84 (2020). https://doi.org/10.1038/s43246-020-00084-0
122. W.W. Xie, C.L. Du, Y.M. Luo, S.Y. Cao, Design and optimization of high responsivity and detectivity of lead-free tin-based perovskite photodetectors by numerical simulation. J. Phys. Chem. Solids **207**, 112919 (2025). https://doi.org/10.1016/j.jpcs.2025.112919
123. H. Xiong, F. Zabihi, H. Wang, Q. Zhang, M. Eslamian, Grain engineering by ultrasonic substrate vibration post-treatment of wet perovskite films for annealing-free, high performance, and stable perovskite solar cells. Nanoscale **10**(18), 8526–8535 (2018). https://doi.org/10.1039/c8nr00540k
124. K. Xu, A. Al-Ashouri, Z.W. Peng, E. Köhnen, H. Hempel, F. Akhundova, J.A. Marquez, P. Tockhorn, O. Shargaieva, F. Ruske, J. Zhang, J. Dagar, B. Stannowski, T. Unold, D. Abou-Ras, E. Unger, L. Korte, S. Albrecht, Slot-die coated triple-halide perovskites for efficient and scalable perovskite/silicon tandem solar cells. ACS Energy Lett. **7**(10), 3600–3611 (2022). https://doi.org/10.1021/acsenergylett.2c01506
125. P. Yadav, S. Khurana, D. Gill, A. Rahman, S. Bhattacharya, S. Sapra, Chemical control of optical band gap and magnetic properties in Fe-alloyed Na–In halide layered perovskites. Chem. Mater. **36**(11), 5456–5464 (2024). https://doi.org/10.1021/acs.chemmater.4c00393
126. W. Yan, W. Yang, K. Zhang, H. Yu, Y. Yang, H. Fan, Y. Qi, H. Xin, Enhancing performance and stability of perovskite solar cells through surface defect passivation with organic bidentate lewis bases. ACS Omega **7**(36), 32383–32392 (2022). https://doi.org/10.1021/acsomega.2c03802
127. B. Yang, K. Han, Ultrafast dynamics of self-trapped excitons in lead-free perovskite nanocrystals. J. Phys. Chem. Lett. **12**(34), 8256–8262 (2021). https://doi.org/10.1021/acs.jpclett.1c01828
128. Y. Yang, Z. Huang, H. Gao, Z. Xu, W. Fang, Y. Chen, Y. Hu, Z. Yi, J. Huang, H. Zhu, Green anti-solvent engineering for high-efficiency and environmentally friendly perovskite solar cells. RSC Adv. **14**(44), 32370–32388 (2024). https://doi.org/10.1039/d4ra05082g
129. F. Yuan, G. Folpini, T. Liu, U. Singh, A. Treglia, J.W.M. Lim, J. Klarbring, S.I. Simak, I.A. Abrikosov, T.C. Sum, A. Petrozza, F. Gao, Bright and stable near-infrared lead-free perovskite light-emitting diodes. Nat. Photon. **18**(2), 170–176 (2024). https://doi.org/10.1038/s41566-023-01351-5
130. Q. Zhang, D. Zhang, B. Cao, S. Poddar, X. Mo, Z. Fan, Improving the operational lifetime of metal-halide perovskite light-emitting diodes with dimension control and ligand engineering. ACS Nano **18**(12), 8557–8570 (2024). https://doi.org/10.1021/acsnano.3c13136

131. D. Zhao, Z. Yang, Z. Mo, D. Wu, H. Li, J. Yang, L. Zeng, X. Yang, Y.H. Tsang, L. He, A self-powered WSe_2/SivdW heterojunction photodiode with high carrier collection efficiency for noninvasive multimodal biosensing applications. ACS Photon. **1**, 2275 (2025). https://doi.org/10.1021/acsphotonics.5c02275
132. D.G. Zheng, D.H. Kim, Degradation mechanisms of perovskite light-emitting diodes under electrical bias. Nanophotonics **12**(3), 451–476 (2023). https://doi.org/10.1515/nanoph-2022-0569
133. Z.H. Zheng, H.B. Lan, Z.H. Su, H.X. Peng, J.T. Luo, G.X. Liang, P. Fan, Single source thermal evaporation of two-dimensional perovskite thin films for photovoltaic applications. Sci. Rep. **9**(1), 609 (2019). https://doi.org/10.1038/s41598-019-53609-0
134. H. Zhong, Y. Wang, L. Hou, X. Song, F. Zhang, Advancing roll-to-roll manufacturing of flexible perovskite solar cells from lab to fab. ACS Appl. Mater. Interf. **17**(36), 50168–50190 (2025). https://doi.org/10.1021/acsami.5c14588
135. Y. Zou, Z. Yuan, S. Bai, F. Gao, B. Sun, Recent progress toward perovskite light-emitting diodes with enhanced spectral and operational stability. Mater. Today Nano **5**, 100028 (2019). https://doi.org/10.1016/j.mtnano.2019.100028

Chapter 2
Synthesis, Properties, and Applications of Lead-Free Perovskites

Abhishek Kumar Srivastava, C. K. Pandey, Omwati Rana, Pratigya Tripathi, Shilpi Srivastava, and Manisha Bajpai

Abstract Lead (Pb)-free perovskite materials (PMs) have gained attention as a safe substitute for conventional lead-based halide PMs for next-generation optoelectronic applications. These materials offer the key benefits of lead perovskites, such as strong absorption in the visible region, less toxic, environmentally friendly nature, tunableband gaps, and long carrier lifetimes. At present there is an advancement in the synthesis route, such as solution processing, hot-injection methods, and solid-state reactions, which enables researchers to fabricate high-quality thin films and nanocrystals. Due to the toxicity associated with the Pb, the Pb^{2+} ion is successfully replaced by the Sn^{2+}, Ge^{2+}, Bi^{3+}, Sb^{3+} ions. Many of these show improved stability against heat, moisture, and photodegradation. Double perovskites are particularly interesting because of their structural properties and tunable optoelectronic behaviour, which make them relevant for solar cells, LEDs, photodetectors, and X-ray detection. But significant hurdles remain. Sn^{2+} and Ge^{2+} oxidize readily, some low-dimensional structures exhibit poor charge transport, and device efficiencies still lag behind lead-based benchmarks. This chapter covers the primary synthesis routes, structural characteristics, and optoelectronic behaviour of lead-free perovskites. Recent device improvements are reviewed in this study, and applications across different optoelectronic platforms are discussed, with particular attention to stability enhancement and future pathways toward environmentally friendly, high-performance materials.

Keywords Lead-free perovskite · Synthesis · Optoelectronic properties · Double perovskite · Photovoltaics · Light-emitting diodes

A. K. Srivastava · M. Bajpai (✉)
Department of Physics, Siddharth University, Pachangawa, Uttar Pradesh, India
e-mail: mansa83@gmail.com

C. K. Pandey
Department of Physics, S.N.S. College, Motihari, BRABU, Muzaffarpur, Bihar, India

O. Rana
Department of Physics, Daulat Ram College, University of Delhi, Delhi, India

P. Tripathi · S. Srivastava
Department of Chemistry, Siddharth University, Pachangawa, Uttar Pradesh, India

V. Chauhan et al. (eds.), *Sustainable Lead-Free Perovskite Materials*, Topics in Applied Physics 157, https://doi.org/10.1007/978-3-032-24873-2_2

2.1 Introduction

Lead-based halide perovskites, represented as APbX3, where A stands for an organic cation and X stands for a halide metal ion, have gained significant research interest for optoelectronic applications [1–4]. Lead-based materials were first discovered in the late nineteenth century. Numerous studies have been done to study these materials until Kojima et al. demonstrated the potential of lead material for solar cells in 2009 and emission of light in 2012 [5]. Tan et al. achieved a remarkable advancement for LEDs in 2014 [6]. The advancement of hybrid perovskite solar cells (PSCs) in 2012 led to an advancement in the perovskite market for future optoelectronics applications [7, 8]. Due to its optical and electronic properties, and chemical flexibility of lead-halide perovskite materials (PMs), it has gained attention in the global market [9–11]. Numerous works have been done in the various fields, including solar cells, LEDs, photodetectors, humidity sensors, lasers, supercapacitors, and X-ray detectors [12–18]. PMs absorb light efficiently in their visible range, have tunable band gaps, mobile charge carriers that exhibit long lifetimes in comparison to other materials, long diffusion lengths, and low-cost solution-based fabrication, making them a sustainable candidate in the field of photovoltaic (PV). The high defect tolerance associated with the lead compounds due to which the researchers are moving towards next-generation devices. The power conversion efficiency of single-junction lead-halide perovskite solar cells (PSCs) has increased significantly, rising from 3.8% to over 25% [19]. Perovskite–silicon tandems have exceeded PCE over 32% [20]. Perovskite based LEDs have made a progress in comparison to previously reported data, with external quantum efficiencies (EQE) exceeding over 25% in red and green emission [21, 22]. Present research efforts led to increase in the performance and stability by optimizing device structure, and device passivation strategies to successfully suppress the trap charge carriers and non-radiative recombination which arises at the interface [23].

Despite huge advancements in both thin-film and lead-halide perovskites, the toxicity and stability are the major concerns due to that they continue to restrict the commercial market as compared to silicon-based technology [24–26]. Various research has been done, 2D/3D mixed perovskites, surface defect passivation [27], but PSCs remain far from the silicon-based devices in terms of stability [28]. Perovskite-based LEDs have degraded within ten hours, which shows their limited longevity [29]. Replacing Pb in $CsPbBr_3$ with low-toxic metal ions offers a resultant path towards sustainable materials. The challenges in the fabrication of perovskite-based structures are very obvious, in which substitution plays an important role which may reduce environmental risks without affecting the optical performance or detecting quality [30]. Lead-halide PMs usually degrade under heat, UV light, moisture, and humidity. While degrading, it releases harmful byproducts like PbI_2, which can leach into water systems and spread into the environment. Researchers have therefore focused on replacing Pb in ABX3 perovskite materials with heavy elements from the s and p blocks in the periodic table [31–33]. Sn^{2+} and Ge^{2+} look promising in recent reported studies, but their reactive 5 s orbitals oxidizes very easily to Sn^{4+} and Ge^{4+} [31]. Bi^{3+} and Sb^{3+}, both have the same electronic configuration as Pb^{2+}, generally form low-dimensional $A_3B_2X_9$ structures with facesharing $[Bi_2I_9]^{3-}$ group [34–36]. These structures oppose charge transport and cause carrier localization [37–40]. These limitations have successfully pushed researchers toward double PMs, where two different metals replace the single lead ion in regular ABX_3 structures.

$A_2BB'X_6$-type double perovskites offer promise as stable alternatives to lead-based halide perovskites. The double perovskite structure allows different combinations of monovalent (1+) and trivalent (3+) metal ions, including Bi^{3+}, Sb^{3+}, In^{3+}, Cu^{+}, and Ag^{+} [41–43]. The oxide analogues ($A_2BB'O_6$) have been studied extensively and display diverse physical properties, including ferromagnetism, ferroelectricity, colossal magnetoresistance, and multiferroicity. The oxide-based double PMs have been widely used in scintillation detection techniques [44, 45]. Double perovskites have been further studied due to their tunable band gaps and broad light-absorbing properties [46]. Among these materials, lead-free $Cs_2AgBiBr_6$comes as a viable candidate with a strong thermal as well as environmental stability [47], it exhibit long charge carrier lifetimes [41], and low effective masses [44]. Early demonstrations of solar cells based on $Cs_2AgBiBr_6$have reported promising efficiencies [48–50], though the highest PCE_S of metal-halide double perovskites ($MHDP_S$) are still below those of Pb-based perovskites [50]. In recent years, a number of review articles have focused mainly on synthesis approaches and crystal structures of metal-halide PM_S [51–55].

This chapter reviews recent progress in lead-free halide perovskite nanomaterials, examining structure engineering approaches, methods for controlling morphology and dimensions, and the resulting effects on optoelectronic properties. Applications in LED_S, solar cells, and photodetectors are also discussed.

2.2 Synthesis Approaches for Lead-Free Perovskite Materials (PMs)

2.2.1 Synthesis and Properties of Tin (Sn^{2+}) Halide PMs

Tin (Sn^{2+}), which belongs to the same Group IVA family as lead (Pb^{2+}), shares several important chemical similarities with it. It has ionic radius (1.35 Å) is close to the ionic radius of Pb^{2+} (1.49 Å), and both ions contain a comparable lone-pair s orbital. Because of these parallels, tin-based perovskites can achieve optoelectronic behaviour very similar to that of lead-based materials [56, 57]. There are several methods for the synthesis of these materials, including hot injection (HI), ligand-assisted reprecipitation (LARP), chemical vapour deposition (CVD), and hydrothermal methods [58].

2.2.1.1 Hot Injection Method

Nanocrystals (NCs) prepared using the HI method were firstly reported by Kovalenko et al. [59] with excellent optical and structural properties. Building on this approach, Jellicoe et al. [60] successfully synthesized $CsSnX_3$ NCs using the same process in an oxygen-free environment. In this method, SnX_2 powder is dissolved in the mildly reducing solvent tri-n-octylphosphine (TOP) to form a Sn–X–TOP solution. This solution is then rapidly injected into a Cs-oleate precursor containing oleic acid (OA) and oleylamine (OAm) at 170 °C. The reaction was completed in less than one minute and was immediately quenched in an ice–water bath. The powder XRD pattern shown in Fig. 2.1a confirms the successful production of colloidal $CsSnX_3$ nanocrystals (NCs) with different halide compositions including Cl (Chloride), Cl0.5Br (Bromide)0.5, Br, Br0.5I0.5, I(Iodide). Chloride-based NCs crystallize in a Pm3m cubic space group, whereas both Br and I counterparts exhibit perovskite-like structures in a Pnma orthorhombic phase with slightly lower symmetry, allowing their optical bandgap can be tuned throughout the visible spectrum to the near-infrared spectrum (as shown in Fig. 2.1b). The resulting NCs, especially $CsSnI_3$ NCs, were cubic and smaller than 20 nm (Fig. 2.1c). While OAm and OA were used as surface ligands to stabilize the $CsSnX_3$'s structure and morphology, the NCs were highly unstable and degraded within minutes under normal conditions. The $CsSnI_3$ NCs also had a very low photoluminescence quantum yield (PLQY) of less than 1%, suggesting numerous structural defects that cause light to be lost as heat rather than emitted. A drawback of this synthesis is that a large quantity of TOP is required. Trioctylphosphine oxide (TOPO), a proven effective alternative for lead-based NCs, could potentially replace TOP. Transmission electron microscopy (TEM) images of $CsSnI_3$ nanocrystals revealed a well-defined cubic shape with a uniform particle size of approximately 10 nm, as shown in Fig. 2.1c. By adjusting the halide ion composition, the bandgap of $CsSnX_3$ shall be finely tuned from 2.8 to 1.3 eV, covering a broad spectral range from the visible spectrum region to the near-infrared region. Additionally, for $CsSnBr_3$ nanocrystals, the optical bandgap can be shifted between 630 and 680 nm by changing the nanocrystal size. However, the maximum photoluminescence quantum yield (PLQY) achieved for these Sn-based nanocrystals is only 0.14%, primarily because of the presence of non-radiative defect states, as confirmed by transient PL measurements. This is largely because Sn-based perovskites have a relatively low defect formation energy (~250 meV), which encourages the generation of high defect densities (up to 10^{19} cm^{-3}). Many of these defects introduce trap states within the bandgap, facilitating non-radiative recombination and significantly reducing photoluminescence efficiency.

Wang et al. [61] developed a TOP-free method for synthesizing $CsSnBr_3$ nanocrystals (NCs) using tin(II) 2-ethylhexanoate and $MgBr_2{\cdot}6H_2O$ as precursors. By injecting Cs-oleate into the Sn–Br solution at 230 °C, they achieved hollow nanocage structures through a self-assembly process. The reaction temperature played a crucial roleat 190 °C; only nanocubes formed, while higher temperatures led to the transformation into hollow nanocages. A distinct hollow morphology appeared only when $Sn(Oct)_2$ was used as the tin precursor, most likely because the bulky,

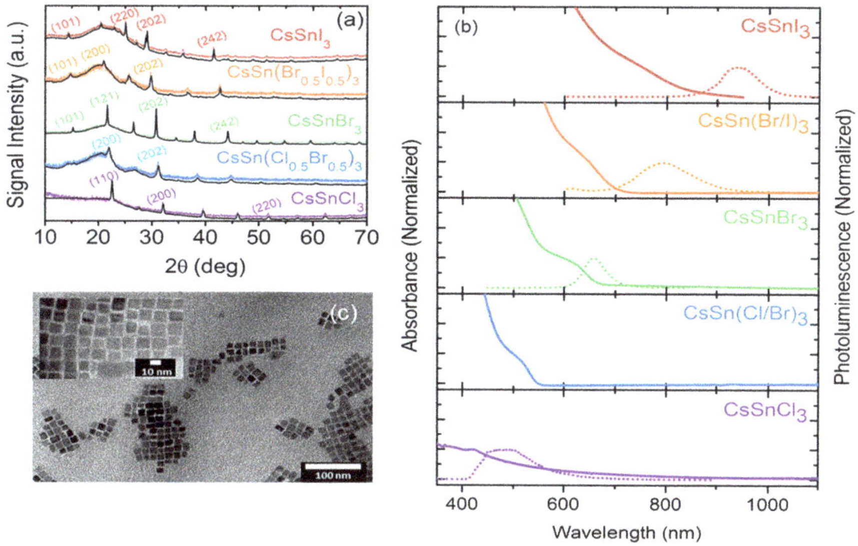

Fig. 2.1 (**a**) XRD patterns showing the crystal structure of $CsSnX_3$ nanocrystals across different halide ratios (X = Cl, $Cl_{0.5}Br_{0.5}$, Br, $Br_{0.5}I_{0.5}$, I). (**b**) Optical absorption and PL emission spectra for nanocrystals with single and mixed halide compositions. (**c**) TEM micrograph of $CsSnI_3$ NCs. Adapted from Ref. [60], © 2016, with permission from American Chemical Society (ACS)

branched ligand influenced the way the crystals formed. This synthesis was carried out entirely in ambient air, and the resulting nanocages showed noticeably better stability than the usual nanocubes. The steric hindrance imposed by the branched fatty-acid chains of phosphatidylcholine affected the growth kinetics of $CsSnX_3$ nanocrystals, allowing controlled formation with a photoluminescence quantum yield (PLQY) of 12%, which is the highest reported result among the previously reported studies [61]. As shown in Fig. 2.2a, $CsSnBr_3$ nanocages displays a broad absorption band which is centred around 655 nm, an emission peak at 685 nm, and a full width to the half maximum (FWHM) of 56 nm. The choice of the coordinates and the reaction temperature is totally responsible for the morphology and optical response of $CsSnBr_3$. The hollow nanocages are more stable in comparison to solid nanocube structures, As shown in Fig. 2.2b. This enhancement in the stability is due to the use of heavy electron-withdrawing ligands such as perfluorooctanoic acid, which act as surface-capping agents and help to protect Sn(II) on the nanocrystal surface [62]. Due to its high surface area and unique architecture, hollow nanostructures are emerging as a promising candidate for applications in drug delivery, energy storage, catalysis, and molecular separation. Using a hot-injection method, ultra-small (2–3 nm) organic–inorganic $MASnBr_2I$ nanocrystals were also synthesized. A deeper understanding of the nucleation and growth behaviour of $CsSnX_3$ nanocrystals formed through the hot-injection method is needed because of the more controlled and rational fabrication approaches for tin-halide perovskite nanocrystals with tunable structural and functional properties.

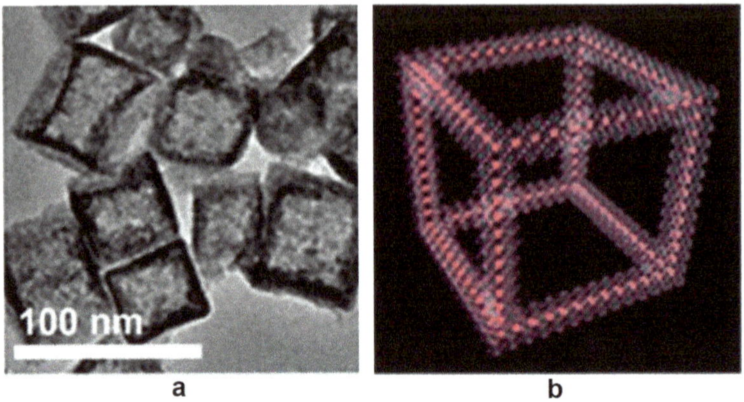

Fig. 2.2 (**a**) Transmission electron microscopy of $CsSnBr_3$ hollow nanostructures. (**b**) Structural model showing the cage-like architecture. Adapted from Ref. [61], © 2017. Reproduced with permission from the ACS

2.2.1.2 Ligand-Assisted Reprecipitation Method

This approach involves injecting a precursor dissolved in polar solvents like DMF or DMSO into nonpolar media such as hexane or toluene. The abrupt mixing triggers fast nucleation and crystal growth due to the sharp difference in how ions dissolve in each solvent. When solubility drops suddenly, the solution becomes supersaturated, causing crystallization to occur within seconds at room temperature.

Researchers have used the LARP method successfully to make different lead-based perovskite nanostructures, nanocubes, nanoplatelets, and nanosheets. For tin-based halide perovskites, though, this technique has mainly produced nanoplatelets. Weidman and colleagues first made Sn-based perovskite nanoplatelets following the formula $L_2[ABX_3]_{n-1}BX_4$ using a non-solvent crystallization approach. In these materials, L stands for long-chain organic ligands like butylammonium (BA) or octylammonium; A is a monovalent cation (Cs^+, FA^+, or MA^+); B is the metal cation (Pb^{2+} or Sn^{2+}); and n tells you how many perovskite layers are present, which sets the nanoplatelet thickness.

Weidman's team made L_2SnI_4 ($n = 1$) and $L_2[FASnI_3]SnI_4$ ($n = 2$) nanoplatelets by gradually adding a DMF solution containing BAI, CsI, and SnI_2 to toluene. By changing precursor ratios, they could control nanoplatelet thickness precisely, as shown in Fig. 2.3a–c. The long butylammonium chains took the place of FA^+ sites and limited crystal growth in one direction, producing thin, layered structures (Fig. 2.3b and d). Measurements of interlayer spacing (d-spacing) for $n = 1$ and for $n = 2$ nanoplatelets, it is approximately 1.7 nm and 2.3 nm, respectively. After accounting for the estimated ~ 0.5 nm length of the BA ligand, the actual thickness of the perovskite layer is about 0.6 nm for $n = 1$ and 1.2 nm for $n = 2$, aligning with the thickness of one and two perovskite unit cells Fig. 2.3d. Although the choice of A-site cation has only a minor influence on the absorption and emission energies

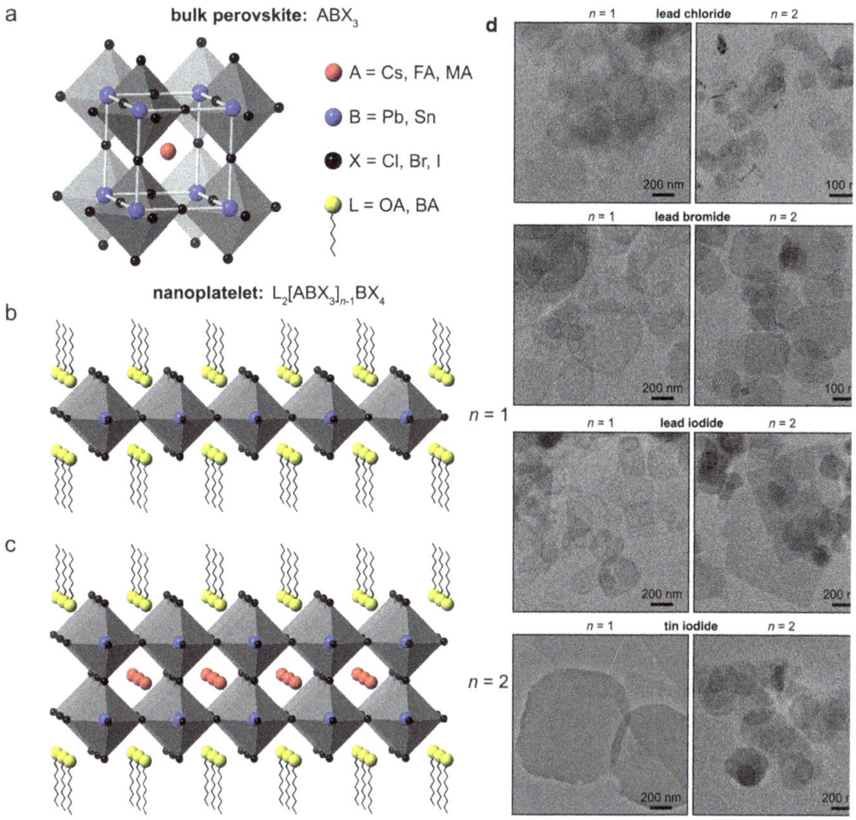

Fig. 2.3 (**a**) Unit cell diagram of the bulk perovskite (white cube) with labelled A, B, X, and L site occupancy (**b**, **c**) Structures of nanoplatelets containing n = 1 and n = 2 octahedral layers formed by metal–halide bonds. (**d**) Electron microscopy images showing $L_2[ABX_3]_{n-1}BX_4$ nanoplatelets with one and two layers (A = FA; B = Pb or Sn; X = Cl, Br, or I). Adapted from Ref. [63], © 2016. Reproduced with permission from the ACS

of nanoplatelets (NPLs), it can have a significant impact on their stability. When the NPLs are closer together, charge carriers can move more easily between them instead of staying trapped on individual platelets, which further improves the carrier mobility. The stronger connections between neighbouring NPLs also help photo generated carriers to last longer, which improves the light emission efficiency.

2.2.1.3 Chemical Vapour Deposition Method

Beside the solution-based synthesis routes, chemical vapour deposition (CVD) operates at high temperatures and generally involves longer growth times. In this method the reaction is very controlled by taking care of the various parameters such as vapour pressure and the flow of carrier gases. CVD is widely used to achieve epitaxial growth of low-dimensional perovskite structures, including nanowires, nanosheets, and other ordered architectures. The application of CVD for producing perovskite nanocrystals began around 2015, when $MAPbI_3$ nanowires of high structural quality were successfully fabricated through a two-step reaction using PbI_2 and MAI precursors. Since then, CVD-derived perovskite nanocrystals with reduced dimensionality have attracted considerable attention owing to their improved stability, superior crystallinity, and reliable performance across a range of optoelectronic devices [64]. Chen et al. developed a vapour-phase approach for growing horizontally aligned $CsSnX_3$ (X = Br, I) nanowires and microwires on mica substrates via high-temperature vapour-phase epitaxy (VPE) [65]. They began by preparing $CsSnX_3$ ingots at 550 °C, which then served as the source material. Fluorophlogopite mica substrates were chosen to promote epitaxial alignment.

The growth process used a custom CVD system (see Supporting Information, Figure S1 for details and schematic). $CsSnX_3$ powders sat at the centre of the heating zone and were brought to 280 °C (Fig. 2.4a). Argon gas carried the vaporized material downstream to a cooler region where freshly cleaved fluorophlogopite mica substrates were positioned $[KAl_2(Si_3Al)O_{10}(OH)_2](001)[KAl_2(Si_3Al)O_{10}(OH)_2](001)[KAl_2(Si_3Al)O_{10}(OH)_2](001)$ was positioned to enable epitaxial growth. It is important to note that the precursors used in this process were pre-melted $CsSnX_3$ ingots synthesized at 550 °C, which differs from earlier CsPbX nanowire growth methods that relied on CsX and PbX_2 mixtures. Attempts to use CsX + SnX_2 mixtures for $CsSnX_3$ nanowires were unsuccessful, producing only SnX_2 or SnX_4 deposits on mica (as shown for SnI_2 in Figure S2), likely due to the low melting point of SnX_2, which prevents proper perovskite phase formation. The $CsSnX_3$ (X = Br, I) perovskite nanowires and microwires prepared through vapour-phase epitaxy (VPE) exhibit high crystallinity, uniform structures, and tunable optical properties. The growth time and crystallization of tin halide PMs are typically very sensitive to Sn^{2+} purity, even small Sn^{4+} impurities degrade their crystal quality. The CVD method reduces this issue because SnI_4 decomposes into SnI_2 and I_2 at higher temperatures. Notably, when CsI and SnI_3 were used to produce $CsSnI_3$ ingots, XRD confirmed that the resulting nanowires retained a pure orthorhombic $CsSnI_3$ phase, comparable to that obtained using CsI and SnI_2, while optical microscopy images showed well-defined $CsSnI_3$ nanowires with morphologies dependent on growth time. After 1 h of reaction, the NWs had diameters ranging from hundreds of nanometres to several micrometres, and lengths of several hundred micrometres. Elemental mapping confirmed the uniform distribution of Cs, Sn, and I with an approximate 1:1:3 ratio. When the growth time was extended to 2 h, the NWs grew longer and wider, eventually forming interconnected hexagonal networks, as shown in Fig. 2.4b, c, d.

Their photoluminescence (PL) characteristics confirm strong emission, with bandgaps of ~ 1.3 eV ($CsSnI_3$) and ~ 1.9 eV ($CsSnBr_3$)including intense photoluminescence at specific wavelengths (650 nm for $CsSnBr_3$ and 950 nm for $CsSnI_3$), with a good carrier lifetimes of 1.1 ns and 0.94 ns, respectively, Fig. 2.4e and f making them ideal for the next-generation optoelectronic applications such as solar cells, LEDs,

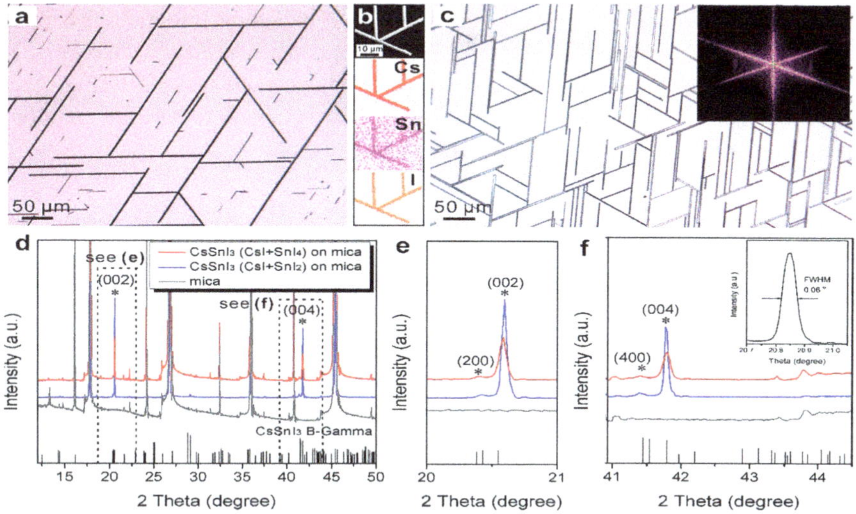

Fig. 2.4 Synthesis and structure of $CsSnI_3$ wire arrays on mica substrates. (**a**) Microscopy of wires prepared at 280 °C (1 h reaction) using CsI/SnI_4 precursors. (**b**) Energy-dispersive X-ray maps confirming homogeneous Cs, Sn, and I distribution throughout the nanowire assembly. (**c**) Wires obtained after 2 h heating at 280 °C from CsI/SnI_2, inset displays FFT analysis revealing hexagonal crystal alignment. (**d**) X-ray diffraction data comparing samples in panels (**a**) and (**c**). (**e, f**) Zoomed-in XRD sections corresponding to panel (**d**), panel (**f**) inset present the rocking curve measurement at the (004) peak for the 2 h growth. Adapted from Ref. [65], © 2019. Reproduced with permission from the ACS

and photodetectors. The wires also show reduced defect density, longer carrier lifetimes, and suppressed nonradiative recombination, particularly at temperatures below 225 K. Moreover, their thermal stability and anisotropic epitaxial growth behavior enhance their material robustness. These properties make CsSnX nanowires a strong lead-free candidate for the next-generation PVs for highly sustainable optoelectronics applications.

The crystallization in CVD occurs more slowly and in a more controlled manner in comparison to solution-based methods, that is why this approach is highly suitable for growing THP NPs with uniform and uniform structures. By tuning the substrate, precursor choice, growth temperature, and pressure, a diversity of well-defined shapes and morphologies can be obtained.

2.2.1.4 Hydrothermal Method

At present, perovskite films for optoelectronic applications are mainly prepared using either solution-based or vapour-based deposition methods. In the solution processing method, metal halide precursors are dissolved in high-boiling polar solvents such as dimethylformamide (DMF) or dimethyl sulfoxide (DMSO). In contrast, vapour-based techniques require controlled environments and well-specialized equipment.

Both methods require strict processing conditions, including low humidity, inert atmosphere, and high-vacuum setups. These requirements, combined with relatively low production output, are a limitation for making large-scale manufacturing. As a result, there is increasing interest in developing new synthesis routes that are more scalable and offer higher yields for perovskite production.

2.2.1.5 Challenges of Tin Halide PMs

Tin-based halide perovskite (THP) nanocrystals emerge as a strong contender in the field of PVs, but several key challenges still limit their ability in comparison with lead-based perovskites [66].

Limited Synthetic Strategies

The available synthesis methods for THP nanocrystals are still quite narrow, especially for organic–inorganic hybrid systems. Most of the previous research has been focused on allinorganic $CsSnX_3$, while hybrid materials such as $FASnI_3$, known for achieving high PCE which remain largely unexplored. Developing more flexible and low-temperature synthesis approaches beyond hot-injection, including LARP or hydrothermal methods, improves scalability and production efficiency.

Sn^{2+} Oxidation

One of the most serious issues in tin halide perovskite nanocrystals is the oxidation of Sn^{2+} to Sn^{4+}, which changes both the structure as well as its optoelectronic properties. This oxidation occurs even under inert atmospheres. Introducing reducing agents or suitable dopants during synthesis may help to prevent Sn^{4+} formation and reduce unwanted self-doping.

Surface and Structural Defects

The soft and highly ionic nature of tin halide perovskites makes them susceptible to surface reconstruction and internal defect formation. Improving stability and performance will require a better understanding of these surface states and the use of ligand engineering or epitaxial strategies to effectively passivate them.

Poor Environmental Stability

Tin-based perovskite NCs degrade rapidly when exposed to moisture, oxygen, heat, or light. Some protective approaches, such as strong surface-binding ligands, can slow degradation. Doping or alloying the perovskite NCs may also help to enhance the crystal lattice of the compounds.

Low Photoluminescence Quantum Yields (PLQYs)

The PL efficiency of tin halide NCsremains significantly lower than that of their lead-based counterparts. Metal doping, for example, with Mn^{2+} or In^{3+} is satisfied for Sn vacancies and improves optical properties, but this area is still not well explored.

Overall, addressing these challenges, the researcher should require better insight into the chemistry, such as crystallization, and defect behaviour of tin halide perovskite nanocrystals. By addressing the above challenges, these materials could become reliable and efficient lead-free candidates for future PV technologies.

2.2.2 *Synthesis and Properties of Lead-Free Double PMs*

The lead-free double PMs$Cs_2AgBiBr_6$ has attracted the global market due to its long carrier lifetime, straightforward synthesis, tunable optoelectronic properties, efficient carrier transport behaviour, and low defect tolerance. However, there are some challenges, including its indirect and wide bandgap, the presence of intrinsic defects, strong electron–phonon coupling, and several issues related to ion migration. Halide double perovskites ($A_2B^+B^{3+}X_6$) have become attractive for various optoelectronic applications because they maintain the benefits of the 3D corner-sharing perovskite structure without the toxicity of lead. These materials have vast potential for combinatorial chemistry because of the wide variety of possible monovalent (B^+) and trivalent (B^{3+}) metal ions. Since 2016, several types of halide double perovskites have been investigated, including Cs_2AgBiX_6 [67], $Cs_2AgSbCl_6$ [54, 67], $Cs_2AgInCl_6$, $(MA)_2AgBiBr_6$ [54, 67], $(MA)_2KBiCl_6$ [54], and $(MA)_2TlBiBr_6$ [54].

2.2.2.1 Hot Injection Method

Creutz et al. [68] successfully performed colloidal hot injection synthesis of heterometallic elpasolite nanocrystals for the first time, specifically Cs_2AgBiX_6 (X = Cl or Br) and further demonstrated a post synthetic ion-exchange strategy to access Cs_2AgBiI_6. This material has not yet been synthesized in any form. Beyond introducing new synthetic pathways for elpasolite nanocrystals with the potential for broad generalization, this work underscores the powerful role of nanocrystal chemistry in accelerating the discovery of novel materials.

The Cs_2AgBiX_6 nanocrystals was synthesised by mixing cesium acetate, silver acetate, and bismuth acetate were dissolved in solvent in appropriate amount. Cube-shaped Cs_2AgBiX_6 nanocrystals were synthesised followed by the reaction, centrifugation, and extraction. The Absorption (blue) and emission (red) spectra of $Cs_2AgBiCl_6$ (A) and $Cs_2AgBiBr_6$ (B) nanocrystal are shown in Fig. 2.5a and b respectively. Cs_2AgBiX_6 nanocrystals (~8–9 nm) exhibit well-defined absorption peaks at 3.39 eV (Cl) and 2.87 eV (Br), indicating indirect band gaps, consistent with bulk values. Tauc plots estimated the gaps as 2.9 eV (Cl) and 2.3 eV (). These nanocrystals show high extinction coefficients (~15,000 cm^{-1} M^{-1} per Bi^{3+}), with weak orange photoluminescence (PL) (~0.3% quantum yield) at room temperature, which brightens significantly at cryogenic temperatures (20 K). At 20 K, $Cs_2AgBiCl_6$ displays a broad emission at 1.68 eV, while $Cs_2AgBiBr_6$ shows two peaks: a broad

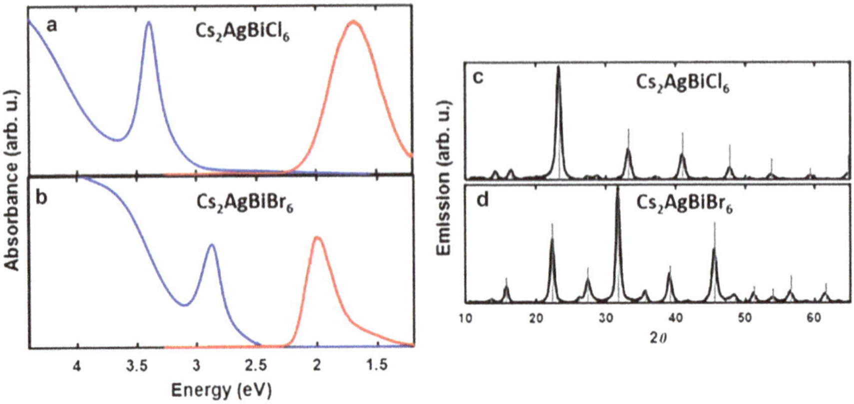

Fig. 2.5 (**a, b**) Optical characterization of 8 nm $Cs_2AgBiCl_6$ (**a**) and $Cs_2AgBiBr_6$ (**b**) nanocrystals showing absorption (blue curves) and photoluminescence (red curves). Absorption measurements were performed on hexane dispersions at room temperature, and emission data were collected from thin films at 20 K under 365 nm excitation for $Cs_2AgBiCl_6$ and 405 nm for $Cs_2AgBiBr_6$. (**c, d**) Diffraction patterns (black traces) obtained from ~ 13 nm $Cs_2AgBiBr_6$ (**c**) and $Cs_2AgBiBr_6$ (**d**) nanocrystals, displayed with reference patterns (grey lines, ICSD codes 239,874 and 239,875). Adapted from Ref. [68], © 2018. Reproduced with permission from the ACS

band at 1.75 eV and a sharper one at 1.97 eV, red shifting to 1.92 eV at room temperature. The broad emissions likely stem from defects or surface states, whereas the sharper peak in the Br variant may arise from a distinct process. PL excitation spectra confirm consistent origins of absorption and emission features.

Powder XRD patterns (black) for nanocrystals of $Cs_2AgBiCl_6$ (C) and $Cs_2AgBiBr_6$ (D) is shown in Fig. 2.5c and d respectively. XRD patterns confirm phase purity and structural identity of the nanocrystals, matching known double perovskite structures. The synthesized Cs_2AgBiX_6 (X = Cl, Br) nanocrystals displayed a uniform cube-shaped morphology with edge lengths around 8 nm and a narrow size distribution. The formation of phase-pure nanocrystals was found to be highly dependent on key parameters, such as precursor stoichiometry, ligand concentration, reaction time, and temperature. Beyond the direct synthesis of $Cs_2AgBiCl_6$ and $Cs_2AgBiBr_6$, a novel Cs_2AgBiI_6 composition was successfully obtained via postsynthetic anion exchange using trimethylsilyl iodide (TMSI) under an inert atmosphere, demonstrating the potential of ion-exchange strategies for expanding the halide double-perovskite library. These results highlight the promising optical properties of the Cs_2AgBiX_6 NCs for next generation PVs.

2.2.2.2 Mechanochemical Synthesis Using High-Energy Ball Milling

The thin films Double PMs are generally prepared by spin coating or via single crystals precipitation method [47, 69, 70], while other approaches such as hydrothermal synthesis, solvent evaporation, and solid-state method, have not been very effective in comparison to these methods [70, 71]. This difficulty arises from the competing formation of the layered $A_3B_2X_9$ phase, which has a comparable formation energy and often leads to mixed phases or the exclusive formation of $A_3B_2X_9$ instead of the desired $A_2BB'X_6$ structure. As a result, careful control of synthesis parameters is crucial for obtaining phase-pure $A_2BB'X_6$ double perovskites.

Among the existing fabrication techniques for Pb-based perovskites, mechanochemical synthesis stands out as a solvent-free method that offers short reaction times, low energy demand, and minimal reagent use. It has already proven effective for producing a wide variety of materials, including supported metal nanoparticles, metal oxides, composites, and MOFs, indicating strong potential for achieving pure double perovskite phases [72]. Garcia et al. [72] introduced a reliable mechanochemical route for synthesizing both inorganic and organic–inorganic hybrid double PMs by ball-milling bromide precursors in a 2:1:1 ratio to obtain polycrystalline $Cs_2AgBiBr_6$, $MA_2TlBiBr_6$, and $Cs_2AgSbBr_6$ under ambient high-energy milling conditions. The reaction yields for $Cs_2AgBiBr_6$ and $MA_2TlBiBr_6$ are very high (~100%), while $Cs_2AgSbBr_6$ showed a slightly lower yield due to some remaining precursors being left after milling. Overall, the results highlight that $Cs_2AgBiBr_6$ exhibits good crystallinity, defect-related optical activity, and favourable carrier dynamics, which make it a potential candidate for optoelectronic applications.

2.3 Applications of Lead-Free PMs

2.3.1 Lead-Free PMs in Solar Cell

Perovskite materials have gained global attention in next-generation PVtechnology because of their excellent optoelectronic properties [73]. Lead-free PMs offers tunableband gaps, broad spectral response, scalability, and low-cost synthesis, making them strong candidates to replace organic dyes in mesoscopic solar cells [54].

In particular, all-inorganic CsSnI QRs increase the optical density of the active layer while providing improved photostability and chemical durability. Their higher surface area compared to bulk counterparts increases interfacial contact, which enhances light absorption and boosts device performance. Solar cells based on CsSnX have already achieved PCE values close to 13% [74].

$Cs_2AgBiBr_6$ also stands out due to its high stability, long photoluminescence lifetime (~660 ns), and tunable bandgap, making it a promising option for LF-PSCs. Greul et al. first introduced it into solar cell architecture using one-step spin coating for thin-film fabrication [75]. Ning et al. later developed the first $Cs_2AgBiBr_6$-based device with the structure ITO/C-TiO_2/$Cs_2AgBiBr_6$/Spiro-OMeTAD/Au, reporting a PCE of 1.11% and identifying the coexistence of free carriers and excitons with a diffusion length of ~ 110 nm [48]. Igbari et al. [76], have done some improvements in their studies that compared vacuum-sublimated and solution-processed $Cs_2AgBiBr_6$ films. The results of the yielded films show better crystallinity, narrow band gaps, high carrier mobility, and longer carrier lifetimes, leading to a PCE of 2.51% by the solution processing method. These devices also exhibited strong ambient stability, making them suitable for long-term operation in next-generation PV.

Previously reported studies have examined the long-term stability of these materials and observed better performance under ambient conditions. The PCE of MHDP-based devices is still less efficient than that of lead-based perovskites, this efficiency gap shows challenges between MHDP and lead-based materials. For MHDPs to become competitive alternatives, the absorber materials must have ideal optoelectronic characteristics, such as a direct bandgap in the range of 1.0–1.8 eV, long carrier diffusion lengths, and minimal trap states. These requirements are essential for improving the overall performance and leading to the advancement in the next-generation of MHDPsPV technology [54, 77].

2.3.2 *Lead-Free PMs in Light Emitting Diode*

Various lead-free PMs, such as organic compounds, colloidal chalcogenide quantum dots, and perovskite quantum dots, have been explored for LED fabrication owing to their narrow emission bandwidths and tunableband gaps. However, organic molecules often result in broad emission lines, while the use of chalcogenides and lead-based perovskites raises concerns owing to the toxicity of cadmium and lead. This has motivated researchers to explore lead-free perovskites for environment-friendly LED applications. In 2016, Chao et al. achieved the first high-performance infrared LED from solution-processed $CsSnI_3$, with 950 nm emission, 40 W sr^{-1} m^{-2} radiance, and 3.8% EQE. However, Sn^{2+} instability limits durability. Inspired by photovoltaics, 2D PEA_2SnI_4 red LEDs (633 nm) were developed, and Sargent et al. improved stability using H_3PO_2, achieving 0.3% EQE and 70 cd m^{-2} brightness—among the best Pb-free red LEDs. However, oxidation remains a problem that leads researchers to explore more stable Sb^{3+}- and Bi^{3+}-based alternatives [78].

In 2018, Creutz et al. [68] successfully synthesized Cs_2AgBiX_6 (X = Cl, Br) nanocrystals via the thermal injection method. At the room-temperature, PLQY remained low (~0.3%), it only improves at cryogenic temperatures. Chen et al. in 2019 [79] incorporated Mn^{2+} and rare-earth ions into double perovskites, which improved their photoluminescence. Later, in 2023, Wang et al. [80] synthesized Na^+-doped $Cs_2AgBiBr_6$ nanocrystals that show an increment in the PLQY of about 9.08%. Sun et al. prepared $Cs_2AgBiBr_6$ and K^+-alloyed $Cs_2AgBiBr_6$ nanocrystals by taking care of the previous studies using a thermal-injection method. In which$Cs_2Ag_{0.64}K_{0.36}BiBr_6$ reached a PLQY of 6.82%, which is a significant improvement over the undoped material (~0.4%). Time-resolved photoluminescence (TRPL) measurements also showed that the exciton lifetime increased from 2.08 ns to 10.64 ns, indicating reduced nonradiative recombination [81]. The above results underline the promise of K^+-alloyed$Cs_2AgBiBr_6$ nanocrystals for future LED applications.

2.3.3 *Photodetectors*

PMs are also widely used in photodetectors, which convert incoming light energy into electrical energy [54]. Lead-free perovskite NCs gain global attraction because they offer high charge carrier mobility, broad and strong light absorption, better environmental stability, and low toxicity, making them suitable candidates for modern photodetectorsin the next generation PV technology. Photodetectors are widely used in X-ray detection for medical imaging, non-destructive testing, and scientific research [55]. The X-ray detectors made from lead-halide PMs have shown excellent sensitivity [82], but the toxicity associated with lead limits their performance in practical applications. Due to this, the researchers successfully moved to further development, which aims to reduce the detection limits while moving toward safer materials. In medical applications, lowering the radiation dose is crucial, as it can help to minimize long-term cancer risks [83]. Achieving a lower detection threshold is necessary for improving X-ray imaging resolution and enhancing the performance of security screening systems [84]. For these reasons, the search for lead-free single-crystal PMs with ultra-low X-ray detection limits has become an important and active area of research in the next generation PV technology [55, 84].

Ghosh et al. demonstrated efficient photodetectors fabricated from Cs_2SnI_6 nanomaterials of various morphologies. These devices were produced using a straightforward spin-coating method on patterned ITO substrates, highlighting the material's suitability for simple solution-processed photodetector architectures [85]. Among these, devices made from Cs_2SnI_6 nanorods (NRs) showed the highest photocurrent gain under white-light illumination, mainly due to the efficient charge transport provided by their one-dimensional structure. In another study, Waleed et al. fabricated photodetectors from $MASnI_3$ nanowire arrays prepared with the assistance of a porous alumina template, showing a responsivity of 0.47 A/W along with a detectivity reaching 8.80×10^{10} Jones [86]. However, oxidation of Sn^{2+} into Sn^{3+} remains a key constraint for device performance. To address this, researchers synthesized 2D PEA_2SnI_4 perovskites, which delivered a much higher responsivity of 16 A/W. Incorporating SnX_2 and suitable reducing agents helped stabilize the perovskite lattice and maintain a Sn^{2+}-rich environment by suppressing the formation of Sn vacancies. Using this approach, an air-stable broadband photodetector with a responsivity of up to 105 A/W, spanning the UV to near-infrared region, was achieved [87]. This improvement was attributed to the addition of hydroxybenzene sulfonic acid along with excess $SnCl_2$, which encapsulated and protected the perovskite grains. Similar enhancements in stability and performance were also observed when SnF_2 and ascorbic acid were introduced into CsSnI-based photodetectors.

Although $Cs_2AgBiBr_6$ possesses an indirect bandgap of around 2.1 eV, which limits its performance in photovoltaic and lighting applications, it shows several notable advantages for X-ray detection. Its high average atomic number (Z = 53.1), higher than that of $MAPbI_3$ and $MAPbBr_3$, allows for stronger X-ray absorption. The indirect bandgap also contributes to a long photoluminescence lifetime, indicating a potentially high $\mu\tau$ product and effective charge transport. Moreover, $Cs_2AgBiBr_6$ exhibits high resistivity (10^9–10^{11} Ω·cm), which helps minimize dark current and reduce noise. In addition, compared with many other halide perovskites, it shows lower halide ion migration, enhancing both device stability and sensitivity [88, 89].

2.4 Conclusions

In this chapter, we have discussed the basic introduction and properties of perovskite materials. This chapter compiled the key fabrication methods, including hot injection, ligand assisted re-precipitation, chemical vapor deposition, hydrothermal processing synthesis, which enable controlled growth of high-quality lead-free perovskite materials. Tin-based halide perovskite remain widely studied but face major challenges such as Sn^{2+} oxidation, structural defects, limited synthetic routes, low photoluminescence quantum yields, and instability under environmental conditions. In contrast, lead-free double perovskite offer enhanced structural robustness and promising optical behavior, making them suitable for solar cells, LEDs, and photo detectors. Although significant progress has been made in material development and device integration, further improvement in stability, charge transport, and efficiency is needed. Continued research in material engineering and scalable synthesis approaches will be crucial for realizing high-performance, environmentally friendly perovskite technologies.

Acknowledgements The authors, Abhishek Kumar Srivastava and Dr. Manisha Bajpai express deep thanks towards Council of Science & Technology, Uttar Pradesh for providing financial assistance (Project ID: 3061). Dr. Manisha Bajpai also express sincere acknowledgment to Siddharth University, Kapilvastu, Siddharth Nagar for providing assistance (Project ID FS-018/R&D/SUK/2024) and Department of Higher Education, Uttar Pradesh (No: 41/2024/336/70-4-2024-001-4(59)2022) to execute this research work.

The authors, Dr. Manisha Bajpai would like to thank Council of Science & Technology, Uttar Pradesh (Project ID—3061) for financial assistance. One of the authors, Dr. Manisha Bajpai would like to thanks Siddharth University, Kapilvastu, Siddharth Nagar for providing assistance (Project ID FS-018/R&D/SUK/2024) and Department of Higher Education, Uttar Pradesh (No: 41/2024/336/70-4-2024-001-4(59)2022) to execute this research work.

References

1. Y. Chen, Z. Feng, A. Pal, J. Zhang, Recent progress on the performance of lead-based halide perovskite APbX3 detectors. Physica Status Solidi (A) **219**(9), 2200018 (2022). https://doi.org/10.1002/pssa.202200018

2. T. Miyasaka, A.K. Jena, Research background and recent progress of perovskite photovoltaics, in *Perovskite Photovoltaics and Optoelectronics: From Fundamentals to Advanced Applications* (Wiley Online Books, 2022), pp. 1–60. ISBN: 978-3-527-82639-1
3. M. Shekarnoush, *Room-Temperature Solid-State Synthesis, Characterization, and Properties of Halide Perovskite Materials* (Doctoral Dissertation) (2024)
4. S. Ahmed, M.A. Gondal, A.S. Alzahrani, M. Parvaz, A. Ahmed, S. Hussain, Recent trends and challenges in lead-free perovskite solar cells: a critical review. ACS Appl. Energy Mater. **7**(4), 1382–1397 (2024). https://doi.org/10.1021/acsaem.3c02327
5. A. Kojima, M. Ikegami, K. Teshima, T. Miyasaka, Highly luminescent lead bromide perovskite nanoparticles synthesized with porous alumina media. Chem. Lett. **41**(4), 397–399 (2012). https://doi.org/10.1246/cl.2012.397
6. Z.K. Tan, R.S. Moghaddam, M.L. Lai, P. Docampo, R. Higler, F. Deschler, M. Price, A. Sadhanala, L.M. Pazos, D. Credgington, F. Hanusch, Bright light-emitting diodes based on organometal halide perovskite. Nat. Nanotechnol. **9**, 687–692 (2014). https://doi.org/10.1038/nnano.2014.149
7. A.K. Jena, A. Kulkarni, T. Miyasaka, Halide perovskite photovoltaics: background, status, and future prospects. Chem. Rev. **119**(5), 3036–3103 (2019). https://doi.org/10.1021/acs.chemrev.8b00539
8. J.Y. Kim, J.W. Lee, H.S. Jung, H. Shin, N.G. Park, High-efficiency perovskite solar cells. Chem. Rev. **120**(15), 7867–7918 (2020). https://doi.org/10.1021/acs.chemrev.0c00107
9. J.S. Manser, J.A. Christians, P.V. Kamat, Intriguing optoelectronic properties of metal halide perovskites. Chem. Rev. **116**(21), 12956–13008 (2016). https://doi.org/10.1021/acs.chemrev.6b00136
10. J. Fu, S. Ramesh, J.W. Melvin Lim, T.C. Sum, Carriers, quasi-particles, and collective excitations in halide perovskites. Chem. Rev. **123**(13), 8154–8231 (2023). https://doi.org/10.1021/acs.chemrev.2c00843
11. G. Wu, R. Liang, M. Ge, G. Sun, Y. Zhang, G. Xing, Surface passivation using 2D perovskites toward efficient and stable perovskite solar cells. Adv. Mater. **34**(8), 2105635 (2022). https://doi.org/10.1002/adma.202105635
12. Q. Tai, K.C. Tang, F. Yan, Recent progress of inorganic perovskite solar cells. Energy Environ. Sci. **12**(8), 2375–2405 (2019). https://doi.org/10.1039/C9EE01479A
13. Q. Shan, J. Song, Y. Zou, J. Li, L. Xu, J. Xue et al., High performance metal halide perovskite light-emitting diode: from material design to device optimization. Small **13**(45), 1701770 (2017). https://doi.org/10.1002/smll.201701770
14. H. Wang, D.H. Kim, Perovskite-based photodetectors: materials and devices. Chem. Soc. Rev. **46**(17), 5204–5236 (2017). https://doi.org/10.1039/C6CS00896H
15. K. Tambwe, N. Ross, P. Baker, T.T. Bui, F. Goubard, Humidity sensing applications of lead-free halide perovskite nanomaterials. Materials **15**(12), 4146 (2022). https://doi.org/10.3390/ma15124146
16. H. Dong, C. Zhang, X. Liu, J. Yao, Y.S. Zhao, Materials chemistry and engineering in metal halide perovskite lasers. Chem. Soc. Rev. **49**(3), 951–982 (2020). https://doi.org/10.1039/C9CS00598F
17. Y. Cao, J. Liang, X. Li, L. Yue, Q. Liu, S. Lu et al., Recent advances in perovskite oxides as electrode materials for supercapacitors. Chem. Commun. **57**(19), 2343–2355 (2021). https://doi.org/10.1039/D0TA09756J
18. F. Yao, K. Dong, W. Ke, G. Fang, Micro/Nano perovskite materials for advanced x-ray detection and imaging. ACS Nano **18**(8), 6095–6110 (2024). https://doi.org/10.1021/acsnano.3c10116
19. J. Park, J. Kim, H.S. Yun, M.J. Paik, E. Noh, H.J. Mun et al., Controlled growth of perovskite layers with volatile alkylammonium chlorides. Nature **616**(7958), 724–730 (2023). https://doi.org/10.1038/s41586-023-05825-y
20. S. Castelletto, F. De Angelis, A. Boretti, Prospects and challenges of quantum emitters in perovskites nanocrystals. Appl. Mater. Today **26**, 101401 (2022). https://doi.org/10.1016/j.apmt.2022.101401

21. D. Ma, K. Lin, Y. Dong, H. Choubisa, A.H. Proppe, D. Wu et al., Distribution control enables efficient reduced-dimensional perovskite LEDs. Nature **599**(7886), 594–598 (2021). https://doi.org/10.1038/s41586-021-03997-z
22. K. Wang, Z.Y. Lin, Z. Zhang, L. Jin, K. Ma, A.H. Coffey et al., Suppressing phase disproportionation in quasi-2D perovskite light-emitting diodes. Nat. Commun. **14**(1), 397 (2023). https://doi.org/10.1038/s41467-023-36118-7
23. R.K. Raman, S.A.G. Thangavelu, S. Venkataraj, A. Krishnamoorthy, Materials, methods and strategies for encapsulation of perovskite solar cells: from past to present. Renew. Sustain. Energy Rev. **151**, 111608 (2021). https://doi.org/10.1016/j.rser.2021.111608
24. A. Babayigit, A. Ethirajan, M. Muller, B. Conings, Toxicity of organometal halide perovskite solar cells. Nat. Mater. **15**(3), 247–251 (2016). https://doi.org/10.1038/nmat4572
25. R. Wang, M. Mujahid, Y. Duan, Z.K. Wang, J. Xue, Y. Yang, A review of perovskites solar cell stability. Adv. Func. Mater. **29**(47), 1808843 (2019). https://doi.org/10.1002/adfm.201808843
26. C. Otero-Martínez, J. Ye, J. Sung, I. Pastoriza-Santos, J. Pérez-Juste, Z. Xia et al., Colloidal metal-halide perovskite nanoplatelets: thickness-controlled synthesis, properties, and application in light-emitting diodes. Adv. Mater. **34**(10), 2107105 (2022). https://doi.org/10.1002/adma.202107105
27. F. Liu, T. Zhang, D. Mondal, S. Teng, Y. Zhang, K. Huang et al., Light-emitting metal–organic halide 1D and 2D structures: near-unity quantum efficiency, low-loss optical waveguide and highly polarized emission. AngewandteChemie **133**(24), 13660–13665 (2021). https://doi.org/10.1002/anie.202017274
28. C.C. Boyd, R. Cheacharoen, T. Leijtens, M.D. McGehee, Understanding degradation mechanisms and improving stability of perovskite photovoltaics. Chem. Rev. **119**(5), 3418–3451 (2018). https://doi.org/10.1021/acs.chemrev.8b00336
29. S.J. Woo, J.S. Kim, T.W. Lee, Characterization of stability and challenges to improve lifetime in perovskite LEDs. Nat. Photonics **15**(9), 630–634 (2021). https://doi.org/10.1038/s41566-021-00863-2
30. L. Clinckemalie, D. Valli, M.B. Roeffaers, J. Hofkens, B. Pradhan, E. Debroye, Challenges and opportunities for $CsPbBr_3$ perovskites in low-and high-energy radiation detection. ACS Energy Lett. **6**(4), 1290–1314 (2021). https://doi.org/10.1021/acsenergylett.1c00007
31. E.T. Hoke, D.J. Slotcavage, E.R. Dohner, A.R. Bowring, H.I. Karunadasa, M.D. McGehee, Reversible photo-induced trap formation in mixed-halide hybrid perovskites for photovoltaics. Chem. Sci. **6**(1), 613–617 (2015). https://doi.org/10.1039/C4SC03141E
32. B. Lee, C.C. Stoumpos, N. Zhou, F. Hao, C. Malliakas, C.Y. Yeh et al., Air-stable molecular semiconducting iodosalts for solar cell applications: Cs_2SnI_6 as a hole conductor. J. Am. Chem. Soc. **136**(43), 15379–15385 (2014). https://doi.org/10.1021/ja508464w
33. C.C. Stoumpos, L. Frazer, D.J. Clark, Y.S. Kim, S.H. Rhim, A.J. Freeman et al., Hybrid germanium iodide perovskite semiconductors: active lone pairs, structural distortions, direct and indirect energy gaps, and strong nonlinear optical properties. J. Am. Chem. Soc. **137**(21), 6804–6819 (2015). https://doi.org/10.1021/jacs.5b01025
34. M. Lyu, J.H. Yun, M. Cai, Y. Jiao, P.V. Bernhardt, M. Zhang et al., Organic–inorganic bismuth (III)-based material: a lead-free, air-stable and solution-processable light-absorber beyond organolead perovskites. Nano Res. **9**(3), 692–702 (2016). https://doi.org/10.1007/s12274-015-0948-y
35. M. Ghasemi, M. Lyu, M. Roknuzzaman, J.H. Yun, M. Hao, D. He et al., Phenethylammonium bismuth halides: from single crystals to bulky-organic cation promoted thin-film deposition for potential optoelectronic applications. J. Mater. Chem. A **7**(36), 20733–20741 (2019). https://doi.org/10.1039/C9TA07454F
36. B. Saparov, F. Hong, J.P. Sun, H.S. Duan, W. Meng, S. Cameron et al., Thin-film preparation and characterization of $Cs_3Sb_2I_9$: a lead-free layered perovskite semiconductor. Chem. Mater. **27**(16), 5622–5632 (2015). https://doi.org/10.1021/acs.chemmater.5b01989
37. T. Singh, A. Kulkarni, M. Ikegami, T. Miyasaka, Effect of electron transporting layer on bismuth-based lead-free perovskite $(CH_3NH_3)_3\ Bi_2I_9$ for photovoltaic applications. ACS Appl. Mater. Interfaces **8**(23), 14542–14547 (2016). https://doi.org/10.1021/acsami.6b02843

38. H. Zhu, M. Pan, M.B. Johansson, E.M. Johansson, High photon-to-current conversion in solar cells based on light-absorbing silver bismuth iodide. Chemsuschem **10**(12), 2592–2596 (2017). https://doi.org/10.1002/cssc.201700634
39. X. Han, Y. Zheng, S. Chai, S. Chen, J. Xu, 2D organic-inorganic hybrid perovskite materials for nonlinear optics. Nanophotonics **9**(7), 1787–1810 (2020). https://doi.org/10.1515/nanoph-2020-0038
40. W. Li, J. Ma, H. Wang, C. Fang, H. Luo, D. Li, Biexcitons in 2D (iso-BA)$_2$PbI$_4$ perovskite crystals. Nanophotonics **9**(7), 2001–2006 (2020). https://doi.org/10.1515/nanoph-2019-0528
41. A.H. Slavney, T. Hu, A.M. Lindenberg, H.I. Karunadasa, A bismuth-halide double perovskite with long carrier recombination lifetime for photovoltaic applications. J. Am. Chem. Soc. **138**(7), 2138–2141 (2016). https://www.x-mol.com/paperRedirect/84074
42. E.T. McClure, M.R. Ball, W. Windl, P.M. Woodward, Cs_2AgBiX_6 (X= Br, Cl): new visible light absorbing, lead-free halide perovskite semiconductors. Chem. Mater. **28**(5), 1348–1354 (2016). https://doi.org/10.1021/acs.chemmater.5b04231
43. G. Volonakis, M.R. Filip, A.A. Haghighirad, N. Sakai, B. Wenger, H.J. Snaith, F. Giustino, Lead-free halide double perovskites via heterovalent substitution of noble metals. J. Phys. Chem. Lett. **7**(7), 1254–1259 (2016). https://doi.org/10.3390/met8090667
44. E.V.D. Van Loef, P. Dorenbos, C.W.E. Van Eijk, K.W. Krämer, H.U. Güdel, Scintillation and spectroscopy of the pure and Ce^{3+}-doped elpasolites: Cs_2LiYX_6 (X= Cl, Br). J. Phys. Condens. Matter **14**(36), 8481 (2002). https://doi.org/10.1088/0953-8984/14/36/307
45. S. Vasala, M. Karppinen, A2B′ B ″O6 perovskites: a review. Prog. Solid State Chem. **43**(1–2), 1–36 (2015). https://doi.org/10.1016/j.progsolidstchem.2014.08.001
46. R.F. Berger, J.B. Neaton, Computational design of low-band-gap double perovskites. Phys. Rev. B Condens. Matter Mater. Phys. **86**(16), 165211 (2012). https://doi.org/10.1103/PhysRevB.86.165211
47. C. Wu, Q. Zhang, Y. Liu, W. Luo, X. Guo, Z. Huang et al., The dawn of lead-free perovskite solar cell: highly stable double perovskite $Cs_2AgBiBr_6$ film. Adv. Sci. **5**(3), 1700759 (2018). https://doi.org/10.1002/advs.201700759
48. W. Ning, F. Wang, B. Wu, J. Lu, Z. Yan, X. Liu et al., Long electron–hole diffusion length in high-quality lead-free double perovskite films. Adv. Mater. **30**(20), 1706246 (2018). https://doi.org/10.1002/ADMA.201706246
49. M. Pantaler, K.T. Cho, V.I. Queloz, I. García Benito, C. Fettkenhauer, I. Anusca et al., Hysteresis-free lead-free double-perovskite solar cells by interface engineering. ACS Energy Lett. **3**(8), 1781–1786 (2018). https://doi.org/10.1021/acsenergylett.8b00871
50. M. Ghasemi, L. Zhang, J.H. Yun, M. Hao, D. He, P. Chen et al., Dual-ion-diffusion induced degradation in lead-free $Cs_2AgBiBr_6$ double perovskite solar cells. Adv. Func. Mater. **30**(42), 2002342 (2020). https://doi.org/10.1002/adfm.202002342
51. R. Wang, J. Wang, S. Tan, Y. Duan, Z.K. Wang, Y. Yang, Opportunities and challenges of lead-free perovskite optoelectronic devices. Trends Chem. **1**(4), 368–379 (2019). https://doi.org/10.1016/j.trechm.2019.04.004
52. I. López-Fernández, D. Valli, C.Y. Wang, S. Samanta, T. Okamoto, Y.T. Huang et al., Lead-free halide perovskite materials and optoelectronic devices: progress and prospective. Adv. Func. Mater. **34**(6), 2307896 (2024). https://doi.org/10.1002/adfm.202307896
53. S.F. Hoefler, G. Trimmel, T. Rath, Progress on lead-free metal halide perovskites for photovoltaic applications: a review. Monatshefte für Chemie Chem. Mon. **148**(5), 795–826 (2017). https://doi.org/10.1007/s00706-017-1933-9
54. F. Zhang, Z. Ma, Z. Shi, X. Chen, D. Wu, X. Li, C. Shan, Recent advances and opportunities of lead-free perovskite nanocrystal for optoelectronic application. Energy Mater. Adv. (2021). https://doi.org/10.1002/adma.202001367
55. M. Ghasemi, M. Hao, M. Xiao, P. Chen, D. He, Y. Zhang et al., Lead-free metal-halide double perovskites: from optoelectronic properties to applications. Nanophotonics **10**(8), 2181–2219 (2021). https://doi.org/10.1515/nanoph-2020-0548
56. P. Umari, E. Mosconi, F. De Angelis, Relativistic GW calculations on $CH_3NH_3PbI_3$ and $CH_3NH_3SnI_3$ perovskites for solar cell applications. Sci. Rep. **4**(1), 4467 (2014). https://doi.org/10.1038/srep04467

57. F. Li, C. Zhang, J.H. Huang, H. Fan, H. Wang, P. Wang et al., A cation-exchange approach for the fabrication of efficient methylammonium tin iodide perovskite solar cells. AngewandteChemie Int. Ed. **58**(20), 6688–6692 (2019). https://doi.org/10.1002/anie.201902418
58. A. Dey, J. Ye, A. De, E. Debroye, S.K. Ha, E. Bladt et al., State of the art and prospects for halide perovskite nanocrystals. ACS Nano **15**(7), 10775–10981 (2021). https://doi.org/10.1021/acsnano.0c08903
59. L. Protesescu, S. Yakunin, M.I. Bodnarchuk, F. Krieg, R. Caputo, C.H. Hendon et al., Nanocrystals of cesium lead halide perovskites ($CsPbX_3$, X= Cl, Br, and I): novel optoelectronic materials showing bright emission with wide color gamut. Nano Lett. **15**(6), 3692–3696 (2015). https://doi.org/10.1021/nl5048779
60. T.C. Jellicoe, J.M. Richter, H.F. Glass, M. Tabachnyk, R. Brady, S.E. Dutton et al., Synthesis and optical properties of lead-free cesium tin halide perovskite nanocrystals. J. Am. Chem. Soc. **138**(9), 2941–2944 (2016). https://doi.org/10.1021/jacs.5b13470
61. A. Wang, Y. Guo, F. Muhammad, Z. Deng, Controlled synthesis of lead-free cesium tin halide perovskite cubic nanocages with high stability. Chem. Mater. **29**(15), 6493–6501 (2017). https://doi.org/10.1021/acs.chemmater.7b02089
62. A. Wang, X. Yan, M. Zhang, S. Sun, M. Yang, W. Shen et al., Controlled synthesis of lead-free and stable perovskite derivative Cs_2SnI_6 nanocrystals via a facile hot-injection process. Chem. Mater. **28**(22), 8132–8140 (2016). https://doi.org/10.1021/acs.chemmater.6b01329
63. M.C. Weidman, M. Seitz, S.D. Stranks, W.A. Tisdale, Highly tunable colloidal perovskite nanoplatelets through variable cation, metal, and halide composition. ACS Nano **10**(8), 7830–7839 (2016). https://doi.org/10.1021/acsnano.6b03496
64. J. Xing, X.F. Liu, Q. Zhang, S.T. Ha, Y.W. Yuan, C. Shen et al., Vapor phase synthesis of organometal halide perovskite nanowires for tunable room-temperature nanolasers. Nano Lett. **15**(7), 4571–4577 (2015). https://doi.org/10.1021/acs.nanolett.5b01166
65. J. Chen, Z. Luo, Y. Fu, X. Wang, K.J. Czech, S. Shen et al., Tin (IV)-tolerant vapor-phase growth and photophysical properties of aligned cesium tin halide perovskite ($CsSnX_3$; X= Br, I) nanowires. ACS Energy Lett. **4**(5), 1045–1052 (2019). https://doi.org/10.1039/D5CC03229F
66. J.K. Chen, B.B. Zhang, Q. Liu, N. Shirahata, O.F. Mohammed, O.M. Bakr, H.T. Sun, Advances and challenges in tin halide perovskite nanocrystals. ACS Mater. Lett. **3**(11), 1541–1557 (2021). https://doi.org/10.1021/acsmaterialslett.1c00444
67. Y. Bekenstein, J.C. Dahl, J. Huang, W.T. Osowiecki, J.K. Swabeck, E.M. Chan et al., The making and breaking of lead-free double perovskite nanocrystals of cesium silver–bismuth halide compositions. Nano Lett. **18**(6), 3502–3508 (2018). https://doi.org/10.1021/acs.nanolett.8b00560
68. S.E. Creutz, E.N. Crites, M.C. De Siena, D.R. Gamelin, Colloidal nanocrystals of lead-free double-perovskite (elpasolite) semiconductors: synthesis and anion exchange to access new materials. Nano Lett. **18**(2), 1118–1123 (2018). https://doi.org/10.1021/acs.nanolett.7b04659
69. A.H. Slavney, T. Hu, A.M. Lindenberg, H.I. Karunadasa, A bismuth-halide double perovskite with long carrier recombination lifetime for photovoltaic applications. J. Am. Chem. Soc. **138**(7), 2138–2141 (2016). https://doi.org/10.1021/jacs.5b13294
70. Z. Deng, F. Wei, S. Sun, G. Kieslich, A.K. Cheetham, P.D. Bristowe, Exploring the properties of lead-free hybrid double perovskites using a combined computational-experimental approach. J. Mater. Chem. A **4**(31), 12025–12029 (2016). https://doi.org/10.1039/C6TA05817E
71. M.R. Filip, X. Liu, A. Miglio, G. Hautier, F. Giustino, Phase diagrams and stability of lead-free halide double perovskites Cs_2BB' X6: B= Sb and Bi, B′= Cu, Ag, and Au, and X= Cl, Br, and I. J. Phys. Chem. C **122**(1), 158–170 (2018). https://doi.org/10.1021/acs.jpclett.6b01041
72. G. García-Espejo, D. Rodríguez-Padrón, R. Luque, L. Camacho, G. de Miguel, Mechanochemical synthesis of three double perovskites: $Cs_2AgBiBr_6$, $(CH_3NH_3)_2$ $TlBiBr_6$ and Cs_2 $AgSbBr_6$. Nanoscale **11**(35), 16650–16657 (2019). https://doi.org/10.1039/C9NR06092H
73. Y. Li, X. Zhang, H. Huang, S.V. Kershaw, A.L. Rogach, Advances in metal halide perovskite nanocrystals: synthetic strategies, growth mechanisms, and optoelectronic applications. Mater. Today **32**, 204–221 (2020). https://doi.org/10.1016/j.mattod.2019.06.007

74. H. Xu, H. Yuan, J. Duan, Y. Zhao, Z. Jiao, Q. Tang, Lead-free $CH_3NH_3SnBr_3$-xIx perovskite quantum dots for mesoscopic solar cell applications. Electrochim. Acta **282**, 807–812 (2018). https://doi.org/10.1039/C8QM00337H
75. E. Greul, M.L. Petrus, A. Binek, P. Docampo, T. Bein, Highly stable, phase pure $Cs_2AgBiBr_6$ double perovskite thin films for optoelectronic applications. J. Mater. Chem. A **5**(37), 19972–19981 (2017). https://doi.org/10.1039/C7TA06816F
76. F. Igbari, R. Wang, Z.K. Wang, X.J. Ma, Q. Wang, K.L. Wang et al., Composition stoichiometry of $Cs_2AgBiBr_6$ films for highly efficient lead-free perovskite solar cells. Nano Lett. **19**(3), 2066–2073 (2019). https://doi.org/10.1021/acs.nanolett.9b00238
77. X.G. Zhao, J.H. Yang, Y. Fu, D. Yang, Q. Xu, L. Yu et al., Design of lead-free inorganic halide perovskites for solar cells via cation-transmutation. J. Am. Chem. Soc. **139**(7), 2630–2638 (2017). https://doi.org/10.1021/jacs.6b09645
78. S. Singh, F. Ahmad, Synthesis, properties and applications of lead-free perovskites: a review. J. Ultra Sci. Phys. Sci. Sect. (Math.) **33**(6), 330601 (2021). https://doi.org/10.22147/jusps-A/330601
79. N. Chen, T. Cai, W. Li, K. Hills-Kimball, H. Yang, M. Que et al., Yb-and Mn-doped lead-free double perovskite Cs_2AgBiX_6 ($X= Cl^-$, Br^-) nanocrystals. ACS Appl. Mater. Interfaces **11**(18), 16855–16863 (2019). https://doi.org/10.1021/acsami.9b02367
80. S. Wang, W. Jiang, W. Cui, W. Jiang, M. Zhao, J. Sun et al., Improved optical properties of lead-free double perovskite $Cs_2AgBiBr_6$ nanocrystals via Na ions doping. Adv. Opt. Mater. **11**(10), 2202745 (2023). https://doi.org/10.1002/adom.202202745
81. R. Sun, W. Jiang, S. Wang, W. Cui, L. Qi, K. Pan, K-alloyed lead-free double-perovskite $Cs_2AgBiBr_6$ nanocrystals for use in light-emitting diodes. ACS Appl. Nano Mater. **6**(16), 15247–15254 (2023). https://doi.org/10.1021/acsanm.3c03020
82. W. Wei, Y. Zhang, Q. Xu, H. Wei, Y. Fang, Q. Wang et al., Monolithic integration of hybrid perovskite single crystals with heterogenous substrate for highly sensitive X-ray imaging. Nat. Photonics **11**(5), 315–321 (2017). https://doi.org/10.1038/nphoton.2017.43
83. H. Wei, Y. Fang, P. Mulligan, W. Chuirazzi, H.H. Fang, C. Wang et al., Sensitive X-ray detectors made of methylammonium lead tribromide perovskite single crystals. Nat. Photonics **10**(5), 333–339 (2016). https://doi.org/10.1038/nphoton.2016.41
84. B.T. Polischuk, Z. Shukri, A. Legros, H. Rougeot, Selenium direct-converter structure for static and dynamic x-ray detection in medical imaging applications, in *Medical Imaging 1998 Physics of Medical Imaging*, vol. 3336 (SPIE, 1998) (pp. 494–504). https://doi.org/10.1117/12.317051
85. S. Ghosh, S. Paul, S.K. De, Control synthesis of air-stable morphology tunable Pb-free Cs_2SnI_6 perovskite nanoparticles and their photodetection properties. Part. Part. Syst. Charact. **35**(9), 1800199 (2018). https://doi.org/10.1002/ppsc.201800199
86. A. Waleed, M.M. Tavakoli, L. Gu, Z. Wang, D. Zhang, A. Manikandan et al., Lead-free perovskite nanowire array photodetectors with drastically improved stability in nanoengineering templates. Nano Lett. **17**(1), 523–530 (2017). https://doi.org/10.1021/acs.nanolett.6b04587
87. F. Yan, Z. Wei, X. Wei, Q. Lv, W. Zhu, K. Wang, Toward high-performance photodetectors based on 2D materials: strategy on methods. Small Methods **2**(5), 1700349 (2018). https://doi.org/10.1002/smtd.201700349
88. V.S. Chirvony, J.P. Martínez-Pastor, Trap-limited dynamics of excited carriers and interpretation of the photoluminescence decay kinetics in metal halide perovskites. J. Phys. Chem. Lett. **9**(17), 4955–4962 (2018). https://doi.org/10.1021/acs.jpclett.8b01241
89. J.A. Steele, W. Pan, C. Martin, M. Keshavarz, E. Debroye, H. Yuan, M.B. Roeffaers, Photophysical pathways in highly sensitive $Cs_2AgBiBr_6$ double-perovskite single-crystal X-ray detectors. Adv. Mater. **30**(46), 1804450 (2018). https://doi.org/10.1002/adma.201804450

Chapter 3
Recent Advancement in Lead-Free Perovskites for Optoelectronic Devices

Manisha Sharma, Anirudh Kumar, Deepak Kumar, Abd. Rashid bin Mohd Yusoffe, and Sanjeev Kumar Sharma

Abstract Lead (Pb) halide perovskite materials are gaining popularity in various areas of optoelectronic devices, including high-performance solar cells, light-emitting diodes (LEDs), photodetectors, and neuromorphic memristor devices, due to their lightweight nature, flexibility, ease of manufacturing, and remarkable optical properties. However, the toxicity of Pb hindered the commercial production of these optoelectronic devices due to health risks. Therefore, for the development of environmentally friendly optoelectronic devices, it is highly desirable to replace Pb^{2+} with other non-toxic elements in perovskite materials without sacrificing the material's functionality and device performance. Recently, researchers have considered Sn, Ge, Bi, and Cu as promising alternatives to replace Pb^{2+} in halide perovskite materials with suitable bandgaps, optical, and electrical properties. In this chapter, we discuss the progress in the performance of Pb-free perovskite-based optoelectronic devices. Furthermore, the challenges and future perspectives to improve the performance of Pb-free perovskite-based optoelectronic devices are also highlighted in detail.

Keywords Lead-free Perovskite · Solar cells · Photodetectors · LEDs · Optical memristors

3.1 Introduction

Lead (Pb) free perovskites have emerged as sustainable and versatile alternatives for optoelectronic systems enabling next-generation devices, including solar cells, light-emitting diodes (LEDs), photodetectors, and lasers. The combination of strong light absorption, tunable bandgap, high carrier mobility, and solution process ability

---

M. Sharma · A. Kumar · D. Kumar · S. K. Sharma (✉)
Biomaterials and Sensor Laboratory, Department of Physics, Ch. Charan Singh University, Meerut Campus, Meerut, Uttar Pradesh, India
e-mail: sksharma18@ccsuniversity.ac.in

Abd. R. bin Mohd Yusoffe
Department of Physics, Faculty of Science, University Teknologi Malaysia, Johor Bahru, Malaysia

V. Chauhan et al. (eds.), *Sustainable Lead-Free Perovskite Materials*, Topics in Applied Physics 157, https://doi.org/10.1007/978-3-032-24873-2_3

has accelerated progress in perovskite research, with solar cell efficiencies already exceeding 26%, comparable to silicon technologies. The dominance of Pb-based Halide perovskites (HP) has been constrained by toxicity and limited durability, necessitating global efforts to develop safer counterparts with comparable performance [1, 2]. Replacing Pb with cations such as Sn^{2+}, Ge^{2+}, Bi^{3+}, Sb^{3+}, Cu^{2+}, Ag^{+}, and Mn^{2+} has opened new design pathways for material design. Beyond simple cation substitution, a range of advanced strategies including dimensional engineering, defect passivation, doping, compositional tuning, nanostructuring, and heterojunction formation have been widely employed to address key challenges such as instability, oxidation, and efficiency losses. These approaches are reshaping the structural and electronic design principles of HPs, yielding eco-friendly materials with wide ranging applicability. Pb-free perovskites have demonstrated promising integration into diverse optoelectronic platforms: Sn and Ge based perovskites underpin more efficient and stable photovoltaics; in LEDs, their tunable emission and defect tolerance enable high-quality light sources; photodetectors leverage their broadband sensitivity and ultrafast response; while emerging applications such as tunable lasers and perovskite based memristors highlight their multifunctionality in neuromorphic and intelligent electronics. Despite notable achievements, persistent challenges including the intrinsic instability of Sn and Ge based materials, defect-mediated nonradiative recombination, and scalability constraints continue to hinder commercial deployment. Overcoming these barriers is essential for moving beyond laboratory demonstrations toward market ready technologies. Interdisciplinary progress in material design and device engineering indicates that these challenges can be overcome, underscoring the promise of Pb-free perovskites for sustainable optoelectronics [3, 4]. As shown in Fig. 3.1, the trajectory of Pb-free perovskite research reveals critical milestones, from early exploration of non-toxic alternatives to their integration across photovoltaics, LEDs, lasers, and photodetectors. The timeline underscores how systematic advances in material design, coupled with device-level innovations, have steadily expanded the functional scope of this class of semiconductors. This chapter aims to provide an in-depth perspective on the evolution of Pb-free perovskites, linking fundamental material innovations with emerging device applications. While Pb-based systems established the foundation of perovskite optoelectronics, the transition toward Pb-free analogues represents both a scientific imperative and an opportunity to establish safer, greener, and more sustainable technologies for future energy and electronic applications [5].

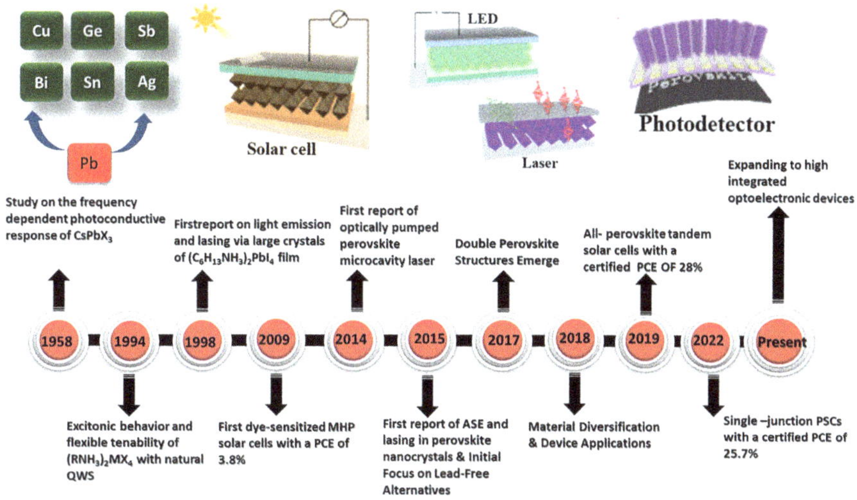

Fig. 3.1 Timeline of major milestones in the development of Pb-free perovskite and their application in solar cells, LEDs, lasers, and photodetectors

3.2 Major Advancements in Pb-Free Perovskites

The advancement of Pb-free perovskites is motivated by the need to eliminate Pb toxicity while ensuring long-term stability [6]. Early systems demonstrated limited efficiency and rapid degradation, restricting their application to preliminary experimental demonstrations. Through systematic innovations, however, their optoelectronic potential has been considerably enhanced. Recent research directions emphasize structural stabilization to suppress phase transitions, defect management to reduce trap assisted recombination, and bandgap engineering to align optical absorption with target device requirements [7]. In parallel, nanoscale design strategies have been employed to control carrier confinement, interface quality, and excitonic behaviour, all of which contribute to improved charge transport and reduced non radiative losses [8, 9]. These refinements collectively increase tolerance against environmental stressors such as moisture, thermal fluctuations, and continuous illumination. The integration of multiple approaches including cation and anion substitution, dimensional tailoring, defect passivation, controlled doping, compositional tuning, and the design of nanostructured heterointerfaces has enabled the realization of Pb-free perovskites with superior optoelectronic characteristics and enhanced operational durability. These materials have advanced from early prototypes to strong candidates for efficient photovoltaics, stable light emitters, and sensitive photodetectors, highlighting their potential as sustainable, non-toxic solutions for next-generation optoelectronics [10, 11].

3.2.1 Cation Substitution for Bandgap Engineering

Cation substitution is a powerful strategy for tuning the bandgap and optimizing the optoelectronic properties of Pb-free perovskites, addressing both toxicity and stability concerns for next-generation devices. Recent research demonstrates that substituting various cations at different lattice sites can significantly alter band structure, absorption, and carrier dynamics, paving the way for high-performance, environmentally friendly optoelectronic materials [10]. Gu et al. [11] employed trivalent Ru^{3+} to partially substitute Bi^{3+} in $Cs_3Bi_2I_9$, systematically investigating the resulting structural, optoelectronic, and environmental stability characteristics. The introduction of Ru^{3+} effectively narrows the bandgap and modifies the band structure by generating defect states and enhancing phonon–electron coupling, thereby boosting optoelectronic performance. Likewise, Fe doping in $Cs_3Bi_2Br_9$ decreases the bandgap from 2.54 to 1.78 eV, although higher doping concentrations promote the formation of secondary perovskite phases. These results emphasize the role of targeted cation substitution in bandgap engineering and stability enhancement of Pb-free perovskites. Keshavarz et al. [12] explored the structural and optoelectronic properties of pristine $Cs_2AgBiBr_6$ and alkali metal substituted single crystals, $(Cs_{1-x}Y_x)_2AgBiBr_6$ ($Y = Rb^+, K^+, Na^+$; $x = 0.02$). While alkali substitution induces minimal changes in the bandgap, it significantly enhances carrier recombination lifetimes and improves X-ray sensitivity, making these materials promising for photodetector applications. Among the variants, Rb-substituted devices exhibit the highest photocurrent density, as confirmed by time-dependent photoresponse measurements at 300 K under a 20 V bias. Replacing Pb^{2+} with Ge^{2+} and varying Br content in $CsGeI_{3-x}Br_x$ altered the bandgap and optical absorption. Higher Br content increased the bandgap, while Ge substitution red-shifted absorption peaks, enabling bandgap tuning for optoelectronic applications [13]. First-principles calculations showed that A site Rb substitution in Pb HPs had little effect on optoelectronic properties but increased halide migration. B site Sr substitution suppressed halide migration but enlarged the bandgap, reducing visible light absorption. Sn substitution improved both stability and optoelectronic performance [14]. Bandgap tuning in triple cation perovskites was achieved via cation and halide substitution, preventing photoinduced segregation and achieving high efficiency (18.3%) with a 1.67 eV bandgap for tandem solar cells [15].

3.2.2 Dimensional Engineering (2D/1D/0D Architectures)

A promising route to modulate the optical response of double perovskites is through dimensional regulation, which enables the exploitation of quantum confinement effects when moving from bulk 3D structures to 2D or even 0D systems [16, 17]. Although achieving precise control over dimensionality remains challenging, several advances have been reported. For instance, in thin-film double perovskites, the incorporation of organic spacer molecules can drive the transformation from a 3D to a 2D framework, thereby converting the indirect bandgap of $Cs_2AgBiBr_6$ into a direct one. Parallel efforts have focused on fabricating double perovskite nanoplatelets (NPls) with controlled thickness. Cs_2AgBiX_6 (X = Cl, Br, I) NPls, synthesized via temperature-regulated hot-injection methods Fig. 3.2a–c, exhibit absorption and photoluminescence (PL) characteristics largely comparable to their bulk counterparts [18]. At ambient conditions, monolayer NPls can also be obtained, showing a blue-shifted emission attributed to direct bandgap exciton recombination. Despite this, their luminescence is weak, restricting their device applicability. To address these shortcomings, compositional engineering has been applied to low-dimensional double perovskites. Notably, 2D $Cs_2AgIn_xBi_{1-x}Cl_6$ ($0 \leq x \leq 1$) NPls have demonstrated a relatively high photoluminescence quantum yield (PLQY) of ~ 40%. Their potential has been validated in electrically driven LEDs Fig. 3.2d–f, though the external quantum efficiency (EQE) achieved (~ 0.01%) remains far below that of Pb-based HPs. Collectively, these findings underscore both the opportunities and persistent limitations of dimensional and compositional engineering strategies in optimizing the optoelectronic behavior of Pb-free double perovskites [18].

3.2.3 Defects and Passivation

Defects remain a critical bottleneck for the optoelectronic performance and long-term stability of perovskite devices, as they act as non-radiative recombination centers. These are broadly classified into intrinsic defects, such as vacancies, grain boundaries, and dangling bonds, and extrinsic defects, which originate from interfaces and electrodes. For instance, the migration of electrode materials, such as Ag or Au, into the perovskite layer can trigger irreversible degradation. To address these challenges, defect passivation strategies are generally categorized into bulk passivation and surface/interface passivation, with significant emphasis placed on the latter to suppress recombination and enhance device durability. Among Pb-free systems, Sn-based perovskites are particularly defect-prone due to the easy oxidation of Sn^{2+} to Sn^{4+}, which produces both Sn and halide vacancies acting as recombination centers. Introducing suitable passivating agents can effectively fill these vacancies, improving carrier transport. However, excessive incorporation of passivators may partially replace A-site cations, thereby reducing carrier pathways and transport efficiency. Thus, optimizing passivator concentration is crucial to achieving high efficiency and long-term device stability Fig. 3.2g. Surface and interface defects are equally important. They introduce unwanted energy states that disrupt band alignment with charge transport layers, accelerating recombination. A typical example is the iodide vacancy-induced dangling bonds on the surface of $FASnI_3$ Fig. 3.2h. To address this, graphitic carbon nitride (g-C_3N_4) has been employed as an effective surface passivator. Its functional groups form hydrogen bonds with FA^+ cations and

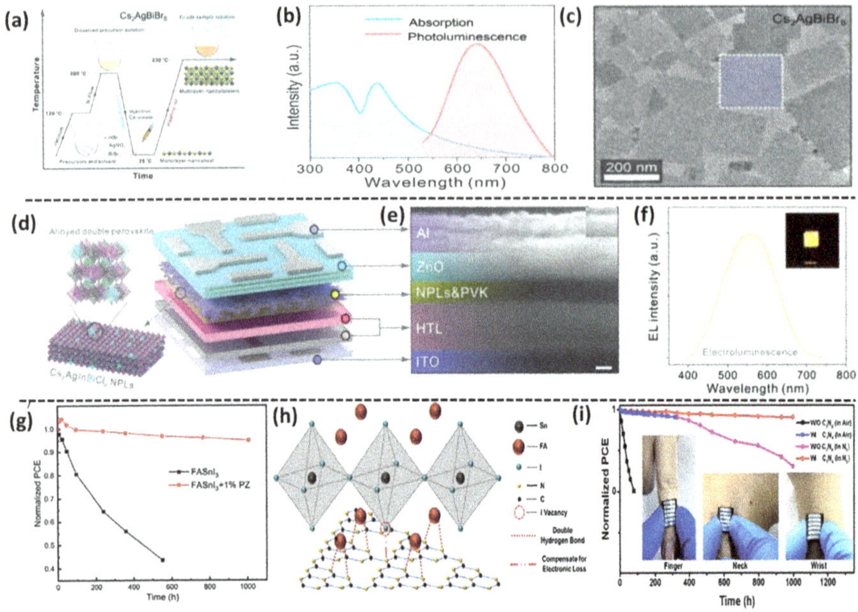

Fig. 3.2 **a** Schematic synthesis route of $Cs_2AgBiBr_6$ double perovskite nanoplates (DP NPls), **b** UV–vis absorption and PL spectra, **c** TEM image of $Cs_2AgBiBr_6$ DP NPls; *Reproduced with permission* © **ACS** [18]. **d** Device architecture of DP-NPl LEDs, **e** cross-sectional SEM image, **f** EL spectra; *Reproduced with permission* ©**Wiley** [19]. **g** Storage stability of $FASnI_3$ and $FASnI_3$ + 1% PZ devices under N_2; Reproduced with permission © **Wiley** [20]. **h** Illustration of g-C_3N_4 interaction with perovskite, **i** Stability of 20 PSCs stored in air and N_2; inset shows a wearable prototype device; *Reproduced with permission* ©**Wiley** [21]

coordinate with iodide vacancies, thereby stabilizing the surface. This dual passivation suppresses recombination, enhances structural robustness, and significantly extends device lifetime (Fig. 3.2i).

3.2.4 Doping

Doping has emerged as a pivotal approach to optimize the optoelectronic performance of Pb-free perovskites by tailoring their structural, electronic, and defect landscapes [22]. Substitution at the A-site (e.g., Cs^+, FA^+, or MA^+ with larger/smaller cations) is employed to tune lattice symmetry, stabilize desirable phases, and enhance tolerance factors, thereby improving film quality and durability. B-site doping is widely explored, where the incorporation of elements such as Bi^{3+}, Sb^{3+}, Ge^{2+}, or transition metals modulates band alignment, reduces trap densities, and facilitates efficient charge transport [23]. At the halide site, partial replacement of Cl^-, Br^-, or I^- is used to adjust bandgap energies and balance carrier mobility with emission wavelength, which is particularly beneficial for light-emitting and photodetector applications. Transition metals such as Mn^{2+}, Fe^{3+}, Cu^{2+}, Ni^{2+}, and Cr^{3+} are incorporated to tune the bandgap, boost photoluminescence (PL), and enable new emission bands, such as visible and near-infrared, by breaking parity-forbidden transitions and enhancing radiative recombination [24].

3.2.5 Compositional Engineering

Compositional engineering represents a holistic strategy in which multiple cations and anions are intentionally combined within the perovskite lattice to achieve properties that cannot be realized through single-ion substitution or low-level doping alone. By designing multi-component frameworks, such as triple-cation systems (Cs/FA/MA), mixed-halide compositions (I/Br/Cl), and dual B-site alloys (Sn-Ge, Bi–Sb), this approach provides a powerful means of balancing structural distortions and mitigating phase segregation, which are critical challenges in Pb-free systems. One of the key advantages of compositional mixing is its ability to regulate crystallization pathways and grain growth, resulting in dense films with fewer defects and more uniform microstructures [25]. The incorporation of multiple ionic species also reduces lattice strain, optimizes tolerance factors, and stabilizes metastable phases that would otherwise degrade under operational conditions. In halide-mixed systems, bandgap tunability is achieved without significantly compromising charge-carrier mobility, allowing fine adjustment of optical absorption across the visible spectrum. Similarly, multi-cation and multi-anion frameworks enhance thermal and moisture resistance, as the different ionic radii and bonding environments distribute stress more evenly within the crystal lattice. These refinements collectively enhance the operational durability of Pb-free perovskites against environmental stressors, including humidity, thermal cycling, and continuous illumination, while simultaneously enabling precise control over optoelectronic properties. As a result, compositional engineering not only bridges the gap between structural stability and electronic functionality but also provides a scalable pathway toward device-grade perovskite platforms. Its adaptability makes it a cornerstone strategy for advancing Pb-free perovskites in photovoltaics, light-emitting devices, and photodetectors [26].

3.2.6 Nanostructuring and Heterojunctions

Nanostructuring and heterojunction engineering have emerged as pivotal approaches to overcome the intrinsic limitations of Pb-free perovskites and unlock their full optoelectronic potential. By tailoring perovskites into nanocrystals, nanoplates, nanowires, quantum dots, or layered architectures, researchers can precisely manipulate carrier confinement, excitonic interactions, and light-matter coupling. Such nanoscale control enhances absorption across a broader spectrum, improves radiative recombination for efficient emission, and promotes directional charge transport [27]. By simultaneously advancing efficiency, stability, and functionality, these strategies are accelerating the transition of Pb-free perovskites from laboratory prototypes to scalable, next-generation optoelectronic technologies [28, 29]. Extensive research has explored the optoelectronic applications of heterojunctions based on II–VI semiconductor nanostructures, including solar cells, photodetectors, and light-emitting diodes. This section highlights the diverse device architectures and performance outcomes of such heterojunctions. Reported efficiencies are often underestimated due to limited light absorption in thin nanowires; however, improvements can be achieved through surface passivation, which suppresses carrier recombination at surface states. In recent years, solar cells based on II–VI semiconductor nanostructures have attracted significant attention and demonstrated promising outcomes. Yang et al. reported CdS/Cu_2S, core–shell nanowire solar cells with superior open-circuit voltage and fill factor compared to their planar counterparts, achieving a ~ 5.4% efficiency despite limited light absorption [30]. They further integrated multiple cells on single nanowires in both series and parallel configurations to enhance output performance. The improved photovoltaic response was attributed to the low-temperature cation-exchange process that formed a heteroepitaxial junction between single-crystalline CdS and Cu_2S. Zhou et al. developed CdSe/ZnTe core–shell nanowire devices using indium and nickel as electrodes for the core and shell, respectively. These devices exhibited rectifying I–V behavior, with the best cell reaching an efficiency of 1.7% (Voc = 0.18 V, Jsc = 38 pA, FF = 0.38). Zhang et al. designed coaxial ZnSe/ZnO p–n junction nanowires, yielding an efficiency of 1.24% and a high Voc of 0.87 V under UV illumination, attributed to the wide bandgap and high-quality interface. In a further study, ZnO/ZnSe type-II core–shell nanowire arrays were fabricated, exhibiting enhanced light absorption due to increased junction area and light-trapping effects, thereby broadening the options for absorber materials in nanostructured solar cells [31]. Xie et al. reported a high-performance zero-power photodetector based on a p-CdTe nanorod/n-Si nanowire array heterojunction, which operates effectively in the visible to near-infrared region with excellent stability, sensitivity, and a fast response. Under white light illumination (62.5 $\mu W\ cm^{-2}$), the device exhibited reproducible performance up to 600 Hz, with rapid response times of 1.2 ms (rise) and 1.58 ms (fall), surpassing previously reported CdTe nanowire and nanoribbon photodetectors [32].

3.3 Pb-Free Perovskites: Unlocking Next-Generation Optoelectronic Devices

The exploration of Pb-free perovskites represents a crucial step toward achieving safe and sustainable optoelectronic technologies. By uniting strong optoelectronic performance with reduced environmental risk, these materials are increasingly viewed as promising alternatives to Pb-based perovskites. Their structural flexibility, solution processability, and tunable band gap enable precise control over light absorption, emission, and carrier dynamics, making them suitable for a wide range of device architectures. In contrast, vacancy-ordered Ge-based halides stand out for their narrower band gaps and strong solar absorption; Bi and Sb-based systems are valued for their remarkable stability. At the same time, vacancy-ordered double perovskites offer non-toxicity and compositional robustness [33]. Each category presents unique advantages and limitations, but collectively they provide a versatile materials platform that can be tailored to specific performance requirements. This versatility opens the door to diverse optoelectronic applications, including high-efficiency solar cells, color-tunable LEDs, broadband photo detectors, and coherent light sources such as lasers, thereby paving the way toward next-generation, environmentally sustainable technologies. The following subsections examine how different classes of Pb-free perovskites are being developed and integrated into these devices, with particular emphasis on recent breakthroughs and the challenges that remain [6].

3.3.1 Perovskite Solar Cells

Perovskite solar cells (PSCs) have rapidly emerged as one of the most promising photovoltaic technologies, offering exceptional optoelectronic properties, solution-based low-cost fabrication, and steadily rising efficiencies. Yet, their path to commercialization remains challenging due to issues of long-term stability, reproducibility, and environmental concerns, particularly the toxicity of Pb-based compositions. To address these challenges, researchers have turned to more sustainable alternatives. Among them, Sn and Ge-based HPs stand out, not only because they eliminate lead but also because of their narrower band gaps, which allow broader solar absorption. This bandgap reduction arises from fundamental factors: the smaller ionic radii of Sn^{2+} and Ge^{2+} compared with Pb^{2+} lead to lattice contraction and more substantial orbital overlap, while differences in electronegativity shift charge distribution and further tune the electronic states. Computational studies reinforce this picture, showing how ionic size and bonding characteristics shape the unique band structures of Sn/Ge perovskites, making them highly appealing for next-generation photovoltaic absorbers. The narrower bandgap makes Sn perovskites excellent light harvesters, but their efficiency still trails behind Pb-based systems due to the easy oxidation of Sn^{2+} to Sn^{4+}, which compromises both stability and device reproducibility. Several strategies have been developed to overcome this drawback. For instance, $CsSn_{0.5}Ge_{0.5}I_3$ synthesized via melt-crystallization produced high-quality films with fewer defects Fig. 3.3a–c. Devices based on these films achieved a PCE of 7.11% and retained stability under 500 h of continuous solar exposure Fig. 3.3d [34]. The use of pentafluorophenoxyethylammonium iodide (FOEI) enabled controlled growth of $FASnI_3$, delivering dense, defect-suppressed films Fig. 3.3e. Compared with other passivators such as FEAI, FOEI offered superior stabilization, yielding devices with a PCE of 10.81% Fig. 3.3f [35]. Beyond absorber engineering, tailoring charge-transport layers has also been critical. Among four regioisomers of diethylmalonate-C_{60} bisadduct

(DCBA) synthesized as electron transport layers, the trans-3 isomer provided the best alignment and extraction, leading to a certified 14.30% efficiency, one of the highest values reported for Sn-based perovskites Figs. 3.3g, h [36]. The overall trajectory of Sn-based PVSCs clearly shows steady improvement, with efficiencies now surpassing 14% Fig. 3.3i. Although stability and scalability remain key hurdles, these materials are widely regarded as the most promising Pb-free alternative to conventional Pb-based devices. In comparison with other Pb-free alternatives, Sn/Ge perovskites hold a distinctive advantage. While Bi- and Sb-based systems provide excellent stability, their wide bandgap and indirect transitions limit photovoltaic efficiency. Vacancy-ordered double perovskites, such as $Cs_2AgBiBr_6$, are non-toxic and structurally robust, but they suffer from poor carrier transport and unfavorable band alignment. By contrast, Sn/Ge perovskites uniquely combine broad solar absorption, structural tunability, and reduced toxicity, positioning them as the most promising Pb-free absorbers, provided their intrinsic instability can be effectively mitigated.

3.3.2 *Light Emitting Diodes (LEDs)*

Pb-free perovskites have emerged as promising materials for next-generation light-emitting diodes due to their tunable bandgap, high color purity, and solution-processability. Substituting Pb with environmentally benign cations such as tin, bismuth, antimony, or copper enables efficient radiative recombination while mitigating toxicity concerns. Their defect-tolerant electronic structure and ability to form low-dimensional phases enhance exciton binding and improve luminescence efficiency. Recent advances in compositional engineering, dimensional control, and surface passivation have further improved device stability and emission quality, positioning Pb-free perovskite LEDs as a sustainable alternative for displays and solid-state lighting. LEDs generate light via photoluminescence (PL) or electroluminescence (EL). Key parameters for evaluating LED performance include luminous efficacy (LE), color rendering index (CRI), correlated color temperature (CCT), and CIE color coordinates. High-energy chips (e.g., InGaN, GaN) excite perovskite emitters, while encapsulation governs stability. Leng et al. [37] achieved white emission using $Cs_3Bi_2Br_9$/silica composites with phosphors under GaN excitation, yielding CIE (0.29, 0.30) and CCT 8477 K (Fig. 3.4a). Warm-white emission was realized by combining Cs_2SnCl_6:Bi^{3+} nanocrystals with yellow phosphors under 365 nm excitation, giving CIE (0.36, 0.37) and CCT 4486 K (Fig. 3.4b). These studies highlight the tunability and stability of Pb-free perovskites for WLEDs.

Wang et al. [38] developed 2D tin HPs, (octylammonium)$_2SnX_4$ (X = Br, I, or mixed), via an aqueous acid-based route, exhibiting near-unity PL quantum yield, broadband 600 nm emission, significant Stokes shift, long-lived luminescence, and remarkable ambient stability for over 240 days. Their emission was tunable from yellow to deep red by halide composition. UV-pumped LEDs based on these phosphors produced warm yellow light (CCT 2500 K, CRI 69; CIE 0.488, 0.415), warm

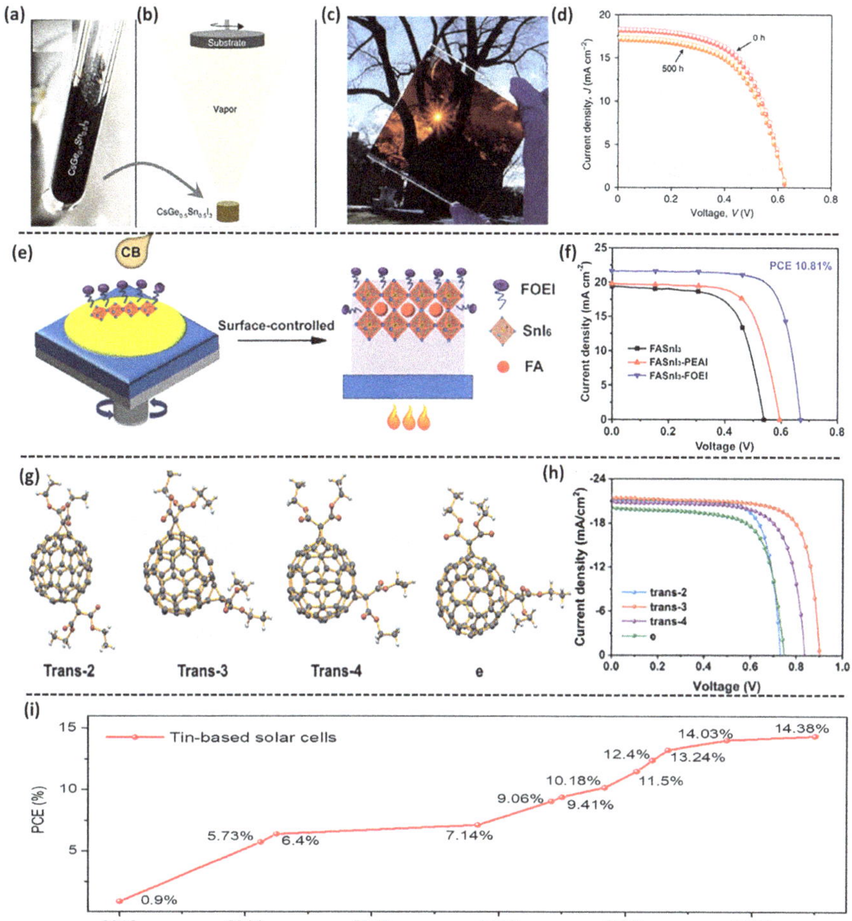

Fig. 3.3 **a** Photograph of $CsSn_{0.5}Ge_{0.5}I_3$ solid prepared via melt-crystallization, **b** Schematic of the single-source evaporation process for depositing smooth $CsSn_{0.5}Ge_{0.5}I_3$ films, **c** Large-area $CsSn_{0.5}Ge_{0.5}I_3$ thin film on glass substrate showing a dark-red appearance, **d** J–V curves of the device at initial measurement and after 500 h of continuous operation; *Reproduced with permission* **©Nature** [34]. **e** Schematic of surface controlled growth in $FASnI_3$-FOEI films, **f** J-V curves of $FASnI_3$, $FASnI_3$-PEAI, and $FASnI_3$-FOEI devices under 1 sun illumination (100 mWcm^{-2}, reverse scan); *Reproduced with permission* **©Elsevier** [35]. **g** Single-crystal structures of trans-2, trans-3, trans-4, and e regioisomers; **h** J–V curves of TPSCs with different ETLs; *Reproduced with permission* **©Wiley** [36]. **i** Efficiency evolution of Sn-based perovskite solar cells

white with yellow/blue mixing (CCT 5635 K, CRI 81; CIE 0.33, 0.26), and optimized white with yellow/blue/green blending (CCT 6530 K, CRI 89; CIE 0.33, 0.31). Figure 3.4c shows the UV pumped WLED fabricated with successive phosphor combinations—yellow-only, yellow/blue, and yellow/blue/green—illustrating the transition from warm yellow to high-quality white emission with enhanced

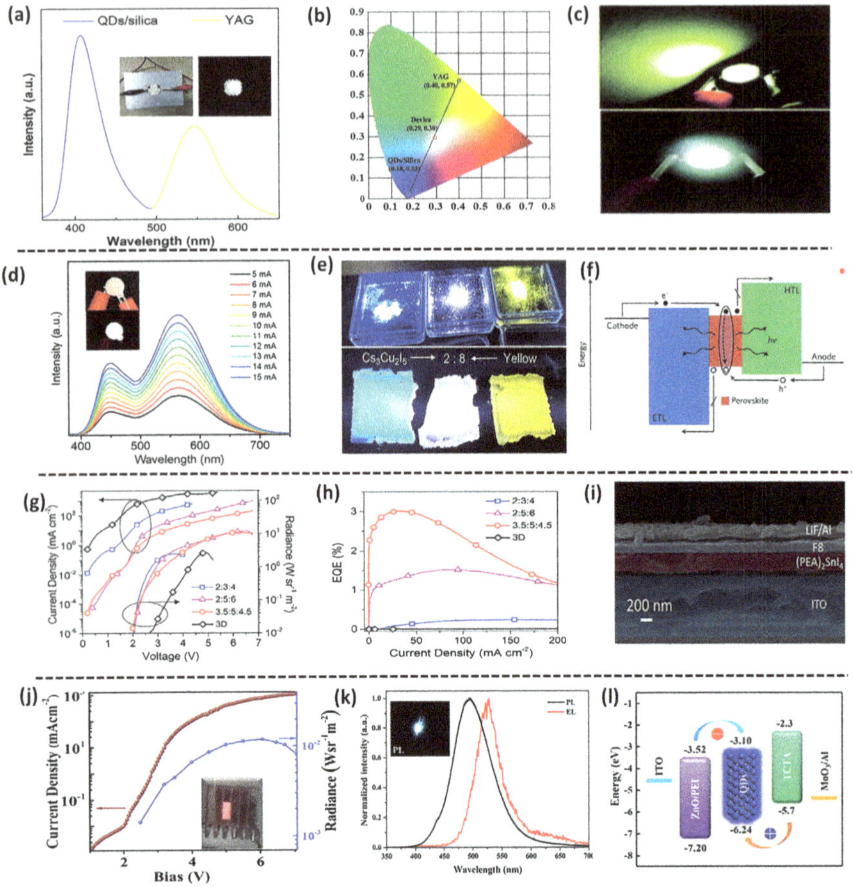

Fig. 3.4 Shows the EL spectrum of a WLED combining blue-emitting $Cs_3Bi_2Br_9$ NCs with yellow $Y_3Al_5O_{12}$ phosphor, **b** presents the CIE coordinates of a WLED based on Cs_2SnCl_6:Bi^{3+} and yellow phosphors on a 365 nm chip, with the operating device inset; *Reproduced with permission* **©Wiley** [37]. **c** displays a 365 nm UV-pumped single yellow $(OCTAm)_2SnBr_4$ LED and a WLED fabricated with $(OCTAm)_2SnBr_4$/$BaMgAl_{10}O_{17}$:Eu^{2+}/G2762/365 nm chip; *Reproduced with permission* **©RSC** [38]. **d** Shows EL spectra at different currents with device on/off insets; *Reproduced with permission* **©RSC** [46]. **e** Shows photographs of $Cs_3Cu_2I_5$ and its mixtures; *Reproduced with permission* **©Wiley** [40]. **f** depicts a typical EL device structure; *Reproduced with permission* **©Nature** [47]. **g-h** present current density, radiance, and EQE of $(PEAI)_x(CsI)_y(SnI_2)z$ films, with a maximum EQE of 3% for $(PEAI)_{3.5}(CsI)_5(SnI_2)_{4.5}$ MQW LED; *Reproduced with permission* **©ACS** [43]. **i** shows the cross-section of ITO/PEDOT:PSS/$(PEA)_2SnI_4$/F8/LiF/Al; *Reproduced with permission* **©ACS**[48]. **j** Shows J–V and radiance voltage for ITO/PEDOT:PSS/$Cs_3Sb_2I_9$/TPBi/LiF/Al; *Reproduced with permission* **©ACS** [44]. **k** Presents PL of $CsCuBr_2$ micro-crosses and EL at 2.8 V; *Reproduced with permission* **©RSC** [45]. **l** illustrates the energy level diagram of a $Cs_3Sb_2Br_9$ nanocrystal LED; *Reproduced with permission* **©ACS** [49]

CRI and stability. A 1:1 mixture of 0D $(C_4N_2H_{14}Br)_4SnBr_6$ and blue phosphor ($BaMgAl_{10}O_{17}:Eu^{2+}$) produced stable white light with CCT 4946 K, CRI 70, and CIE (0.35, 0.39). Figure 3.4d shows consistent color stability across varying operation currents, attributed to the absence of energy transfer between the phosphors [39]. Similarly, blending blue-emitting $Cs_3Cu_2I_5$ with a yellow phosphor in a PDMS matrix yielded bright white emission under UV excitation Fig. 3.4e [40].

Electroluminescent (EL) devices emit light through radiative recombination of injected electrons and holes within the active layer. While Pb-based perovskite LEDs have achieved remarkable efficiencies (EQEs exceeding 20% for green, red, and infrared, and up to 9.5% for blue), Pb-free alternatives are being actively explored due to environmental concerns [41, 42]. A typical EL device structure is illustrated in Fig. 3.4f, which shows the layered architecture that enables charge injection and recombination. Early demonstrations employed Sn-based 2D perovskites, where uniform thin films of $(PEAI)_x(CsI)_y(SnI_2)_z$ delivered tunable EL spectra by adjusting precursor ratios. The best-performing device, $(PEAI)_{3.5}(CsI)_5(SnI_2)_{4.5}$, reached an EQE of 3% at 920 nm with a radiance of 40 W sr^{-1} m^{-2} (Fig. 3.4g, h) [43]. Sb-based perovskites were introduced by Singh et al. [44] using $Cs_3Sb_2I_9$ thin films in ITO/PEDOT:PSS/perovskite/TPBi/LiF/Al devices. Figure 3.4i shows a rapid current increase at 2 V and radiance of 0.012 W sr^{-1} m^{-2} at 6 V (Fig. 3.4j). Similarly, $CsSnX_3$ thin films produced red-emitting EL with current densities up to 915 A cm^{-2}. Copper halide systems enabled green EL at 527 nm via $CsCuBr_2$ micro-crosses integrated into Ag/$CsCuBr_2$/p-Si/In-Ga structures (Fig. 3.4k), where the PL–EL peak shift reflected surface versus interfacial recombination [45]. Finally, Pb-free perovskite nanocrystals $Cs_3Sb_2Br_9$ quantum dots with TCTA as a hole-transporting and electron-blocking layer exhibited violet EL at 408 nm with EQE ~ 0.206% Fig. 3.4l. These devices remain rare due to low PLQY, broad emission, and limited stability.

3.3.3 *Photodetectors*

Pb-free perovskites have emerged as promising materials for photodetectors (PDs) in optoelectronic applications, offers a combination of high photoresponsivity, broad spectral sensitivity, and environmental safety. By substituting toxic Pb with elements such as tin, bismuth, copper, or silver, these materials address health and ecological concerns while maintaining excellent optoelectronic properties [50, 51]. Notably, devices based on Pb-free perovskites, such as $Cs_3Bi_2I_9$ and $Cs_3Cu_2I_5$, demonstrate high responsivity, fast response times, and remarkable long-term stability, even under ambient conditions. Integration with silicon and the development of heterojunction architectures further enhance device performance, enabling broadband detection from ultraviolet to near-infrared regions. Although challenges remain in optimizing film quality and device architecture, ongoing advances in material synthesis and interface engineering are rapidly closing the performance gap with Pb-based counterparts, positioning Pb-free perovskite photodetectors as strong candidates for next-generation, environmentally friendly optoelectronic devices [52, 53].

Therefore, researchers working on PDs have taken great efforts to optimize Pb-free perovskite devices. However, the stability of Sn-based perovskite materials is even more problematic than that of Pb-based perovskites, presenting a significant obstacle for further device development. Waleed et al. [54] reported the first fabrication of 3D $CH_3NH_3SnI_3$ nanowire (NW) arrays using nanoengineered templates, addressing critical challenges of nanowire integration and stability. Photo detectors based on $MASnI_3$ NWs exhibited visible-light sensitivity with a responsivity of 0.47 A W^{-1}, increasing monotonically with incident light intensity, and a specific detectivity of 8.8×10^{10} Jones (Fig. 3.5a–d). Periodic light on/off switching revealed response times of 1500 ms (rise) and 400 ms (decay), which can be further improved by shortening NW lengths. Notably, the light current in NW devices decayed nearly 500 times slower than in thin-film counterparts, highlighting superior charge retention. These results demonstrate a promising strategy for stable, high-performance Sn-based perovskite optoelectronics. Han et al. [55] demonstrated the fabrication of high-quality $CsSnI_3$ perovskite nanowire (NW) arrays with a narrow bandgap of 1.34 eV, enabling their application as active channel materials for photodetectors (PDs) operating from the visible to the near-infrared (NIR) region. The devices exhibited a rapid photoresponse of 8.38/243.4 ms, which can be attributed to the reduced surface states and lower density of trap centers in the $CsSnI_3$ NWs. As illustrated in Fig. 3.5e, the photodetectors achieved a maximum responsivity of 54 mA W^{-1} at 940 nm. Furthermore, the devices maintained excellent stability and reproducibility, highlighting the promise of Pb-free all-inorganic perovskite NW arrays for reliable NIR optical detection.

Double perovskites (A_2AgBiX_6), derived from the ABX_3 framework by substituting two B^{2+} cations with B^+ and Bi^{3+}, offer long carrier recombination lifetimes and enhanced stability (Fig. 3.5f). In 2018, Shi et al. [56] reported $Cs_2AgBiBr_6$ PDs fabricated by one-step spin-coating. Owing to efficient Au/$Cs_2AgBiBr_6$ interfacial charge transfer, the devices achieved a detectivity of 5.66×10^{11} Jones and an EQE of 2146% (Fig. 3.5g), though both declined with increasing incident power. Response dynamics measured at 100 Hz revealed rise and decay times of 956 and 995 μs (Fig. 3.5h, i). Notably, the devices retained performance after 36 h of continuous operation without encapsulation and preserved photodetection capability following two weeks in ambient air, highlighting their durability for stable optoelectronic applications.

Li et al. [58] demonstrated a $Cs_3Cu_2I_5$/GaN heterojunction photodetector (PD) (Fig. 3.5j) designed for narrowband UV detection within 300–370 nm. The strong interfacial charge interaction between $Cs_3Cu_2I_5$ and GaN improved performance, resulting in rapid response times of 95 μs (rise) and 130 μs (decay) under UV illumination. The I–V characteristics (Fig. 3.5k), reveal device behavior under dark conditions and varying irradiation intensities at 320 nm. Furthermore, the PD was successfully employed as a sensing pixel in a UV imaging system, enabling high-resolution image acquisition (Fig. 3.5l, m). Collectively, these results position $Cs_3Cu_2I_5$ as a stable, Pb-free material capable of spectrum-selective detection without the need for external filters.

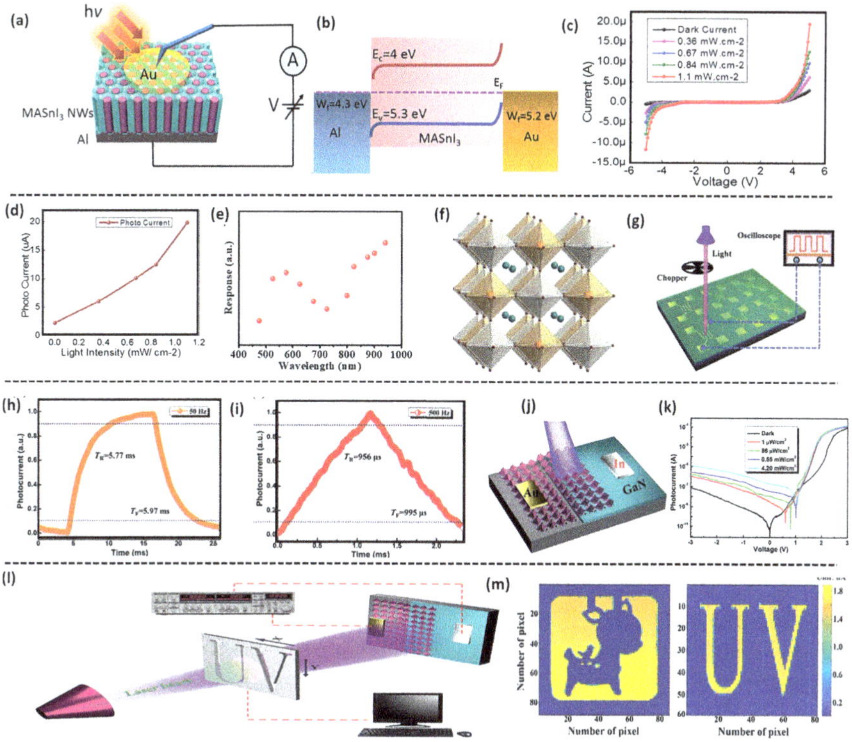

Fig. 3.5 **a** Schematic of the $MASnI_3$ nanowire device structure, **b** Energy band alignment between $MASnI_3$ and the electrodes, **c** I–V characteristics under varying light intensities, **d** Photocurrent response at different illumination levels; *Reproduced with permission* ©**ACS** [54]. **e** Spectral responsivity of the $CsSnI_3$ nanowire array photodetector*; Reproduced with permission* ©**ACS** [55]. **f** Representative morphology of $Cs_2AgBiBr_6$ crystals; *Reproduced with permission* ©**ACS** [57]. **g** System for measuring photodetector response speed. **h, i** Single response cycle showing rise and fall times at 50 Hz and 500 Hz; *Reproduced with permission* ©**RSC** [56]. **j** Schematic of the $Cs_3Cu_2I_5$/GaN heterojunction device. **k** I–V curves under varying light intensities. **l** Diagram of the image sensing setup using the $Cs_3Cu_2I_5$/GaN device. **m** Captured high-resolution patterns; *Reproduced with permission* ©**RSC** [58]

3.3.4 Lasers

LASER (Light Amplification by Stimulated Emission of Radiation) devices produce highly coherent and monochromatic light through the process of optical amplification, where incident photons stimulate the emission of additional photons from an active medium. Charles Townes and Arthur Schawlow proposed the theoretical foundation for this phenomenon, while the first operational laser was successfully demonstrated by Theodor Maiman in 1960. Since then, laser technologies have undergone remarkable advancements, with innovations in active materials, resonator architectures, and excitation mechanisms, leading to their integration in diverse applications, including optical data storage, biomedical imaging, precision fabrication, and high-resolution microscopy.

In recent years, Pb-free perovskites have attracted significant research interest as next-generation laser gain media due to their tunable bandgap, strong light absorption, low exciton binding energies, and enhanced photoluminescence efficiency. While traditional Pb-based perovskites exhibit excellent optoelectronic performance, their toxicity and environmental concerns limit large-scale deployment. To address this, researchers have explored non-toxic alternatives based on tin (Sn), bismuth (Bi), antimony (Sb), and copper (Cu), which provide eco-friendly compositions without compromising critical properties. These features make Pb-free perovskites highly suitable for continuous-wave and pulsed laser operations. Furthermore, the structural flexibility of perovskite materials enables precise bandgap engineering, allowing tunable emission across a wide spectral range from the visible to the near-infrared region. Recent advances in nanostructuring strategies, including the design of nanowires, nanoplates, microcavities, and photonic crystals, have further enhanced light confinement, leading to lower lasing thresholds, higher quantum efficiencies, and improved operational stability. Their combination of environmental safety, superior optoelectronic characteristics, and scalable manufacturing potential positions them as a cornerstone for the development of next-generation photonic and optoelectronic devices, paving the way for advanced applications in integrated optics, optoelectronics, and nanophotonics. The application of (HPs) in laser technologies was first highlighted by Kondo et al. [59] who reported biexciton lasing in the layered perovskite system $[C_6H_{13}NH_3]_2PbI_4$. This discovery sparked significant interest, leading to extensive research on Pb-based perovskite thin films and micro/nanostructures as potential optical gain media. In 2014, Deschler et al. [60] demonstrated a vertical-cavity laser using a thin film of $MAPbI_{3-x}Cl_x$, integrated between a dielectric mirror, a PMMA spacer layer, and a gold mirror. At high excitation fluences, the device exhibited amplified spontaneous emission (ASE), marking a breakthrough in the development of perovskite-based light sources and confirming their immense potential for next-generation optoelectronic applications.

Despite the remarkable progress in Pb-based perovskite lasers, the exploration of Pb-free perovskites has been comparatively limited. Xing et al. [61] addressed this gap by investigating $CsSnI_3$, an inorganic Pb-free perovskite, which demonstrated optical gain in the near-infrared (NIR) region. They discovered that optimizing the orthorhombic phase through SnF_2 doping (5, 10, and 20%) significantly improved crystallinity and photophysical properties. The emission peak remained stable at approximately 950 nm, indicating excellent spectral stability even with increased dopant concentrations (Fig. 3.6a). Importantly, they observed a transition from spontaneous emission to ASE when the photoexcitation fluence exceeded approximately 70 μJ cm^{-2}. At an optimal SnF_2 concentration of 20%, a remarkably low lasing threshold of 6 ± 1 μJ cm^{-2} was achieved, comparable to Pb-based perovskites, showcasing the potential of Sn-based systems for efficient lasing. To further enhance optical confinement, a natural photonic crystal, the microstructured surface of a butterfly wing, was used as a template for depositing $CsSnI_3$ films. This bio-inspired architecture facilitated single-mode lasing with a low threshold of ~ 15 μJ cm^{-2}, producing a narrow emission line with a FWHM ≈ 1.9 nm and a quality factor (Q) ≥ 500 (Fig. 3.6b, c). Additionally, bandgap tunability was demonstrated using $CsSnBr_xI_{3-x}$ compositions (x = 0, 1, 2, 3), enabling ASE peak tuning from 700 to 950 nm while maintaining competitive thresholds. These findings confirm that free–electron–hole bimolecular recombination predominantly drives coherent light generation in these systems. Collectively, this work provides compelling evidence that Sn-based Pb-free perovskites can deliver high optical gain, tunable emission, and low lasing thresholds, positioning them as strong candidates for the development of environmentally friendly, efficient, and versatile perovskite laser sources. Future advancements integrating nanophotonic architectures, compositional engineering, and stability enhancement strategies are expected to unlock further the potential of Pb-free HPs for scalable photonic and optoelectronic applications.

3.3.5 Sensors

Pb-free perovskites have emerged as promising materials for high-performance gas and humidity sensors due to their tunable optoelectronic properties, environmental stability, and eco-friendly composition. Materials such as $Cs_2AgBiBr_6$ and $MASnI_3$ demonstrate excellent sensitivity and fast response times. $Cs_2AgBiBr_6$ thin-film sensors exhibit a linear humidity response (15–78% RH) with high stability, while $MASnI_3$-based devices achieve ultra-low NO_2 detection limits (~ 25 ppb) and show distinct resistance changes for oxidizing and reducing gases. These advancements highlight the potential of Pb-free perovskites for developing sensitive, stable, and sustainable sensor technologies. Hien et al. [62] explored the gas-sensing behavior of $MASnI_3$ thin films under both illumination and dark conditions. The sensors exhibited exceptional NO_2 sensitivity with a detection limit of 25 ppb. Exposure to reducing gases, such as H_2S and NH_3 led to an increase in resistance, while oxidizing gases, such as NO_2 caused a decrease. This characteristic response reflects the p-type conduction mechanism, governed by surface interactions between adsorbed oxygen species and gas molecules (Fig. 3.6d, e).

3.3.6 Memristors

Pb-free perovskite memristors have emerged as promising candidates for next-generation non-volatile memory and neuromorphic computing due to their eco-friendly composition, tunable electronic properties, and low fabrication cost. Materials such as Sn, Bi, and Cu-based perovskites exhibit excellent resistive switching characteristics, enabling stable, reversible transitions between high and low-resistance states. These devices offer low operating voltages, high endurance,

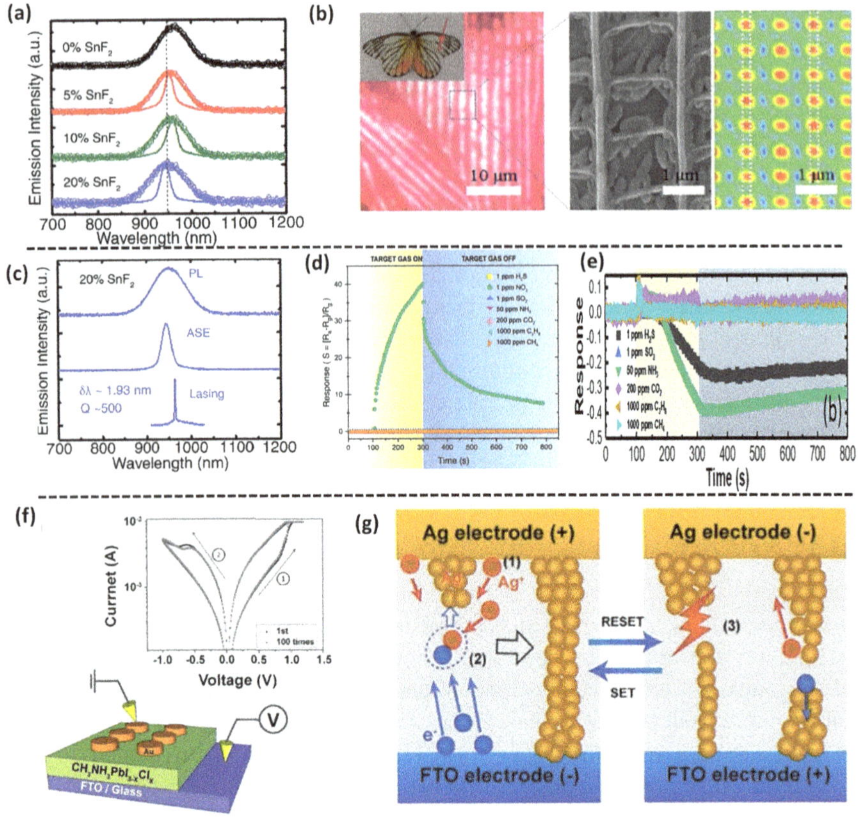

Fig. 3.6 **a** Emission spectra of $CsSnI_3$ films with different SnF_2 concentrations at low and high pump fluences (≈ 70 μJ·cm^{-2}), **b** Optical and SEM images of a butterfly wing scale, showing vertically aligned lamellae, **c** Simulated electric field distribution at 950 nm and comparison of PL, ASE, and single-mode lasing for $CsSnI_3$ (20% SnF_2); *Reproduced with permission* **©Advanced materials** [61]. **d** Gas-sensing response of $MASnI_3$ films to various gases **e** and to gases excluded NO_2; *Reproduced with permission* **©Elsevier** [62]. **f** I–V curves of the Au/$MAPbI_{3-x}Cl_x$/FTO memristor showing initial and stable switching after 100 cycles; *Reproduced with permission* **©Advanced materials** [63]. **g** Mechanism of Ag filament formation and rupture in the Ag/$MAPbI_{3-x}Cl_x$/FTO device under applied bias; *Reproduced with permission* **©Wiley** [64]

and good retention performance, making them suitable for applications in data storage, artificial synapses, and low-power computing architectures. The combination of environmental safety and scalable fabrication positions Pb-free perovskite memristors as a key platform for future energy efficient and sustainable memory technologies [65, 66]. One of the significant benefits of memristors is their potential for miniaturization to the nanoscale, which allows seamless integration into a wide range of applications while also enhancing their overall performance at reduced dimensions. Chua et al. [67] introduced the memristor as a unique two-terminal device that defines a direct relationship between magnetic flux (φ) and electric charge (q), a concept later extended to memristive systems. These devices exhibit the ability to switch between different resistance states in response to an applied voltage or current, enabling the combination of data storage and logic operations within a single architecture. A key signature of memristor behavior is the pinched hysteresis loop in their I–V characteristics, which reflects their switching mechanism and non-volatile memory properties, making them strong candidates for advanced memory devices, neuromorphic computing, and reconfigurable electronics. HPs have gained attention as potential materials for memristive devices due to their exceptional opto-electronic characteristics. The initial demonstration of a memristor using perovskite was conducted by Yoo et al. [63] where a thin film of $MAPbI_{3-x}Cl_x$ was placed between Au and fluorine-doped tin oxide (FTO) electrodes. This device exhibited bipolar resistive switching (RS), impressive endurance (Fig. 3.6f), low operating voltages, and stable non-volatile memory with a retention time of approximately 10^4 s, underscoring the promise of HPs in resistive switching applications. The primary mechanism responsible for resistive switching in HP-based memristors is generally believed to be the creation and breaking of conductive filaments (CFs), mainly driven by the movement of halide ions, vacancies, or active metal ions within the perovskite structure. An example of this process, involving the formation and disruption of Ag filaments, is depicted in Fig. 3.6g. Furthermore, some research suggests an interface-type mechanism, where resistive switching results from changes in the Schottky barrier at the electrode-perovskite interface [64, 68]. HPs possess remarkable properties, such as high carrier mobility, long diffusion lengths, solution processability, and low production costs, which have led to the creation of various device architectures. These are often enhanced with metal oxides and polymeric interlayers, resulting in improved performance and stability [69]. However, the presence of Pb poses a significant challenge, prompting research into Pb-free HPs as more sustainable options. Recently, several Pb-free perovskite-based memristors have been investigated, using Sn, Bi, Sb, and Cu-based perovskites, which have shown practical resistive switching and compatibility with neuromorphic computing architectures. These developments not only enhance environmental safety but also pave the way for designing next-generation, low-power, and multifunctional memristive systems [70, 71].

3.4 Challenges in Pb-Free Perovskites for Optoelectronic Devices

Despite various advantages, such as a tunable bandgap, strong light absorption, high photoluminescence efficiency, and low fabrication costs. Several critical challenges hinder their widespread commercialization and practical applications in photodetectors, light-emitting diodes (LEDs), lasers, solar cells, memristors, and sensors. These challenges primarily arise from material limitations, stability concerns, device performance constraints, and integration complexities, which must be addressed for advancing next-generation optoelectronic technologies. One of the major challenges lies in the intrinsic material stability of Pb-free perovskites. Unlike Pb-based counterparts, tin (Sn)-, bismuth (Bi), antimony (Sb), and copper (Cu)-based perovskites are prone to structural degradation when exposed to moisture, oxygen, light, and elevated temperatures. For instance, Sn^{2+} ions in Sn-based perovskites readily oxidize to Sn^{4+}, resulting in the formation of defects and poor long-term operational stability. Similarly, double perovskites, such as $Cs_2AgBiBr_6$, exhibit superior environmental tolerance but suffer from intrinsic deep defect states that restrict efficient charge transport and limit device performance. Another significant limitation involves defect density and charge-carrier dynamics. Pb-free perovskites often exhibit high trap densities at grain boundaries and interfaces, which accelerate non-radiative recombination and reduce the power conversion efficiency (PCE) and optical gain of devices.

Achieving high-quality, defect-free thin films with uniform crystal growth remains a substantial fabrication challenge, especially when utilizing solution-processing techniques for large-area and flexible device manufacturing. Moreover, improper energy-level alignment between the perovskite layer and charge transport layers can lead to interfacial recombination losses, severely affecting optoelectronic efficiency. Furthermore, while Pb-free perovskites are less toxic and environmentally friendly, their device performance still lags behind that of Pb-based systems in terms of efficiency, optical gain, and operational lifetime. For example, Sn-based perovskite solar cells have reported efficiencies exceeding 14%, but these remain significantly lower than the > 25% achieved by Pb-based perovskites. Similarly, Pb-free perovskite-based LEDs and lasers face challenges in achieving high brightness, narrow emission linewidths, and low lasing thresholds due to incomplete control over crystallinity, exciton confinement, and energy transfer mechanisms. In summary, while Pb-free perovskites hold immense potential for next-generation optoelectronic devices, challenges related to stability, defect control, phase transitions, device performance, large-scale fabrication, and integration must be comprehensively addressed. Overcoming these barriers will require multidisciplinary approaches combining material engineering, surface passivation, nanostructuring, and device optimization. Continued advancements in these areas will pave the way for environmentally sustainable, high-performance, and commercially viable Pb-free perovskite technologies.

3.5 Summary

Recent advancements in Pb-free perovskites position them as promising, sustainable alternatives to toxic Pb-based materials for next-generation optoelectronic devices. Driven by environmental and health concerns, significant research has focused on developing compositions that retain key optoelectronic properties, such as high absorption coefficients, tunable band gaps, long carrier lifetimes, and solution processability. Multiple material engineering strategies have been employed to address the challenges of instability and efficiency losses. Cation substitution using elements like Sn^{2+}, Bi^{3+}, and Sb^{3+} enables bandgap tuning, phase stabilization, and defect suppression. Dimensional engineering, particularly the development of 2D and quasi-2D perovskite structures, improves moisture resistance and structural stability while maintaining effective charge transport. Defect passivation techniques, including the use of halide additives and molecular passivating agents, significantly reduce trap densities and extend carrier lifetimes. Doping provides control over carrier concentration, ion migration, and the optimization of electronic structure.

Compositional engineering through tailored halide and cation ratios allows precise tuning of optoelectronic behaviour, while nanostructuring and heterojunction formation enhance light absorption, charge separation, and device multifunctionality. These strategies have led to notable progress across various device platforms. Pb-free perovskite solar cells demonstrate continuous improvements in both efficiency and operational stability, gradually approaching the performance of Pb-based systems. In LEDs, Pb-free compositions offer efficient and colour-pure emission with scalable fabrication methods suitable for display and lighting technologies. Photodetectors based on these materials exhibit broadband sensitivity, high detectivity, and fast response times, underscoring their potential in advanced sensing applications. Emerging laser devices based on Pb-free perovskites offer high optical gain and narrow emission linewidths. At the same time, their integration into optoelectronic memristors opens up pathways toward multifunctional devices that can combine memory and optical functionalities for neuromorphic computing. Despite being in the early stages of development, Pb-free perovskites show significant potential across a wide range of optoelectronic applications. Critical challenges remain, including long-term stability, scalability, and device efficiency, but the outlined strategies offer a solid foundation for future advancements. With continued interdisciplinary research in materials innovation and device integration, Pb-free perovskites are poised to become key components in the development of safe, efficient, and scalable optoelectronic technologies.

References

1. R. Wang et al., Opportunities and challenges of lead-free perovskite optoelectronic devices. Trends Chem. **1**(4), 368–379 (2019)
2. I. López-Fernández et al., Lead-free halide perovskite materials and optoelectronic devices: progress and prospective. Adv. Func. Mater. **34**(6), 2307896 (2024)
3. H. Wang et al., Research progress of ABX 3-type lead-free perovskites for optoelectronic applications: materials and devices. Phys. Chem. Chem. Phys. **24**(45), 27585–27605 (2022)
4. A. Bibi et al., Lead-free halide double perovskites: toward stable and sustainable optoelectronic devices. Mater. Today **49**, 123–144 (2021)
5. M. Roknuzzaman et al., Insight into lead-free organic-inorganic hybrid perovskites for photovoltaics and optoelectronics: a first-principles study. Org. Electron. **59**, 99–106 (2018)
6. M.H. Miah, et al., Lead-free alternatives and toxicity mitigation strategies for sustainable perovskite solar cells: a critical review. Mater. Adv. (2025)
7. C.-Z. Ning, L. Dou, P. Yang, Bandgap engineering in semiconductor alloy nanomaterials with widely tunable compositions. Nat. Rev. Mater. **2**(12), 1–14 (2017)

8. I. Ahmed, Prakash, K., S.M. Mobin, Lead-free perovskites for solar cells applications: recent progress, ongoing challenges, and strategic approaches. Chem. Commun. (2025)
9. S. Ahmed et al., Recent trends and challenges in lead-free perovskite solar cells: a critical review. ACS Appl. Energy Mater. **7**(4), 1382–1397 (2024)
10. A.D. Khan et al., Advancements in the stability, protection and lead-free strategies of perovskite solar cells: a critical review. Environ. Sci.: Adv. **3**(7), 1004–1029 (2024)
11. P.V. Kamat, J. Bisquert, J. Buriak, Lead-free perovskite solar cells. ACS Energy Lett. **2**(4), 904–905 (2017)
12. M. Keshavarz et al., Tuning the structural and optoelectronic properties of Cs2AgBiBr 6 double-perovskite single crystals through alkali-metal substitution. Adv. Mater. **32**(40), 2001878 (2020)
13. S.S. Almishal, O. Rashwan, First-principles investigation of lead-free trigonal CsGeI3-xBrx mixed-halide perovskite system for optoelectronic applications: electronic and optical properties. Mater. Sci. Semicond. Process. **151**, 107017 (2022)
14. L. Xiao, L. Chen, S. Zhao, Effect of partial cation substitution on optoelectronic properties and ion diffusion behavior of inorganic lead halide perovskite: a theoretical investigation. J. Phys. Chem. Solids **193**, 112193 (2024)
15. J. Sala et al., Compositional investigation for bandgap engineering of wide bandgap triple cation perovskite. ACS Appl. Energy Mater. **4**(7), 6377–6384 (2021)
16. A. Dey et al., State of the art and prospects for halide perovskite nanocrystals. ACS Nano **15**(7), 10775–10981 (2021)
17. C. Otero-Martínez et al., Colloidal metal-halide perovskite nanoplatelets: thickness-controlled synthesis, properties, and application in light-emitting diodes. Adv. Mater. **34**(10), 2107105 (2022)
18. Z. Liu et al., Synthesis of lead-free Cs2AgBiX6 (X= Cl, Br, I) double perovskite nanoplatelets and their application in CO_2 photocatalytic reduction. Nano Lett. **21**(4), 1620–1627 (2021)
19. Z. Liu et al., Two-dimensional Cs2AgInxBi1-xCl6 alloyed double perovskite nanoplatelets for solution-processed light-emitting diodes. Adv. Mater. **35**(19), 2211235 (2023)
20. B. Chang et al., Efficient bulk defect suppression strategy in FASnI3 perovskite for photovoltaic performance enhancement. Adv. Func. Mater. **32**(12), 2107710 (2022)
21. L. Rao et al., Wearable tin-based perovskite solar cells achieved by a crystallographic size effect. Angew. Chem. **133**(26), 14814–14821 (2021)
22. S. Jiang et al., Doping strategies for inorganic lead-free halide perovskite solar cells: progress and challenges. Phys. Chem. Chem. Phys. **26**(6), 4794–4811 (2024)
23. Y. Lu et al., Doping up the light: a review of A/B-site doping in metal halide perovskite nanocrystals for next-generation LEDs. J. Phys. Chem. C **128**(24), 10084–10107 (2024)
24. D. Cortecchia et al., Lead-free MA2CuCl x Br 4–x hybrid perovskites. Inorg. Chem. **55**(3), 1044–1052 (2016)
25. Y. Sun et al., Compositional and structural engineering of inorganic nanowires toward advanced properties and applications. InfoMat **1**(4), 496–524 (2019)
26. Y. Zhang, Y. Liu, S. Liu, Composition engineering of perovskite single crystals for high-performance optoelectronics. Adv. Func. Mater. **33**(9), 2210335 (2023)
27. Z.U. Abideen et al., Structural engineering three-dimensional nano-heterojunction networks for high-performance photochemical sensing. ACS Appl. Mater. Interfaces. **15**(48), 56464–56477 (2023)
28. X. Xu, X. Wang, Perovskite nano-heterojunctions: synthesis, structures, properties, challenges, and prospects. Small Struct. **1**(1), 2000009 (2020)
29. Q. Su et al., Heterojunction photocatalysts based on 2D materials: the role of configuration. Adv. Sustain. Syst. **4**(9), 2000130 (2020)
30. S. Martinuzzi, Trends and problems in CdS/CuxS thin film solar cells: a review. Solar Cells **5**(3), 243–268 (1982)
31. X. Zhang et al., ZnSe nanowire/Si p–n heterojunctions: device construction and optoelectronic applications. Nanotechnology **24**(39), 395201 (2013)

32. Y. Zhang et al., ZnO/ZnSe type II core-shell nanowire array solar cell. MRS Online Proc. Libr. **1396**(1), 158–163 (2012)
33. C. Li et al., Wide-bandgap lead halide perovskites for next-generation optoelectronics: current status and future prospects. ACS Nano **18**(52), 35130–35163 (2024)
34. M. Chen et al., Highly stable and efficient all-inorganic lead-free perovskite solar cells with native-oxide passivation. Nat. Commun. **10**(1), 16 (2019)
35. X. Meng et al., Surface-controlled oriented growth of FASnI3 crystals for efficient lead-free perovskite solar cells. Joule **4**(4), 902–912 (2020)
36. C. Sun et al., Well-defined fullerene bisadducts enable high-performance tin-based perovskite solar cells. Adv. Mater. **35**(9), 2205603 (2023)
37. M. Leng et al., All-inorganic bismuth-based perovskite quantum dots with bright blue photoluminescence and excellent stability. Adv. Func. Mater. **28**(1), 1704446 (2018)
38. A. Wang et al., Aqueous acid-based synthesis of lead-free tin halide perovskites with near-unity photoluminescence quantum efficiency. Chem. Sci. **10**(17), 4573–4579 (2019)
39. C. Zhou et al., Facile preparation of light emitting organic metal halide crystals with near-unity quantum efficiency. Chem. Mater. **30**(7), 2374–2378 (2018)
40. T. Jun et al., Lead-free highly efficient blue-emitting Cs3Cu2I5 with 0D electronic structure. Adv. Mater. **30**(43), 1804547 (2018)
41. S. Chen, H. Xu, Electroluminescent materials toward near ultraviolet region. Chem. Soc. Rev. **50**(15), 8639–8668 (2021)
42. J.S. Konthoujam et al., Electrical control of electroluminescence in 2D semiconductor light emitting device through synchronous injection of electrons and holes. Adv. Mater. Technol. **10**(7), 2401743 (2025)
43. Y. Wang et al., Tin-based multiple quantum well perovskites for light-emitting diodes with improved stability. J. Phys. Chem. Lett. **10**(3), 453–459 (2019)
44. A. Singh et al., Lead-free antimony-based light-emitting diodes through the vapor–anion-exchange method. ACS Appl. Mater. Interfaces. **11**(38), 35088–35094 (2019)
45. T. Li et al., Distinct green electroluminescence from lead-free CsCuBr 2 halide micro-crosses. Chem. Commun. **55**(31), 4554–4557 (2019)
46. C. Zhou et al., Luminescent zero-dimensional organic metal halide hybrids with near-unity quantum efficiency. Chem. Sci. **9**(3), 586–593 (2018)
47. B.R. Sutherland, E.H. Sargent, Perovskite photonic sources. Nat. Photonics **10**(5), 295–302 (2016)
48. L. Lanzetta et al., Two-dimensional organic tin halide perovskites with tunable visible emission and their use in light-emitting devices. ACS Energy Lett. **2**(7), 1662–1668 (2017)
49. Z. Ma et al., Electrically-driven violet light-emitting devices based on highly stable lead-free perovskite Cs3Sb2Br 9 quantum dots. ACS Energy Lett. **5**(2), 385–394 (2019)
50. Y. Zhang et al., Lead-free perovskite photodetectors: progress, challenges, and opportunities. Adv. Mater. **33**(26), 2006691 (2021)
51. F. Cao, L. Li, Progress of lead-free halide perovskites: from material synthesis to photodetector application. Adv. Func. Mater. **31**(11), 2008275 (2021)
52. S.A. Behera, P.G.R. Achary, Lead free perovskite based heterojunction photodetectors: a mini review. Appl. Surf. Sci. Adv. **14**, 100393 (2023)
53. W. Sun et al., Integration of one-dimensional (1D) lead-free perovskite microbelts onto silicon for ultraviolet–visible–near-infrared (UV-vis-NIR) heterojunction photodetectors. J. Phys. Chem. Lett. **15**(9), 2359–2368 (2024)
54. A. Waleed et al., Lead-free perovskite nanowire array photodetectors with drastically improved stability in nanoengineering templates. Nano Lett. **17**(1), 523–530 (2017)
55. M. Han et al., Controllable growth of lead-free all-inorganic perovskite nanowire array with fast and stable near-infrared photodetection. J. Phys. Chem. C **123**(28), 17566–17573 (2019)
56. L.-Z. Lei et al., High-efficiency and air-stable photodetectors based on lead-free double perovskite Cs 2 AgBiBr 6 thin films. J. Mater. Chem. C **6**(30), 7982–7988 (2018)
57. A.H. Slavney et al., A bismuth-halide double perovskite with long carrier recombination lifetime for photovoltaic applications. J. Am. Chem. Soc. **138**(7), 2138–2141 (2016)

58. Y. Li et al., Highly stable and spectrum-selective ultraviolet photodetectors based on lead-free copper-based perovskites. Mater. Horiz. **7**(2), 530–540 (2020)
59. T. Kondo et al., Biexciton lasing in the layered perovskite-type material (C6H13NH3) 2PbI4. Solid State Commun. **105**(4), 253–255 (1998)
60. F. Deschler et al., High photoluminescence efficiency and optically pumped lasing in solution-processed mixed halide perovskite semiconductors. J. Phys. Chem. Lett. **5**(8), 1421–1426 (2014)
61. G. Xing, et al., *Solution-Processed Tin-Based Perovskite for Near Infrared Lasing* (2016)
62. V.X. Hien et al., Growth and gas sensing properties of methylammonium tin iodide thin film. Scripta Mater. **178**, 108–113 (2020)
63. E.J. Yoo, et al., Resistive switching behavior in organic-inorganic hybrid CH_3 NH_3 $PbI_{3\text{-}x}$ Clx perovskite for resistive random access memory devices. Adv. Mater. (Deerfield Beach, Fla.) **27**(40), 6170–6175 (2015)
64. E. Yoo et al., Bifunctional resistive switching behavior in an organolead halide perovskite based Ag/CH 3 NH 3 PbI 3–x Cl x/FTO structure. J. Mater. Chem. C **4**(33), 7824–7830 (2016)
65. K. Sun, J. Chen, X. Yan, The future of memristors: materials engineering and neural networks. Adv. Func. Mater. **31**(8), 2006773 (2021)
66. N. Youngblood et al., Integrated optical memristors. Nat. Photonics **17**(7), 561–572 (2023)
67. L. Chua, Memristor-the missing circuit element. IEEE Trans. Circuit Theory **18**(5), 507–519 (2003)
68. X. Xiao et al., Recent advances in halide perovskite memristors: materials, structures, mechanisms, and applications. Adv. Mater. Technol. **5**(6), 1900914 (2020)
69. H. An et al., Highly-stable memristive devices based on poly (methylmethacrylate): CsPbCl3 perovskite quantum dot hybrid nanocomposites. Org. Electron. **56**, 41–45 (2018)
70. Y. Wu et al., Capping CsPbBr 3 with ZnO to improve performance and stability of perovskite memristors. Nano Res. **10**(5), 1584–1594 (2017)
71. Z. Xu et al., To probe the performance of perovskite memory devices: defects property and hysteresis. J. Mater. Chem. C **5**(23), 5810–5817 (2017)

Chapter 4
Role of Lead Free Perovskite Materials in Gas Sensors

Aruna Raj, J. S. Anaswara, R. Adithya Nath, and R. Jayakrishnan

Abstract Assessing air quality has become increasingly vital in today's world, where climate change, pollution, and the demand for sustainable environments pose escalating challenges. In response, there is a growing need for gas sensors that are both selective and highly sensitive. Lead halide perovskites have emerged as promising candidates for such applications due to their exceptional optoelectronic properties. However, the toxicity of lead has raised significant environmental concerns, prompting an urgent search for safer alternatives. This urgency is further amplified by rising awareness of harmful pollutants—such as nitrogen dioxide, carbon monoxide, and hydrogen sulphide—and by stringent health and safety regulations. Lead-free perovskites (LFPs) have gained attention as a viable class of materials for advanced gas sensing. Compositions based on tin, bismuth, antimony, and germanium offer comparable optoelectronic characteristics with reduced toxicity. Nonetheless, their practical deployment is hindered by challenges related to lower stability and diminished efficiency. LFP nanoparticles, in particular, exhibit enhanced sensing capabilities owing to their high surface-to-volume ratio, quantum confinement effects, and strong exciton binding energy. Under external stimuli, these materials undergo reversible changes in properties through mechanisms such as intercalation, surface adsorption, trap passivation, and ion exchange. This chapter presents a detailed analysis of key factors influencing the sensing performance of LFPs, including band gap characteristics, intrinsic defect tolerance, and the role of electronic surface traps at the nanocrystal interface.

Keywords: Lead free perovskites · Gas sensors · Defect tolerance · Oxygen vacancy

A. Raj · J. S. Anaswara · R. Adithya Nath · R. Jayakrishnan (✉)
Photovoltaic Research Laboratory, Department of Physics, University of Kerala, Karyavattom, Thiruvananthapuram, India
e-mail: rjk@keralauniversity.ac.in

V. Chauhan et al. (eds.), *Sustainable Lead-Free Perovskite Materials*, Topics in Applied Physics 157, https://doi.org/10.1007/978-3-032-24873-2_4

4.1 Introduction

Atmospheric contamination from the release of harmful gases, even at trace levels, can cause serious health hazards. As we know, the air we breathe has a mixture of gases such as oxygen (O_2), Nitrogen (N_2), Carbon monoxide (CO), Carbon dioxide (CO_2), several volatile organic compounds (VOCs) etc. When the concentration of such gases exceeds the safe limits, they become hazardous, posing serious environmental threats. This factor necessitates the need for highly sensitive, selective, environmentally safe, and economically viable sensors for sensing gases. Seiyama et al. [56] has developed a thin film based gas chromatographic sensor for the first time. Thereafter, several innovative designs came up using nanomaterials such as metal oxides, graphenes, carbon nanotubes, MXenes etc. The common limitation on the requirement of high material processing temperatures for efficient sensing has opened up the urge to look for alternate materials with improved performance. The basic working principle of metal oxide semiconductor based gas sensors was based on change in resistance of the gas sensor material in air and gas atmosphere. Thus the nature as well as the structure of the sensing material plays a crucial role in gas sensing performance. Perovskite nano crystals provides a major prerequisite on high binding energy sites for high performance of gas sensors at room temperature. These materials are highly promising optoelectronic class of materials possessing additional features such as high and fast response, selectivity etc. Fang et al. [19] has demonstrated the first perovskite based sensor to detect oxygen and water vapor. Recent reports has shown the huge potential of perovskite materials in sensing various gases.

Among the perovskite based materials, lead halide perovskites outperform all other compounds due to their exceptional optoelectronic features as well as chemical tunability. The inherent toxicity as well as the environmental instability associated with lead has raised significant health concerns. This factor has raised the need for the development of lead free perovskites thereby offering a sustainable platform for gas sensors. Lead free perovskite materials utilizes the exceptional combination of high surface activity and favourable electronic properties, enhancing their efficient response towards gas molecule interactions. The fast as well as high conductive and optical (optical absorption and photoluminescence) response of these materials under gas exposure has formed a wide range of mechanisms for gas sensing. For optimising gas sensing efficiency, synthesis techniques, structural parameters, defect analysis as well as surface passivations need to be assessed before engineering the device.

This chapter widely explores the role of lead free perovskites in gas sensing starting from giving descriptions on various mechanisms involved, several classes of materials, various associated synthesis and processing techniques, and strategies for improving the performances. The chapter also highlights different gas sensor architectures, finally addressing the challenges as well as future direction towards sustainable device integration in upcoming world technologies.

4.2 Fundamentals

In the current era of automation, there is a huge necessity of precise instruments for measuring various aspects. The capability enhancement of measurement and detection systems can be achieved through integrating sensor units at input points of domestic and scientific devices. This should be implemented on a cost effective and compact approach.

Gas sensor is a device that detects as well as quantifies the amount of a particular gas or a mixture of gases present in the atmosphere. Two major components present with in a gas sensor are a receptor and a transducer. The receptor would be composed of materials that would undergo a change in parameters such as dielectric constant, work function, mass, electrode potential etc. or will be emitting energy in the form of heat or light. A transducer would convert this into an electrical response in the form of current, voltage, capacitance, frequency etc. [70]. A model of a typical gas sensor is illustrated in Fig. 4.1.

4.2.1 Properties

The performance of a gas sensor is evaluated by a number of parameters. For a gas sensor to be an excellent one, it should perform well in such key parameters as listed below [51, 64, 65].

Stability: A consistent response should be maintained to a particular concentration of gas over some duration, which in fact shows the reliability of the device.

Optimum Operating Temperature: A critical parameter where the rate of adsorption and desorption are at equilibrium theoretically. The determination of this

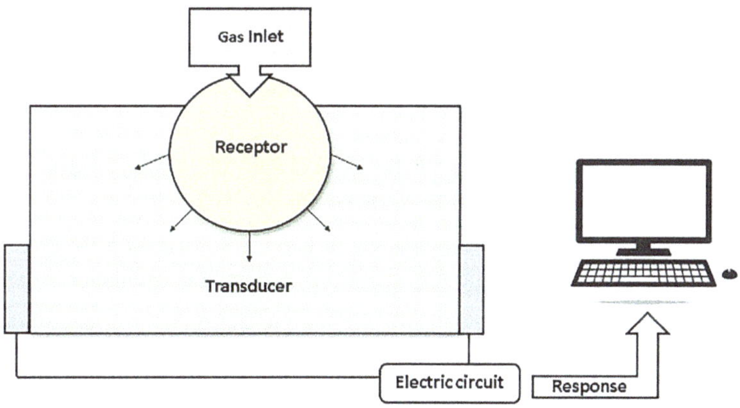

Fig. 4.1 Gas sensor model

parameter requires an analysis of the sensor's performance over a range of temperatures.

Responsiveness and Sensitivity: Gas sensitive responsiveness is measured by noting the change in electrical signals when the gas is passed on. This value for an n-type semiconductor material is opposite to that of p-type. Sensitivity is a measure on how effectively does the sensor detects the gas. It is the ratio of responsiveness to the concentration of the particular gas.

Selectivity: This parameter defines the ability to distinguish between gases. A good sensor with great selectivity should resist interference from non-target gases by giving a stronger response to the gas under consideration.

Recovery Time and Response Time: This parameter determine how fast the sensor detects the target gas which is influenced by the concentration of the gas. Higher the concentration, higher will be the rates of recovery and response. Recovery time is defined as the time required for the sensor to get back to the initial state after the exposure to gas. Response time is the time taken by the sensor to detect change in concentration of the gas and then to respond to it quantitatively.

Repeatability: Repeatability determines the consistency of the measurement results obtained from the gas sensor under identical atmosphere and conditions.

Limit of Detection: This is the lowest concentration where the sensor could effectively detect the gas. This parameter is influenced by the sensitivity, responsiveness, temperature, humidity, pressure etc.

4.2.2 Classification of Gas Sensors

Based on the mode of conversion of gas interaction into measurable signals, various classification of gas sensors exist [51].

Electrochemical Sensors: These sensors use chemical reactions occurring at the electrodes immersed within the electrolyte to produce current which is proportional to the concentration of the gas.

Infrared Sensors: The absorption of the gases within the infrared wavelengths is utilized to measure the concentration of the gases.

Catalytic Bead Sensors: Combustible gases can be sensed using such sensors through a process of catalysis, thereby producing a change in resistance.

Photoionization Detectors: An electric current proportional to that of the gas concentration is produced as a result of the ionization of gas due to the incidence of ultraviolet radiation.

Other than this commonly used gas sensors, Magnetic sensors, calometric sensors, acoustic wave sensors, thermal conductivity sensors, optical fibre sensors are recently used for specific gas sensing applications.

4.3 Lead Free Perovskite Based Gas Sensors

Lead based perovskite gas sensors, despite of their tunable crystal structure, optoelectronic properties, defect chemistry and responsiveness, fail in the aspect of non-toxicity. This has made researchers to move from conventional lead based perovskite gas sensors to environmentally benign alternatives. In that context, Lead free Perovskites such as single component halide perovskites, mixed halide perovskites, mixed cation perovskites, metal oxide perovskites has emerged that aligns with the toxicity regulations without compromising the sensing performance.

Room temperature gas sensing was made possible with single component halide perovskites such as $CsSnBr_3$ and $CsBi_3I_{10}$.The peculiarity of the material that contributes towards efficient gas sensing are intrinsic semiconductivity and high surface activity. Mixed component Lead free Perovskite materials like $CH_3NH_3AgBiBr_6$ and $Cs_2AgBiCl_6$ were also researched for gas sensor applications. The enhancement of selectivity and chemical stability was observed with the heterovalent cations at B-site and layered morphology. The sensing capability for certain gases like NO_2, H_2S with such mixed component systems provides better defect tolerance, control over the band structure, adsorption phenomena etc. Metal oxide perovskites with ABO_3 structure are another broad class of materials having huge applications in gas sensors. $SrTiO_3$, $BaTiO_3$, $LaFeO_3$ are researched and found to be thermally robust and stable materials. They were commonly use in chemiresistive type sensors. Several metal oxide perovskites can be tailored on their porosity, oxygen vacancies, particle size which could enhance parameters such as selectivity, sensitivity and response/recovery times in gas sensors [64, 65].

4.3.1 Categories of Lead Free Perovskites in Gas Sensing

The immense research during the recent times on perovskite gas sensors has explored different classes of Lead free perovskite systems as halide based, oxide based, double perovskites, Organic Inorganic perovskites and two dimensional perovskites.

Halide Based Lead Free Perovskites: Halide perovskites based on Bismuth, antimony and tin were used actively in optical and chemiresistive gas sensors. Narrow band gap materials like $Cs_3Bi_2Br_9$ [20], as well as $(CH_3NH_3)_3Bi_2I_9$ [7] has shown excellent NO_2 as well as NH_3 sensing. These perovskite materials possess high carrier mobility. Most of the materials in this category shows the sensing through changes in conductivity and photoluminescence. The limitations associated with certain Sn based compounds are the oxidative nature of Sn would necessitate the composite or encapsulation strategy which will otherwise lead to phase instability affecting the sensitivity of gas sensors.

Oxide Based Lead Free Perovskites: Most of the ABO_3 structured perovskites are known for their chemical and thermal stability. They have rich vacancies of oxygen and different doping strategies can be implemented. These materials are good candidates for high temperature gas sensing.The mechanisms associated with this class of materials are usually based on chemiresistive mechanism as in the case of $LaFeO_3$, which could sense NO_2, H_2, CO etc. [52], oxygen adsorption/desorption as in the case of $LaFeO_3$ co doped with Sr or Co showing sensitivity to CO_2 and acetone and vacancy mediated charge transfer mechanisms [15].

Lead Free Double Perovskites: These materials with the general structure $A_2B'B''X_6$ often are indirect band gap materials. They could be utilized in optical as well as resistive gas sensors due to their favourable defect tolerance with enhanced stability. Also, a greater extend of optoelectronic tuning is possible with this class of materials. $Cs_2AgBiBr_6$ detects NO_2 as well as acetone with its photoconductive changes [4].

Organic Inorganic Hybrid Lead Free Perovskites: The advantage of these materials over others are two. One is the possibility of low temperature processing and other one is solution processing. A combination of structural flexibility from the organic part and semiconducting properties from the inorganic part would help in the good performance of these materials in gas sensing. The common gas sensing mechanism associated with these materials is optical sensing.$CH_3NH_3Bi_2I_9$ [7] and $CH_3NH_3SnI_3$ has been used in NH_3 sensing sensing [46].

Two Dimensional (2D) Lead Free Perovskites: A_2BX_4 structures (Ruddlesden-Popper type) would offer structural and phase stability. These material exhibits anisotropic transport properties. The advantage of this class of materials over others is that they produce a high surface area as well as active sites for the gaseous species. (Heterojunctions with 2D and 0D materials, such as $Cs_3Bi_2Br_9$ with Bi_2WO_6, have demonstrated CO_2 sensing [20].

4.4 Gas Sensing Mechanisms in Perovskites

Many widely accepted models of gas sensor mechanisms have evolved with huge research on perovskite based gas sensors recently. The working principle of the gas sensor materials can be explained using such models.

Oxygen Adsorption Model is a widely recognized model that is used mostly by chemiresistive oxide and halide perovskite gas sensors. As per this model, oxygen molecules from the atmosphere gets adsorbed on to the perovskite surface, taking electrons, forming ionic species such as O^-, O_2^- or O^{2-} [65].When a gas that is capable of reducing or oxidizing any material interacts with the surface, reaction with oxygen species occur thereby releasing or trapping the electrons by itself. $BaTiO_3$ [14], $LaFeO_3$ [69] are some of the lead free perovskites that works on oxygen adsorption model.

Grain Boundary Model is a common model in polycrystalline perovskite gas sensors.The importance of microstructure control with the optimisation of grain size as well as porosity is understood with this model. Adsorption of gas by the perovskite facilitates or hinders the charge transport across the grains, causing a change in the potential barrier height. This changes in the potential barrier causes changes in the conductivity of the material [65]. $LaFeI_3$ detects ethanol [8] Silver (Ag)-doped double perovskite Lanthanum iron tin oxide ($LaFe_{1-x}Sn_xO_3$) detects methanol using this model.

Oxygen Vacancy Model: This model is used in cases where the addition of dopants or post deposition treatments are made on the material where the chance of increasing the oxygen vacancy concentration is high. Oxygen vacancies serves as sites for the adsorption of gases increasing the reactivity as well as sensitivity. $LaFe_{1-x}PxO_{3-\delta}$ nanosheets for acetone detection also works as per the oxygen vacancy model [65].

Photophysical Models such as **Fluorescence Resonance Energy Transfer (FRET)** and **Photoinduced Electron Transfer (PET)** are commonly used in optical gas sensors. In FRET, donor-acceptor pair gets excited and the energy is transferred on a non-radiative mode to the nearby acceptor molecule. This gets reflected in the fluorescence wavelength or intensity. $CsPbBr_xI_{(3-x)}$ perovskite nanocrystals sense ascorbic acid using this model. PET involves the transfer of excited electrons from a donor to an acceptor, showing modulations on the optical signal.

Intramolecular Charge Transfer Model (ICT) is a commonly used model in hybrid perovskites having structural flexibility. The sensing material on interacting with the gas produces a change in dipole moment and electronic structure [65].

Excimer Association Model is mostly used in the case where excimer or exciplex complexes are formed between the perovskite and the ligand. Gas molecules disturb these complexes, showing a change in photoluminescence [65].

4.5 Synthesis and Processing Approaches of Lead-Free Perovskites for Gas Sensors

Various strategies of synthesis for the development of lead free perovskite materials has been formulated through a vast research on gas sensors. Techniques based on solution based as well as vapour based methods were observed to enhance the quality of lead free perovskite materials. The process of synthesis as well as post processing, needs to be designed carefully as the quality of the synthesized material to be high, is central in determining the performance of gas sensors. Figure 4.2 illustrates the

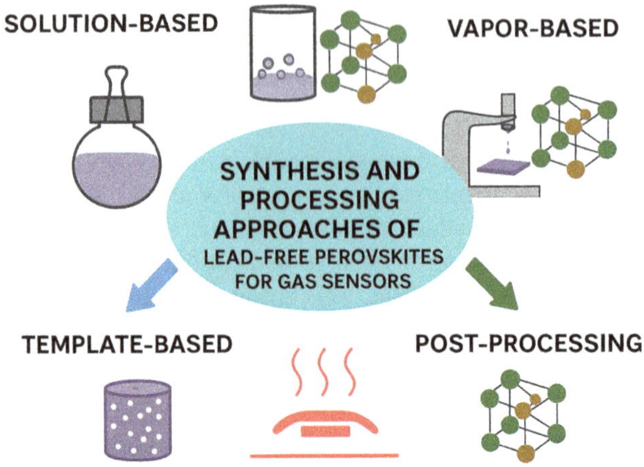

Fig. 4.2 Synthesis approaches of Lead free perovskite materials towards Gas sensors

various approaches on synthesis and processing of lead free perovskite materials towards gas sensors.

4.5.1 *Solution-Based Methods*

- **Spin Coating**: This is one of the most widely used techniques for thin-film fabrication due to its ease and suitability with large-area substrates. A common procedure is to spin a substrate at a high speed after depositing a tiny puddle of a fluid material in the middle of the substrate. The resin will spread across the substrate due to centripetal acceleration, leaving behind a thin layer of substance [21]. The qualities of the fluid material (viscosity, drying rate, solids percentage, surface tension, etc.) and the spin process parameters (rotation speed, acceleration, and fume exhaust) will determine the final film thickness. Repeatability is one of the most crucial aspects of spin coating since even little changes to the parameters that determine a process can have a significant impact on the coated film [21]. This process creates a homogeneous thin film by depositing a precursor solution (such as $MASnI_3$ in DMF or DMSO) over the substrate and spinning it quickly. After that, the film is annealed to increase its crystallinity. Particularly for Sn-based systems, the technique is susceptible to ambient oxygen and moisture, therefore regulated conditions are required [66].
- **Solvothermal/Hydrothermal Synthesis**: This method involves the growth of perovskite crystals in a closed vessel at elevated temperatures and pressures. According to solvothermal or hydrothermal, a solvent is used to promote chemical reactions during the synthesis of materials in a closed system at high temperatures

and pressures [57]. By adjusting reaction parameters including temperature, pressure, and solvent type, this method enables the production of a variety of materials, such as metal-organic frameworks, ceramics, and nanoparticles. Using solvents like ethylene glycol or water, materials such as $Cs_2AgBiBr_6$, $Cs_4MnBi_2Cl_{12}$ nanocrystals can be synthesized with improved crystallinity and morphology. This process allows for better control over shape and particle size, beneficial for tailoring gas interaction surfaces [53].

- **Ligand-Assisted Reprecipitation (LARP)**: A dual-solvent system is employed where a precursor solution is rapidly injected into a poor solvent containing capping ligands. This results in the formation of nanocrystals with controlled size and photoluminescent properties. This is a simple process using the principle of supersaturation in the presence of ligands. The procedure starts when the polar solvent's precursor solution is transferred dropwise into a weak solvent that contains ligands. The miscibility of the solvents is affected by the difference in their polarity. Consequently, the precursor experiences a nonequilibrium condition and the two solvents are unable to mix. Perovskites are nucleated and grow via instantaneous supersaturation, which eliminates the need for high temperatures. The LARP approach can be easily implemented to large-scale manufacturing with a high production yield of stable QDs because room temperature (RT) is sufficient for the synthesis of QDs. For example, $CsSnBr_3$ nanocrystals produced via LARP demonstrate PL sensitivity to oxidative gases like NO_2. Bi-based $Cs_3Bi_2I_9$ and $CsBi_3I_{10}$ perovskites via the LARP synthesis were also studied extensively [32, 61].

4.5.2 *Vapor-Based Approaches*

- **Chemical Vapor Deposition (CVD)**: CVD enables the growth of uniform, high-quality films over complex geometries. In the case of lead-free perovskites, CVD is used to deposit thin films like $CsCu_2I_3$, $Cs_3Sb_2Br_9$, from vapor-phase precursors [50, 59]. In this procedure, volatile precursors are introduced into a reaction chamber, where they undergo a reaction and produce a solid substance that is deposited on the substrate surface. Because CVD is flexible, it can be used to create a variety of materials with adjustable characteristics. This method ensures superior film adherence and uniformity but requires careful handling of precursors and optimization of deposition conditions.
- **Thermal Evaporation**: A popular technique for applying a thin layer is thermal evaporation, in which the source material evaporates in a vacuum as a result of heating at a high temperature. This allows the vapour particles to move and reach a substrate, where they again solidify. This technique uses a resistive coil or charge holding boat that is either solid or powdered. To achieve the high melting temperatures required for metals, a high direct current (DC) is applied to the resistive boat or coil, and the high vacuum ($<10^{-4}$ Pa) facilitates the metal's evaporation and subsequent transport to the substrate. This method works particularly well with

materials that have low melting points [6]. Bi- and Sb-based perovskite thin films have been successfully fabricated by this method. Lead-free halide perovskites like $CH_3NH_3SnI_3$ and (phenethylammonium)$_3Bi_2Br_9$ (PEA-Bi–Br) has attained more attention for ammonia gas sensors [43, 46]. Post-deposition annealing is critical for phase purity.
- **Atomic Layer Deposition (ALD)**: Though still under development for halide perovskites, ALD offers atomic-level control of film thickness and uniformity. This method is promising for fabricating ultra-thin perovskite films in sensor architectures requiring precise dimension control.

4.5.3 Template-Assisted and Nano Structuring Techniques

The process of creating materials with precise, well-defined shapes and structures, especially at the nanoscale, is called template-assisted synthesis or template synthesis. In order to direct the growth or assembly of another material, a template typically a material with an established structure, such as a self-assembled structure or a nanoporous membrane is used. The size, shape, and morphology of the final nanomaterials can be precisely controlled using this technique [55].

- **Anodic Alumina Templates**: It has been known since the mid-1900s that anodising aluminium produces porous alumina, as long as an electrolyte is used in which the alumina is somewhat soluble. Masuda and Fukuda showed that, under the right conditions, self-assembly can occur during the electrochemical etching of aluminium. Repulsion between adjacent pores happens as a result of the volume expansion that takes place during the alumina production process [17]. The pores are compelled to form a hexagonal lattice in order to reduce them. The 10% porosity rule must be followed when adjusting process parameters like the applied voltage, temperature, and electrolyte concentration in order to achieve self-ordered pore growth. The aluminium substrate thus displays regularly spaced pits following the removal of the alumina layer, which serve as seeds for the formation of pores during a subsequent etching step. Highly organised pore arrays are produced by this two-step procedure [17, 74]. These templates provide vertical channels that guide the growth of one-dimensional structures such as nanorods or nanotubes. For instance, $Cs_3Bi_2I_9$ nanorods synthesized using anodic alumina templates exhibit enhanced gas interaction due to increased surface area. Synthesis of $Ba_{0.5}Sr_{0.5}Co_{0.8}Fe_{0.2}O_{3-\delta}$ (BSCF) perovskite nanowires [24] and Perovskite quantum dot oxygen sensors [27] are the notable works.
- **Nanoimprint Lithography (NIL)**: NIL uses a patterned stamp to create micro- and nanostructures on perovskite films, beneficial for fabricating periodic structures with optical resonance effects. A prospective solution to the drawbacks of chemical prepattern-based methods is the micro/nanoimprinting of a perovskite film using a topologically prepatterned mould. Developing a soft and moldable precursor film and delaying the rapid crystallization of the precursor film to a

perovskite one in the amorphous adduct state for as long as possible to guarantee adequate patterning time are two of the most important factors for the successful nanoimprinting of a thin perovskite film [29]. These structures enhance gas detection through guided light-matter interaction.

- **Soft Template Approaches**: Using surfactants or block copolymers, mesoporous structures of perovskites can be synthesized. These materials show improved gas sensing due to enhanced surface area and diffusion paths. Cs_2PdBr_6 hollow nanospheres for trace CO detection [62] and High performance porous $LaFeO_3$ gas sensor with embedded p–n junctions enabling ppb-level formaldehyde detection [67] were the notable work.

4.6 Post-Synthesis Treatments

Thermal annealing increases grain size, eliminates residual solvents, and improves film crystallinity all of which are essential for gas-sensing functionality. One of the main problems with Sn-based perovskites like $MASnI_3$ is that when exposed to air, Sn^{2+} oxidises to Sn^{4+}, introducing deep-level trap states. High conductivity and device stability are maintained by preserving Sn^{2+} by controlled annealing in inert or vacuum atmospheres. Annealing Bi-based perovskite thin films such as $(CH_3NH_3)_3Bi_2I_9$ in vacuum or nitrogen improves phase purity and charge transport, which are essential for resistive or capacitive-type sensors. Annealed films exhibit lower defect densities and increased surface uniformity, which facilitate reproducible gas adsorption and desorption kinetics [54]. The effectiveness of gas reaction is decreased by surface and grain boundary imperfections, which serve as non-radiative recombination centres. These flaws can be lessened and environmental stability improved using passivation treatments. It has been demonstrated that halide salt passivation with either phenylethylammonium iodide (PEAI) or formamidinium iodide (FAI) enhances film quality and prolongs device life. Additionally, undercoordinated sites can be passivated and grain boundary recombination can be decreased by organic ligands like polyethylenimine or thiocyanate. For long-term monitoring, these treatments enhance reproducibility and lessen baseline drift in sensor responses.

4.7 Mechanistic Approaches for Enhancing Gas Sensing in Lead Free Perovskites

To enhance the performance of lead free perovskite-based gas sensors, various mechanistic approaches have been developed. These strategies can optimize the material's properties at atomic composition, nanostructure morphology and interfacial interactions.

4.7.1 Defect Engineering

Defect engineering is the creation or controlling of point defects within the crystal lattice of perovskite materials. These defects, particularly oxygen vacancies, play a significant role in influencing the electronic structure and surface reactivity of the material, thereby affecting its gas sensing properties [60]. The concept of defect tolerance is very crucial in metal halide perovskites, where intrinsic defects induce shallow transition levels, allowing the material to maintain good electronic quality. These defects can act as electronic traps, increasing the interaction with gas compounds [2, 10, 60].

Defects can increase active sites for gas adsorption and improve charge carrier dynamics. A- or B-site deficiencies along with oxygen vacancies can affect the adsorption and dissociation energy of oxygen species and target analytes which in turn improves selectivity. Defects, dangling bonds, and organic ligands can form trap states that intensify interaction with target gases [2, 10, 60].

The defect tolerance of halide perovskites is attributed to the lack of bonding-antibonding interactions between the valence band (VB) and the conduction band (CB) [25]. The relatively high density of surface defects in lead-free perovskites, is a drawback in optical applications [72] and is beneficial for gas sensing as these defects act as electronic traps for interacting with gas compounds, potentially explaining their superior performance compared to lead-based alternatives [9]. The contrasting orbital characteristics (spin-orbit coupling in CB and antibonding nature in VB) cause the formation of shallow trap states, enhancing sensitivity to target gases [33].

Defect engineering can directly influence the active sites for gas interaction by improving both sensitivity and selectivity [2]. It can also improve intrinsic properties of perovskites like defect tolerance in lead-free halides [30]. But precise control over the type and concentration of defects is still challenging for a lead free perovskite [9].

4.7.2 Catalyst Surface Functionalization

This strategy includes loading of catalytic particles or forming of spontaneous metal catalysts on the surface of perovskite sensing layers. These catalysts act as "reaction promoters" by enhancing specific gas adsorption and dissociation, or by facilitating the transfer of activated species to the sensing layer, thereby improving overall sensor performance. The presence of catalytic particles will increase the number of gas reaction sites and promotion of specific gas adsorption. Catalytic effects can considerably improve specific gas adsorption and dissociation. They can also enhance analyte adsorption or desorption, causing faster response and recovery times. Ex-solution catalysts offer exceptional durability under harsh conditions due to their strong adhesion on the perovskite surface [31, 28, 36, 37].

Introducing highly catalytic PO particles onto conventional sensing layers acts as reaction promoters, sensitizing the sensing layer and promoting specific gas adsorption and dissociation, thus enhancing selectivity. $La_{0.8}Sr_{0.2}FeO_3$ (LSFO) nanoparticles used to decorate Ga_2O_3 nanorod gas sensors improved CO sensitivity at high temperatures due to the catalytic spillover effect of LSFO particles [44]. Adding $La_{0.75}Sr_{0.25}Cr_{0.5}Mn_{0.5}O_{3-\delta}$ (LSCM) particles onto highly porous SnO_2 fiber-in-tubes (FITs) improved formaldehyde (HCHO) detection. LSCM particles, rich in oxygen vacancies, increased oxygen spillover on the SnO_2 surface, enhancing sensitivity and selectivity towards HCHO [34].

Ex-solution, the spontaneous growth of metal nanoparticles (NPs) on the surface of metal oxides, can be used for enhancing catalytic properties. Kim et al. functionalized ex-solved alloy NPs (CoFe, NiFe, CoNiFe) on $La_{0.6}Ca_{0.4}FeO_{3-\delta}$ (LCF)-loaded porous SnO_2 nanofibers [37]. The ex-solved CoNiFe NPs on the LCF PO surface dissociate oxygen molecules and transport them to adjacent oxide supports and cause 148.3-fold enhanced DMS (dimethyl sulfide) response [37, 44]. Single-atom catalysts (SACs), such as Pt SACs on SnO_2 nanosheets, have demonstrated high catalytic activity for chemiresistive sensing of acetone gas, with a 7.6-fold enhancement compared to pristine SnO_2 [60].

Catalyst surface functionalization also creates additional reaction sites, enhancing sensitivity and selectivity [31]. Cost-effective substitution to noble metals are highly needed [35]. Ex-solution catalysts are stable and are stable over a long time due to strong adhesion [28, 36, 37].

4.7.3 Heterojunction Engineering (Including Encapsulation and Composite Approaches)

This is the combination of different materials, often semiconductors having different electronic properties (e.g., p-type and n-type, or perovskites combined with other nanostructures like graphene), to create heterojunctions or composites [2, 11, 9, 37]. These interfaces can form electron depletion layers that are highly sensitive to gas interactions, or they can increase synergistic effects between the components. Encapsulation and composite approaches are specific forms of heterojunction engineering where one material is integrated within or supported on another to improve stability or sensing characteristics [2, 37]. Increased electron depletion layers and catalytic effects lead to greater resistance changes on gas exposure. Specific interactions at the heterojunction can also improve the ability to discriminate target gases. Composites are designed to maximize active surface area and gas penetration [2, 11, 62, 37].

The use of composite materials and polynary compounds (like $LaFeO_3/SnO_2$) is a promising strategy for enhancing conductometric sensor performance [2, 22, 52, 65]. $LaFeO_3/In_2O_3$ composites improved ethanol sensing by increasing oxygen adsorption, creating a heterojunction with unique Fermi level effects, and catalyzing redox reactions [71]. LSCM@SnO_2 FITs (fiber-in-tubes) exhibited exceptional sensitivity towards formaldehyde and outstanding stability over repeated tests [34]. TiO_2–$LaFeO_3$ nanocomposites for acetone sensing [8]. Cotton-modified $LaFeO_3$/MXene composites are engineered for NO_2 sensing [16]. Lead-free perovskite nanocrystals (Cs_3CuBr_5 and $Cs_2AgBiBr_6$) supported on graphene were successfully used for the first time to detect H_2 and H_2S, and achieved ultrasensitive detection of NO_2 at room temperature [9, 62]. Graphene acts as a transducing element with high and efficient carrier transport, leveraging the quantum confinement effect and exciton binding energy of the nanocrystals [11, 9]. The $Cs_3Cu_2Br_5$ NCs supported on graphene demonstrated higher sensitivity than $Cs_2AgBiBr_6$ NCs for all tested gases, partially due to its direct band gap favoring faster carrier recombination and interaction [9]. A $CH_3NH_3SnI_3/SnO_2$/Pd/Au nanocomposite was developed for formaldehyde gas detection, showing good response and low detection limits [73]. A MXene/perovskite-based bionic human odor sensor array also has been developed [26].

4.7.4 Doping

Minor quantities of foreign elements are introduced into the perovskite crystal lattice and electronic and chemical properties are modified [2, 30]. As a result, carrier concentration, defect formation (e.g., oxygen vacancies), and surface reactivity changes, which are highly requisite for gas sensing [2]. A or B site doping of the ABO_3 structure can influence the material's composition [2, 30]. Changes in carrier concentration and number of active sites can significantly alter resistance when exposed to gas [2]. Cation substitution helps tailoring physicochemical properties for selective sensing [2]. Doping enhances the interaction with target gases due to modified electronic properties or increased active sites [2]. Modified surface reactivity can facilitate faster adsorption and desorption processes [2]. Figure 4.3 depicts the gas sensimg mechanism of boron-doped graphene quantum dots/Ag–$LaFeO_3$ in air as well as in benzene. Table 4.1 summarizes recent works based on doped lead free perovskite based gas sensors.

4.7.5 Substitution

A or B sites within the ABO_3 perovskite are replaced by other elements thereby causing changes in cationic/anionic stoichiometry, surface reactivity, and redox properties. It will improve the gas response, especially for p-type materials, by changing the type of semiconductor (p to n or vice versa) or creating favorable reaction sites. New elemental combinations can provide tunable interactions for specific analytes. It can lead to a transition from p-type to n-type semiconducting behavior, resulting in the development of a material which can detect different types of gases (e.g., volatile organic compounds vs. oxidizing gases). The A and B sites in the PO structure are highly flexible and can accommodate most of the elements from the periodic

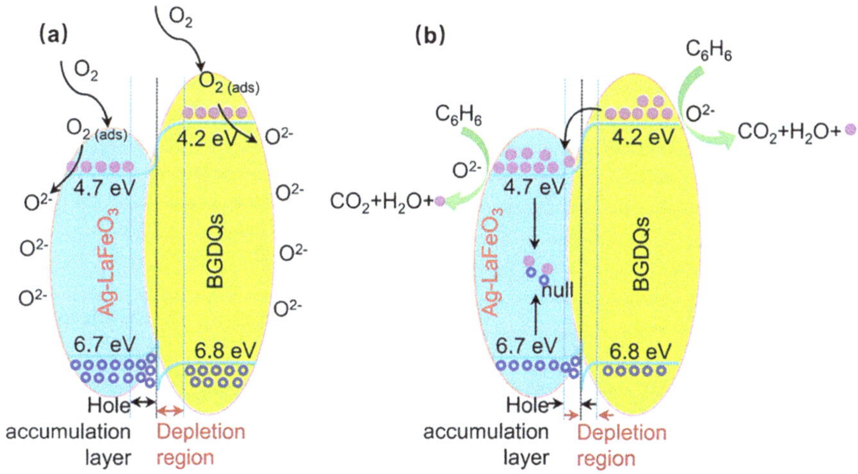

Fig. 4.3 Gas sensing mechanism of boron-doped graphene quantum dots/Ag–$LaFeO_3$ **a** in air and **b** in benzene. Reprinted from ref. [41]. Copyright 2018, with permission from the royal society of chemistry

table in various combinations and proportions [2, 37]. Sr substitution in $LaFeO_3$ ($La_{1-x}Sr_xFeO_3$) for hydrogen sensing, Co substitution in $LaFeO_3$ ($LaFe_{0.8}Co_{0.2}O_3$) for CO sensing and replacement of Cs^+ with long-chain alkyl chains in $Cs_2AgBiBr_6$ to form 2D hybrid double perovskites like $(BA)_4AgBiBr_8$, $(HA)_4AgBiBr_8$, and $(OA)_4AgBiBr_8$ for enhanced environmental stability of the sensors (stable for over 270 days) and improved CO detection performance are examples [5, 52, 62].

Substitution helps achieve control over the fundamental properties of the perovskite, including its semiconducting nature. It can also overcome inherent limitations of certain perovskite types (e.g., low response of p-type materials to VOCs).

4.7.6 Galvanic Replacement Reaction (GRR)

Galvanic Replacement Reaction (GRR) is an effective chemical method used to control the elemental composition and morphology of parent oxides. It is a redox process driven by differences in electrochemical potentials of metal species, leading to the dissolution of one species and the simultaneous deposition of another onto the parent structure as oxides. This method allows for partial or complete substitution, creating novel composite structures.

GRR significantly increases response, especially by forming p–n heterojunctions. Response speed is improved through the creation of porous structures. GRR can induce porous structures, enhancing the active surface area for gas interaction. It

Table 4.1 Doped lead free perovskite gas sensors

Dopant/ modifier	Base material	Target gas(es)	Reported effect/ mechanism	Source
Yttrium (Y^{3+})	$LaFeO_3$ (LFO)	General gas sensing	Increased free electrons and surface charge exchange, creating more active sites	Liu et al.
Calcium (Ca^{2+})	$GdFeO_3$	Methanol	Generated oxygen vacancies and promoted hole production	Xiaofeng et al. [68]
Calcium (Ca^{2+})	$La_{0.8}Ca_{0.2}FeO_2$	Sulfur dioxide (SO_2)	Improved SO_2 sensing	Ma et al. [45]
Phosphorus (P^{5+})	$LaFe_{1-x}P_xO_3$-δ	Acetone	Created oxygen vacancies by replacing a B-site cation	Qin et al. [23]
Magnesium (Mg^{2+})	$LaFeO_3$ (LFO)	Methanol	Created oxygen vacancies, facilitated oxygen absorption	Ahmad et al. [2]
Strontium (Sr)	$La_{1-x}Sr_xFeO_3$	Hydrogen (H_2)	Used for hydrogen sensing	Murade et al. [48], Koonsaeng et al. [49]
Strontium (Sr)	$La_{0.8}Sr_{0.2}FeO_3$ (LSFO) on Ga_2O_3	Carbon monoxide (CO)	Improved CO sensitivity	Lin et al. [44]
Strontium (Sr)	$La_{0.75}Sr_{0.25}Cr_{0.5}Mn_{0.5}O_{3-\delta}$	Formaldehyde	Improved formaldehyde detection.	Kang et al. [34]
Strontium (Sr)	$BaSnO_3$	Hydrogen sulfide (H_2S)	Improved H_2S sensitivity.	Marikutsa et al. [47]
Strontium (Sr)	$SrTi_{0.65}Fe_{0.35}O_{3-\delta}$ (STFO)	Oxygen (O_2)	Robust temperature independence and enhanced sensitivity.	Jong Won Baek et al. [31]
Silver (ag)	$LaFeO_3$	Formaldehyde (HCHO)	Improved HCHO sensing.	Zhang et al. [73], Wei et al. [65]

(continued)

Table 4.1 (continued)

Dopant/ modifier	Base material	Target gas(es)	Reported effect/ mechanism	Source
Silver (ag)	$LaFeO_3$	Acetylene	Enhanced response to acetylene.	Alharbi [1]
Silver (ag)	$LaFeO_3$	Toluene, xylene	Used for sensing toluene and xylene.	Chen et al. [12]

enables precise control over elemental composition, including transitioning from p-type to n-type materials. GRR has been successfully applied to p-type PO to induce porous structures and facilitate a transition from p-type to n-type composition. Baek et al. demonstrated a systematic design approach to fabricate p-type $LaFeO_3$/n-type SnO_2 composite nanotubes ($LaFeO_3$/SnO_2 NTs) by using GRR on $LaFeO_3$ NTs. This resulted in a remarkable 31.2-fold increase in response to acetylene with an improved response speed. The process created a highly coherent interface between SnO_2 and $LaFeO_3$, forming heterogeneous p–n junctions. Jang et al. produced SnO_2 (n-type) hollow nanocubes functionalized with catalytic Co_3O_4 (p-type) components by combining metal–organic frameworks (MOF) templating and GRR. This led to a 21.9-fold improvement in acetone response, attributed to the catalytic effect and the highly porous and hollow multicomponent nanostructures [2, 5].

4.7.7 *Morphology Tuning*

Morphology tuning involves engineering the physical shape, size, and porosity of the perovskite sensing material. Nanostructured architectures with high surface-to-volume ratios and high porosity are effective for gas sensing. High surface area provides more active sites for gas adsorption, amplifying sensing signals. Porous structures cause efficient molecular diffusion to the sensing surface, leading to rapid response and recovery. Optimized morphology can improve the access of gas molecules into the sensing layer. Certain morphologies, like suspended web structures, can reduce cross-sensitivity to temperature fluctuations [5].

The fabrication of nanostructured architectures with large surface area and well-developed porosity is a primary strategy for achieving better chemical sensing performance. Honeycomb-like periodic porous $LaFeO_3$ thin films demonstrated enhanced gas-sensing performances, including high response and fast response (~4 s). The electrospinning technique is widely used to fabricate 1D nanostructures such as nanofibers. For example, electrospun p-type $LaNiO_3$ nanofibers and $SrTi_{0.65}Fe_{0.35}O_{3-\delta}$ (STFO) nanofibers achieved highly porous nanostructured PO with a high surface area. STFO NFs showed enhanced sensitivity to oxygen

(R_{O_2}/R_{N_2} = 4.08 at 20% O_2) due to high surface area and porous structure facilitating efficient molecular diffusion. Precise control of NF morphology by adjusting electrospinning parameters can lead to smaller diameters and nanobelts, further enhancing gas penetration. A graphene-cellulose templating route was developed for the universal synthesis of various inorganic porous nanosheets (NSs), including perovskite $LaCoO_3$ (LCO), $LaMnO_3$ (LMO), and $La_xSr_{1-x}Mn_yNi_{1-y}O_3$ (LSMN) NSs. This sacrificial template approach results in highly porous perovskite NSs with a high surface area. Hollow $LaFeO_3$ (LFO-HS) synthesized via a hydrothermal method with CTAB resulted in a hollow, porous structure with a large surface area, enhancing ethanol sensing performance and achieving an ultralow detection limit of 1 ppb. Mesoporous hollow $PrFeO_3$ nanofibers prepared by electrospinning were studied for acetone sensing [5, 11].

The use of lead-free perovskite nanocrystals supported on graphene achieves excellent sensitivity due to the quantum confinement effect and exciton binding energy. These confined structures increase the exciton binding energy above thermal energy at room temperature, promoting radiative recombination and suppressing exciton dissociation. The nanoscale, with its relatively high surface-to-volume ratio, enhances the importance of defects (acting as electronic traps) for gas sensing. It significantly increases the active surface area for gas interaction. It also causes faster diffusion of gas molecules, leading to quicker response and recovery times. It can contribute to low-power consumption devices due to improved efficiency.

Progress in research related to the mechanistic enhancement strategies of improving sensor performance parameters like sensitivity, response time, sensing range and selectivity are appreciable while a significant achievement in stability is worth mentioning (e.g., 270 days for $(BA)_4AgBiBr_8$) [62]. Sensor research also aims to develop lead free perovskites which can remain stable under extreme conditions.

4.8 Types of Gas Sensors Using Lead Free Perovskites

Based on the gas sensing mechanism, Lead free perovskite based sensors can be broadly classified into chemiresististive sensors and optical sensors. Even though specific lead-free examples in gas sensing are rare, capacitance changes are also explored for perovskite gas sensors.

4.8.1 Chemiresistive Gas Sensors

Chemiresistive gas sensors show sensor response to analyte as molecules as chane in electrical resistivity or conductivity. The core mechanism identified by many of the sensor researchers involves a redox reaction between the gas and the sensor surface, causing change in resistance. Lead-free perovskites are non-toxic replacement in this area offering high sensitivity, selectivity, and room-temperature operation.

$CsCu_2I_3$-based sensors: [2, 50, 52].

A. NO_2 sensing: $CsCu_2I_3$ sensors possess high sensitivity (1509% to 5 ppm) and high selectivity for nitrogen dioxide (NO_2) detection at room temperature with appreciable recovery. They exhibit high stability in ambient air and high-humidity environments, a characteristic often not achievable with traditional halide perovskite sensors [38].
B. H_2S sensing: A lead-free $CsCu_2I_3$ film was developed as an H_2S gas sensor, demonstrating excellent sensing properties, including high response and selectivity toward H_2S, even at low concentrations (0.2 ppm) at room temperature. This sensor showed a response of 0.95 at 1 ppm H_2S at 25 °C, with response and recovery times of 185 and 106 s, respectively. It did not respond to H_2 or CO_2, and only exhibited a weak response to O_2 and N_2, with a significantly lower negative response to NO_2 (approximately −0.11 at 5 ppm). The sensor maintained high stability over a 45-day period under ambient moisture and reactive gases [50].
C. Ammonia (NH_3) sensing: A $CsCu_2I_3$ film has been identified for its potential in ammonia sensing and stimuli-responsive anti-counterfeiting applications, [3, 9, 30–46, 65].

For $CsCu_2I_3$, which is a direct band-gap p-type semiconductor exhibiting self-trapped excitons, defects play a critical role. Iodine vacancies are commonly present in $CsCu_2I_3$. In the presence of a reducing gas like H_2S (an electron donor), these defects act as electronic traps. This interaction leads to an increase in minority carriers, causing a decrease in the number of available holes in the p-type semiconductor, which results in a reduction of the current. Conversely, an oxidizing gas like NO_2 causes the current to increase.

$Cs_2AgBiCl_6$ and $Cs_2AgBiBr_6$-Based Sensors: Lead-free halide double perovskites like $Cs_2AgBiCl_6$ and $Cs_2AgBiBr_6$ have emerged as non-toxic alternatives for gas sensing [9, 43, 50, 58, 60].

A. H_2S sensing ($Cs_2AgBiCl_6$): A sensor based on $Cs_2AgBiCl_6$ showed excellent H_2S gas sensing at room temperature over a wide humidity range. It exhibited a high response (90.6 for 10 ppm H_2S), a fast response speed (99.6 s for 400 ppb H_2S), a low detection limit (5 ppb H_2S), and excellent selectivity.
B. H_2, H_2S, NO_2, NH_3 sensing ($Cs_2AgBiBr_6$ and $Cs_3Cu_2Br_5$ on Graphene): $Cs_3Cu_2Br_5$ and $Cs_2AgBiBr_6$ supported on graphene have been developed, and they marked the first time H_2 and H_2S detection. Ultrasensitive NO_2 detection was also achieved at room temperature [9].

$CH_3NH_3SnI_3$/$MASnI_3$ sensors:

A. Ammonia (NH_3) sensing: A textile-based sensor using $CH_3NH_3SnI_3$ was developed for ammonia gas detection at room temperature. It achieved a maximum calibrated sensitivity of ~85% upon exposure to 100 ppm ammonia gas and is compatible with low-power wearable applications (operating at 2 V bias with ~2 nA current).

B. NO_2 sensing (Au/$MASnI_3$/SnO_2): A novel NO_2 sensing material, Au/$MASnI_3$/SnO_2, was fabricated by combining light-sensitive Au-decorated organic-inorganic hybrid perovskite (Au/$MASnI_3$) with tin dioxide (SnO_2). This composite material shows high sensitivity to NO_2 at room temperature under UV illumination, including high sensitivity ($R_g/R_a = 240$ up to 5 ppm NO_2), ultra-fast recovery time (about 12 s), and excellent selectivity. This is attributed to a synergistic effect, including the high light absorption of $MASnI_3$, the surface plasmon resonance, active catalytic effect of Au nanoparticles, and special interactions at the p–n heterojunction between p-$MASnI_3$ and n-SnO_2 [40, 42]

Other Chemiresistive Lead-Free Perovskites

A. Nitric Oxide (NO) Detection: Lead-free halide Cs_2PtI_6 perovskite has shown potential for trace NO detection as low as 100 ppb, with ultrahigh selectivity and stability for over 2 months at room temperature. The strong Pt-N bonding interaction is key to its high adsorption energy and conductive change [9].

4.8.2 *Optical Gas Sensors*

Optical gas sensors show sensing properties as changes in light absorption, emission (luminescence), or other optical properties when it interacts with target gases. Lead free perovskites are promising in this area with their high optical absorption coefficient, adjustable band gap, and low exciton binding energy.

Changes in photoluminescence (PL) of graphene-supported $Cs_3Cu_2Br_5$ and $Cs_2AgBiBr_6$ nanocrystals under different gases have been observed, supporting the importance of exciton-phonon coupling strength in their sensing mechanism. Textile-based lead-free halide perovskite $CH_3NH_3SnI_3$ ($MASnI_3$ or MASI) has been utilized for ammonia gas sensing at ambient temperature. This material can trace ammonia through visual color change methods in addition to electrical readout, indicating a direct optical sensing component in its mechanism [60].

$CsCu_2I_3$ (Copper-based Perovskites): Lead-free $CsCu_2I_3$ films have been developed as H_2S gas sensors at room temperature. It shows notable optical properties and sensing mechanisms that involve direct light interaction. It displays a broad yellow photoluminescence (PL) emission at 577 nm with a large Stokes shift. The absorption edge is at 332.4 nm and a direct band gap of 3.74 eV, showing that the absorption range is at the ultraviolet (UV) region. A reversible color change is observed in the $CsCu_2I_3$ film when exposed to H_2S gas. This color change indicates a reversible chemical reaction between H_2S and $CsCu_2I_3$. Exposure to H_2S can cause a red shift in the absorption band edge, suggesting the formation of new compounds (e.g., Cs_2S and Cu_2S) and it can also reduce the optical band gap of the mixture. The sensing mechanism is due to the defects in the perovskite material (acting as electronic traps) and the interaction of H_2S as an electron donor, which influences the material's electrical and optical behavior [50].

- PeroGaS Project (Perovskite/Graphene Nanocomposites) aims to develop gas sensors using lead-free perovskite crystals conjugated to reduced graphene oxide (rGO) to detect various air pollutants, including CO_2, CO, O_3, NO_x, and SO_x. They intend to identify the specific vibrational features of targeted gases via light modulation, utilizing Stimulated Raman Scattering and other similar methods.

4.8.3 Self-Powered Gas Sensors

Self-powered gas sensors are a class of lead free perovskite gas sensors which are desirable for portable and wearable applications, as they eliminate the need for external power sources. Halide perovskites have seen significant advancements in this area.

- NO_2 Detection (FMCPIB): A triple cation perovskite, $FA_{0.80}MA_{0.15}Cs_{0.05}PbI_{2.55}Br_{0.45}$ (FMCPIB) nanocrystalline, has been successfully applied in both self-powered and externally-powered modes for NO_2 gas detection at room temperature, with an estimated detection limit of 0.2 ppm. [12]. Figure 4.4 illustrates self charging and self-powered chemical sensing of FMCPIB.
- Perovskite/Graphene Nanocomposites: The integration of perovskites with graphene in nanocomposites is being explored for self-powered gas sensing, leveraging graphene's excellent carrier transport properties [9].

4.9 Challenges and Future Prospects

Increasing concerns about the toxicity and environmental impacts associated with lead-based perovskites have made the development of lead-free perovskite gas sensors a primary area of research. However, lead free perovskite gas sensors face numerous challenges that prevent their common usage and commercialisation. Change in humidity and oxygen concentration results in rapid oxidation which have a major effect on their sensing performance, resulting in decreased sensitivity, poor reproducibility, and short operational lifetimes. Apart from that, lack of mechanistic knowledge regarding the interactions between gas molecules and perovskite materials is another factor [64]. This is because of the knowledge gap that results in material design and optimisation the adsorption dynamics, charge transfer mechanisms, and structural alterations that take place during gas detection. As the synthesis of lead-free perovskites frequently involves intricate, expensive processes that are not reproducible, and the conversion of laboratory-scale prototypes into industrially scalable devices is still uncertain, manufacturing and scalability pose additional challenges in addition to stability and mechanistic limitations [65].

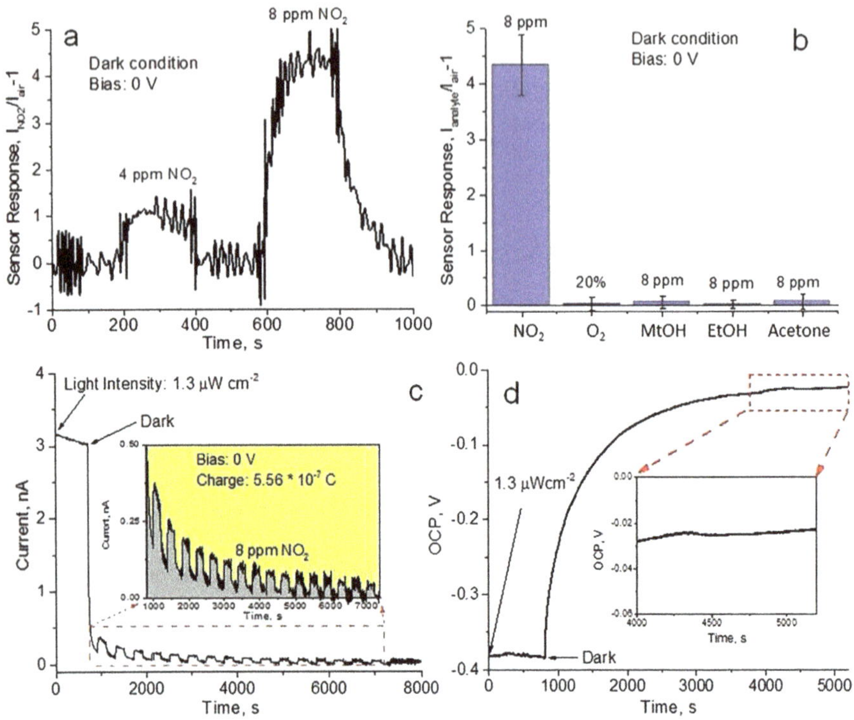

Fig. 4.4 Demonstration of the self-charging and self-powered chemical sensing of FMCPIB in dark conditions. Reprinted from ref. [8] with permission © 2020 WILEY-VCH Verlag GmbH and Co. KGaA, Weinheim

New studies have also shown a number of ways to get over these obstacles and realise the perovskites' potential in environmentally friendly monitoring systems. Engineering advanced material systems, such as incorporating self-healing capabilities into perovskite structures, is one promising strategy. This enables sensors to partially recover their performance following environmental deterioration, thus prolonging their operational lifetimes [64]. For instance, it has been demonstrated that combining lead-free perovskites with conductive nanostructures such as graphene or carbon nanotubes improves adsorption kinetics, speeds up charge transfer, and increases sensitivity, resulting in quicker and more accurate sensor responses [9]. Additionally, studies on organic-inorganic hybrids indicate that the addition of organic cations to tin-based frameworks enhances selectivity and lowers detection limits by increasing the active surface area for gas interactions and improving environmental stability. Another significant area of innovation is protective strategies. Encapsulation techniques is one of the effective ways to prevent lead free perovskite sensors from moisture and reactive gases [65]. Silica coatings, hydrophobic layers, and polymer barriers can all greatly extend sensor lifetimes while preserving sensitivity. These protective coatings can function as molecular filters, allowing specific analytes to reach the sensor surface, improving selectivity thereby slowing down deterioration.

Focussing on both performance and environmental issues, the use of sustainable synthesis routes such as green solvents, low-cost-energy fabrication techniques, and non-toxic precursors further increases the attractiveness of lead-free perovskite gas sensors. Another frontier is the investigation of new hybrid architectures, which integrates lead-free perovskites with nanostructured materials like carbon nanomaterials, metal-organic frameworks, or two-dimensional semiconductors to create synergistic effects that lower detection thresholds, increase selectivity, and speed up response times [9]. Research on graphene-perovskite hybrids has shown that the effective charge transfer channels offered by graphene networks significantly enhance the sensor performance [9]. However, the application of encapsulation and protective coatings has demonstrated that it is feasible to create sensors that not only endure but also function well in challenging environmental circumstances, such as high humidity and temperature fluctuations, where traditional perovskite sensors usually malfunction [3]. Altogether, advancements in self-healing lead-free perovskites, hybrid materials, encapsulation techniques, and sustainable methods will offer a path forward for resolving current issues and realising the full potential of lead-free perovskite gas sensors as dependable, scalable, and ecologically conscious technologies.

4.10 Conclusions and Outlook

The rapid advancement of gas sensor technologies necessitates materials that are not only selective and sensitive but also environmentally benign. Lead-free perovskites (LFPs) have emerged as promising candidates in this regard, offering reduced toxicity without compromising on performance. Their diverse structural frameworks and favourable optoelectronic properties position them well for sustainable integration into next-generation gas sensing platforms.

This chapter began by introducing the relevance of LFPs in gas sensor development, followed by a discussion of the fundamental parameters governing sensor performance. It explored the various sensing mechanisms and theoretical models applicable to LFPs, and highlighted the different material classes suited for gas detection. The role of synthesis strategies—ranging from solution-based and vapor-phase methods to template-assisted approaches—was examined in detail, with emphasis on how these techniques influence key material attributes such as porosity, crystallinity, and surface area. Enhancements in sensing performance were further achieved through mechanistic interventions including doping, heterojunction engineering, and surface functionalization. Innovations in device architecture—such as chemiresistive, field-effect transistor (FET)-based, and self-powered configurations—have expanded the functional applicability of LFP-based sensors.

Despite these advances, certain challenges persist. Environmental safety concerns remain for some hybrid compositions, and the complete elimination of toxicity is still constrained by the nature of substituents and halide components like iodide.

Moreover, issues related to long-term reproducibility and scalability continue to limit broader deployment.

Future research must focus on developing LFP materials with enhanced stability under humidity and temperature fluctuations, while adopting green synthesis techniques that align with global sustainability goals. Such efforts are essential to fully realize the potential of lead-free perovskites in delivering high-performance, eco-friendly gas sensors.

References

1. A. Alharbi, Gas sensors based on perovskite structured material [doctoral dissertation, University of Tübingen]. DSpace Repositorium (2020). http://hdl.handle.net/10900/100916
2. N. Ahmad, P. Kanjariya, G.P. Priya, A. Kumar, R. Thakur, R. Sharma, M. Kumari, S. Kaur, M.K. Mishra, Recent advances on the gas-sensing properties and mechanism of perovskite oxide materials: a review. ACS Omega **10**(14), 13780–13796 (2025). https://doi.org/10.1021/acsomega.4c11667
3. S. Ahmed, M.A. Gondal, A.S. Alzahrani, M. Parvaz, A. Ahmed, S. Hussain, Recent trends and challenges in lead-free perovskite solar cells: a critical review. ACS Appl. Energy Mater. **7**(4), 1382–1397 (2024). https://doi.org/10.1021/acsaem.3c02327
4. A. Argyrou, R.M. Giappa, E. Gagaoudakis, V. Binas, I. Remediakis, K. Brintakis, A. Kostopoulou, E. Stratakis, Cs2AgBiBr6 perovskites: designing stable, sensitive and selective eco-friendly ozone sensors. Adv. Sensor Res. **2**, 18 (2025). https://doi.org/10.1002/adsr.202500018
5. J.W. Baek, I.-D. Kim, Advances in perovskite oxides for chemiresistive sensors. Acc. Mater. Res. **4**(12), 1108–1120 (2023). https://doi.org/10.1021/accountsmr.3c00206
6. A. Bashir, T.I. Awan, A. Tehseen, M.B. Tahir, M. Ijaz, Interfaces and surfaces. Chem. Nanomater. **12**, 51–87 (2020). https://doi.org/10.1016/B978-0-12-818908-5.00003-2
7. M.K. Bhosale, A.I. Kazi, K.K. Pawar, R.S. Shingate, A.D. Kadam, N.J. Patil, A.D. Sheikh, Eco-friendly $MA_3Bi_2I_9$ perovskite thin films based ammonia sensor. Nanotechnology **34**(6), 065501 (2023). https://doi.org/10.1088/1361-6528/aca0f7
8. E. Cao, Y. Yang, T. Cui, Y. Zhang, W. Hao, L. Sun, H. Peng, X. Deng, Effect of synthesis route on electrical and ethanol sensing characteristics for $LaFeO_{3-\delta}$ nanoparticles by citric sol–gel method. Appl. Surf. Sci. **393**, 134–143 (2017). https://doi.org/10.1016/j.apsusc.2016.10.013
9. J. Casanova-Chafer, R. Garcia-Aboal, P. Atienzar, E. Llobet, Unraveling the gas-sensing mechanisms of lead-free perovskites supported on graphene. ACS Sens. **7**(12), 3753–3763 (2022). https://doi.org/10.1021/acssensors.2c01581
10. J. Casanova-Chafer, R. Garcia-Aboal, K. Mego, S.B. Malik, P. Atienzar, E. Llobet, Lead-free perovskite nanocrystals decorating graphene for detecting nerve agents. ACS Appl. Electr. Mater. **6**(9), 6974–6981 (2024). https://doi.org/10.1021/acsaelm.4c01220
11. J. Casanova-Cháfer, R. García-Aboal, P. Atienzar, E. Llobet, Gas sensing properties of perovskite decorated graphene at room temperature. Sensors **19**(20), 4563 (2019). https://doi.org/10.3390/s19204563
12. H. Chen, M. Zhang, B. Xing, X. Fu, R. Bo, H.K. Mulmudi, e al., Superior self-charged and-powered chemical sensing with high performance for NO_2 detection at room temperature. Adv. Opt. Mater. **8**(11), 1901863 (2020). https://doi.org/10.1002/adom.201901863
13. A. Cheng, J. Zhao, X.-A. Wang, Z. Lu, Y. Qi, J. Sun, Hybrid 1D/3D-structured perovskite as a highly selective and stable sensor for NO_2 detection at room temperature. Molecules **28**(6), 2615 (2023). https://doi.org/10.3390/molecules28062615

14. X. Chu, Z. Gan, L. Bai, Y. Dong, M.N. Rumyantseva, The acetic acid vapor sensing properties of $BaSnO_3$ microtubes prepared by electrospinning method. Mater. Sci. Eng. B **259**, 114606 (2020). https://doi.org/10.1016/j.mseb.2020.114606
15. A. Cyza, Ł. Cieniek, T. Moskalewicz, W. Maziarz, J. Kusiński, K. Kowalski, A. Kopia, The effect of strontium doping on $LaFeO_3$ thin films deposited by the PLD method. Catalysts **10**(9), 954 (2020). https://doi.org/10.3390/catal10090954
16. Dhariwal et al (2024). https://doi.org/10.1021/acssensors.4c02597
17. J.W. Diggle, T.C. Downie, C.W. Goulding, Anodic oxide films on aluminum. Chem. Rev. **69**(3), 365–405 (1969). https://doi.org/10.1021/cr60259a005
18. T. Du, A.T. John, H. Chen, H. Pham, K. Murugappan, T. Tran-Phu, A. Tricoli, K. Catchpole, Mixed-dimensional organic–inorganic metal halide perovskite (OIMHP) based gas sensors with superior stability for NO_2 detection. Mater. Adv. **3**(3), 1263–1271 (2022). https://doi.org/10.1039/D1MA00976A
19. H.-H. Fang, S. Adjokatse, H. Wei, J. Yang, G.R. Blake, J. Huang, J. Even, M.A. Loi, Ultrahigh sensitivity of methylammonium lead tribromide perovskite single crystals to environmental gases. Sci. Adv. **2**(7), 534 (2016). https://doi.org/10.1126/sciadv.1600534
20. Y. Feng, D. Chen, Y. Zhong, Z. He, S. Ma, H. Ding, W. Ao, X. Wu, M. Niu, A lead-free 0D/2D $Cs_3Bi_2Br_9/Bi_2WO_6$ S-scheme heterojunction for efficient photoreduction of CO_2. ACS Appl. Mater. Interf. **15**(7), 9221–9230 (2023). https://doi.org/10.1021/acsami.2c19703
21. M. Foster, J. Mendoza Cortes, H.C. Mayer, Spin coating experiments and theory for undergraduate physics and engineering students—a connection to microfabrication. Am. J. Phys. **92**(7), 545–558 (2024). https://doi.org/10.1119/5.0169090
22. G. Jayanthi, S. Sumathi, K. Kannan, V. Andal, & Murugan, S. (2022). A review on synthesis, properties, and environmental application of Fe-based perovskite. Adv. Mater. Sci. Eng. 2022, 6607683. https://doi.org/10.1155/2022/6607683
23. Gin et al. (2022). https://doi.org/10.1016/j.cej.2021.134280
24. A.S. Habiballah, A.M.M. Jani, A.H. Mahmud, N. Osman, S. Radiman, Facile synthesis of $Ba_{0.5}Sr_{0.5}Co_{0.8}Fe_{0.2}O_{3-\delta}$ (BSCF) perovskite nanowires by templating from nanoporous anodic aluminium oxide membranes. Mater. Chem. Phys. **177**, 371–378 (2016). https://doi.org/10.1016/j.matchemphys.2016.04.041
25. High Defect Tolerance in Lead Halide Perovskite CsPbBr3
26. Hu et al. (2023). https://doi.org/10.1016/j.cej.2023.143752
27. J. Iskandar, C.-Y. Liu, C.-C. Lee, K.-Y. Ke, M.R.A. Septian, R. Estrada, H. Humaidi, S. Biring, C.-S. Chu, Z.-L. Tseng, S.-W. Liu, Stabilizing perovskite quantum dot oxygen sensors through ultra-long 2 mm horizontally aligned nanopores in anodic alumina oxide templates. J. Mater. Chem. C **13**(2), 709–717 (2025). https://doi.org/10.1039/D4TC03851G
28. Jang et al. (2020). https://doi.org/10.1021/jacs.7b05246
29. B. Jeong, H. Han, H.H. Kim, W.K. Choi, Y.J. Park, C. Park, Polymer-assisted nanoimprinting for environment- and phase-stable perovskite nanopatterns. ACS Nano **14**(2), 1645–1655 (2020). https://doi.org/10.1021/acsnano.9b06980
30. S. Jiang, M. Liu, D. Zhao, Y. Guo, J. Fu, Y. Lei, Y. Zhang, Z. Zheng, Doping strategies for inorganic lead-free halide perovskite solar cells: progress and challenges. Phys. Chem. Chem. Phys. **26**, 4794–4811 (2024). https://doi.org/10.1039/D3CP05444F
31. John wong baek et al. (2023). https://doi.org/10.1021/accountsmr.3c00206
32. A. KV, S. Subashchandran, P. Mariyappan, Investigation on optical properties of lead-free $Cs_3Bi_2Br_9$ perovskite derivative quantum dots synthesised via modified LARP method. Opt. Quant. Electron. **56**(6), 1044 (2024). https://doi.org/10.1007/s11082-024-06975-7
33. Kang, Wang (2017). https://doi.org/10.1021/acs.jpclett.6b02800
34. Kang et al. (2018). https://doi.org/10.1039/C8TA02725K
35. Kim et al. (2017). https://doi.org/10.1016/j.snb.2017.04.029
36. Kim et al. (2022). https://doi.org/10.1016/j.cej.2022.136025
37. Kim et al. (2023). https://doi.org/10.1021/acsnano.2c12580
38. Kim et al. (2025). https://doi.org/10.1021/acs.nanolett.4c06149

39. H. Lee, D. Lee, H. Jin, D. Baek, M.K. Kim, J. Cha, S. Kim, M. Kim, Optical humidity sensors based on lead-free cu-based perovskite nanomaterials. Nanoscale Adv. **4**(16), 3309–3317 (2022). https://doi.org/10.1039/D2NA00168C
40. B. Li, L. Qiao, H. Yang, Y. Wang, W. Gu, B. Du, Perovskite materials empower sensors and batteries: environmental health monitoring and flexible wearable innovation. J. Mater. Chem. C **31**, 1739 (2025a). https://doi.org/10.1039/D5TC01739D
41. X. Li, J. Shi, Lead-free halide double perovskite for high-performance photodetectors: progress and perspective. Materials **16**(12), 4490 (2023). https://doi.org/10.3390/ma16124490
42. M. Li, W. Ye, J. Ruan, Q. Ren, S. Dong, D. Chen, N. Li, Q. Xu, H. Li, J. Lu, Lead-free halide double perovskite $Cs_2AgBiCl_6$ for H_2S trace detection at room temperature. ACS Sens. **10**(3), 2224–2233 (2025b). https://doi.org/10.1021/acssensors.4c03532
43. G. Li, Y. Zhang, X. Zhao, J. Lin, C. She, S. Liu, C. Jing, Y. Cheng, J. Chu, Bismuth-based lead-free perovskite film for highly sensitive detection of ammonia gas. Sens. Actuat. B Chem. **345**, 130298 (2021). https://doi.org/10.1016/j.snb.2021.130298
44. Lin et al. (2016). https://doi.org/10.1021/acsami.6b01709
45. Ma et al. (2018). https://doi.org/10.1016/j.snb.2017.09.194
46. A. Maity, S. Mitra, B. Ghosh, Textile based lead-free halide perovskite $CH_3NH_3SnI_3$ ammonia gas sensor working at room temperature. ACS Appl. Electr. Mater. **6**(4), 2677–2682 (2024). https://doi.org/10.1021/acsaelm.4c00230
47. Marikutsa et al. (2023). https://doi.org/10.1016/j.jallcom.2023.169141
48. Murade et al. (2011). https://doi.org/10.1016/j.cap.2010.08.020
49. N. Koonsaeng, W.A. Thaweechai, Preparation of Sr-doped Lafeo3 by thermal decomposition of metal organic complex and their gas-sensing properties. SOJ Mater. Sci. Eng. **6**(1), 1–9 (2018). https://doi.org/10.1039/C8TA02725K
50. K. Ou, Y. Wang, W. Zhang, Y. Tang, Y. Ni, Y. Xia, H. Wang, Highly sensitive H_2S gas sensor based on a lead-free $CsCu_2I_3$ perovskite film at room temperature. ACS Omega **8**(50), 48326–48335 (2023). https://doi.org/10.1021/acsomega.3c07694
51. S. Panda, S. Mehlawat, N. Dhariwal, A. Kumar, A. Sanger, Comprehensive review on gas sensors: unveiling recent developments and addressing challenges. Mater. Sci. Eng. B **308**, 117616 (2024). https://doi.org/10.1016/j.mseb.2024.117616
52. S.S. Patil, B.M. Babar, La–Fe–O perovskite based gas sensors: recent advances and future challenges. ACS Omega **9**(28), 29994–30014 (2024). https://doi.org/10.1021/acsomega.4c00334
53. X.-F. Qi, F. Zhang, Z.-P. Chen, X. Chen, M.-C. Jia, H.-F. Ji, Z.-F. Shi, Hydrothermal synthesis of stable lead-free $Cs_4MnBi_2Cl1_2$ perovskite single crystals for efficient photocatalytic degradation of organic pollutants. J. Mater. Chem. C **11**(11), 3715–3725 (2023). https://doi.org/10.1039/D3TC00185G
54. S. Rana, M. Akteruzzaman, A comprehensive systematic literature review on perovskite solar cells: advancements, efficiency optimization, and commercialization potential for next-generation photovoltaics. SSRN Electron. J. **23**, 7693 (2025). https://doi.org/10.2139/ssrn.5237693
55. M. Rycenga, P.H.C. Camargo, Y. Xia, Template-assisted self-assembly: a versatile approach to complex micro- and nanostructures. Soft Matter. **5**(6), 1129–1136 (2009). https://doi.org/10.1039/B811021B
56. T. Seiyama, A. Kato, K. Fujiishi, M. Nagatani, A new detector for gaseous components using semiconductive thin films. Anal. Chem. **34**(11), 1502–1503 (1962). https://doi.org/10.1021/ac60191a001
57. M. Shandilya, G.A. Kaur, Low temperature crystal growth of lead-free complex perovskite nano-structure by using sol–gel hydrothermal process. J. Solid State Chem. **280**, 120988 (2019). https://doi.org/10.1016/j.jssc.2019.120988
58. M. Shellaiah, K.W. Sun, Review on sensing applications of perovskite nanomaterials. Chemosensors **8**(3), 55 (2020). https://doi.org/10.3390/chemosensors8030055
59. S.K. Shil, F. Wang, K.O. Egbo, Z. Lai, Y. Wang, Y. Wang, D. Zhao, S.-W. Tsang, J.C. Ho, K.M. Yu, Two-step chemical vapor deposition-synthesized lead-free all-inorganic $Cs_3Sb_2Br_9$

perovskite microplates for optoelectronic applications. ACS Appl. Mater. Interf. **13**(30), 35930–35940 (2021). https://doi.org/10.1021/acsami.1c07839
60. P.V. Shinde, A. Patra, C.S. Rout, A review on the sensing mechanisms and recent developments on metal halide-based perovskite gas sensors. J. Mater. Chem. C **10**(28), 10196–10223 (2022). https://doi.org/10.1039/D2TC01980A
61. S. Vijaya, J. Subbiah, D.J. Jones, S. Anandan, LARP-assisted synthesis of $CsBi_3I_{10}$ perovskite for efficient lead-free solar cells. RSC Adv. **13**(15), 9978–9982 (2023). https://doi.org/10.1039/D3RA00365E
62. W. Ye, & H.-Z. Lin (2025). Dimensional reduction in $Cs_2AgBiBr_6$ enables long-term stable perovskite-based gas sensing. Nat. Commun., 16(1), 4820. https://doi.org/10.1002/adma.202100674
63. X. Wang, K. Huang, W. Ma, Y. Cong, C. Ge, S. Feng, Defect engineering, electronic structure, and catalytic properties of perovskite oxide $La_{0.5}Sr_{0.5}CoO_{3-\delta}$. Chem. A Eur. J. **23**(5), 1093–1100 (2017). https://doi.org/10.1002/chem.201604065
64. Y. Wang, J. Liu, Y. Liu, S. Li, X. Xu, Z. Lou, Recent advances in lead-free halide perovskites: from synthesis to applications. J. Mater. Chem. C **12**, 10267–10329 (2024b). https://doi.org/10.1039/D4TC01556H
65. C. Wei, Z. Guo, H. Wang, S. Zhang, D. Hao, J. Huang, Recent progress of gas sensors based on perovskites. Mater. Horiz. **12**, 317–342 (2025). https://doi.org/10.1039/D4MH01306A
66. J. Wu, X. Xu, Y. Zhao, J. Shi, Y. Xu, Y. Luo, D. Li, H. Wu, Q. Meng, DMF as an additive in a two-step spin-coating method for 20% conversion efficiency in perovskite solar cells. ACS Appl. Mater. Interf. **9**(32), 26937–26947 (2017). https://doi.org/10.1021/acsami.7b08504
67. C. Xiao, Z. Tang, Z. Ma, X. Gao, H. Wang, L. Jia, High performance porous $LaFeO_3$ gas sensor with embedded p–n junctions enabling ppb-level formaldehyde detection. Sens. Actuat. B Chem. **397**, 134670 (2023). https://doi.org/10.1016/j.snb.2023.134670
68. Xiaofeng et al (2017). https://doi.org/10.1016/S1002-0721(17)60965-7
69. A.K. Yadav, P. Singh, R. Pandey, Advances in lanthanum ferrite-based sensors: a comprehensive review. Anal. Sens. **5**(4), 86 (2025). https://doi.org/10.1002/anse.202400086
70. N. Yamazoe, K. Shimanoe, Fundamentals of semiconductor gas sensors. Semicond. Gas Sens. **49**, 3–38 (2020). https://doi.org/10.1016/B978-0-08-102559-8.00001-X
71. Yuan et al. (2025). https://doi.org/10.1016/j.snb.2024.135415
72. F. Zhang, Z. Ma, Z. Shi, X. Chen, D. Wu, X. Li, C. Shan, Recent advances and opportunities of lead-free perovskite nanocrystal for optoelectronic application. Energy Mater. Adv. **2021**, 5198145 (2021)
73. Y. Zhang, Q. Rong, J. Zhao, J. Zhang, Z. Zhu, Q. Liu, Boron-doped graphene quantum dot/ag–$LaFeO_3$ p–p heterojunctions for sensitive and selective benzene detection. J. Mater. Chem. A **6**(26), 12647–12653 (2018). https://doi.org/10.1039/C8TA03425G
74. L. Zhao, M. Steinhart, M. Yosef, S.K. Lee, S. Schlecht, Large-scale template-assisted growth of $LiNbO_3$ one-dimensional nanostructures for nano-sensors. Sens. Actuat. B Chem. **109**(1), 86–90 (2005). https://doi.org/10.1016/j.snb.2005.03.093

Chapter 5
Lead-Free Perovskites: Crystallographic Classification and Photocatalysis Insights

Narendra Kumar, Niwesh Ojha, and Kamal Kishore Pant

Abstract Developing advanced materials to meet the ever-growing global energy demand has accelerated research for renewable energy conversion. Perovskite materials, whether employed as active layers in solar cells or catalysts for valuable chemical conversion, have emerged as promising materials. Lead-based perovskites have achieved remarkable power conversion efficiencies and exhibit excellent optoelectronic properties. However, their intrinsic toxicity and limited long-term stability present significant obstacles to widespread commercialization. These limitations have driven an intensive research shift toward the design and synthesis of environmentally benign, lead-free perovskites (LFPs) alternatives with enhanced structural and chemical stability. This chapter comprehensively discusses the synthetic strategies, crystallographic classifications, and mechanistic insights relevant to lead-free perovskite materials. Special emphasis is placed on Sn^{-}, Ge^{-}, Sb^{-}, and Bi-based perovskites, double perovskites, and mixed-metal frameworks. The classification of these materials is presented systematically based on their structural dimensionality (0D, 1D, 2D, and 3D), followed by a detailed analysis of their various synthesis methodologies, including composition, oxidation state, etc. The correlation between crystal structure, defect chemistry, and optoelectronic properties is highlighted to establish key structure–property relationships for photocatalytic applications. Finally, the chapter critically examines stability challenges, processing limitations, and scalability concerns, offering future perspectives to optimize synthetic routes and enable the development of efficient, durable, and commercially viable LFPs for various photocatalytic applications.

Keywords Lead-free perovskites · Photocatalysis · Perovskite crystals

---

N. Kumar · N. Ojha (✉) · K. K. Pant
Department of Chemical Engineering, Indian Institute of Technology Roorkee, Roorkee, Uttarakhand, India
e-mail: niwesh.ch@sric.iitr.ac.in

V. Chauhan et al. (eds.), *Sustainable Lead-Free Perovskite Materials*, Topics in Applied Physics 157, https://doi.org/10.1007/978-3-032-24873-2_5

5.1 Introduction

Perovskite-based materials have recently emerged due to their optoelectronic properties, structural tunability, thermal stability, and broad light absorption capabilities. Generally characterized by the formula ABC_3, where A is a bigger cation, usually an alkaline earth metal, such as Ba^{2+}, La^{3+}, Sr^{2+}. B denotes the smaller transition metal cation, such as Fe^{3-}, Co^{3-}, Mn^{3-}, Ni^{3-}. C represents an anion, commonly oxygen or larger ions like oxides, halides, nitrides, and sulfides. Such atoms can be arranged in various forms, such as a cubic structure where the A-cation is positioned at the corners of the cube with 12-fold coordination. At the same time, the B-cation occupies the center, and the C-anion atoms are arranged at each face center [22]. Initially, Lev Perovski discovered such structures in 1872 [19]. After that, Wenk and Bulakh provided a detailed description in 1926 and published it in 1945 [103]. Despite many advantages, perovskite material for photocatalysis faces several challenges, like instability under operational conditions, rapid charge recombination, and, in some cases, environmental concerns related to lead (Pb) content. Lead-based perovskite may leach into soil and water streams during synthesis, operation, or disposal. It can create a significant threat to the marine aquatic life and affect the ecological balance. As a result, an urgent search for LFPs is prompted without compromising the structural stability and desired properties for photocatalysis and other applications. To improve the efficient separation and migration of photogenerated carriers with an enhanced number of catalytically active sites, and adequate long-term stability [112, 117] Addressing this limitation has become a focal point in recent research, leading to developing doped perovskites, composite heterojunctions, and lead-free alternatives with enhanced stability and activity performance.

5.2 Lead-Free Perovskites Classification

Perovskites were classified into different types based on the cell structure, composition, and dimensionality, which provides an essential understanding of these materials' different properties and potential applications. Researchers gain valuable information by examining the arrangement of atoms in the LFP crystal lattice.

5.2.1 Based on Composition

(a) Oxide Perovskites: Oxide Perovskites are a class of perovskites in which oxygen atoms replace the C atom, so the general formula becomes ABO_3 (Fig. 5.1a). Some examples of oxide perovskites are $LaCoO_3$, $SrTiO_3$, and $BaFeO_3$. They offer higher thermal stability, and each atom acts as an active site to interact with the reactant. Their tunable size allows for the creation of a heterostructure with other catalysts to improve the overall catalytic performance synergically. For example, Xu et al. optimized $LaNiO_3$/CdS photocatalyst for hydrogen evolution reaction (HER) without any rare earth metal cocatalyst (e.g., Pt). They observed an H_2 production rate of 74 μmol h^{-1} (e.g., 3700 μmol h^{-1} g^{-1}), significantly better than the individual $LaNiO_3$ and CdS. The composite photocatalyst also provides higher stability. The $LaNiO_3$/CdS catalyst surface facilitates photoinduced charge carriers' dynamics, as indicated by photoelectrochemical (PEC) analyses, which improved the overall HER efficiency.

(b) Halide Perovskites: Halide perovskites are a class of perovskites in which oxygen atoms are replaced by the halide group members' atoms (like Cl^-, Br^-, I^-), so the general formula becomes ABX_3 (Fig. 5.1b). Some examples of the halide perovskites are methylammonium lead iodide ($CH_3NH_3PbI_3$), cesium lead iodide ($CsPbI_3$), cesium lead bromide $CsPbBr_3$ etc. Lead halide perovskite has gained significant attention due to its large absorption coefficients, high mobility, high photoluminescence quantum efficiency, and long carrier diffusion lengths. Due to these characteristics, these materials are widely used in optoelectronics applications, such as photodetectors, solar cells, lasers,

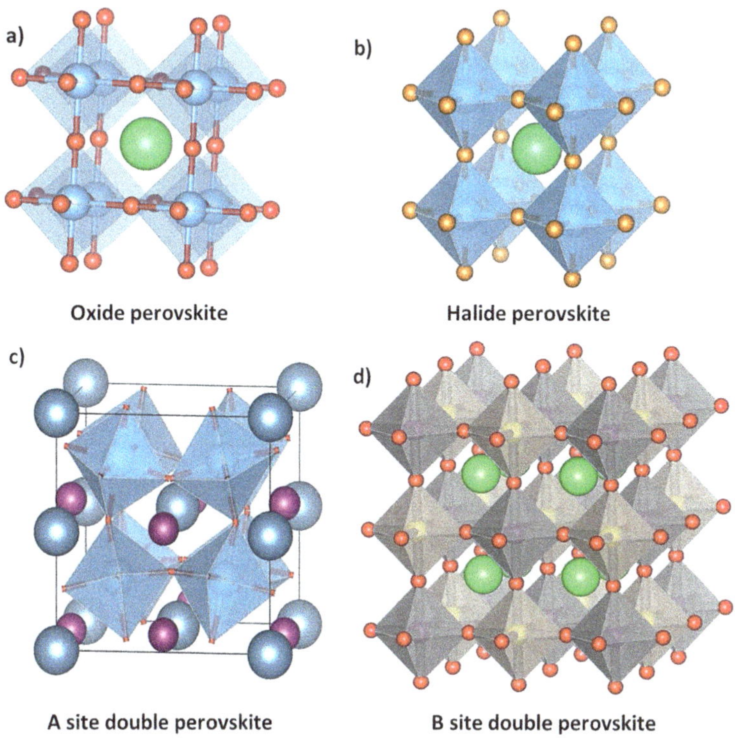

Fig. 5.1 Schematic representation of different types of perovskites based on composition: **a** Oxide perovskite; **b** Halide perovskite; **c**–**d** double perovskites

and light-emitting diodes. Specifically, the power efficiency conversion has uniformly increased from 3.8 to 26.1%. If we compared organic–inorganic hybrid perovskites, we found that all inorganic perovskite materials have higher stability. Benhafsa et al. focus on predicting the structural, electronic, optical properties, and photocatalytic performance of the $BaAgX_3$ (where X = I, Br, Cl) using the full potential linearized augmented plane wave method. [8] The lattice parameter values are 5.23 Å ($BaAgCl_3$), 5.85 Å ($BaAgI_3$), and 5.49 Å ($BaAgBr_3$). Calculated band structure reveals that halide perovskite compounds are indirect band gap semiconductors. Additionally, linear optical analysis indicates strong absorption in the ultraviolet spectrum. Exciton stability was examined through binding energy calculations to determine the efficiency of photogenerated charge carrier (electron and holes) separation. The binding energy values were obtained as 34.9 meV for $BaAgI_3$, 78.1 meV for $BaAgBr_3$, and 120.8 meV for $BaAgCl_3$. From this result, it is indicated that exciton stability is enhanced from iodide to bromide to chloride. Band gap values for $BaAgI_3$, $BaAgBr_3$, and $BaAgCl_3$ were determined to be 1.525 eV, 2.184 eV, and 2.436 eV, respectively. Photocatalytic studies suggest $BaAgI_3$ is a promising candidate for CO_2 reduction, $BaAgBr_3$ is well-suited for photocatalytic water splitting, and $BaAgCl_3$ shows potential for pollutant degradation due to its robust oxidative properties.

(c) Double Perovskites: It is similar to other perovskites but instead of using one cation at A-site and B-site, they are using two different cation at the A and B site, so that generally formula of double perovskites becomes $AA'B_2O_6$(double A-site) or $A_2BB'O_6$ (Double B-site) as shown in (Fig. 5.1c, d), some examples of the double perovskites are Strontium Iron Molybdenum Oxide (Sr_2FeMoO_6), Lanthanum Cobalt Manganese Oxide (La_2CoMnO_6), they were mainly used in the fuel cells, catalysis, magnetoresistance [23]. For example, Kobayashi et al. show an adverse tunnelling magnetoresistance effect at room temperature, which is of interest for a wide range of applications in magnetic devices[47]. Some reactions that need excessive H^+ ions, such as hydrogen evolution or the CO_2 reduction reaction, must be conducted in controlled reaction conditions to avoid the decomposition of perovskites. For example, Zhang et al. utilized an alcohol-based photocatalytic system to achieve dye degradation using $Cs_2AgBiBr_6$ photocatalysts, simultaneously enhancing lead-free perovskite materials' stability [131]. The reaction rate on $Cs_2AgBiBr_6$ is analogous to that of CdS, and the reaction follows pseudo-zeroth-order kinetics, indicating that the structure of $Cs_2AgBiBr_6$ is also a function of the reaction sites for photocatalysis. Next, it is deposited with Pt and Au. Average lifetime decreases to 0.71 ns and 1.52 ns, compared to the $Cs_2AgBiBr_6$ lifetime of 3.18 ns. In the same line, Sato et al. focus on illustrating crystal structures, optical, magnetic, and photocatalytic properties of the B-site substituted double perovskite $Ba_2Pr(Bi_{1-x}Sb_x)O_6$ ($x = 0, 0.1, 0.2, 0.5,$ and 1.0) [92]. All the samples show magnetic moments are concentrated around three μ_B, when measured by the magnetization, indicating a valence mixing state between Pr^{3+} and Pr^{4+}. Band gap energy for the sample was estimated at $x = 0$ as $Eg = 1.06$ eV and at $x = 1.0$

as $E_g = 2.71$ eV. As a result, samples of $Ba_2Pr(Bi_{1-x}Sb_x)O_6$ (x = 0 and 0.1) show enhanced photocatalytic activities compared to the solid-state method for the same compounds. Their result indicated that photocatalytic activities strongly depend on catalyst preparation methods, band gaps, and photogenerated charge separation.

5.2.2 Classification Based on Goldschmidt's Tolerance Factor

The structural stability and crystal symmetry of perovskite (ABC_3) are crucial for determining its physicochemical properties and photocatalytic performance. One of the most widely used criteria to predict the structural formability of perovskites is Goldschmidt's tolerance factor (t) [20]. Usually, evaluates the geometric compatibility of the A, B, and X ions based on their ionic radii, and is defined as given below,

$$t = \frac{r_a + r_c}{\sqrt{2}(r_b + r_c)} \tag{5.1}$$

where r_A and r_B are the radii of the cations of A and B, and r_c is the radius of the anion.

A tolerance factor value close to 1 indicates a good geometrical fit. It typically results in a stable cubic structure, as observed in $SrTiO_3$, favoring efficient charge mobility and photocatalytic activity. In contrast, values significantly lower or higher than 1.0 lead to structural distortions, like orthorhombic or rhombohedral phases (Fig. 5.2), such as octahedral tilting and symmetry lowering. These distortions can affect band structure, oxygen vacancy orientation, and surface reactivity including usage in optoelectronics and photocatalysis applications, with a tolerance factor calculated using Eq. 5.1 in the range $0.813 < t < 1.107$, depending on the ionic radius [70]. For a better understanding of complex perovskite, especially double perovskites of the type $A(B'_{1/2}B''_{1/2})\,C_3$, the above equation can be modified to account for the average size of the B-site cation, as modified as,

$$t = \frac{R_A + R_C}{\sqrt{2\left(\frac{R'_B + R''_B}{2} + R_C\right)}} \tag{5.2}$$

However, indicating the stability of perovskites can't be predicated on a single parameter, so other parameters that also help in estimating the stability of the perovskite structure are the octahedral factor (μ), defined as,

$$\mu = \frac{r_B}{r_C} \tag{5.3}$$

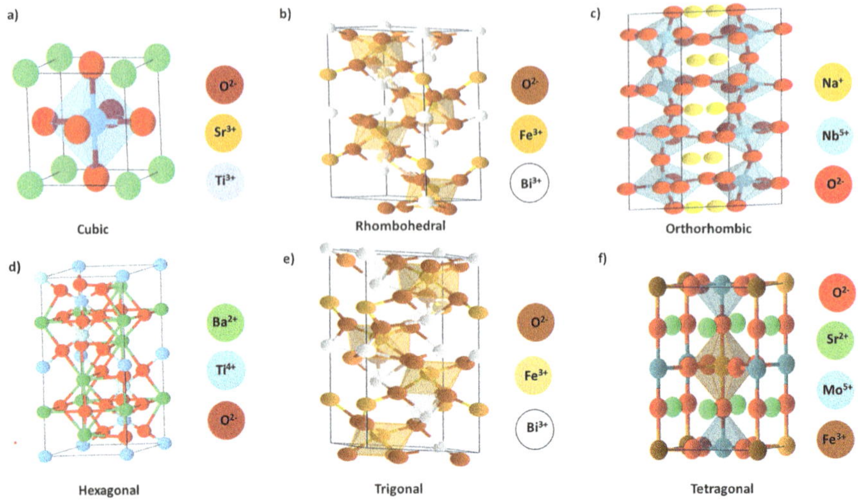

Fig. 5.2 Illustrates the different types of perovskites based on the tolerance factor

Generally, the stability range for μ-value for the perovskite is between $0.442 < \mu < 0.895$ [70]. Researchers have adopted material informatics and machine learning (ML) methods to overcome geometric model limitations. For instance, Pilania et al. developed a classification model using a dataset of 185 experimentally labelled ABX_3 compounds, applying machine learning algorithms to identify perovskites from non-perovskites [84]. Although the model achieved a high classification accuracy of 92.5% on the training set and 91.5% on the test set. It did not generate a new quantitative descriptor that could universally predict perovskite stability. Building on this, Lu et al. employed a gradient boosting classification algorithm trained on a much larger dataset of 1109 compounds, considering 132 different features [65]. The ML prediction accuracy of energy difference and bandgap regressions is over 90%. ML produces results comparable to those of density functional theory calculations. A significant advancement came from Bartel et al., who proposed a new tolerance factor (τ) that incorporates both geometric constraints and chemical nature through a statistical learning framework [6]. New tolerance factor defined as,

$$\tau = \frac{r_X}{r_B} - n_A\left(n_A - \frac{r_A/r_B}{\ln(r_A/r_B)}\right) \tag{5.4}$$

An et al. synthesized Cs $(Mn_{0.2}Ni_{0.2}Sn_{0.2}Co_{0.2}Bi_{0.1})$ Br_3 by the high-energy ball milling method for different contents of Bi^{3+}. The ionic radii of each ion are shown in Table 5.1, from which they calculated t and μ were 0.958 and 0.447, respectively [3].

Table 5.1 Ionic radius of different in $Cs(Mn_{0.2}Ni_{0.2}Sn_{0.2}Co_{0.2}Bi_{0.1})Br_3$

Ion	Cs^+	Mn^{2+}	Ni^{2+}	Sn^{2+}	Co^{2+}	Bi^{3+}	Br^-
Ionic radii (Å)	1.88	0.83	0.69	1.18	0.65	1.03	1.96

5.2.3 Classification Based on Perovskite Dimension

Perovskite material can also be classified based on dimensionality, which determines the number of perovskite structures that can be extended. Figure 5.3 indicates that four main types of perovskite materials are classified on this basis: zero-dimensional perovskites (0D), one-dimensional perovskites (1D), two-dimensional perovskites (2D), and three-dimensional perovskites (3D), which have been used recently by researchers to evaluate various unique properties of perovskites.

(a) **Zero-dimensional lead-free perovskites (0D)**:

The individual octahedral anions are completely isolated and surrounded by the organic cations [60]. These structures are confined to all the surrounding space and exist as quantum dots or discrete nanocrystals, where the crystal lattice cannot be extended beyond individual nanocrystals (Fig. 5.3a). The general chemical formula for the 0D LFP is A_4BX_6, where A denotes a monovalent organic cation and BX_6 denotes metal halide octahedra. Although OD perovskites exhibited limited charge transport due to their structural isolation, they show promise in photocatalysis due to their high surface-to-volume ratio. Cs_4SnBr_6 is a well-known 0D perovskite that

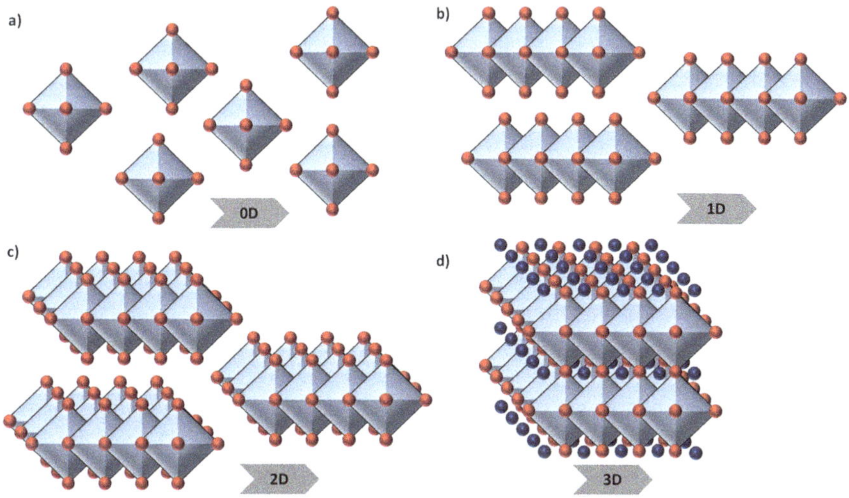

Fig. 5.3 Illustrate the different types of perovskites based on the dimensionalities: **a** zero-dimensional perovskite; **b** one-dimensional perovskite; **c** two-dimensional perovskite; **d** three-dimensional perovskite

has been investigated for photocatalytic and sensing applications. For example, Huang et al. synthesized a series of Mn^{2+}/Sb^{3+} co-doped Cs_4SnBr_6 powders via a water-assisted wet ball milling method [37]. The incorporation of Mn^{2+} and Sb^{3+} ions enhanced the light emission originating from self-trapped excitons (STEs) in Cs_4SnBr_6, resulting in a broadband emission centered at 525 nm and a high photoluminescence quantum yield of up to 67.7%. In addition, the Mn^{2+}/Sb^{3+} co-doped Cs_4SnBr_6 exhibited good thermal stability, making it more promising for practical optoelectronic applications. This improvement can be attributed to the Mn^{2+}/Sb^{3+}-induced distortion of the $[SNBr_6]^{4-}$ octahedra, which enhances the electron-phonon coupling and increases the STE binding energy.

(b) **One-dimensional perovskite (1D)**:

In 1D perovskite, octahedral anions are connected only in one dimension, like a chain, and surrounded by the organic cations as shown in the Fig. 5.2b, having a general formula $A'_2A_mM_mX_{3m+2}$ [30]. Mainly, Size-dependent effects emerge when structures are below characteristic length scales such as the Bohr radius, the electron mean free path, or the Coulomb charging length. As a result, quantum confinement becomes significant when crystallite dimensions are below ~ 10 nm, although notable deviations from bulk behaviour occur. However, reduced dimensionality can hinder charge mobility, which must be compensated for by surface engineering. For example, Zou et al. synthesized stable $CsCu_2I_3$ by optimizing the annealing treatment parameters [139]. High-quality, pinhole-free $CsCu_2I_3$ films with strong crystallinity were achieved, exhibiting excitation binding energy of 116 meV and carrier mobility ($\mu_e = 23.29\ cm^2\ V^{-1}\ s^{-1}$ and $\mu_h = 12.83\ cm^2\ V^{-1}\ s^{-1}$). The synthesized $CsCu_2I_3$-based photodetectors demonstrated excellent photo-response performance under 365 nm UV irradiation, with a detectivity of 1.1 * 1011 Jones, responsivity of 42 mA W^{-1}, and EQE (External quantum efficiency) of 12.26%. The device showed the fastest response time ($t_{rise} = 1.49$ ms, $t_{decay} = 1.33$ ms) and a large on/off ratio exceeding 7.5×10^5. Furthermore, $CsCu_2I_3$ films also exhibited high operational stability, with only 8% degradation observed in two months in open air while maintaining an EQE of 11.2%.

(c) **Two-dimensional perovskites (2D)**:

In the 2D perovskites' structure, nucleation grows in two dimensions instead of only one (Fig. 5.3c), usually consisting of layered structures where alkaline earth metals are separated by organic cations or inert layers, forming a quantum well structure. The formula of 2D perovskite structure is $(A')2(A)_{n-1}B_nX_{3n+1}$ where, A is organic cations with large size (such as phenylethylammonium (PEA), butylammonium (BA), iso-butylammonium (iso-BA), n-propaneammonium (PA), and propane-1,3-diammonium (PDA)), B is lead, germanium or tin, X is halide anion, and n is the number of perovskites layer [73, 121]. These materials offer better environmental stability due to hydrophobic organic layers and can be engineered for anisotropic charge transport. Next, M. Masri et al. addressed the water instability issues by synthesizing 2D $Cs_3Bi_2Br_9$ (CBB) nanosheets via an anti-solvent reprecipitation method [74]. Four samples were prepared using varying volumes of isopropanol (75, 125, 250, 500 ml) under vigorous stirring and denoted as CBB75, CBB125, CBB250, and CBB500, respectively. The result showed that increasing the volume of isopropanol led to reduced agglomeration and facilitated the formation of 2D nanosheets. The optical band energy of CBB75, CBB125, CBB250, and CBB500 samples is 2.57, 2.57, 2.55, and 2.52 eV, respectively, and they have the lowest charge rate of e−/h+ recombination owing to their 2D nanosheet properties. CBB250 shows the highest efficiency for MB dye removal compared to the other samples. Approximately 80% dye is removed in the first 90 min, and within 150 min, complete (100%) removal was achieved. In contrast, CBB500 has approximately 70% dye removal in 90 min, while CBB75 and CBB125 demonstrated less than 40% dye degradation after 150 min. Furthermore, the catalyst dosage influenced the photocatalytic degradation efficiency, and an optimal performance was observed at 30 mg/L. In the same line, Sun et al. synthesized lead-free perovskite $Cs_3Bi_2Br_9$ via an antisolvent precipitation method [100]. They demonstrated that constructing a heterojunction structure significantly enhanced the photocatalytic activity. In particular, a 12.4% $Cs_3Bi_2Br_9$/TiO_2 composite achieved a degradation rate of 74.7% within the first 10 minutes, reaching 99.9% within 20 min. In contrast, pure TiO_2 and $Cs_3Bi_2Br_9$ show only 2.8 and 30.9% degradation rates within 10 min. Additionally, the degradation rate for MBT reaches 99.9% in 20 min over the 12.4% $Cs_3Bi_2Br_9/TiO_2$. The Production rate achieved by the $Cs_3Bi_2Br_9/TiO_2$ for the selective oxidation of benzyl alcohol is 1465 μmol g^{-1} h^{-1} as compared to inorganic catalysts without a zero-valent metal co-catalyst.

(d) **Three-dimensional perovskites (3D)**:

In 3D perovskites, the octahedral anion (BX_6) can be extended to all three spatial directions, forming a continuous network, as indicated in the Fig. 5.3d. Owing to halide anion connectivity, efficient charge carrier transport, and strong light absorption are generally observed. Hence, highly suitable for photocatalytic reactions such as water splitting, CO_2 reduction, and pollutant degradation. Examples include $SrTiO_3$, $CH_3NH_3PbI_3$, and $BaTiO_3$. However, changing the precursor concentration often regulates 3D halide perovskites' morphology or particle size. Zhou et al. focus on developing a strategy to enhance the growth of $Cs_2AgInCl_6$ crystal size from several micrometers to several millimeters [135] $Cs_2AgInCl_6$ was synthesized by the hydrothermal method, resulting in a crystal structure in a cubic unit cell with space group Fm3m. Structurally, this perovskite comprises [$AgCl_6$] and [$InCl_6$] octahedra alternating in an ordered rock-salt structure. By varying experimental parameters, such as the hydrothermal reaction time, the crystal size would be tuned from 5 to 1.5 mm. $Cs_2AgInCl_6$ behaves as a direct gap semiconductor with a wide band gap of 3.23 eV obtained experimentally and 3.33 eV obtained by DFT calculation. In addition to tunable properties, $Cs_2AgInCl_6$ perovskites show excellent moisture, light, and heat stability, indicating their significant potential for photovoltaic and other optoelectronic applications.

Table 5.2. Summarizes the crystal structure, synthesis method, properties, and the application of halide/oxide perovskites with potential in optoelectronic and energy-related applications.

5.2.4 Classification Based on the Cation Oxidation States

In perovskite structures, the oxidation states of the cations were generally observed to be +2 at the A-site and +4 at the B-site. However, deviations from this trend were also reported. Instead of the typical +2 (A-site) and +4 (B-site) combination, oxidation states such as +3 for both cations or +1 at the A-site and +5 at the B-site were found. These variations were observed when the B-site cation was maintained in a six-fold coordination, i.e., surrounded by six oxygen atoms. To preserve the charge neutrality of the ABO_3 lattice, the oxidation states of both A-site and B-site cations were adjusted accordingly. Perovskites were classified according to the oxidation states of the cations involved, as shown in the Fig. 5.4.

(a) **Oxides of $A^{1+}B^{5+}O_3$ type**:

It is a subgroup of the ternary oxides in which the oxidation state of cations is +1 for A and +5 for B. They mainly contain tantalates and niobates at site B, and for A, Na, K, Ag, and Li [107]. Alkali tantalate-based perovskite semiconductors have been utilised in photocatalytic reactions, modulating superconductivity, electrochemical reduction reactions, and electromagnetic applications. Twelve oxygen anions coordinate a-site in A1+B5+O3 and is generally occupied by relatively bigger cations (like Li, Na, K) [46]. The B-site is taken up by smaller cations (e.g., Ta or Nb) with a coordination of six. Tantalates possess conduction bands consisting of Ta5d orbitals at a more negative position than titanates (Ti_3d) and niobates (Nb_4d). Such a high potential of the conduction band is advantageous for reducing water to produce H_2.

(b) **$A^{2+}B^{4+}O_3$ type Oxides**:

Perovskite oxides with the general formula $A^{2+}B^{4+}X_3$, where $X = O^{2-}$, are ideal perovskites. In this group, the elements that often occupy the B-site are tin (Sn), titanium (Ti), and hafnium (Hf), except Mn^{4+}; these compounds typically have cubic symmetry and exhibit metallic behavior. Transition metals, which have a + 4-oxidation state, like iron (Fe), Chromium (Cr), and manganese (Mn), can also be suitable at thc B-site. However, it isn't easy to maintain the +4 valence in 3d transition metals, limiting the stability of these kinds of perovskite material. Conversely, transition metal elements that are 4d and 5d are more suited for these kinds of structures since their +4 oxidation states quickly stabilise.

The B-site cation can affect the magnetic ordering in these types of perovskite oxides. For example, Co^{4+} produces the strong ferromagnetic ordering, Mn^{4+} and Cr^{4+} show weak ferromagnetism, and Fe^{4+} produces a complicated magnetic state. Substituting different elements at the B-site can significantly change the material's

Table 5.2 List of LFP Crystal structures, space group, properties, and applications

Perovskite	Crystal structure	Space group	Synthesis method	Properties	Application	References
$Cs_2AgInCl_6$	Cubic	Fm-3 m	Hydrothermal	Sensitivity, high luminescent, Electro-optical	Photodetector, photovoltaic	[66]
$CsPb(Sn)X_3$ (X: Cl, Br,I)	Cubic	Pnma	Solid state	Electro-optical	Energy generation	[87]
$BiFeO_3$	Rhombohedral	R3c	Sol–gel	Photocatalyst, small bandgap, Ferromagnetic, Electro-optical, high Curie temperature	Treatment and decomposition of wastewater	[33]
$CH_3NH_3PbI_3$	Tetragonal	14/mcm	Solid state-modified	Electro-optical, Excitonic	Solar cell	[97]
$Pb(Zr,Ti)O_3$	Tetragonal	P4mm	Solid state	Ferroelectricity, Piezoelectricity	Solar cells	[https://doi.org/10.1080/10584580590898730]
$NdFeO_3$	Orthorhombic	P63/mmc	Co-precipitation	Ferromagnetic, dielectric	Gas sensors	[35]
$BaTiO_3$	Hexagonal	P63/mmc	Solid state	Piezoelectricity, high dielectric constant, ferroelectricity	Solid oxide fuel cell (SOFC)	[54, 109]
$ReFeO_3$ (Re: La, Sm, Eu, Gd)	Orthorhombic	Pbma	Hydrothermal Sol–gel	Catalytic, photocatalytic, ferroelectricity, Piezoelectricity	Gas sensors, data storage, and spintronic devices	[7, 29]
Sr_2FeMoO_6	Tetragonal	I4/mcm	Sol–gel	Ferromagnetism	Memory devices	[15]
$NaNbO_3$	Orthorhombic	Pbma	Hydrothermal/ solvothermal	Ferroelectricity, piezoelectricity	Electrochemical systems	[42]

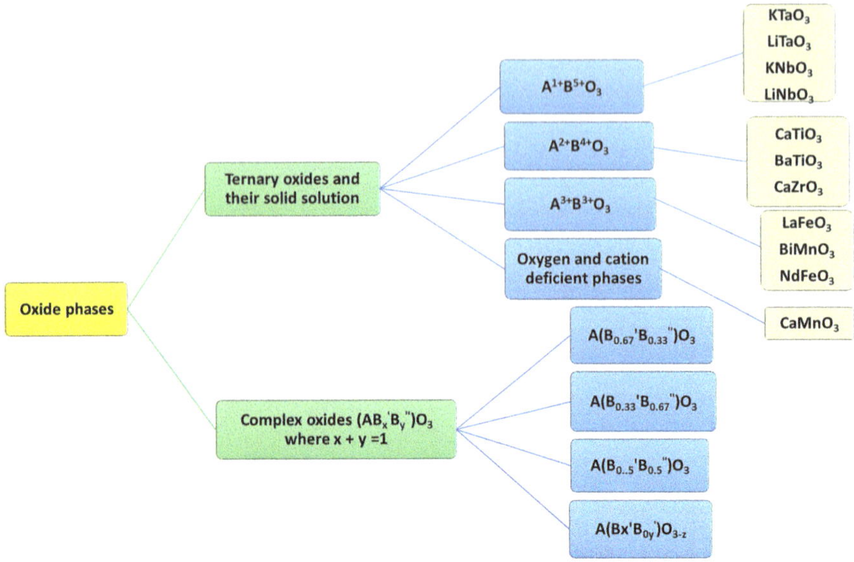

Fig. 5.4 Illustrates the classification of perovskites based on the oxidation state of the cation involved in them

physical and chemical behavior, leading to defects and lattice distortions. The purpose of this substitution is to improve specific functional characteristics. For instance, the photocatalytic performance of $SrTiO_3$ is improved by substituting chromium (Cr), and when manganese (Mn) is substituted in $BaTiO_3$, it improves their ferroelectric and piezoelectric properties.

(c) **$A^{3+}B^{3+}O_3$ type oxide**:

These types of Perovskite oxides have a general formula $A^{3+}B^{3+}O_3$, where A-site is generally of a rare-earth metal like La, Y, Nd, Gd, Bi, Sm, and at the B-site is a transition metal like Fe, Cr, Ga, Sc, In. Among these, Fe is used most of the time at the B-site; two essential members of this subclass are bismuth ferrite ($BiFeO_3$) and lanthanum orthoferrite ($LaFeO_3$). Because of its various physical and chemical characteristics, $LaFeO_3$ can be used in multiple cutting-edge technologies, such as gas sensors, solid oxide fuel cells (SOFCs), electromagnetic devices, and heterogeneous catalysis. However, it shows limited photocatalytic performance due to poor bulk surface area, recombination of photogenerated charge carriers, insufficient solar spectrum utilisation, and an inappropriate conduction band (CB) minimum oriented too positively to drive photoreduction processes efficiently.

With a high Curie temperature (T_C) of 1103 K and a Néel temperature (T_N) of 643 K, $BiFeO_3$ (BFO) is a multiferroic material that has been extensively studied. It possesses both room-temperature ferroelectricity and G-type antiferromagnetism, which makes it a promising material for magnetoelectric and spintronic applications. With a rhombohedral angle of 89.45° and a lattice constant of 5.63 Å, it crystallises in a rhombohedral form (space group R3c). The structure comprises FeO_6 octahedra with a Fe^{3+} centre and Bi^{3+} ions at the corners that are joined along the direction. Bi^{3+} displacement and polarisation result from the hybridisation of Bi^{3+} (6s) and O^{2-} (2p) orbitals, which gives rise to ferroelectricity. The antiferromagnetic ordering forms a cycloidal spin structure with a period of roughly 62 nm from Fe^{3+}, which is perpendicular to the polarization direction.

(d) **Complex oxides**:

$A(B_x{}' B\gamma'')O_3$ is a general formula of complex perovskite with x + y= 1 to maintain their charge neutrality and structure stability. $A(B_{0.67}{}' B_{0.33}{}'')O_3$ perovskites the amount of lower valence cation is twice that of higher valence cation. In contrast, $A(B_{0.33}{}' B_{0.67}{}'')O_3$ perovskite, the lower valence is half the higher valence cation. This is a balanced example in which both B-site cations are present in equal proportion, such as $A(B_{0.5}{}' B_{0.5}{}'')O_3$. Many of these complicated oxides show oxygen-deficient phases, in which many oxygen or cation vacancies causes the divergence from stoichiometry and a distinct chemical formula. Generally, A and B cations have an oxidation state of +2 and +4. Still, in some structures, both have a +3 oxidation state, particularly when the B3+ ion can maintain a six-fold coordination, i.e., octahedral.

Furthermore, there are some perovskites that contain only one type of B-site cation but in two different oxidation states. Rather than being mixed-composition compounds, they are best characterised as non-stoichiometric or defect-containing phases, which set them apart from complex perovskites with two separate B elements.

5.3 Lead-Free Perovskite Synthesis Methods

Perovskite synthesis techniques have rapidly evolved over the years [22]. This advancement has aimed chiefly to enhance structural quality, morphology, crystallinity, and defect density of perovskite materials to influence their photocatalytic performance. As a result of this, various chemical and physical methods have been developed to prepare perovskites with controlled composition, dimensionality, and particle size, such as sol-gel, hydrothermal, solvothermal, coprecipitation, solid-state reaction, ball milling, Pechini method, etc. [50]. There are mainly two approaches: the top-down and the bottom-up, each with its own benefits and drawbacks. These approaches are discussed below in detail.

5.3.1 Top-Down Approach

In this approach, large bulk material is broken into smaller particles with sizes up to the nanoparticle range (< 2 nm) through chemical or physical methods (Fig. 5.5a). Techniques such as ball milling, lithography, etching, high-pressure torsion, and laser ablation are used to do it. This method used techniques like carving and sculpting a larger material into a desired shape. This approach is generally used when bulk material is required in a short period, and exact control over size and morphology is essential to generate surface defects and concentration distribution, but such a process consumes high energy. Furthermore, impurities may be introduced, or the purity of the crystal could be damaged by the high-energy or temperature operations.

(a) **Ball milling method**

Ball milling method was widely employed owing to its simplicity, scalability, cost effectiveness, and ability to produce fine particles with uniform homogeneity. In this approach, bulk materials were broken down into nanoscale particles using mechanical milling devices. The fundamental principle of the method relied on the impact and attrition forces generated by the movement of balls within the milling chamber. As the chamber rotated, the grinding balls repeatedly collided with the precursor particles, reducing their size and promoting solid-state reactions between them. An et al. synthesised high-entropy LFP nanocrystals $Cs(Mn_{0.2}Ni_{0.2}Sn_{0.2}Co_{0.2}Bi_{0.1})Br_3$ (MNSCB) via a solvent-free synthetic route of high-energy ball milling as shown in (Fig. 5.6h) [3]. The high-entropy material has provided a triple advantages of prolonging the carrier lifespan, improving the availability of active sites, and modulating the bandgap energy level, for improving the Photocatalytic activity of Rhodamine B. As a result, they achieved a removal rate of 95% within 60 min. Furthermore, the unequal molar ratio of Bi led to shallow surface defects, which induced lower carrier recombination rate and the energy band structure showed the lattice distortion effect due to doping of various ions, narrowed the bandgap (1.61 eV), improved the visible light absorption juxtaposed with other single-phase perovskite, while upshifting the conduction band edges in more negative potential region. This encourages electron transfer to oxygen species, generating more superoxide anion radicals. Next, the multi-ion introduction provided abundant active sites, expediting the absorption of more organic pollutants and improved stability while affecting Mn-Mn spacing to enhance carrier lifespan provided eminent electron transport capacity, enhancing the transported efficiency of photoexcited carriers within the structure, and also result in lowering carrier recombination rate, extending the lifespan of both

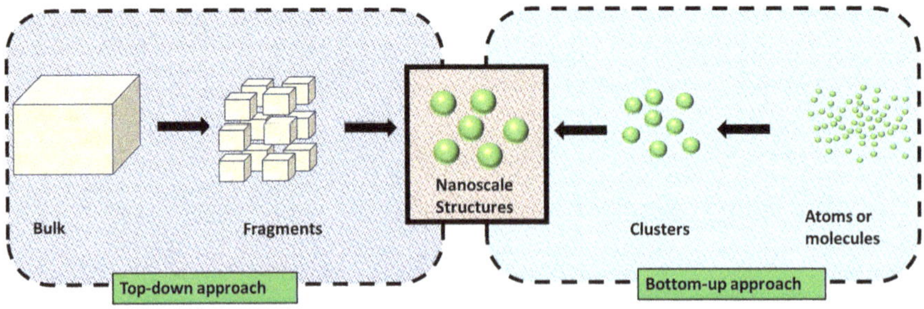

Fig. 5.5 Illustrates the different types of approaches for synthesizing materials: **a** Top-down approach; **b** bottom–up approach

holes and electrons by becoming a trap energy level for trapping and storing carriers, which strengthens the chances for carriers to relocate to the surface to participate in the photocatalytic reaction.

(b) **Laser ablation and Pulsed Laser**

In the laser ablation method, a high-energy pulsed laser beam is focused on a solid material, typically a metal, immersed in a liquid medium such as water, ethanol, acetone, chloroform, or isopropanol. Molecules were removed from the substrate surface to create nano-/microstructures, such as perovskites. The substance was eliminated from the substrate when the laser beam was focused on it, and ablation occurred once sufficient energy was absorbed to induce evaporation or melting. In laser machining applications, including laser beam milling, laser cutting, and high-precision drilling, the overall process of laser ablation was considered constant.

5.3.2 *Bottom-Up Approach*

In this approach, instead of breaking the material into smaller nanoparticles. Herein, self-assemble the atoms or molecules at nanoparticles agglomerate together, just like nature builds crystals or proteins (Fig. 5.5b) [26, 50]. Various methods such as sol-gel and co-precipitation. Hydrothermal and solvothermal, etc, are used. This approach provides greater control over particle size, shape, and composition, allowing for the controlled synthesis of perovskites with specific functionality, high purity and defect-free perovskites. However, scaling up this method for industrial production would be expensive and complex, as they are more sensitive to process variables like temperature, pH, and precursor selection. It is generally suitable for applications requiring nanostructures such as quantum dots, nanowires, thin films, and functional catalysts.

(a) **Sol–gel method**:

The sol–gel method was recognized as a bottom-up approach, where the transformation of colloidal suspension metal precursor particles in a liquid (sol) into a gel-like network was achieved through hydrolysis and condensation reactions, followed by drying and heating to obtain a solid catalyst [12]. Initially, a sol was prepared by dissolving the precursor in a solvent (water or alcohol) and mixing with a magnetic stirrer while the pH was maintained at the desired values. Subsequently, the precursors were subjected to hydrolysis and polymerization to form hydroxides, as indicated by the reaction in Eq. 5.5 [2]. After undergoing condensation processes, these hydrolysed species generated a continuous gel structure. The gel was dried at 60–100 °C for 12–48 h, during which the solvent (moisture) trapped inside was eliminated, leading to the formation of xerogel, as represented in Eq. 5.6. Finally, calcination was carried out, which yielded a porous xerogel with enhanced crystallinity and free from residual organics. The sol–gel method was widely employed due to its excellent

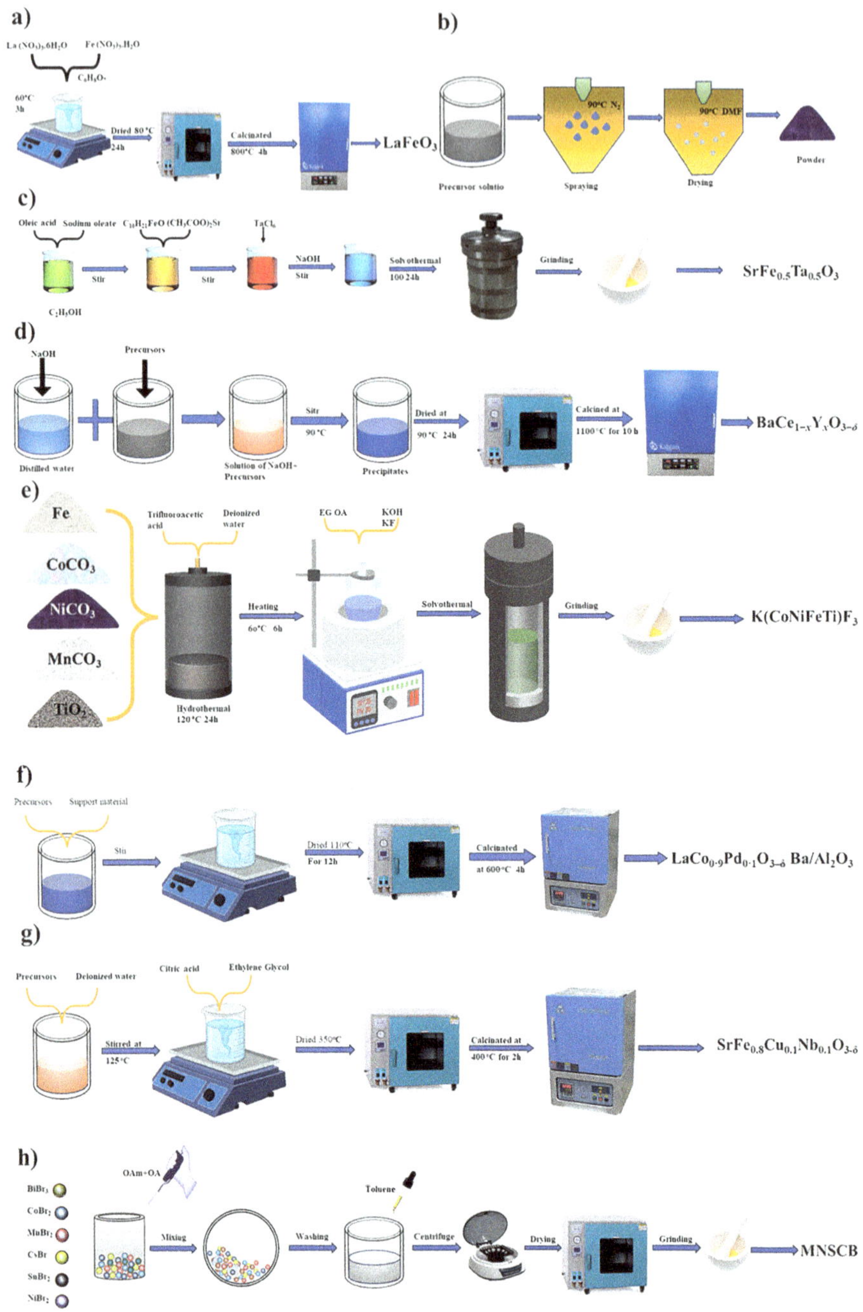

Fig. 5.6 Illustrates the different types of synthesis methods: **(a)** Sol–gel method; **(b)** Spray drying method; **(c)** solvothermal method; **(d)** Co-precipitation method; **(e)** Hydrothermal method; **(f)** Wet impregnation method; **(g)** Pechini method; **(h)** Ball milling method

homogeneity with 99.99% purity, low processing temperature, and adaptability in fabricating thin films, powders, fibres, and aerogels [45].

$$\mathrm{M} - OR + \mathrm{H_2O} \rightarrow \mathrm{M} - (\mathrm{OH}) + \mathrm{R} - \mathrm{OH} \quad \text{Hydrolysis reaction} \tag{5.5}$$

$$\mathrm{M} - OR + \mathrm{M} - \mathrm{OH} \rightarrow \mathrm{M} - \mathrm{O} - \mathrm{M} + \mathrm{R} - \mathrm{OH} \quad \text{Condensation reaction} \tag{5.6}$$

Upendranath et al. synthesized a visible light-harvesting perovskite oxide, $LaFeO_3$, having a band gap of 2.35 eV via the sol-gel method, as illustrated in the Fig. 5.6a [108]. The reaction involved in the preparation of $LaFeO_3$ was represented in Eqs. 5.7–5.10. The material exhibited luminescence, emitting a yellowish-orange colour, which rendered it suitable for optical display applications. Furthermore, the powder dusting method was applied to both porous and non-porous surfaces, and it was found that $LaFeO_3$ is used for forensic science because researchers found distinct ridges of fingerprints without hindrance by the background with precise sensitivity.

a. **Hydrolysis reaction**:

$$\mathrm{La(OR)_3} + 3\mathrm{H_2O} \rightarrow \mathrm{La(OH)_3} + 3\mathrm{ROH} \tag{5.7}$$

$$\mathrm{Fe(OR)_3} + 3\mathrm{H_2O} \rightarrow \mathrm{Fe(OH)_3} + 3\mathrm{ROH} \tag{5.8}$$

b. **Condensation reaction**:

$$\mathrm{La(OR)_3} + \mathrm{Fe(OH)_3} \rightarrow \mathrm{La} - \mathrm{O} - \mathrm{Fe} + \mathrm{R} - \mathrm{OH} \tag{5.9}$$

c. **Dehydration/Calcination**

$$\mathrm{La(OH)_3} + \mathrm{Fe(OH)_3} \rightarrow \mathrm{LaFeO_3} + 3\mathrm{H_2O} \tag{5.10}$$

(b) **Hydrothermal method**:

The hydrothermal method was considered a bottom-up approach, and the perovskite structure was synthesized through chemical reactions under controlled pressures (1–100 MPa) and temperatures (160–250 °C) in a sealed autoclave.[110] This method dissolves precursors, typically water or organic compounds, in a solvent. Then the prepared solution is sealed in an autoclave and heated to the desired temperature, generating solvent vapour pressure which exerts pressure of several atmospheres, i.e., facilitating the precursors' dissolution and reaction for 6–48 h. After the reaction, the autoclave was cooled, and the obtained product was filtered, washed, and dried. This technique promoted low-temperature crystallization compared to other conventional synthesis methods. For example, Li et al. synthesized a medium-entropy perovskite electrocatalyst of $K(CoNiFeTi)F_3$ by the hydrothermal method for the reduction of the NO_3^- to NH_3 (Fig. 5.6e) [58]. The optimized electrocatalyst performance of $K(CoNiFeTi)F_3$ was better as compared to $KCoF_3$, $K(CoNi)F_3$, and

$K(CoNiFe)F_3$ in reduction NO_3^- to NH_3 conversion with approximately 100% NH_3 Feand has 13.9 mg h^{-1} cm^{-2} ammonia stabile field rate at 0.6V vs RHE in alkaline medium because various transition metal is introducing which leads to increase electron transfer, which prevent the transition of Co 1s-3d and Co-F bond strengthened, which decrease the catalyst symmetry. Next, $LaFeO_3$ is known for good thermal stability, oxygen mobility, redox behaviour, and electronic and ionic conductivity. Due to this, they are applicable to catalysts in oxidation processes, sensors, electrodes, etc., but the preparation is time-consuming (up to 14 days) and involves several synthesis steps. This limitation can be solved by using microwave-heating conditions. Next, it was modified by Kostyukhin et al. via one-step hydrothermal microwave-assisted at a temperature of 240 °C and pressure of 60 bar without including an annealing [49]. Result shows that using microwave irradiation as a heating source increases the crystallization rate of nanoparticles. $LaFeO_3$ prepared by the modified method reduces the total synthesis time to 3 h, which is approximately 16 times faster than conventional hydrothermal synthesis.

(c) **Solvothermal method**:

Herein, solvent is used in place of water as discussed in the hydrothermal methods for synthesizing various LFPs materials, such as rare-earth-doped, Pb-based, fluoride, and oxide nanomaterials. Generally, nanosized halide double perovskites are not synthesized through the solvothermal method [71]. Because they are less stable in high temperatures and high pressures. Properly utilize organic solvents in a sealed autoclave under elevated temperature and autogenous pressure. Dissolve metal precursors in an organic solvent (like oleic acid, ethanol, or butanol), then add surfactants to tailor the size and shape of the nanoparticles and prevent aggregation in a solvent medium. Then transfer the mixture into a sealed autoclave and heat at 120–200 °C for 12–48 h after cooling, washing with appropriate solvents (ethanol, hexane, toluene), then dry it to get the desired product [110]. Cui et al. synthesised a double perovskite oxide $SrF_{0.5}Ta_{0.5}O_3$ by the solvothermal method at 100 °C, as shown in the Fig. 5.6c) [21]. A researcher reveals that the interaction between precursor cations and oleate ions is the key parameter in decreasing the solvothermal temperature. The $SrF_{0.5}Ta_{0.5}O_3$ has high crystallinity with a particle size of ca. 31 nm. $SrF_{0.5}Ta_{0.5}O_3$ exhibits photocatalytic NO removal efficiency of ca. 35.1% under visible light irradiation in 80 min and the AQE of 22.02% at 420 nm, greater than that of $SrFTaO_6$.

(d) **Co-Precipitation method**:

In the co-precipitation method, more than one metal precursor forms a homogeneous solution. Initially, the metal precursor is dissolved in the deionized water then added solvent until pH reaches desired value under constant stirring after that mixed metal hydroxides forms then left the suspension for 12–24 h at room temperature to improve crystallinity, then filtrated and washed it with deionized water several times to remove residual nitrates or ammonium ions then dry it 100–200 °C for 12–16 h. Finally, calcined at 600–800 °C for 4–6 h in air to form the perovskite phase. For example,

Zybert et al. synthesized a Y-modified perovskite-type oxide, $BaCe_{1-x}Y_xO_{3-\delta}$ (x = 0–0.030), as a support for the cobalt catalyst by the co-precipitation method, as shown in Fig. 5.6d [141]. Study concludes that a small amount of yttrium in barium cerate increased the basicity and specific surface area. They also tested the industrial scale (T = 400–470 °C, P = 6.3 MPa, H_2/N_2 = 3) and found that the most active catalyst containing 10 mol% Y is deposited on a support. The synthesis reaction rate of ammonia is 15–20% more than that of undoped $Co/BaCeO_3$. Also, it reveals that the activity of the catalyst decreases, greatly increasing the Y content in the Support by up to 30 mol% %. Researchers show that the Co/ $BaCe_{1-x}Y_xO_{3-\delta}$ catalyst has improved ammonia synthesis results than the commercial iron catalyst (ZA-5).

(e) **Wet impregnation**

In the wet impregnation method, an excess amount of the solution is applied, and next is pore volume impregnation (PVI), in which the volume of the solution is adjusted to match the pore volume of the support [76]. A solution of metal precursors was then employed to deposit active metal ions on the porous support. Common precursors included nitrates, chlorides, sulfates, carbonates, acetates, and metal complexes such as acetylacetonates. Water was commonly used as the solvent due to its high solubility for many inorganic precursors, while organic solvents were preferred for organometallic species. The concentration was maintained below saturation or supersaturation to avoid premature deposition of the precursor in the bulk solution. For uniform dispersion and adsorption of the metal ions within the pores and on the surface of the support, the impregnated system was either left undisturbed or stirred for several hours. The material was subsequently dried to remove the solvent. Calcination was then performed to decompose the precursor and generate the desired metal oxide or catalyst phase anchored on the support. For instance, Lyu et al. synthesized a Pd-doped perovskite $LaCo_{0.9}Pd_{0.1}O_3$ to enhance the reduction of NO_x to NH_3, as shown in (Fig. 5.6f) [68]. The catalysts are synthesized by the Sequential impregnation of 15 wt% of $LaCo_{1-x}Pd_xO_3$-type perovskites and 20 wt% of barium over alumina. Researchers claim that $LaCo_{0.9}Pd_{0.1}O_3$-Ba/Al_2O_3 catalyst has the best activity interval around 100°C, as compared to $LaCoO_3$-Ba/Al_2O_3, which is around 450–550 °C, and increases the selectivity of ammonia. NO_x reduction conversion by $LaCo_{0.9}Pd_{0.1}O_3$-Ba/Al_2O_3 catalyst rose from 72.9% to around 100% at H_2/NO ratio equal to 1, and production of NH_3 increased from 20.1% to 58.6% at H_2/NO ratio equal to 3.

(f) **Spray drying method**:

Spray drying was recognized as a versatile industrial technique categorized as a bottom-up approach and employed to synthesize perovskite materials rapidly.[52] In this process, a homogeneous precursor solution was prepared in a solvent such as dimethylformamide. It was subsequently transferred into an atomizer, which was converted into fine droplets (10–100 μm in diameter) using a nozzle or nebulizer. These droplets were then introduced into a hot drying chamber (typically 100–200 °C), where the solvent evaporated within a few seconds, forming solid particles.

The resulting powders were collected either at the bottom of the chamber or through a cyclone separator, while the evaporated solvent was removed with the exhaust gas (Fig. 5.6b. This technique was suitable even for heat-sensitive materials due to the short residence time of 2–10 seconds. This approach successfully synthesized catalyst powders such as $LaFeO_3$ and NiO/Al_2O_3 battery electrode materials, and ceramic oxides [reference]. However, the process was often accompanied by the formation of secondary phases and particle size and morphology variations. These limitations were mitigated by employing more stable precursor salts and by lowering the drying temperature, which allowed a slower evaporation process and produced perovskite powders with reduced crystallinity. Owing to these advantages, spray drying was extensively utilized for the preparation of $CaMnO_3$ structures in research applications [5, 11, 48].

(g) **Hot injection method**

The hot injection method produced monodisperse nanocrystals with controlled morphology, including LFP nanocrystals. After the publication of the hot-injection method by the Kovalenko group, this is one of the most used methods for synthesizing all-inorganic halide perovskites.[85] For example, LFP nanocrystals were synthesized by Jellicoe et al. through the reaction of SnX_2 (where X = Cl, $Cl_{0.5}Br_{0.5}$, Br, $Br_{0.5}I_{0.5}$ and I) [41]. SnX_2 was dissolved in of tri-n-octylphosphine to prevent precipitation, then injected into a mixture of Cs_2CO_3 with oleyl-amine and oleic acid at 170 °C. $CsSnCl_3$ nanocrystals were obtained with a cubic crystal structure (space group Pm3m), whereas $CsSnBr_3$, and $CsSnI_3$ nanocrystals were found to crystallize in orthorhombic structures (space group Pnma). The optical bandgap was tuned across the visible and near-infrared regions by employing different halides in the perovskite lattice as well as by exploiting quantum confinement effects. Bandgap values of 2.8–1.3 eV were achieved for $CsSnCl_3$, $CsSnBr_3$, and $CsSnI_3$ nanocrystals. Furthermore, two different approaches were highlighted to produce mixed-halide perovskite nanostructures: (i) direct synthesis using mixed-halide precursors, and (ii) post-synthetic anion exchange, in which as-synthesized pure-halide perovskite nanocrystals were subjected to halide exchange. The temporal and spectral evolution of the photoluminescence (PL) of Sn-based perovskite nanocrystals was resolved, revealing two luminescent decay channels. These were assigned to a fast band-to-band emission and a slow radiative recombination process at shallow intrinsic defect sites. Wu et al. synthesized a lead-free halide perovskite $Cs_2AgBiBr_6$ quantum dot via the hot injection method [119]. Photocatalytic activity of $Cs_2AgBiBr_6$ was demonstrated for the degradation of organic dyes and antibiotic degradation in ethanol under visible-light irradiation. Specifically, about 97% of rhodamine B and methyl orange were removed within 10 min and 30 min, respectively. Additionally, 60% tetracycline hydrochloride residue was degraded within 30 min, 7 times more than commercial titania (P25).

(h) **Pechini method**:

The Pechini method, first proposed by M.P. Pechini in 1967, was developed to form a polymeric network enabling the atomic-level dispersion of metal cations. The process began by dissolving metal salts or nitrates of the target elements in a suitable solvent [24]. Subsequently, chelating agents like ethylene glycol and citric acid were introduced into the precursor solution. These agents formed stable complexes with the metal ions, as outlined by Eq. 5.11. Following chelation, the poly-esterification step was initiated by combining ethylene glycol with citric acid (Eq. 5.12). Under constant stirring and progressive heating (temperature range 150–250 °C), ester bonds were established between the components. This resulted in a polyester polymeric gel wherein metal ions were uniformly distributed throughout the organic polymer network. Continued heating transformed the mixture into a viscous gel, yielding a resin-like mass. Once formed, the gel was dried or subjected to pre-decomposition steps to remove excess moisture, yielding a dry precursor. Calcination in air followed, serving to thermally break down the organic matrix and convert the metal complexes into metal oxides, as in Eq. 5.13. Gradual heating facilitated further decomposition of the chelating agents, promoting the thorough incorporation of metal ions into the matrix. Finally, the calcined gel underwent a controlled, multi-cycle heat treatment regimen at elevated temperatures. This mechanism induced thermal decomposition and complete elimination of organic content, including chelating agents and residual polymers. As a result, the desired perovskite ceramic powder was obtained, with the metal ions evenly dispersed at the atomic scale within the final product

$$M^{n+} + n(C_6H_8O_7) \rightarrow [M(C_6H_8O_7)]_n \quad (5.11)$$

$$n(C_6H_8O_7) + n(C_2H_6O_2) \rightarrow [C_{10}H_{12}O_8]_n + nH_2O \quad (5.12)$$

$$[M(C_6H_8O_7)]_n \rightarrow MO_x + CO_2 + H_2 \quad (5.13)$$

Zou et al., synthesised Shuttle-shaped perovskite oxide $SrFe_{0.8}Cu_{0.1}Nb_{0.1}O_{3\text{-}\delta}$ by the Pechini method at 1000 °C as shown in (Fig. 5.6g) [140]. Study reveals that when $SrFe_{0.8}Cu_{0.1}Nb_{0.1}O_{3\text{-}\delta}$ is added with active carbon to make a composite electrode for an ammonia fuel cell based on an alkaline membrane electrolyte and Pt/C anode. When a dilute 0.02 M ammonia solution (340 ppm) was used as fuel, the fuel cell showed a power density of 0.03mW/cm^2 and an open circuit voltage of 0.35 V, which has the potential to be used for a stationary application, such as the removal of ammonia from wastewater. For a further increase in the cell's OCV, NaOH is added to the fuel when the fuel has a low concentration of ammonia.

To provide clear understanding of the advantages and disadvantages of the various synthesis method of perovskite, a comparative summary of some standard techniques is presented in the (Table 5.3). In which the outlines the typical precursors/additives, relative cost, yield, achievable purity, particle size range, and degree of shape/size control reported in the literature for each method. Such a comparison facilitates the selection of an appropriate synthesis route based on the targeted physicochemical properties and application requirements.

Table 5.3 Comparison of synthesis methods employed for perovskite material preparation

S. No.	Synthesis method	Precursors/Additives	Cost	Yield	Purity	Particle size (nm)	Shape/size control	References
1	Co-precipitation	Oxides, nitrate salts, acrylamide, ammonia/ammonium carbonate, EDTA, polyvinylpyrrolidone	Low	High	High	50–500	Poor	[38]
2	Sol–gel	Oxides, NaOH, nitrate salts, HNO3, ethylene glycol, citric acid	High	High	Excellent	20–150	Good	[34]
3	hydrothermal	Nitrate salts, polyethylene glycol, NaOH/KOH, acetone, citric acid	Medium	High	Very high	40–80	Good	[78]
4	Solid-state reaction	Metal oxides, ethanol/acetone	High	High	Poor	50–200	Good	[38]
5	Ball milling	Metal oxides, alumina/the valuables	Low	Medium	Medium	100–500	Poor	[3]
6	Spray drying	Perovskite composition, solvent, air	High	High	High	4–200	Good	[102]
7	Solvothermal	oleic acid, ethanol, butanol, metal precursors	Medium	High	High	10–100	Excellent	[21]
8	Wet Impregnation	Metal nitrates/chlorides, water, alcohol	Low	Medium	Medium	50–300	Poor	[68]
9	Pechini	Nitrate salts, ethyl glycol, citric acid	Medium	High	Excellent	20–100	Good	[140]

(continued)

Table 5.3 (continued)

S. No.	Synthesis method	Precursors/Additives	Cost	Yield	Purity	Particle size (nm)	Shape/size control	References
10	Hot-injection method	Caesium salts, ethyl glycol,	Medium	High	Excellent	5–100	Good	[119]
11	Laser ablation and Pulsed laser	Metal precursor salt	Very high	High	Excellent	2–20	Excellent	[36]

5.4 Photocatalytic Applications

Usually, photocatalysts or semiconductors are considered intermediate between conductors and insulators. Due to a suitable band gap, semiconductors can be widely employed in photocatalysis or harvesting renewable solar energy. UV or visible light incident on the surface of a semiconductor eventually generates photoexcited electrons that gain energy from photons (Fig. 5.7a). However, from the perspective of solar utilization, electrons in the valence band (VB) could transition to the conduction band (CB). In this process, excited electrons were promoted to the CB, leaving behind vacancies in the VB, which were called holes. These photogenerated electron–hole pairs initiated simultaneous reduction reactions, i.e., gain of electron and oxidation reaction on the photocatalyst's surface [4, 39]. Photogenerated electron–hole pairs were utilized in diverse applications (Fig. 5.7b), depending on the redox potential of the semiconductor system.

Selecting perovskite materials for specific applications was crucial for achieving improved performance [75]. In photocatalytic systems, the materials were selected by balancing the band edge potentials of the semiconductors with the corresponding reaction potentials. However, their practical utilization was hindered by certain intrinsic limitations. The major obstacles were identified as the valuable recombination lifetime of the photoinduced electron–hole pairs and the mismatch between the absorption wavelength of the photocatalyst and the solar spectrum. Since most perovskite-based photocatalysts were activated predominantly under UV light, their efficiency was limited, as the UV portion of the solar spectrum accounted for only ~ 5%. Therefore, to achieve enhanced photocatalytic activity, the material band gap must be tuned toward visible and near-infrared (NIR) regions while simultaneously prolonging the recombination lifetime of charge carriers [84].

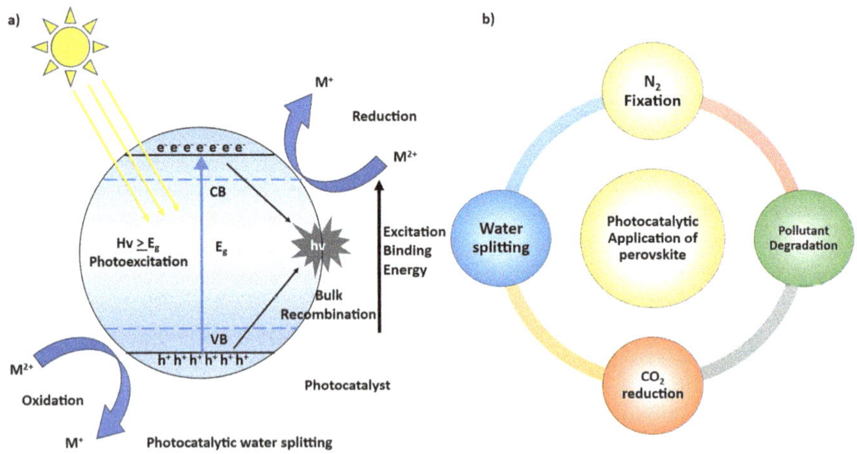

Fig. 5.7 **(a)** Detailed illustration of the mechanism of photocatalysis; **(b)** illustration of the photocatalytic application of the perovskite

The photogenerated electron–hole pairs were also required to possess a favorable band alignment to drive redox reactions effectively. Specifically, the valence band maximum (VBM) had to be more positive than the oxidation potential, while the conduction band minimum (CBM) had to be more negative than the reduction potential. Consequently, a sufficiently large band gap was necessary for photocatalysis. However, from the perspective of solar utilization, a huge band gap was undesirable, as it would limit absorption of low-energy photons. By balancing these considerations, the ideal band gap for photocatalytic semiconductors was estimated to lie within the 1.8–2.8 eV range [137]. A schematic diagram illustrates lead-free perovskite materials' band gaps and energy level alignment (CBM and VBM). Next, the photocatalytic applications of lead-free perovskites are discussed.

5.4.1 Water Splitting

The demand for clean, sustainable, and environmentally friendly fuels has intensified globally. Hydrogen (H_2) emerges as an alternative fuel because of its high gravimetric energy density (120 MJ/Kg) and absence of carbon content. Table 5.4 indicates LFP catalysts utilised for photocatalytic HER production. As shown in Eq. 5.14. The energy required to break the H–O–H bonds is supplied from light (electromagnetic radiation) to initiate the chemical reaction.

$$2H_2O \rightarrow 2H_2 + O_2 \Delta G^o = +237 \text{ kJ mol}^{-1}, E^o = 1.23\ V \tag{5.14}$$

Water splitting processes are shown in Fig. 5.8. Herein, the holes generated in the valence band act as an oxidizing agent and are responsible for the oxidation of water and the release of oxygen, and the electrons from the conduction band act as reducing agents to produce hydrogen from in-contact water molecules.

Apart from the HER using LFP, it is also explored for the photocatalytic CO_2 reduction reaction, as discussed in the next section.

5.4.2 CO_2 Reduction

Photocatalytic CO_2 conversion to valuable chemicals such as CO, CH_4, etc., generally called solar fuel, offers a promising sustainable solution for fulfilling the energy crisis [82]. Fundamentally, the key factor is dictated by both thermodynamic and kinetic constraints. Thermodynamically, the position of the photocatalyst's conduction band must be sufficiently negative relative to the $CO_2/CO_2\bullet^-$ redox potential to enable electron transfer and initiate the reduction process. The bandgap energy and light absorption capacity thus play a central role in determining whether the photogenerated charge carriers possess adequate driving force. Kinetically, the overall efficiency and selectivity of CO_2 conversion depend on the number of electrons and protons (H^+

Table 5.4 Lead free Perovskite for photocatalytic performance of hydrogen evolution

S. No.	Photocatalyst	Light source	H_2 evolution rate (μmol g^{-1} h^{-1})	Medium	References
1	PtSA/Cs_2SnI_6	100 mW cm^{-2} $\lambda \geq 420$ nm	430	HI	[137]
2	$Cs_2AgBiBr_6$-RGO	300 W Xe Lamp $\lambda > 420$ nm	489	–	[113]
3	$Cs_2AgBiBr_6$-NC140	300 W Xe lamp $\lambda > 420$ nm	380	HBr	[44]
4	$DMASnBr_3$-gC_3N_4	500 Wm^{-2}	> 1700	TEOA aqueous solution	[89]
5	Cs_3BiBr_9/gC_3N_4	500 Wm^{-2}	1050	TEOA aqueous solution	[90]
6	$Cs_3Bi_2I_9$	–	1504.5	HI/ethyl acetate	[56]
7	$Cs_3Bi_{0.6}Sb_{1.4}I_9$	$\lambda \geq 420$ nm	157.2	Aqueous HI	[18]
8	$Cs_2AgBiBr_6$/Pt	Visible light irradiation ($\lambda \geq$ 420 nm)	54.7	Aqueous HBr	[95]
9	$[(CH_3)_2NH_2]_3/PtI_x$	Visible light irradiation at 465 nm	47	Aqueous HI	[133]
10	$(CH_3NH_3)_3Bi_2I_9$/Pt	300 W Xe-lamp $\lambda > 400$ nm	169.21	Aqueous HI	[31]
11	Cs_2SnI_6/PtSA	300 W Xe-lamp $\lambda > 400$ nm, 100 mW cm^{-2}	430	Aqueous HI with 20 vol% H_3PO_2 as stabilizer	[137]
12	$Cs_2AgBiBr_6$/RGO	300 W Xe-lamp $\lambda \geq 420$ nm	48.9	Aqueous HBr	[113]
13	$MA_3Bi_2I_9$/DMA_3BiI_6	300 W Xenon lamp $\lambda \geq 420$ nm	198.2	Aqueous HI	[105]
14	PEA_2SnBr_4/g-C_3N_4	Solar light 500 W m^{-2}	1613	H_2O/ TEOA	[88]
15	$DMASnBr_3$@gC_3N_4/ Pt	Xenon lamp, 500 W m^{-2} $\lambda \geq$ 300 nm	1730	H_2O/ TEOA	[89]
16	$Cs_2Pt_{0.05}Sn_{0.95}Cl_6$	300 W Xe lamp $\lambda \geq 460$ nm	16.11	H_2O/ TEOA	[124]

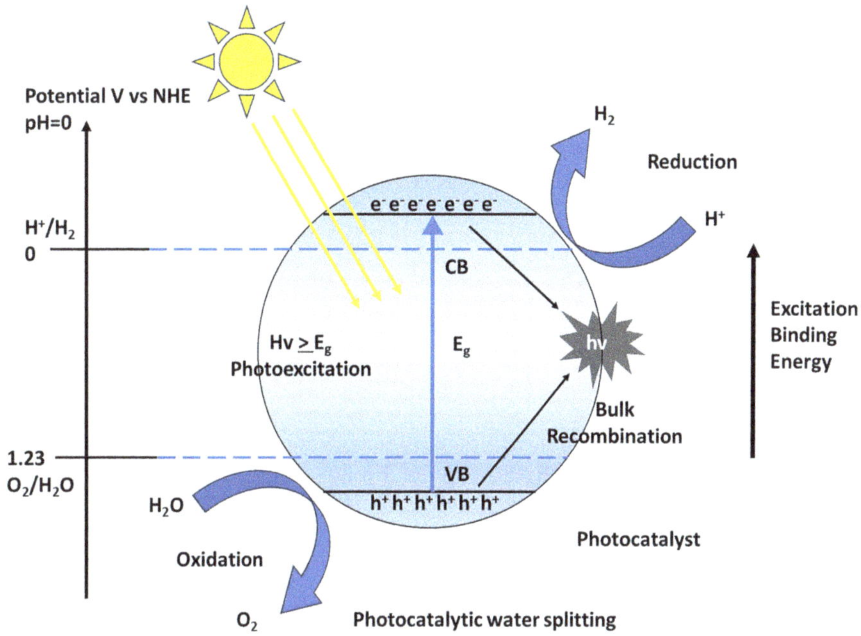

Fig. 5.8 Illustrates the photocatalytic water splitting mechanism

ions) involved in the reduction pathway, which dictates the formation rate of products such as CO, CH_4, CH_3OH, or HCOOH. Furthermore, the separation and migration of photoexcited electron–hole pairs, surface adsorption of reactants, and stabilization of key intermediates strongly influence the reaction kinetics and product distribution. Therefore, the interplay between these thermodynamic and kinetic parameters is critical for designing photocatalysts capable of selectively steering multi-electron, multi-proton reduction pathways with high efficiency under solar illumination. The utilization of various LFP photocatalysts is summarised in Table 5.5.

5.4.3 *Ammonia Production*

Photocatalytic N_2 fixation to NH_3 using water as a proton source has recently emerged as an important sustainable approach. Achieving N_2 fixation under mild conditions remains a significant challenge in catalysis, as the conventional Haber–Bosch process is highly energy intensive, requiring elevated temperatures and pressures and contributing substantially to global CO_2 emissions [81]. Photocatalytic N_2 reduction provides a promising alternative by utilizing solar energy to drive the reaction under near-ambient temperature and pressure. Recent advances in photocatalytic supplying electrons for N_2 activation have demonstrated the feasibility of converting N_2 into NH_3 under simulated natural conditions. Nevertheless, a major

Table 5.5 Summary on Photocatalytic CO_2 Reduction Performance using LFP photocatalyst

S. No.	Photocatalyst	Product and activity (μmol g^{-1} h^{-1})	Strategy	Stability (H)	Reaction environment	Morphology	References
1	$Cs_3Bi_2I_9$	CO (7.76), CH_4 (1.49)	–	> 10	CO_2/H_2O vapor	Nanocrystal	[10]
2	$Cs_3Bi_2Br_9$	CO (26.95)	–	> 20	CO_2/H_2O vapor	Quantum dot	[93]
3	Cs_2SnI_6/SnS_2	CH_4 (2.03)	Heterojunction construction	> 9	CO_2/H_2O vapor/ CH_3OH	Nanocrystal	[115]
4	$CsSnCl_3/ZnSe$	CO (53.96), CH_4 (2.55)	Heterojunction construction	–	Toluene/ isopropanol	Nanocrystal	[57]
5	$Cs_2AgBiBr_6$/ Bi_2WO_6	CO (42.19), CH_4 (0.41)	Heterojunction construction	> 24	Ethyl acetate/ isopropanol	Nanocrystal	[120]
6	$Cs_2AgBiBr_6$/ Sr_2FeNbO_6	CO (50.0), CH_4 (8.12)	Heterojunction construction	> 20	Ethyl acetate/H_2O	Nanocrystal	[69]
7	$Cs_2AgBiBr_6$/ g-C_3N_4	CH_4 (2.0)	Heterojunction construction	> 12	Ethyl acetate/ methanol	Nanoparticle	[116]
8	$Cs_3Bi_2Br_9/TiO_2$	CH_4 (24.2)	Heterojunction construction	> 35	Isopropanol	Nanodot	[100]
9	$Cs_2AgBiBr_6/TiO_2$	CH_4 (32.9)	Heterojunction construction	> 35	Isopropanol	Nanodot	[101]
10	$Cs_2AgBiBr_6$	CO (–), CH_4 (–)	Morphology regulation	> 6	Ethyl acetate	Nanoplatelet	[62]
11	$CsSb_2(Br_{0.7}I_{0.3})_9$	CO (9.23)	Morphology regulation	> 6	CO_2/H_2O vapor	Nanoplatelet	[117]
12	$Cs_3Bi_2Br_9$/ MCM-41	CO (17.24)	Morphology regulation	> 16	CO_2/H_2O vapor	Nanoparticle	[117]
13	$Cs_3Sb_2Br_9$	CO (127.5)	Surface engineering	> 4	Octadecene/H_2O	Nanocrystal	[64]

(continued)

Table 5.5 (continued)

S. No.	Photocatalyst	Product and activity (μmol g^{-1} h^{-1})	Strategy	Stability (H)	Reaction environment	Morphology	References
14	$Cs_2AgBiBr_6$	CO (2.35), CH_4 (1.6)	Surface engineering	> 6	Ethyl acetate	Nanocrystal	[136]
15	$Cs_2NaBiBr_6$	CO (30.22), CH_4 (1.12)	Surface and defect engineering	> 12	CO_2/H_2O vapor	Porous microsphere	[83]

limitation lies in the relatively low energy of photogenerated electrons, which is insufficient to cleave the N≡N triple bond directly. As a result, photocatalytic N_2 fixation generally proceeds through associative pathways, either alternating or distal, where N≡N bond cleavage occurs after stepwise hydrogenation to form N–H intermediates. Although these associative mechanisms reduce the overall energy input compared to direct dissociation, developing photocatalysts capable of activating and cleaving the N≡N bond under ambient conditions remains a key research priority, as this would significantly improve the efficiency of solar-driven ammonia synthesis [80]. Tao et al. synthesized a ternary Perovskite heterojunction nanofiber (TiO_2/$SrTiO_3$/g-C_3N_4) with gradient energy bands with cascade-driven charge separation and transfer properties [106]. The enhanced photocurrent response (about 1.4 and 1.8 times as compared to TiO_2/g-C_3N_4 and $SrTiO_3$/g-C_3N_4 nanofibers), decreased lifetime (about 0.75 and 0.79 times compared to TiO_2/g-C_3N_4 and $SrTiO_3$/g-C_3N_4 nanofibers), and their decrease in photoluminescence intensity suggested that the ternary nanofibers with the gradient energy band alignment had accomplished highly effective cascade charge transfer and separation. Under the Simulated sunlight irradiation, the photocatalytic reduction reaction rate for H_2 and N_2 is 1304 and 2192, which are 2.1 and 1.9 times higher as compared to TiO_2/large amounts of sludge being generated$SrTiO_3$/g-C_3N_4. The orthorhombic phase of $KNbO_3$ was used as a perovskite for the photoreduction of N_2 to NH_3. However, this material provides less surface area and low ammonia production efficiency under UV light irradiation. So, to overcome this problem, M. Chamack et al. used a metal–organic framework (MOF), named as TMU-5 ($[Zn(OBA)(BPDH)_{0.5}]_{n.}$1.5DME, where BPDH = 2,5-bis(4-pyridyl)-3,4-diaza-2,4-hexadiene) and H_2OBA = 4,4′-oxybis- (benzoic acid) for the synthesis of the $KNbO_3$@TMU-5 hybrid material [16]. The ammonia production rate of $KNbO_3$@TMU-5 is 39.9 μmol L^{-1} h^{-1} g^{-1} as compared to the KNbO3 production rate of 20.5 μmol L^{-1} h^{-1} g^{-1} upon UV light irradiation. Furthermore, an increase in the photoactivity of $KNbO_3$@TMU-5 in NH_3 production is observed because of the high surface area, high density of negative charges on Nb sites, and high electron–hole separation efficiency as compared to $KNbO_3$. Further, various LFP photocatalysts are shown in Table 5.6.

5.4.4 *Pollutant Degradation*

With the advancement of modern industry, organic chemical pollution was recognized as a significant environmental challenge, since various organic pollutants were toxic and harmful to ecosystems. Among them, organic dyes discharged into wastewater from textile and related industrial processes were identified as one of the main contributors. These dyes in water were reported to impede plant growth, impair photosynthesis, and disturb the food chain through their accumulation in soil. In humans, dye-contaminated water causes genetic problems, cancer, infertility, immune suppression, and disruption of glandular functions. Conventional methods employed to remove textile dyes, such as adsorption, often generated large amounts of

Table 5.6 LFP photocatalyst for photocatalytic N_2 fixation reaction

S. No.	Photocatalysts	Incident light	Band gap (eV)	Sacrificial agent	NH_3 produced (μmol g^{-1} h^{-1})	References
1	$MCeO_2$-$BiFeO_3$	UV–vis	–	None	117.77	[72]
2	TiO_2/$SrTiO_3$/ g-C_3N_4	Simulated solar light	2.75–3.1	Ethanol	2192	[106]
3	Ag-$KNbO_3$ (0.5% Ag)	Simulated solar light	3.13	Ethanol	385	[122]
4	$LaCoO_3$:Er^{3+}/ATP	Visible	2.88–3.45	Ethanol	71.51	[126]
5	Bi-Bi_2WO_6	Simulated solar light	–	None	86	[138]
6	c-PAN-Bi_2WO_6	Simulated solar light	–	None	160	[125]
7	$BaTiO_3$	UV–visible	3.2		1.93 mg L^{-1} h^{-1}	[134]
8	Defective La_2TiO_5 (R-LTO)	Simulated solar light	4.07	None	158.13	[99]
9	$LaCoO_3$: $Re^{3+/}$ATP	Visible	2.88–3.45		71.51	[126]
10	NiS/$KNbO_3$ (5% NiS) (Air as nitrogen source)	Simulated solar light	3.11	Ethanol	155.6	[128]
11	NiS/$KNbO_3$ (N_2 as nitrogen source)	Simulated solar light	3.11	Ethanol	218	[128]
12	$CaTiO_3$	Natural sunlight irradiation	3.49	Methanol	236.12	[51]
13	$NiSnO_3$-g-C_3H_4	500 W halogen lamp	–	Methanol	566	[32]
14	V_o-Bi_2MoO_6-OH	300 W Xe lamp ($\lambda \geq$ 420 nm)	2.55	Methanol	795	[53]
15	Cu-doped Bi_2MoO_6	300 W Xe lamp	2.41–2.50	Methanol	302 μmol L^{-1} g^{-1} h^{-1}	[111]
16	Co-doped Bi_2MoO_6	300 W Xe lamp ($\lambda >$ 420 nm)	2.59–2.65	None	130.07	[63]
17	Bi/$ZnSnO_3$ Composite	300 W Xe lamp	–	Methanol	988 μmol L^{-1} g^{-1} h^{-1}	[132]
18	ZnO/$ZnSnO_3$/ Carbon dots	500 W Xe lamp	–	Ethanol	4600	[77]

sludge. This sludge was associated with secondary environmental issues, including bacterial infections, skin problems, and the release of secondary pollutants upon recycling of adsorbents, which also required harsh conditions. In contrast, photocatalytic degradation of these contaminants was considered more reliable and efficient, providing an energy-saving approach to eliminate organic dyes from wastewater without generating harmful byproducts. Therefore, developing efficient perovskite-based photocatalysts was extensively pursued as a sustainable strategy for organic dye degradation [13, 39, 91, 96].

Photocatalysts were reported to degrade and mineralize organic pollutants through a series of reactions known as advanced oxidation processes (AOPs). In this process, reactive oxygen species (ROS), such as hydroxyl radicals (OH^*) and superoxide radicals (O_2-), were generated as shown by Eq. 5.15, which subsequently oxidized and mineralized the pollutants. When photocatalysts absorbed light of an appropriate wavelength, charge carriers (electron–hole pairs) were generated and separated. The photogenerated holes in the valence band (VB) participated in the formation of hydroxyl radicals, indicated by Eq. 5.16. In contrast, the electrons in the conduction band (CB) facilitated the generation of superoxide radicals through secondary reactions, indicated by the equation. These ROS species were then involved in the degradation and complete mineralization of organic contaminants. Figure 5.9 shows the schematic illustration of the pollutant degradation (Table 5.7).

$$h^+ + OH^- \rightarrow OH^* \tag{5.15}$$

$$O_2 + e^- \rightarrow O_2^- \tag{5.16}$$

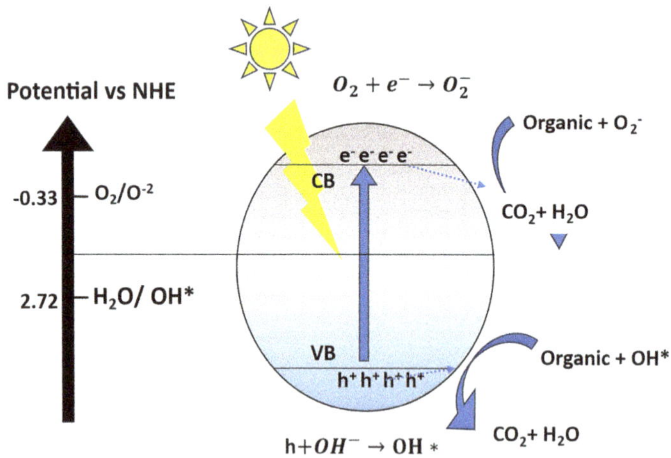

Fig. 5.9 Schematic illustration of pollutant degradation

Table 5.7 Comparison of different perovskite-based photocatalysts for the degradation of organic pollutants

S. No.	Photocatalysts	Contaminants	Morphology	Band gap (eV)	Light source	Time (min)	Degradation rate (%)	References
1	$Cs_3Bi_2I_9$	RhB	Hexagonal particles,3D grains, nano-powder	1.95	Visible light irradiation	180	93	[14]
2	$Cs_2AgBiBr_6$	RhB	Crystalline powder	2.06		120	98	[131]
3	$Cs_2AgInCl_6$	RhB	Microparticle	3.33	300 W Mercury lamp	140	90.8	[55]
4	$CsPbBr_3$	MO	Quantum dots	~ 2.3		100	89	[28]
5	$CsPbCl_3$	MO	Quantum dots	~ 3.0		100	90	[28]
6	MNSCB30	RhB	Nanocrystalline powder	3.3	300 W Xenon lamp λ > 420 nm	60	97	[3]
7	$Cs_3Bi_2Br_9$	MD	Composite	1.88	visible-light irradiation	90/150	80/100	[74]
8	12.4% $Cs_3Bi_2Br_9/TiO_2$	MBT	Composite	–	300 W Xenon lamp λ > 420 nm	10/20	74.7%/99.9%	[100]
9	$CaTiO_3$	MO	Bare	~ 3	UV light	60	54%	[25]
10	$CaTiO_3$-Graphene	MO	Composites	~ 3	UV light	60	98%	[25]
11	C-doped $SrTiO_3$	MB, Mo, RhB, phenol, and BPA	Cubic particle, nanorod, nanotube	< 3.2	Visible light	180	95% of MB, Mo, RhB, and 70% of phenol and BPA	[17]
12	S-$SrTiO_3$	2-propanol	Powder	< 3.2	500 W-Xenon lamp	60		[79]

(continued)

Table 5.7 (continued)

S. No.	Photocatalysts	Contaminants	Morphology	Band gap (eV)	Light source	Time (min)	Degradation rate (%)	References
13	Cu-doped $SrTiO_3$	Methyl Blue	Nanoparticles	2.96	Visible light	120	66%	[86]
14	Fe-doped $SrTiO_3$	Tetra-cyclin TC	–	2.6	Visible light	80	71.6%	[1]
15	Mn-doped $SrTiO_3$	Tetra-cyclin TC	Nano-cubes	2.76	Visible light	60	66.7%	[118]
16	N-doped $NaTaO_3$	MB, MO	Cubic	2.48	UV–visible light	60	95.%	[61]
17	$Ag/AgGaO_2$	MB	Composite	–	Visible light	180	95%	[129]
18	$LaFeO_3$	MB	Nanoparticles	2.36	Visible light	60	100%	[104]
19	Z-scheme $MoS_2/CaTiO_3$	TC	Nanospheres	3.23	Simulated solar light	60	70%	[43]
20	p-n type (30–60% Ag_3PO_4)/$NaTaO_3$	RhB	Crystalline	2.32–3.78	Visible light	25	87%	[114]
21	$BiFeO_3/BiVO_4$	RhB	Nanocomposites	2.23	Visible light	120	69%	[98]
22	$In_2S_3/NaTaO_3$	TC	Composite	2.1–4.0	Stimulated solar irradiation	180	53.2% for 20 wt% $In_2S_3/NaTaO_3$	[123]
23	10 wt% $LaFeO_3/SnS_2$	TC	Composite Z-scheme	2.11	Visible light	120	28.8%	[67]
24	1.7 wt% Ag-$KNbO_3$	RhB	Nanowires	2.2–3.35	UV–Visible	90 min for UV and 120 min for 120	95% with UV and 65% with VIS	[127]
25	7 wt% Ni-doped $BiFeO_3$	MB	Nanoparticle	2.28	Visible light	60	92%	[9]

(continued)

Table 5.7 (continued)

S. No.	Photocatalysts	Contaminants	Morphology	Band gap (eV)	Light source	Time (min)	Degradation rate (%)	References
26	$LaNiO_3$	MO	–	2.26	Visible light	300	74.5%	[59]
27	$NaTaO_3$/rGo	MB	Composite	3.87	8 W UV lamp	90	95%	[27]
28	Sm-doped Bi_2WO_6	RhB	–	2.4–2.5	Visible light	30	98.4%	[130]
29	50 wt% $BiFeO_3$/V_2O_5	MB	Nanoplates	2.05–2.19	Visible light	120	96%	[40]
30	CuS/Bi_2WO_6	RhB	Composite	1.76–2.69	Visible light	50	90%	[130]

RhB—Rhodamine B ($C_{28}H_{31}ClN_2O_3$), MO—Methyl Orange ($C_{14}H_{14}N_3NaO_3S$), MBT—2-mercaptobenzothiazole

Shi et al. synthesized a $BiFeO_3/ZnO$ heterostructure piezo-photocatalyst via a two-step hydrothermal method to construct a $BiFeO_3/ZnO$ heterojunction by decorating ZnO nanoparticles on the surface of $BiFeO_3$ microcubes for the degradation of tetracycline hydrochloride (TCH), exposed under integrated visible-light and ultrasonic vibration [94]. The $BiFeO_3$/35% ZnO catalyst exhibited the highest piezo-photocatalytic removal rate, 0.0142 min^{-1}, which was 6.3 times greater than that of the pristine $BiFeO_3$ removal rate (0.00224 min^{-1}). This enhancement in activity was primarily attributed to the piezoelectric characteristics of both $BiFeO_3$ and ZnO. Their coupling generated an internal polarization field within the heterojunction, facilitating charge carrier migration and reducing recombination of photoinduced electron–hole pairs. Moreover, the S-scheme charge transfer pathway ensured the selective preservation of highly energetic electrons and holes, thereby strengthening the system's redox capability.

5.5 Conclusion and Outlook

Developing lead-free perovskites (LFPs) for photocatalytic applications is a rapidly advancing field, and several research directions are critical for achieving their full potential.

(a) **Bandgap Optimization for Visible-Light Utilization**: A primary focus should be on tuning the bandgap of LFPs, including Sn-, Bi-, Sb-, and double perovskite systems, to maximize their response to the visible-light spectrum. Approaches such as halide alloying (Cl/Br/I mixing), heterovalent cation substitution, and controlled defect engineering can enhance photon absorption while suppressing charge carrier recombination.

(b) **Surface and Interface Engineering**: Since surface states and interfacial charge transfer strongly influence photocatalytic activity, future efforts should prioritize defect passivation, co-catalyst integration (e.g., transition-metal nanoparticles, carbon-based materials), and heterojunction construction with complementary semiconductors to facilitate efficient charge separation and improve catalytic selectivity.

(c) **Improving Stability under Aqueous and Illumination Conditions**: One principal obstacle is halide perovskite degradation in aqueous environments and under prolonged light exposure. The design of intrinsically water-stable LFPs and the use of protective surface coatings and core–shell architectures will be essential to extend the operational lifetimes of photocatalysts for key reactions such as hydrogen evolution and CO_2 reduction.

(d) **Design of Double Perovskites and Mixed-Metal Frameworks**: Chemically robust materials such as $Cs_2AgBiBr_6$ and $Cs_2AgInCl_6$ provide a promising starting point but suffer from wide band gaps that limit visible-light absorption. Targeted cation or anion substitutions at the A-, B-, B'-, or X-sites can reduce bandgap energies and optimize redox potentials for specific photocatalytic processes.

(e) **Mechanistic Insights through In Situ Characterization**: Advanced characterization techniques, including transient absorption spectroscopy, operando X-ray photoelectron spectroscopy (XPS), and time-resolved photoluminescence, are indispensable tools for probing charge carrier dynamics, reaction intermediates, and degradation pathways, ultimately guiding rational material and device design.

(f) **Scalable and Sustainable Synthesis**: Finally, developing environmentally friendly, scalable synthetic approaches is necessary for future deployment. Low-temperature, solvent-minimized techniques such as mechanochemical synthesis and aqueous precursor routes can reduce environmental impact while enabling large-scale production of high-quality LFP photocatalysts.

Lead-free perovskites hold tremendous promise as photocatalysts due to their tunable band gaps, defect-tolerant charge transport, and potential for cost-effective synthesis. Despite these advantages, several fundamental challenges hinder their widespread use, including limited aqueous stability, suboptimal band alignment for redox reactions, and insufficient charge carrier lifetimes. Recent progress in compositional engineering, defect passivation, and heterojunction design has demonstrated that it can enhance photocatalytic efficiency while maintaining structural stability significantly. Integrating co-catalysts, protective layers, and novel synthetic techniques has yielded promising results for photocatalytic reactions. With continued advances in bandgap tuning, stability enhancement, mechanistic understanding, and scalable synthesis, LFPs are poised to emerge as a versatile and sustainable material platform for next-generation photocatalytic applications. Their development will enable efficient solar-to-chemical energy conversion and address the global demand for clean fuels and green chemical synthesis.

References

1. M. Abdi, V. Mahdikhah, S. Sheibani, Visible light photocatalytic performance of La-Fe co-doped SrTiO3 perovskite powder. Opt. Mater. **102**, 109803 (2020). https://doi.org/10.1016/j.optmat.2020.109803
2. N. Abid, A.M. Khan, S. Shujait, K. Chaudhary, M. Ikram, M. Imran, M. Maqbool et al., Synthesis of nanomaterials using various top-down and bottom-up approaches, influencing factors, advantages, and disadvantages: a review. Adv. Coll. Interface. Sci. **300**, 102597 (2022). https://doi.org/10.1016/j.cis.2021.102597
3. X. An, X. Guo, S. Zhu, B. Fan, J. Zhu, G. Shao, H. Lu et al., High-entropy lead-free perovskite nanoparticles as efficient photocatalysis with a wide visible-light absorption band. J. Alloy. Compd. **1036**, 181680 (2025). https://doi.org/10.1016/j.jallcom.2025.181680
4. H. Anwer, A. Mahmood, J. Lee, K.-H. Kim, J.-W. Park, A.C.K. Yip, Photocatalysts for degradation of dyes in industrial effluents: opportunities and challenges. Nano Res. **12**(5), 955–972 (2019). https://doi.org/10.1007/s12274-019-2287-0

5. E.A.R. Assirey, Perovskite synthesis, properties and their related biochemical and industrial application. Saudi Pharm. J. **27**(6), 817–829 (2019). https://doi.org/10.1016/j.jsps.2019.05.003
6. C.J. Bartel, C. Sutton, B.R. Goldsmith, R. Ouyang, C.B. Musgrave, L.M. Ghiringhelli, M. Scheffler, New tolerance factor to predict the stability of perovskite oxides and halides. Sci. Adv. **5**(2), eaav0693 (2019). https://doi.org/10.1126/sciadv.aav0693
7. A. Benali, S. Azizi, M. Bejar, E. Dhahri, M.F.P. Graça, Structural, electrical and ethanol sensing properties of double-doping $LaFeO_3$ perovskite oxides. Ceram. Int. **40**(9), 14367–14373 (2014). https://doi.org/10.1016/j.ceramint.2014.06.029
8. L.M. Benhafsa, Y. Azzaz, H.B. Abdallah, I.K. Bensafa, W. Ouerghui, K. Bencherif, F.Z.Z. Bellahcene, Exploring linear/nonlinear optical responses and photocatalytic efficiency of halide perovskites BaAgX3 (X = I, Br, and cl): toward sustainable energy applications. Comput. Theor. Chem. **1253**, 115416 (2025). https://doi.org/10.1016/j.comptc.2025.115416
9. L.G. Betancourt-Cantera, A.M. Bolarín-Miró, C.A. Cortés-Escobedo, L.E. Hernández-Cruz, F. Sánchez-De Jesús, Structural transitions and multiferroic properties of high Ni-doped $BiFeO_3$. J. Magn. Magn. Mater. **456**, 381–389 (2018). https://doi.org/10.1016/j.jmmm.2018.02.065
10. S.S. Bhosale, A.K. Kharade, E. Jokar, A. Fathi, S. Chang, E.W.-G. Diau, Mechanism of photocatalytic CO_2 reduction by bismuth-based perovskite nanocrystals at the gas-solid interface. J. Am. Chem. Soc. **141**(51), 20434–20442 (2019). https://doi.org/10.1021/jacs.9b11089
11. C.G. Bischak, C.L. Hetherington, H. Wu, S. Aloni, D.F. Ogletree, D.T. Limmer, N.S. Ginsberg, Origin of reversible photoinduced phase separation in hybrid perovskites. Nano Lett. **17**(2), 1028–1033 (2017). https://doi.org/10.1021/acs.nanolett.6b04453
12. D. Bokov, A. Turki Jalil, S. Chupradit, W. Suksatan, M. Javed Ansari, I.H. Shewael, E. Kianfar et al., Nanomaterial by Sol-Gel method: synthesis and application. Adv. Mater. Sci. Eng. **2021**(1), 5102014 (2021). https://doi.org/10.1155/2021/5102014
13. O. Bozorg-Haddad, (ed.), *Water Resources: Future Perspectives, Challenges, Concepts and Necessities* (WA Publishing, 2021). https://doi.org/10.2166/9781789062144
14. B.-M. Bresolin, C. Günnemann, D.W. Bahnemann, M. Sillanpää, Pb-Free Cs3Bi2I9 perovskite as a visible-light-active photocatalyst for organic pollutant degradation. Nanomaterials **10**(4), 763 (2020). https://doi.org/10.3390/nano10040763
15. M. Cernea, F. Vasiliu, C. Plapcianu, C. Bartha, I. Mercioniu, I. Pasuk, L. Pintilie et al., Preparation by sol–gel and solid state reaction methods and properties investigation of double perovskite Sr2FeMoO6. J. Eur. Ceram. Soc. **33**(13–14), 2483–2490 (2013). https://doi.org/10.1016/j.jeurceramsoc.2013.03.026
16. M. Chamack, M. Ifires, S.A. Akbar Razavi, A. Morsali, A. Addad, A. Larimi, R. Boukherroub et al., Photocatalytic performance of perovskite and metal-organic framework hybrid material for the reduction of N_2 to Ammonia. Inorg. Chem. **61**(3), 1735–1744 (2022). https://doi.org/10.1021/acs.inorgchem.1c03622
17. C.-W. Chang, C. Hu, Graphene oxide-derived carbon-doped SrTiO3 for highly efficient photocatalytic degradation of organic pollutants under visible light irradiation. Chem. Eng. J. **383**, 123116 (2020). https://doi.org/10.1016/j.cej.2019.123116
18. G. Chen, P. Wang, Y. Wu, Q. Zhang, Q. Wu, Z. Wang, B. Huang, et al., Lead-free halide perovskite $Cs_3\ Bi_{2x}\ Sb_{2-2x}\ I_9$ ($x \approx 0.3$) possessing the photocatalytic activity for hydrogen evolution comparable to that of $(CH_3\ NH_3)PbI_3$. Adv. Mater. **32**(39), 2001344 (2020). https://doi.org/10.1002/adma.202001344
19. Z. Cheng, J. Lin, Layered organic–inorganic hybrid perovskites: Structure, optical properties, film preparation, patterning and templating engineering. CrystEngComm **12**(10), 2646 (2010). https://doi.org/10.1039/c001929a
20. B. Conings, L. Baeten, C. De Dobbelaere, J. D'Haen, J. Manca, H. Boyen, Perovskite-based hybrid solar cells exceeding 10% efficiency with high reproducibility using a thin film sandwich approach. Adv. Mater. **26**(13), 2041–2046 (2014). https://doi.org/10.1002/adma.201304803

21. E. Cui, G. Hou, Z. Wang, Q. Sun, Y. Wu, J. Wan, Y. Wu, et al., Nanoscale SrFe0.5Ta0.5O3 double perovskite photocatalyst: Low-temperature solvothermal synthesis and photocatalytic NO oxidation performances. Appl. Surface Sci. **531**, 147324 (2020). https://doi.org/10.1016/j.apsusc.2020.147324
22. T. Dawa, B. Sajjadi, Exploring the potential of perovskite structures for chemical looping technology: a state-of-the-art review. Fuel Process. Technol. **253**, 108022 (2024). https://doi.org/10.1016/j.fuproc.2023.108022
23. M.A. De Vries, A.C. Mclaughlin, J.-W.G. Bos, Valence bond glass on an fcc lattice in the double perovskite Ba 2 YMoO 6. Phys. Rev. Lett. **104**(17), 177202 (2010). https://doi.org/10.1103/PhysRevLett.104.177202
24. L. Dimesso, Pechini Processes: An alternate approach of the sol–gel method, preparation, properties, and applications, in *Handbook of Sol-Gel Science and Technology* ed. by L. Klein, M. Aparicio, A. Jitianu (Springer International Publishing, Cham, 2016), pp. 1–22. https://doi.org/10.1007/978-3-319-19454-7_123-1
25. L. Ernawati, R.A. Wahyuono, H. Widiyandari, D.D. Risanti, A.W. Yusariarta, Rebeka, Sitompul, V. Experimental data of CaTiO3 photocatalyst for degradation of organic pollutants (Brilliant green dye)—green synthesis, characterization and kinetic study. Data Brief **32**, 106099 (2020). https://doi.org/10.1016/j.dib.2020.106099
26. P. Fu, Q. Shan, Y. Shang, J. Song, H. Zeng, Z. Ning, J. Gong, Perovskite nanocrystals: synthesis, properties and applications. Sci. Bull. **62**(5), 369–380 (2017). https://doi.org/10.1016/j.scib.2017.01.006
27. Z. Fu, S. Zhang, Z. Fu, Hydrothermal preparation of NaTaO3/rGO composite photocatalyst to enhance UV photocatalytic activity. Results Phys. **15**, 102669 (2019). https://doi.org/10.1016/j.rinp.2019.102669
28. G. Gao, Q. Xi, H. Zhou, Y. Zhao, C. Wu, L. Wang, J. Xu et al., Novel inorganic perovskite quantum dots for photocatalysis. Nanoscale **9**(33), 12032–12038 (2017). https://doi.org/10.1039/C7NR04421F
29. H.T. Giang, H.T. Duy, P.Q. Ngan, G.H. Thai, D.T.A. Thu, D.T. Thu, N.N. Toan, Hydrocarbon gas sensing of nano-crystalline perovskite oxides LnFeO3 (Ln=La, Nd and Sm). Sens. Actuators, B Chem. **158**(1), 246–251 (2011). https://doi.org/10.1016/j.snb.2011.06.013
30. S. González-Carrero, R.E. Galian, J. Pérez-Prieto, Organometal halide perovskites: bulk low-dimension materials and nanoparticles. Part. Part. Syst. Charact. **32**(7), 709–720 (2015). https://doi.org/10.1002/ppsc.201400214
31. Y. Guo, G. Liu, Z. Li, Y. Lou, J. Chen, Y. Zhao, Stable lead-free (CH3 NH3)3 Bi2 I9 perovskite for photocatalytic hydrogen generation. ACS Sustain. Chem. Eng. **7**(17), 15080–15085 (2019). https://doi.org/10.1021/acssuschemeng.9b03761
32. R. Gupta, U. Alam, N. Verma, Efficient photofixation of nitrogen to ammonia over an S-scheme-based NiSnO3-g-C3N4 heterojunction. Chem. Eng. J. **479**, 147644 (2024). https://doi.org/10.1016/j.cej.2023.147644
33. A. Haruna, I. Abdulkadir, S.O. Idris, Synthesis, characterization and photocatalytic properties of Bi0.85−XMXBa0.15FeO3 (M = Na and K, X = 0, 0.1) perovskite-like nanoparticles using the sol-gel method. J. King Saud Univ. Sci. **32**(1), 896–903 (2020). https://doi.org/10.1016/j.jksus.2019.05.005
34. A. Haruna, I. Abdulkadir, S.O. Idris, Visible light induced photodegradation of methylene blue in sodium doped bismuth barium ferrite nanoparticle synthesized by sol-gel method. Avicenna J. Environ. Health Eng. **5**(2), 120–126 (2018). https://doi.org/10.15171/ajehe.2018.16
35. T.G. Ho, T.D. Ha, Q.N. Pham, H.T. Giang, T.A. Thu Do, N.T. Nguyen, Nanosized perovskite oxideNdFeO$_3$ as material for a carbon-monoxide catalytic gas sensor. Adv. Nat. Sci.: Nanosci. Nanotechnol. **2**(1), 015012 (2011). https://doi.org/10.1088/2043-6262/2/1/015012
36. S. Hoffmann-Urlaub, Y. Zhang, Z. Wang, B. Kressdorf, T. Meyer, Fabrication of tin-based halide perovskites by pulsed laser deposition. Appl. Phys. A **126**(7), 553 (2020). https://doi.org/10.1007/s00339-020-03699-9

37. Y. Huang, X. Lu, H. Wu, J. Xu, Z. Lin, Y. Guo, R. Huang, Improving photoluminescence properties of lead-free Cs4SnBr 6 zero-dimensional perovskite via Mn2+/Sb3+ Co-doping. J. Lumin. **277**, 120930 (2025). https://doi.org/10.1016/j.jlumin.2024.120930
38. A. Ibrahim, B.A. Abubakar, Some wet routes for synthesis of hematite nanostructures. Afr. J. Pure Appl. Chem. **7**(3), 114–121 (2013). https://doi.org/10.5897/AJPAC12.002
39. M. Irshad, Q.T. Ain, M. Zaman, M.Z. Aslam, N. Kousar, M. Asim, M. Imran et al., Photocatalysis and perovskite oxide-based materials: a remedy for a clean and sustainable future. RSC Adv. **12**(12), 7009–7039 (2022). https://doi.org/10.1039/D1RA08185C
40. V. Jayaraman, D. Sarkar, R. Rajendran, B. Palanivel, C. Ayappan, M. Chellamuthu, A. Mani, Synergistic effect of band edge potentials on BiFeO3/V2O5 composite: enhanced photo catalytic activity. J. Environ. Manage. **247**, 104–114 (2019). https://doi.org/10.1016/j.jenvman.2019.06.041
41. T.C. Jellicoe, J.M. Richter, H.F.J. Glass, M. Tabachnyk, R. Brady, S.E. Dutton, M.L. Böhm et al., Synthesis and optical properties of lead-free cesium tin halide perovskite nanocrystals. J. Am. Chem. Soc. **138**(9), 2941–2944 (2016). https://doi.org/10.1021/jacs.5b13470
42. S. Ji, H. Liu, Y. Sang, W. Liu, G. Yu, Y. Leng, Synthesis, structure, and piezoelectric properties of ferroelectric and antiferroelectric $NaNbO_3$ nanostructures. CrystEngComm **16**(32), 7598–7604 (2014). https://doi.org/10.1039/C4CE01116C
43. E. Jiang, N. Song, G. Che, C. Liu, H. Dong, L. Yang, Construction of a Z-scheme MoS2/CaTiO3 heterostructure by the morphology-controlled strategy towards enhancing photocatalytic activity. Chem. Eng. J. **399**, 125721 (2020). https://doi.org/10.1016/j.cej.2020.125721
44. Y. Jiang, K. Li, X. Wu, M. Zhu, H. Zhang, K. Zhang, Q.-H. Xu et al., In Situ synthesis of lead-free halide perovskite Cs_2 AgBiBr 6 supported on nitrogen-doped carbon for efficient hydrogen evolution in aqueous HBr solution. ACS Appl. Mater. Interfaces. **13**(8), 10037–10046 (2021). https://doi.org/10.1021/acsami.0c21588
45. K. Kajihara, Recent advances in sol–gel synthesis of monolithic silica and silica-based glasses. J. Asian Ceram. Soc. **1**(2), 121–133 (2013). https://doi.org/10.1016/j.jascer.2013.04.002
46. H. Kato, A. Kudo, Photocatalytic water splitting into H_2 and O_2 over various tantalate photocatalysts. Catal. Today **78**(1–4), 561–569 (2003). https://doi.org/10.1016/S0920-5861(02)00355-3
47. K.-I. Kobayashi, T. Kimura, H. Sawada, K. Terakura, Y. Tokura, Room-temperature magnetoresistance in an oxide material with an ordered double-perovskite structure. Nature **395**(6703), 677–680 (1998). https://doi.org/10.1038/27167
48. K. Korshunova, L. Winterfeld, W.J.D. Beenken, E. Runge, Thermodynamic stability of mixed Pb:Sn methyl-ammonium halide perovskites. Physica Status Solidi (b) **253**(10), 1907–1915 (2016). https://doi.org/10.1002/pssb.201600136
49. E.M. Kostyukhin, A.L. Kustov, L.M. Kustov, One-step hydrothermal microwave-assisted synthesis of LaFeO3 nanoparticles. Ceram. Int. **45**(11), 14384–14388 (2019). https://doi.org/10.1016/j.ceramint.2019.04.155
50. S.A. Kulkarni, S.G. Mhaisalkar, N. Mathews, P.P. Boix, Perovskite nanoparticles: synthesis, properties, and novel applications in photovoltaics and LEDs. Small Methods **3**(1), 1800231 (2019). https://doi.org/10.1002/smtd.201800231
51. A. Kumar, M. Kumar, V. Navakoteswara Rao, M.V. Shankar, S. Bhattacharya, V. Krishnan, Unraveling the structural and morphological stability of oxygen vacancy engineered leaf-templated $CaTiO_3$ towards photocatalytic H_2 evolution and N_2 fixation reactions. J. Mater. Chem. A **9**(31), 17006–17018 (2021). https://doi.org/10.1039/D1TA04180K
52. I. Kuzmanovski, S. Dimitrovska-Lazova, S. Aleksovska, Classification of perovskites with supervised self-organizing maps. Anal. Chim. Acta **595**(1–2), 182–189 (2007). https://doi.org/10.1016/j.aca.2007.04.062
53. G. Li, W. Yang, S. Gao, Q. Shen, J. Xue, K. Chen, Q. Li, Creation of rich oxygen vacancies in bismuth molybdate nanosheets to boost the photocatalytic nitrogen fixation performance under visible light illumination. Chem. Eng. J. **404**, 127115 (2021). https://doi.org/10.1016/j.cej.2020.127115

54. J.-H. Li, X.-Z. Fu, J.-L. Luo, K.T. Chuang, A.R. Sanger, Application of BaTiO3 as anode materials for H2S-containing CH4 fueled solid oxide fuel cells. J. Power Sources **213**, 69–77 (2012). https://doi.org/10.1016/j.jpowsour.2012.03.106
55. K. Li, S. Li, W. Zhang, Z. Shi, D. Wu, X. Chen, X. Li et al., Highly-efficient and stable photocatalytic activity of lead-free Cs2AgInCl6 double perovskite for organic pollutant degradation. J. Colloid Interface Sci. **596**, 376–383 (2021). https://doi.org/10.1016/j.jcis.2021.03.144
56. M. Li, S. Xu, L. Wu, H. Tang, B. Zhou, J. Xu, K. Yan et al., Perovskite Cs_3 Bi_2 I_9 hexagonal prisms with ordered geometry for enhanced photocatalytic hydrogen evolution. ACS Energy Lett. **7**(10), 3370–3377 (2022). https://doi.org/10.1021/acsenergylett.2c01856
57. N. Li, X. Chen, J. Wang, X. Liang, L. Ma, X. Jing, Z. Li et al., ZnSe Nanorods–$CsSnCl_3$ perovskite heterojunction composite for photocatalytic CO_2 reduction. ACS Nano **16**(2), 3332–3340 (2022). https://doi.org/10.1021/acsnano.1c11442
58. R. Li, S. Wang, Y. Li, L. Zi, S. Zhao, L. Qi, J. Zhang et al., The breaking of charge symmetry at the M-site in octahedral units of KMF3 perovskite induces enhanced electrochemical nitrate reduction in ammonia. J. Colloid Interface Sci. **682**, 1175–1184 (2025). https://doi.org/10.1016/j.jcis.2024.11.219
59. Y. Li, S. Yao, W. Wen, L. Xue, Y. Yan, Sol–gel combustion synthesis and visible-light-driven photocatalytic property of perovskite LaNiO3. J. Alloy. Compd. **491**(1–2), 560–564 (2010). https://doi.org/10.1016/j.jallcom.2009.10.269
60. H. Lin, C. Zhou, Y. Tian, T. Siegrist, B. Ma, Low-dimensional organometal halide perovskites. ACS Energy Lett. **3**(1), 54–62 (2018). https://doi.org/10.1021/acsenergylett.7b00926
61. D.-R. Liu, Y.-S. Jiang, G.-M. Gao, Photocatalytic degradation of an azo dye using N-doped NaTaO3 synthesized by a one-step hydrothermal process. Chemosphere **83**(11), 1546–1552 (2011). https://doi.org/10.1016/j.chemosphere.2011.01.033
62. Z. Liu, H. Yang, J. Wang, Y. Yuan, K. Hills-Kimball, T. Cai, O. Chen et al., Synthesis of lead-free Cs_2 $AgBiX_6$ ($X = Cl$, Br, I) double perovskite nanoplatelets and their application in CO_2 photocatalytic reduction. Nano Lett. **21**(4), 1620–1627 (2021). https://doi.org/10.1021/acs.nanolett.0c04148
63. Z. Liu, X. Li, S. Su, W. Ding, L. Meng, Y. Wang, M. Luo, et al., Enhancing photocatalytic nitrogen fixation performance of Co-doped bismuth molybdate through band engineering tuning. Appl. Surface Sci. **611**, 155627 (2023). https://doi.org/10.1016/j.apsusc.2022.155627
64. C. Lu, D.S. Itanze, A.G. Aragon, X. Ma, H. Li, K.B. Ucer, S.M. Geyer et al., Correction: synthesis of lead-free Cs_3 Sb_2 Br 9 perovskite alternative nanocrystals with enhanced photocatalytic CO_2 reduction activity. Nanoscale **14**(22), 8200–8201 (2022). https://doi.org/10.1039/D2NR90107B
65. S. Lu, Q. Zhou, L. Ma, Y. Guo, J. Wang, Rapid discovery of ferroelectric photovoltaic perovskites and material descriptors via machine learning. Small Methods **3**(11), 1900360 (2019). https://doi.org/10.1002/smtd.201900360
66. J. Luo, S. Li, H. Wu, Y. Zhou, Y. Li, J. Liu, J. Tang et al., Cs_2 $AgInCl_6$ double perovskite single crystals: parity forbidden transitions and their application for sensitive and fast UV photodetectors. ACS Photonics **5**(2), 398–405 (2018). https://doi.org/10.1021/acsphotonics.7b00837
67. J. Luo, R. Li, Y. Chen, X. Zhou, X. Ning, L. Zhan, L. Zhang et al., Rational design of Z-scheme LaFeO3/SnS2 hybrid with boosted visible light photocatalytic activity towards tetracycline degradation. Sep. Purif. Technol. **210**, 417–430 (2019). https://doi.org/10.1016/j.seppur.2018.08.028
68. Y. Lyu, G. Lyu, C. Song, A novel Pd-doped perovskite-based LaCo0·9Pd0·1O3–Ba/Al2O3 catalyst enhances NOx to ammonia reaction. Int. J. Hydrogen Energy **69**, 223–233 (2024). https://doi.org/10.1016/j.ijhydene.2024.05.055
69. A. Mahmoud Idris, S. Zheng, L. Wu, S. Zhou, H. Lin, Z. Chen, Z. Li et al., A heterostructure of halide and oxide double perovskites Cs2AgBiBr 6/Sr2FeNbO6 for boosting the charge separation toward high efficient photocatalytic CO_2 reduction under visible-light irradiation. Chem. Eng. J. **446**, 137197 (2022). https://doi.org/10.1016/j.cej.2022.137197

70. H. Mai, D. Chen, Y. Tachibana, H. Suzuki, R. Abe, R.A. Caruso, Developing sustainable, high-performance perovskites in photocatalysis: design strategies and applications. Chem. Soc. Rev. **50**(24), 13692–13729 (2021). https://doi.org/10.1039/D1CS00684C
71. J.-I. Makiura, S. Kakihara, T. Higo, N. Ito, Y. Hirano, Y. Sekine, Efficient CO_2 conversion to CO using chemical looping over Co–In oxide. Chem. Commun. **58**(31), 4837–4840 (2022). https://doi.org/10.1039/D2CC00208F
72. S. Mansingh, S. Sultana, R. Acharya, M.K. Ghosh, K.M. Parida, Efficient photon conversion via double charge dynamics CeO_2–$BiFeO_3$ p–n heterojunction photocatalyst promising toward N_2 fixation and Phenol–Cr(VI) detoxification. Inorg. Chem. **59**(6), 3856–3873 (2020). https://doi.org/10.1021/acs.inorgchem.9b03526
73. L. Mao, C.C. Stoumpos, M.G. Kanatzidis, Two-dimensional hybrid halide perovskites: principles and promises. J. Am. Chem. Soc. **141**(3), 1171–1190 (2019). https://doi.org/10.1021/jacs.8b10851
74. M. Masri, K.B. Girisha, A. Hezam, T.F. Qahtan, K. Alkanad, F. Masri, K. Byrappa et al., Enhanced photocatalytic activity and stability of 2D Cs3Bi2Br 9 perovskite nanosheets synthesized via modified antisolvent method. Colloids Surfaces C: Environ. Asp. **2**, 100024 (2024). https://doi.org/10.1016/j.colsuc.2023.100024
75. R. Molinari, C. Lavorato, P. Argurio, Visible-light photocatalysts and their perspectives for building photocatalytic membrane reactors for various liquid phase chemical conversions. Catalysts **10**(11), 1334 (2020). https://doi.org/10.3390/catal10111334
76. P. Munnik, P.E. De Jongh, K.P. De Jong, Recent developments in the synthesis of supported catalysts. Chem. Rev. **115**(14), 6687–6718 (2015). https://doi.org/10.1021/cr500486u
77. V.-H. Nguyen, M. Mousavi, J.B. Ghasemi, Q.V. Le, S.A. Delbari, M. Shahedi Asl, A. Sabahi Namini et al., High-impressive separation of photoinduced charge carriers on step-scheme ZnO/$ZnSnO_3$/Carbon dots heterojunction with efficient activity in photocatalytic NH_3 production. J. Taiwan Inst. Chem. Eng. **118**, 140–151 (2021). https://doi.org/10.1016/j.jtice.2021.01.012
78. S. Ohara, A. Kondo, H. Shimoda, K. Sato, H. Abe, M. Naito, Rapid mechanochemical synthesis of fine barium titanate nanoparticles. Mater. Lett. **62**(17–18), 2957–2959 (2008). https://doi.org/10.1016/j.matlet.2008.01.083
79. T. Ohno, T. Tsubota, Y. Nakamura, K. Sayama, Preparation of S, C cation-codoped $SrTiO_3$ and its photocatalytic activity under visible light. Appl. Catal. A **288**(1–2), 74–79 (2005). https://doi.org/10.1016/j.apcata.2005.04.035
80. N. Ojha, A. Bajpai, S. Kumar, Enriched oxygen vacancies of Cu2O/SnS2/SnO2 heterostructure for enhanced photocatalytic reduction of CO_2 by water and nitrogen fixation. J. Colloid Interface Sci. **585**, 764–777 (2021). https://doi.org/10.1016/j.jcis.2020.10.056
81. N. Ojha, S. Kumar, Tri-phase photocatalysis for CO_2 reduction and N2 fixation with efficient electron transfer on a hydrophilic surface of transition-metal-doped MIL-88A (Fe). Appl. Catal. B **292**, 120166 (2021). https://doi.org/10.1016/j.apcatb.2021.120166
82. N. Ojha, K. Thakkar, A. Bajpai, K. Joshi, S. Kumar, Photoinduced CO_2 and N_2 reductions on plasmonically enabled gallium oxide. J. Colloid Interface Sci. **629**, 654–666 (2023). https://doi.org/10.1016/j.jcis.2022.09.097
83. J. Pi, X. Jia, Z. Long, S. Yang, H. Wu, D. Zhou, J. Qiu et al., Surface and defect engineering coupling of halide double perovskite Cs_2 $NaBiCl_6$ for efficient CO_2 photoreduction. Adv. Energy Mater. **12**(43), 2202074 (2022). https://doi.org/10.1002/aenm.202202074
84. G. Pilania, P.V. Balachandran, C. Kim, T. Lookman, Finding new perovskite halides via machine learning. Front. Mater. **3** (2016).https://doi.org/10.3389/fmats.2016.00019
85. L. Protesescu, S. Yakunin, M.I. Bodnarchuk, F. Krieg, R. Caputo, C.H. Hendon, M.V. Kovalenko et al., Nanocrystals of cesium lead halide perovskites (CsPbX3, X = Cl, Br, and I): novel optoelectronic materials showing bright emission with wide color gamut. Nano Lett. **15**(6), 3692–3696 (2015). https://doi.org/10.1021/nl5048779
86. Q.I. Rahman, M. Ahmad, S.K. Misra, M. Lohani, Efficient degradation of methylene blue dye over highly reactive Cu doped strontium titanate (SrTiO3) nanoparticles photocatalyst under visible light. J. Nanosci. Nanotechnol. **12**(9), 7181–7186 (2012). https://doi.org/10.1166/jnn.2012.6494

87. P. Ramasamy, D.-H. Lim, B. Kim, S.-H. Lee, M.-S. Lee, J.-S. Lee, All-inorganic cesium lead halide perovskite nanocrystals for photodetector applications. Chem. Commun. **52**(10), 2067–2070 (2016). https://doi.org/10.1039/C5CC08643D
88. L. Romani, A. Bala, V. Kumar, A. Speltini, A. Milella, F. Fracassi, L. Malavasi et al., PEA_2 SnBr 4: a water-stable lead-free two-dimensional perovskite and demonstration of its use as a co-catalyst in hydrogen photogeneration and organic-dye degradation. J. Mater. Chem. C **8**(27), 9189–9194 (2020). https://doi.org/10.1039/D0TC02525A
89. L. Romani, A. Speltini, F. Ambrosio, E. Mosconi, A. Profumo, M. Marelli, L. Malavasi et al., Water-Stable DMASnBr 3 lead-free perovskite for effective solar-driven photocatalysis. Angew. Chem. Int. Ed. **60**(7), 3611–3618 (2021). https://doi.org/10.1002/anie.202007584
90. L. Romani, A. Speltini, C.N. Dibenedetto, A. Listorti, F. Ambrosio, E. Mosconi, L. Malavasi et al., Experimental strategy and mechanistic view to boost the photocatalytic activity of Cs_3 Bi_2 Br 9 lead-free perovskite derivative by g-$_C$3 $_N$4 composite engineering. Adv. Func. Mater. **31**(46), 2104428 (2021). https://doi.org/10.1002/adfm.202104428
91. J.J. Rueda-Marquez, I. Levchuk, P. Fernández Ibañez, M. Sillanpää, A critical review on application of photocatalysis for toxicity reduction of real wastewaters. J. Clean. Prod. **258**, 120694 (2020). https://doi.org/10.1016/j.jclepro.2020.120694
92. A. Sato, M. Matsukawa, H. Taniguchi, S. Tsuji, K. Nishidate, S. Aisawa, K. Zhang et al., Structural, physical and photocatalytic properties of mixed-valence double-perovskite Ba2Pr(Bi,Sb)O6 semiconductor synthesized by citrate pyrolysis technique. Solid State Sci. **107**, 106352 (2020). https://doi.org/10.1016/j.solidstatesciences.2020.106352
93. J. Sheng, Y. He, J. Li, C. Yuan, H. Huang, S. Wang, F. Dong et al., Identification of halogen-associated active sites on bismuth-based perovskite quantum dots for efficient and selective CO_2-to-CO photoreduction. ACS Nano **14**(10), 13103–13114 (2020). https://doi.org/10.1021/acsnano.0c04659
94. L. Shi, T. Jia, C. Sun, F. Fu, J. Huang, Q. Zhang, J.H. Lee et al., Rational construction of BiFeO3-ZnO composite piezo-photocatalysts for synergistically enhanced catalytic performance toward the removal of antibiotics. J. Alloy. Compd. **1038**, 182714 (2025). https://doi.org/10.1016/j.jallcom.2025.182714
95. M. Shi, G. Li, W. Tian, S. Jin, X. Tao, Y. Jiang, C. Li et al., Understanding the effect of crystalline structural transformation for lead-free inorganic halide perovskites. Adv. Mater. **32**(31), 2002137 (2020). https://doi.org/10.1002/adma.202002137
96. M.L. Sikosana, K. Sikhwivhilu, R. Moutloali, D.M. Madyira, Municipal wastewater treatment technologies: a review. Procedia Manuf. **35**, 1018–1024 (2019). https://doi.org/10.1016/j.promfg.2019.06.051
97. A. Singh, R. Chatterjee, Multiferroic properties of La-Rich $BiFeO_3$ -$PbTiO_3$ solid solutions. Ferroelectrics **433**(1), 180–189 (2012). https://doi.org/10.1080/00150193.2012.722462
98. T. Soltani, A. Tayyebi, B.-K. Lee, BiFeO3/BiVO4 p−n heterojunction for efficient and stable photocatalytic and photoelectrochemical water splitting under visible-light irradiation. Catal. Today **340**, 188–196 (2020). https://doi.org/10.1016/j.cattod.2018.09.030
99. M. Song, L. Wang, J. Li, D. Sun, R. Guan, H. Zhai, Z. Sun et al., Defect density modulation of La2TiO5: an effective method to suppress electron-hole recombination and improve photocatalytic nitrogen fixation. J. Colloid Interface Sci. **602**, 748–755 (2021). https://doi.org/10.1016/j.jcis.2021.06.055
100. Q.-M. Sun, J. Xu, F. Tao, W. Ye, C. Zhou, J. He, J. Lu, Boosted inner surface charge transfer in perovskite Nanodots@Mesoporous Titania frameworks for efficient and selective photocatalytic CO_2 reduction to methane. Angew. Chem. Int. Ed. **61**(20), e202200872 (2022). https://doi.org/10.1002/anie.202200872
101. Q. Sun, W. Ye, J. Wei, L. Li, J. Wang, J.-H. He, J.-M. Lu, Lead-free perovskite Cs3Bi2Br 9 heterojunctions for highly efficient and selective photocatalysis under mild conditions. J. Alloy. Compd. **893**, 162326 (2022). https://doi.org/10.1016/j.jallcom.2021.162326
102. S. Sun, D. Yuan, Y. Xu, A. Wang, Z. Deng, Ligand-mediated synthesis of shape-controlled cesium lead halide perovskite nanocrystals *via* reprecipitation process at room temperature. ACS Nano **10**(3), 3648–3657 (2016). https://doi.org/10.1021/acsnano.5b08193

103. P. Szuromi, B. Grocholski, Natural and engineered perovskites. Science **358**(6364), 732–733 (2017). https://doi.org/10.1126/science.358.6364.732
104. P. Tang, Y. Tong, H. Chen, F. Cao, G. Pan, Microwave-assisted synthesis of nanoparticulate perovskite LaFeO3 as a high active visible-light photocatalyst. Curr. Appl. Phys. **13**(2), 340–343 (2013). https://doi.org/10.1016/j.cap.2012.08.006
105. Y. Tang, C.H. Mak, R. Liu, Z. Wang, L. Ji, H. Song, H. Hsu et al., In Situ formation of bismuth-based perovskite heterostructures for high-performance cocatalyst-free photocatalytic hydrogen evolution. Adv. Func. Mater. **30**(52), 2006919 (2020). https://doi.org/10.1002/adfm.202006919
106. R. Tao, X. Li, X. Li, C. Shao, Y. Liu, TiO_2 /$SrTiO_3$ /g-C_3 N_4 ternary heterojunction nanofibers: gradient energy band, cascade charge transfer, enhanced photocatalytic hydrogen evolution, and nitrogen fixation. Nanoscale **12**(15), 8320–8329 (2020). https://doi.org/10.1039/D0NR00219D
107. S. Tasleem, M. Tahir, Recent progress in structural development and band engineering of perovskites materials for photocatalytic solar hydrogen production: a review. Int. J. Hydrogen Energy **45**(38), 19078–19111 (2020). https://doi.org/10.1016/j.ijhydene.2020.05.090
108. K. Upendranath, R.S. Vishwanath, G. Ravitheja, A.R. Lamani, G. Sriram, T. Hwan Oh, T. Altalhi, et al., Sol-gel synthesis of LaFeO3 perovskite oxide for distinct ridges detection of level II and III latent fingerprints. Inorg. Chem. Commun. **170**, 113210 (2024). https://doi.org/10.1016/j.inoche.2024.113210
109. R. Varghese, S.M. Jose, J.S. Lakshmi, J.K. Thomas, Investigation of the structural, optical, electrical, and photovoltaic characteristics of nanostructured Ce-doped BaTiO3 by a modified combustion method. Opt. Quant. Electron. **56**(1), 118 (2024). https://doi.org/10.1007/s11082-023-05686-9
110. R.I. Walton, Perovskite oxides prepared by hydrothermal and solvothermal synthesis: a review of crystallisation, chemistry, and compositions. Chem. A Eur. J. **26**(42), 9041–9069 (2020). https://doi.org/10.1002/chem.202000707
111. J. Wang, C. Zhao, S. Yuan, X. Li, J. Zhang, X. Hu, Y. He et al., One-step fabrication of Cu-doped Bi2MoO6 microflower for enhancing performance in photocatalytic nitrogen fixation. J. Colloid Interface Sci. **638**, 427–438 (2023). https://doi.org/10.1016/j.jcis.2023.02.005
112. M. Wang, D. Wu, X. Tang, Optimizing lead-free halide perovskites: strategies for enhanced performance and selectivity in photocatalytic CO_2 reduction. DeCarbon **9**, 100120 (2025). https://doi.org/10.1016/j.decarb.2025.100120
113. T. Wang, D. Yue, X. Li, Y. Zhao, Lead-free double perovskite Cs2AgBiBr 6/RGO composite for efficient visible light photocatalytic H2 evolution. Appl. Catal. B **268**, 118399 (2020). https://doi.org/10.1016/j.apcatb.2019.118399
114. X. Wang, J. Ma, Y. Kong, C. Fan, M. Peng, S. Komarneni, Synthesis of p-n heterojunction Ag3PO4/NaTaO3 composite photocatalyst for enhanced visible-light-driven photocatalytic performance. Mater. Lett. **251**, 192–195 (2019). https://doi.org/10.1016/j.matlet.2019.05.078
115. X.-D. Wang, Y.-H. Huang, J.-F. Liao, Y. Jiang, L. Zhou, X.-Y. Zhang, D.-B. Kuang et al., In Situ construction of a Cs_2 SnI_6 perovskite nanocrystal/SnS_2 nanosheet heterojunction with boosted interfacial charge transfer. J. Am. Chem. Soc. **141**(34), 13434–13441 (2019). https://doi.org/10.1021/jacs.9b04482
116. Y. Wang, H. Huang, Z. Zhang, C. Wang, Y. Yang, Q. Li, D. Xu, Lead-free perovskite Cs2AgBiBr 6@g-C3N4 Z-scheme system for improving CH4 production in photocatalytic CO_2 reduction. Appl. Catal. B **282**, 119570 (2021). https://doi.org/10.1016/j.apcatb.2020.119570
117. D. Wu, B. Huo, Y. Huang, X. Zhao, J. Yang, K. Hu, X. Tang et al., Synthesis of stable lead-free Cs3 Sb2 (Br x I1–x)9 ($0 \leq x \leq 1$) perovskite nanoplatelets and their application in CO_2 photocatalytic reduction. Small **18**(12), 2106001 (2022). https://doi.org/10.1002/smll.202106001
118. G. Wu, P. Li, D. Xu, B. Luo, Y. Hong, W. Shi, C. Liu, Hydrothermal synthesis and visible-light-driven photocatalytic degradation for tetracycline of Mn-doped SrTiO3 nanocubes. Appl. Surf. Sci. **333**, 39–47 (2015). https://doi.org/10.1016/j.apsusc.2015.02.008

119. J. Wu, X. Wang, J. Hu, C. Li, L. Shi, S.-S. Xia, L. Li et al., Ultra-small cesium silver bismuth bromide quantum dots fabricated by modified hot-injection method for highly efficient degradation of contaminants in organic solvent. J. Environ. Sci. **152**, 577–583 (2025). https://doi.org/10.1016/j.jes.2024.06.011
120. L. Wu, S. Zheng, H. Lin, S. Zhou, A. Mahmoud Idris, J. Wang, Z. Li et al., In-situ assembling 0D/2D Z-scheme heterojunction of Lead-free Cs2AgBiBr 6/Bi2WO6 for enhanced photocatalytic CO_2 reduction. J. Colloid Interface Sci. **629**, 233–242 (2023). https://doi.org/10.1016/j.jcis.2022.08.152
121. H. Xiang, P. Liu, W. Wang, R. Ran, W. Zhou, Z. Shao, Towards highly stable and efficient planar perovskite solar cells: materials development, defect control and interfacial engineering. Chem. Eng. J. **420**, 127599 (2021). https://doi.org/10.1016/j.cej.2020.127599
122. P. Xing, S. Wu, Y. Chen, P. Chen, X. Hu, H. Lin, Y. He, et al., New application and excellent performance of Ag/$KNbO_3$ nanocomposite in photocatalytic NH_3 synthesis. ACS Sustain. Chem. Eng. Acssuschemeng 9b01938 (2019). https://doi.org/10.1021/acssuschemeng.9b01938
123. J. Xu, B. Luo, W. Gu, Y. Jian, F. Wu, Y. Tang, H. Shen, Fabrication of In_2 S_3/$NaTaO_3$ composites for enhancing the photocatalytic activity toward the degradation of tetracycline. New J. Chem. **42**(7), 5052–5058 (2018). https://doi.org/10.1039/C7NJ05123A
124. H. Yin, J. Chen, P. Guan, D. Zheng, Q. Kong, S. Yang, K. Han et al., Controlling photoluminescence and photocatalysis activities in lead-free Cs_2 Pt_x Sn_{1-x} Cl_6 perovskites via ion substitution. Angew. Chem. Int. Ed. **60**(42), 22693–22699 (2021). https://doi.org/10.1002/anie.202108133
125. C. Zhang, G. Chen, C. Lv, Y. Yao, Y. Xu, X. Jin, Q. Meng, Enabling nitrogen fixation on Bi_2 WO_6 photocatalyst by c-PAN surface decoration. ACS Sustain. Chem. Eng. **6**(9), 11190–11195 (2018). https://doi.org/10.1021/acssuschemeng.8b02236
126. H. Zhang, X. Li, H. Su, X. Chen, S. Zuo, X. Yan, C. Yao et al., Sol–gel synthesis of upconversion perovskite/attapulgite heterostructures for photocatalytic fixation of nitrogen. J. Sol-Gel. Sci. Technol. **92**(1), 154–162 (2019). https://doi.org/10.1007/s10971-019-05071-7
127. T. Zhang, W. Lei, P. Liu, J.A. Rodriguez, J. Yu, Y. Qi, M. Liu et al., Organic pollutant photodecomposition by Ag/$KNbO_3$ nanocomposites: a combined experimental and theoretical study. J. Phys. Chem. C **120**(5), 2777–2786 (2016). https://doi.org/10.1021/acs.jpcc.5b11297
128. W. Zhang, P. Xing, J. Zhang, L. Chen, J. Yang, X. Hu, Y. He et al., Facile preparation of novel nickel sulfide modified KNbO3 heterojunction composite and its enhanced performance in photocatalytic nitrogen fixation. J. Colloid Interface Sci. **590**, 548–560 (2021). https://doi.org/10.1016/j.jcis.2021.01.086
129. X. Zhang, A. Tang, Y. Jia, Y. Wang, H. Wang, S. Zhang, Enhanced visible-light-driven photocatalytic performance of Ag/AgGaO2 metal semiconductor heterostructures. J. Alloy. Compd. **701**, 16–22 (2017). https://doi.org/10.1016/j.jallcom.2017.01.085
130. X. Zhang, M. Zhang, K. Cao, Hydrothermal synthesis of Sm-doped Bi_2 WO_6 flower-like microspheres for photocatalytic degradation of rhodamine B. CrystEngComm **21**(41), 6208–6218 (2019). https://doi.org/10.1039/C9CE01043B
131. Z. Zhang, Y. Liang, H. Huang, X. Liu, Q. Li, L. Chen, D. Xu, Stable and highly efficient photocatalysis with lead-free double-perovskite of Cs_2 AgBiBr 6. Angew. Chem. Int. Ed. **58**(22), 7263–7267 (2019). https://doi.org/10.1002/anie.201900658
132. C. Zhao, L. Yue, S. Yuan, X. Ren, Z. Zeng, X. Hu, Y. He et al., Enhanced photocatalytic N2 fixation and water purification using Bi/ZnSnO3 composite: mechanistic insights and novel applications. J. Ind. Eng. Chem. **132**, 135–147 (2024). https://doi.org/10.1016/j.jiec.2023.11.007
133. H. Zhao, Y. Li, B. Zhang, T. Xu, C. Wang, PtI /[(CH3)2NH2]3[BiI6] as a well-dispersed photocatalyst for hydrogen production in hydroiodic acid. Nano Energy **50**, 665–674 (2018). https://doi.org/10.1016/j.nanoen.2018.06.033
134. Z. Zhao, D. Wang, R. Gao, G. Wen, M. Feng, G. Song, Z. Chen et al., Magnetic-field-stimulated efficient photocatalytic N_2 fixation over defective $BaTiO_3$ perovskites. Angew. Chem. Int. Ed. **60**(21), 11910–11918 (2021). https://doi.org/10.1002/anie.202100726

135. J. Zhou, Z. Xia, M.S. Molokeev, X. Zhang, D. Peng, Q. Liu, Composition design, optical gap and stability investigations of lead-free halide double perovskite Cs_2 $AgInCl_6$. J. Mater. Chem. A **5**(29), 15031–15037 (2017). https://doi.org/10.1039/C7TA04690A
136. L. Zhou, Y. Xu, B. Chen, D. Kuang, C. Su, Synthesis and photocatalytic application of stable lead-free Cs_2 AgBiBr 6 perovskite nanocrystals. Small **14**(11), 1703762 (2018). https://doi.org/10.1002/smll.201703762
137. P. Zhou, H. Chen, Y. Chao, Q. Zhang, W. Zhang, F. Lv, S. Guo et al., Single-atom Pt-I3 sites on all-inorganic Cs2SnI6 perovskite for efficient photocatalytic hydrogen production. Nat. Commun. **12**(1), 4412 (2021). https://doi.org/10.1038/s41467-021-24702-8
138. S. Zhou, C. Zhang, J. Liu, J. Liao, Y. Kong, Y. Xu, G. Chen, Formation of an oriented Bi_2 WO_6 photocatalyst induced by *in situ* Bi reduction and its use for efficient nitrogen fixation. Catal. Sci. Technol. **9**(20), 5562–5566 (2019). https://doi.org/10.1039/C9CY00972H
139. J. Zou, J. Liu, X. Wang, Z. Han, Y. Gu, Z. He, Y. Zou et al., Stable and sensitive ultraviolet photodetectors based on high-quality $CsCu_2$ I_3 Films. ACS Appl. Electron. Mater. **5**(5), 2829–2837 (2023). https://doi.org/10.1021/acsaelm.3c00258
140. P. Zou, S. Chen, R. Lan, J. Humphreys, G. Jeerh, S. Tao, Investigation of perovskite oxide SrFe 0.8 Cu 0.1 Nb 0.1 O 3—δ as cathode for a room temperature direct ammonia fuel cell. Int. J. Hydrog. Energy **44**(48), 26554–26564 (2019). https://doi.org/10.1016/j.ijhydene.2019.08.097
141. M. Zybert, H. Ronduda, W. Patkowski, A. Ostrowski, K. Sobczak, W. Raróg-Pilecka, Improving the catalytic performance of Co/$BaCeO_3$ catalyst for ammonia synthesis by Y-modification of the perovskite-type support. RSC Adv. **14**(49), 36281–36294 (2024). https://doi.org/10.1039/D4RA06251E

Chapter 6
Lead Free Perovskites for Energy Storage Applications

Parika Mahajan and Aman Mahajan

Abstract To fulfil the growing global demand for energy, numerous efforts have been directed towards developing efficient materials for energy storage applications. Lead-free perovskites (LFPs), which has the structural formula ABX_3, have emerged as next-generation electrode materials due to their extraordinary conductivity, structural tunability, thermal/chemical stability, excellent ion transport kinetics, and reduced environmental toxicity as compared to their lead-based counterparts and could be employed in various energy storage systems, including supercapacitors, rechargeable batteries, and hybrid devices. Furthermore, their versatile crystal structure allows for compositional modifications that can optimize key performance parameters such as capacity, rate ability, and stability. Thus, this chapter provides a comprehensive overview of LFPs, highlighting their structural and physicochemical properties, and synthesis approaches along with existing challenges and future perspective for developing high-performance and environmentally friendly energy storage devices.

Keywords: Energy storage · Lead-free perovskites · Supercapacitors · Battery · Hybrid system

6.1 Introduction

Energy demand has increased swiftly due to rapid industrialization and growing population all over the world, which can't be met by non-renewable fossil fuels alone [1]. In this context, to address the rising energy crisis, the use of renewable sources of energy, such as, sunlight, water, air, etc. have been increased day by day [66]. Moreover, renewable sources not only help to meet energy demand but are also clean and environment friendly, unlike non-renewable sources which releases toxic gases into the environment and are a major cause of global warming, nowadays [18]. In this direction, scientists have worked extensively to develop systems that

P. Mahajan · A. Mahajan (✉)
Department of Physics, Guru Nanak Dev University, Amritsar, India
e-mail: aman.phy@gndu.ac.in

V. Chauhan et al. (eds.), *Sustainable Lead-Free Perovskite Materials*, Topics in Applied Physics 157, https://doi.org/10.1007/978-3-032-24873-2_6

can harvest energy from these renewable sources using technologies like windmills, solar panels, etc. However, these renewable sources are not available all the time; e.g. solar energy is available only in the daytime, whereas wind energy is also dependant on weather conditions [51]. So, the biggest challenge that still persists in using this harvested renewable energy in daily life is that they are not always readily available when required [52]. Therefore, it is crucial to develop new kind of technology and improving existing energy storage devices to store the energy harvested from these renewable sources effectively, so that it can be used later while maintaining balance between energy supply and demand [33].

To address this, energy could be stored in five different forms: electrochemically, mechanically, electrically (electrostatic & electromagnetic), thermally, and chemically; each of which plays an important role in energy systems used nowadays [7]. Electrochemical energy storage devices are the most common type of energy storage technology these days, especially for electric vehicles (EVs), which use them to transmit power since they are scalable, reliable, and very energy efficient. However, the development of energy storage devices faces several challenges related to optimizing shape, size, cycle stability, and lifespan to maintain cost-effectiveness, high performance and scalability in real-world applications; addressed by carefully selecting the materials, optimization of device geometry, use of advanced fabrication techniques, and adopting strategies that would help in enhancing its conductivity, mechanical integrity, and electrochemical activity [33]. From this point of view, perovskite materials, which have the general structural formula ABX_3 (A and B are cations and X is an anion of oxygen, nitrogen, carbon, or halogen), are particularly promising candidates for energy storage devices of the future generation because of their unique three-dimensional ion transport channels, versatile structural architectures, excellent physio-chemical properties, abundant oxygen vacancies, multiple accessible oxidation states that enables reversible redox activity, robust structural stability, chemical flexibility that allows desired tuning and doping, tunable band gaps and electronic structure, good thermal stability and simple and cost-effective synthesis routes [18, 66]. However, even while lead-based perovskites have great qualities, their poisonous nature and unstable nature remains significant bottleneck which limits their practical applicability and commercialization [42]. In particular, lead (Pb) is recognised as a very toxic element, which can gradually accumulate in the body of a human being and can cause severe illness related to brain hindering its natural growth while disrupting normal neurotransmission and causing headache, stomach problems like constipation, abdominal pain, etc., making it dangerous to human beings, further underscoring its toxicity and raising significant concerns regarding its use in energy storage devices [16]. In addition to this, lead is well-known to affect the reproductive systems of both the male as well as female, causing a reduction in the sperm count in males and increased risks of miscarriage in case of females. Therefore, taking into consideration the aforementioned hazardous and toxic nature of lead, it is crucial to find an alternative to lead which can retain desirable properties of lead without compromising the device performance in energy storage applications. In this perspective, metals such as, tin (Sn), germanium (Ge), titanium (Ti), and bismuth (Bi) have been explored as potential substitutes for lead in storage platforms for energy because of their electronic structures and oxidation states that are comparable to lead [16, 42]. Thus, in response to the growing environmental and health concerns related to lead, nowadays, the researchers are shifting their focus towards lead free perovskite materials for energy storage applications, offering an environment friendly, sustainable pathway and non-toxic approach towards high performing energy storage devices. Although, there are several reports published on lead-based perovskites due to their high efficiency and high performance, however, the field of LFPs are limited and is still yet to be explored [18].

In this direction, herein, in this chapter, at first, we will be discussing the basic structural and chemical properties of lead-free perovskites (LFPs) along with a brief historical background that led to their development in energy storage devices, along with various synthesis route, such as, sol-gel, microwave, hydrothermal, and solvothermal, etc. will be discussed. Further, we will explore the energy storage properties of these LFPs in various classified energy storage systems (electrochemical, mechanical, thermal, electrical, and chemical), while emphasizing its role particularly in electrochemical-based storage solutions, namely, supercapacitors, batteries, and fuel cells. Finally, the chapter will conclude with the current challenges such as, structural instability and long-term performance along with the rising opportunities and possible future perspectives in detail, providing a consolidated perspective on the viability of LFPs as exceptional materials for energy storage. Figure 6.1 outlines the schematic framework of the chapter, providing a structured overview of the key sections and thematic flow.

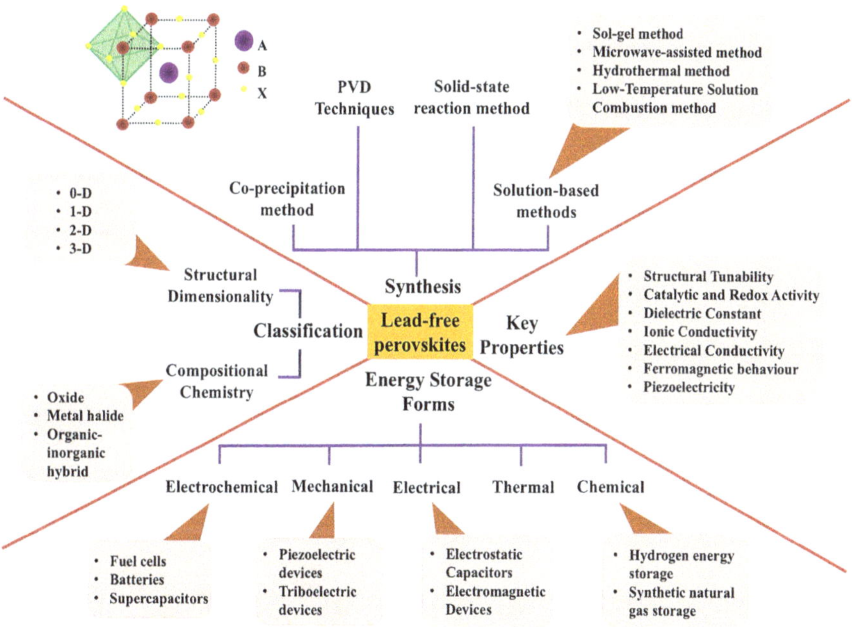

Fig. 6.1 Schematic layout of the chapter

6.2 Fundamentals and Classification of Lead-Free Perovskites for Energy Storage Applications

6.2.1 Introduction to Lead-Free Perovskites: Origin, Basic Structural Framework and Properties

Gustav Rose who was a native of Berlin, discovered the mineral perovskite ($CaTiO_3$) in 1839, during his geological visit to the Ural Mountains in Russia [29]. Further, in the nineteenth century, it was named by Gustav Rose after the famous mineralogist from Russia and statesman Count Lev Alekseyevich von Perovski, symbolizing a unique German-Russian scientific collaboration [66]. In the recent years, different compositions and variants of perovskites were developed linked to the structural stability of perovskites arising from the different combinations of anions and cations, typically denoted as ABX_3, in which A and B are cations (A being larger) and X is usually an anion of nitrogen, oxygen, carbon, or halogen. The structurally optimal crystal form of perovskite is cubic with the space group Pm33m in which cation B is bounded octahedrally by six X anions forming BX_6 while A cation occupies the cuboctahedral void position created between such eight BX_6 octahedra (Fig. 6.2) [16].

The most abundant naturally occurring lead free perovskites found within the earth's crust are magnesium silicate ($MgSiO_3$) and iron silicate ($FeSiO_3$) [29]. In 1926, a Swiss-Norwegian scientist Victor Goldschmidt achieved a remarkable milestone in investigating the atomic-level arrangement and assess the geometric stability of such materials via the introduction of two concepts (1) tolerance factor (t; assesses the geometric fit between the two A & B-site cations and the X-site anion within the perovskite lattice) and (2) octahedral factor (μ; evaluates the size compatibility formed between the B-site cation and the encapsulating X anions within the BX_6 octahedron) [66]. Further, it was found that the whole lattice symmetry is distorted by any partial or complete substitution of either of the A, B or X ions, thereby, leading to change in its ideal cubic geometry to different crystallographic phases such as tetragonal, orthorhombic, or rhombohedral, which in turn influences their

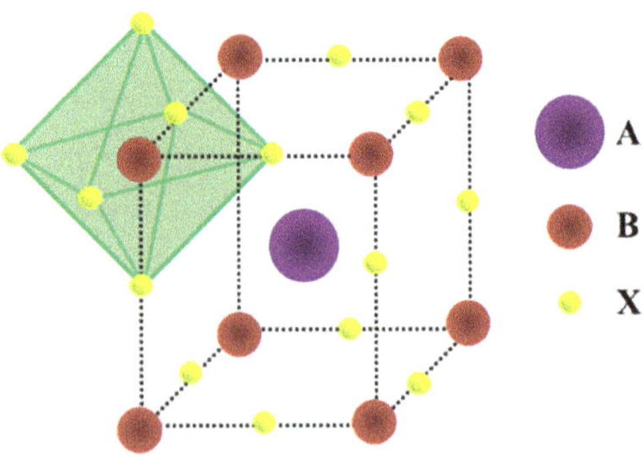

Fig. 6.2 Crystal structure of perovskite

physical and chemical properties. In fact, many of the astonishing electrical, optical, and electrochemical characteristics of perovskite materials arise from such structural flexibility and phase transitions. For instance, the most studied bismuth-based LFPs i.e. $BiFeO_3$ adopts a distorted rhombohedral structure, leading to spontaneous polarization, while the redox-active Fe^{3+}/Fe^{2+} couple enables faradaic charge storage, thus making it a promising material for electrochemical capacitors and battery electrodes [2]. Similar distortions in other LFPs such as sodium niobate ($NaNbO_3$) and potassium sodium niobate (KNN, $K_{0.5}Na_{0.5}NbO_3$) arise from A-site cation size mismatch and off-centre dislocation of Nb^{5+} ions in the BO_6 octahedra, leading to enhanced polarization and local electric fields, which improve both charge storage through dipole-assisted ion transport and piezoelectric performance, highlighting the pivotal role of crystal symmetry and lattice distortion in tuning energy storage functionalities [44]. Moreover, the tunable features arising due to structural and phase transitions, LFPs materials have gained considerable attention in solid-state physics and materials sciences, particularly energy storage as their composition and structure can be deliberately engineered according to the required specific functional applications.

6.2.2 Structural and Chemical Classification of Lead-Free Perovskites for Energy Storage Applications

LFPs can be classified based on either (i) compositional chemistry (ii) structural dimensionality discussed in detail as follows:

1. **Based on compositional chemistry**: LFPs can further be classified as: oxide, metal halide, and organic-inorganic hybrid perovskites based on its compositional chemistry, each of which offers different structural and functional properties tailored by its ionic constituents primarily determined by the type of cations occupying the A and B lattice sites, along with the anion positioned at the X site. The following subsections discuss each class in detail:

   (i) **Lead-free oxide perovskites**: They are generally represented by the structural formula ABO_3, and are typically a class of transition metal oxides but usually having a complex structure with A being any alkaline earth or rare-earth cation (e.g., Na^+, K^+, Bi^{3+}, La^{3+}, etc.) and B cation is commonly a small-sized transition metal element (e.g., Fe^{3+}, Mn^{3+}, Ti^{4+}, Co^{3+}, etc.) [8]. These oxide LFPs can adopt complex architectures, such as, layered perovskites (A_2BO_4), double perovskites ($A_2BB'O_6$), and triple perovskites ($A_2A'B_2B'O_9$), which allows a higher number of redox reactions by introducing oxygen vacancy, structural stability and flexibility as well as modified and tuned electronic structure, suitable for next-generation energy storage applications. For instance, $BiFeO_3$, $NaNbO_3$, and $K_{0.5}Na_{0.5}NbO_3$ (KNN) are among the widely studied oxide LFPs due to their high dielectric constants, ion transport capabilities, and redox-active frameworks, which contribute to enhanced energy and power density, mechanical stability and cycling stability [5].

(ii) **Lead-free metal halide perovskites**: They are generally denoted by the structure ABX'_3, and are typically a class of halide-based compounds, where the A-site is occupied by a monovalent cation, which can be either inorganic or organic in nature (MA^+, FA^+, or Cs^+, etc.), B is either divalent or trivalent metal cation (Sn^{2+}, Ge^{2+}, Bi^{3+}, Sb^{3+}, Cu^+, Mn^{2+}, etc.) and X' is an anion of halogen (I^-, Br^-, etc.) [17, 50]. Significant research efforts have focused on synthesizing either low-dimensional perovskites like Cs_2PdX_6, layered perovskite like $Cs_4ZnSb_2Cl_{12}$, vacancy-ordered or double perovskites like $Cs_2AgBiBr_6$, and even triple perovskites like $BaSr_2MgTa_2O_9$, because these structural variants offered enhanced chemical stability, improved optoelectronic activity, as well as reduced toxicity as demonstrated in recent literature [17]. For instance, Zhou et al. synthesized Cs_2PdX_6 ($X = Br, I$) having thickness of individual unit cell and enhanced cycle life via a simple antisolvent method, demonstrating outstanding resistance to light, moisture, and heat [76].

(iii) **Lead-free organic-inorganic hybrid perovskites**: In the general structure of organic-inorganic hybrid LFPs (i.e. ABX_3), cationic organic molecule such as methylammonium (MA^+), formamidinium (FA^+), or ethyl ammonium, etc. occupies the A-site while a non-toxic metal cation (e.g., Sn^{2+}, Mn^{2+}, etc.), occupies the B-site position and halide ions like I^- and Br^- generally occupy the X-site [37, 49, 54]. Organic-inorganic hybrid LFPs integrates the benefit of both the organic and inorganic cations with structural tunability of organic components and the unique functional properties of inorganic frameworks, offering a unique platform for designing multifunctional materials. Moreover, hydrogen bonding interactions existing between the organic cation and the surrounding metal halide framework provides structural integrity and stability to the material [49]. He et al. synthesized a three-dimensional hybrid molecular solid solution of $[(EPCF)_x(EPCCl)_{1-x}][Mn(dca)_3]$ ($0 \le x \le 1$) (where, $EPCF = [Et_3PCH_2CH_2F]^+$ and $EPCCl = [Et_3PCH_2CH_2Cl]^+$) and required properties like, emergent chirality, ferroelastic behaviour, and nonlinear optical switching were obtained via the compositional control of the organic cations, demonstrating that organic cation engineering plays a pivotal role in tuning multifunctional properties in hybrid perovskites [19].

2. **Based on structural dimensionality**: By understanding the dimensionality we can realise the full potential of LFPs for sustainable and high-performance energy storage devices, as discussed in detail below:

(i) **0-D lead-free perovskites**: They typically include isolated octahedra of metal halide/oxide ions or their clusters which are not interconnected within the lattice due to the incorporation of heavy organic or inorganic cations thus blocking the entire octahedral connectivity as well as the formation of electronic band structure [36]. As a result, 0-D perovskites generally display strong quantum confinement, localized charge storage behaviour and large exciton binding energy, thereby retaining the optical as well as chemical properties of the individual metal halide/oxide, including broad photoluminescence with large Stokes shifts and high emission efficiencies driven by self-trapped exciton (STE) processes. For instance, in a recent review article by Wang et al., Cs_4PbBr_6, a zero-dimensional perovskite composed of isolated $[PbBr_6]^{4-}$ octahedra, was analysed deeply with the observation that Cs_4PbBr_6 has unusual high green photoluminescence efficiency (>95% in bulk crystals), large Stokes shift, and strong environmental stability; suggesting that it is a good candidate for high-efficiency luminescent solar concentrators, which is closely linked to the concept of strong exciton localization and large binding energies [68].

(ii) **1-D lead-free perovskites**: They are composed of linearly aligned chains of metal halide/oxide octahedra interconnected along a single axis, generally through edges or corners, thereby forming a nanowire or nanotube like architecture and exhibiting superior properties like enhanced charge transport and improved structural flexibility, essential for energy storage applications [36]. For example, hybrid structures like $C_4N_2H_{14}PbBr_4$, exhibited emission of strongly luminescent bluish white light due to its well organised 1-D framework consisting of linearly aligned chains of $[PbBr_4{}^{2-}]_\infty$, enclosed by insulating shells of organic ions [6]. Further, in another recent study by Mondal et al., lead-free 1D $CsCuCl_3$ perovskites were used as filler in poly-vinylidene fluoride (PVDF) matrix to fabricate piezoelectric nanogenerators, achieving an output voltage of ~ 73.2 V and demonstrating its practical potential for self-powered mechanical energy harvesting application [47].

(iii) **2-D lead-free perovskites**: 2-D LFPs are composed of alternate layers of inorganic metal oxide & halide sheets along with organic cations having large ionic radius, creating a configuration like quantum well thus offering ionic conductive pathways and environmental stability to the material, making them advantageous for storage-oriented energy technologies (X. [36]). Yadav et al. synthesized a quasi-2-D lead free perovskite (PEA-$Cs_2AgBiBr_6$) with the controlled growth of Ag^+/Bi^{3+} ions and it was found to exhibit an impressive energy density as well as the cyclic stability while retaining 93% capacitance over 2000 cycles, far outperforming typical 3-D analogue [72]. Further, Tewari et al. explored 2-D lead-free $Cs_3Bi_2I_9$ nanosheets with a hexagonal structure to convert sunlight into energy and simultaneously store it in a lithium-ion battery, exhibiting an energy release capacity of 410 mAh/g which increased to 975 mAh/g upon solar illumination, demonstrating a photo-conversion efficiency of ~ 0.43% under illumination [64].

(iv) **3-D lead-free perovskites**: They are composed of 3-D fully connected framework formed by BX_6 octahedra sharing corners in which A-site cation occupies the interstitial voids, which promotes directional charge transport as well as enhances ionic transport capability of the material, making it appropriate for storage applications [36].. However, in terms of energy storage systems, 3D vacancy-ordered double perovskites like $Cs_2AgBiBr_6$ and Cs_2SnI_6 have also been investigated for their stability and redox activity in both supercapacitor and battery electrodes, owing to their robust 3-D frameworks and defect-tolerant nature [38]. Further, Zhang et al. fabricated solid-state photo-rechargeable batteries based on $Cs_3Bi_2I_9$, a 3-D LFPs, in which $Cs_3Bi_2I_9$ simultaneously acted as photo absorbing and the charge storage component and was able to power an LED in the dark with open-circuit voltage of ~ 0.28 V; demonstrating direct solar to chemical energy storage [74].

6.3 Synthesis Strategies of Lead-Free Perovskites for Energy Storage Applications

The choice of synthesis method significantly impacts the resulting structure, morphology, and physicochemical behaviour of LFPs as they directly influence their electrochemical behaviour and performance in energy storage applications. The synthesis of LFPs can be done by either top-down or bottom-up approaches and can be engineered into various architectures, including, nanocrystals, bulk powders, thin films, nanowires, nanotubes, nanocubes, and nanorods, utilizing a wide range of synthetic methods, which are briefly discussed as follows.

6.3.1 Solution-Based Methods

(i) **Sol-gel method**: It is versatile method to produce high-purity and homogeneous oxide materials and several LFPs such as, $BiFeO_3$, $BaTiO_3$, $K_{0.5}Na_{0.5}NbO_3$ (KNN), and $Cs_2AgBiBr_6$ that includes hydrolysis and subsequent polycondensation of metal salts/metal alkoxides in solution to obtain a sol, which forms a gel-like web with both liquid and solid phases, followed by drying and controlled thermal treatment to convert the gel into crystalline perovskite structures [30]. For LFPs, the sol-gel route provides homogeneity, and nanostructure construction, critical parameters for tuning dielectric, ferroelectric, and electrochemical properties of a material. For example, nanostructured (K, Na)NbO_3-based ceramics was fabricated by sol-gel synthesis approach and an improvement in its energy storage capacity was noted due to reduction in grain size, uniform nanostructure formation, and altered domain structure [75].

(ii) **Microwave-assisted method**: This method has gained significant attention in solid state chemistry for synthesizing LFP materials due to its fast-processing time and reduced energy consumption compared to other commonly used methods [3]. $La_{1-x}Ca_xMnO_3$ ($x = 0, 0.2, 0.4$), a LFPs oxide was synthesized by using this method by Tabari et al., which lessened its fabrication time while enhancing its electrochemical properties and specific capacitance, highlighting its potential in advanced energy storage technologies [63].

(iii) **Hydrothermal method**: This method involves reaction in a solvent either aqueous or organic solvent under high temperature and pressure in a sealed autoclave allowing direct crystallization of nanomaterials without the need for post-calcination, thereby, producing high-quality of pure powder with controlled particle size and different morphology, depending upon the temperature [15]. Depending upon the solvent and heating mechanism, this synthesis method can be divided into types as: hydrothermal and solvothermal method. Microwave hydrothermal method involves the synthesis of LFPs via the solid-state reactions in a sealed autoclave. In this direction, Fauziyah et al. reviewed the potential of this method for the synthesis of perovskite-based phase change materials (PCMs) and demonstrated that how this method enables the synthesis of structurally stable and comp ositionally tunable materials with improved thermal and structural performance, making them promising candidates for energy storage applications [15]. However, microwave solvothermal method is also quite similar but it involves the use of organic solvents instead of water for the synthesis of LFP materials. The use of organic solvent can help to stabilize specific oxidation state of the material, enabling the formation of metastable perovskite phases with desired properties for energy storage. A relevant study by Lu et al. synthesized lead-free ternary $NiTiO_3$ perovskite utilized as an electrode in lithium-ion batteries, delivering higher specific capacity and improved cycling durability attributed to its morphology and phase purity enabled by the microwave-assisted solvothermal technique [41].

(iv) **Low-Temperature Solution Combustion method**: It is a facile and cost-effective route to prepare nano- and micro-structured materials which employs mixing the metal precursors, generally nitrates, in aqueous solution with fuels derived from organic compounds, including urea, glycine, or citric acid, which acts both as complexing agents and reducing fuels [67]. A self-sustaining exothermic redox reaction occurs upon heating the mixture at relatively low temperatures (generally below 600 °C), releasing gases and forming highly porous and fine powder with large surface area and controlled stoichiometry. Madhale et al. synthesized granular-like $La_{1-x}Sr_xMnO_3$ LFPs using a facile solution combustion method for energy storage application with significant electrochemical performance, highlighting its potential across various energy storage platforms [43].

6.3.2 Solid-State Reaction Method

It involves different steps starting with accurate stoichiometric mixing of the precursor's powders followed by grinding get uniform powder and then calcinating at high-temperatures and controlled cooling to obtain the final product with desired phase, purity, particle size and morphology which is greatly influenced by the selection and reactivity of the precursors, heating rate, pressure and reaction atmosphere. To synthesize LFPs with structural formula ABX_3, appropriate combination and ratio of the metal-site A and B cations are selected which is then treated at high temperatures, often exceedingly around 2/3rd of the melting point of the respective compound components A, B and X, and is then maintained for a time greater than 10 h to achieve the desired composition [34]. Further, to enhance the uniformity, at first, the mixture is ball-milled using either isopropanol or acetone as milling medium, which is then dried, followed by calcination at a certain temperature for particular time with controlled heating and cooling rate to obtain the desired phase and morphology of the perovskite. Finally, the resultant obtained powder is ground, sieved and is kept for re-calcination at higher temperature than before ranging from 1300–1600 °C for greater times to ensure complete synthesis of a single-phase perovskite structure exhibiting desired characteristics [14].

6.3.3 Co-precipitation Method

This method relies on the coordinated precipitation of multiple cations from a homogenous solution, thus forming insoluble salts such as metal hydroxides, spinel nanoferrites, etc., containing more than one metal cation, making it highly suitable method to synthesize usually complex perovskites like double, triple and hybrid organic-inorganic LFPs [34]. For example, this method has been widely employed to synthesize LFP oxides such as $K_{0.5}Na_{0.5}NbO_3$ (KNN), $BaTiO_3$, and $BiFeO_3$, etc., where several parameters such as, temperature, solution pH, reactant concentration, compositions and phase purity are to be taken care of properly for tuning the ferroelectric, dielectric, and electrochemical properties of the materials, especially for electrochemical energy storage devices.

6.3.4 Physical Vapour Deposition Techniques

Physical vapour deposition (PVD) techniques are used to fabricate high quality thin films of LFP materials with precise control over film thickness, phase purity, crystallographic orientation, grain size, and stoichiometry. This can be done in three ways: (1) direct deposition at crystallization temperature under ambient atmosphere, (2) deposition at an intermediate temperature ranging from 800–1100 K, followed by post-annealing the films, and (3) deposition on a substrate carried out under low-temperature conditions, with subsequent high-temperature post-annealing [14]. In this direction, Yadav et al. has comprehensively reviewed various vacuum based techniques for manufacturing thin-film of perovskites, highlighting physical vapor deposition methods such as PLD, MBE, sputtering, and evaporation, as well as their influence on film quality, crystallinity, and device performance [73].

6.4 Classification Based on Energy Storage Forms

Energy can be stored in five major forms, i.e. electrochemical, mechanical, electrical (electrostatic & electromagnetic), chemical, and thermal; each offering unique advantages and limitations depending upon the physical or chemical principles governing their operation, forming the backbone of modern energy storage technologies from large-scale grid storage systems to portable electronics and wearable systems as shown in Fig. 6.3.

6.4.1 Electrochemical Energy Storage Systems

Among all the available energy storage technologies, electrochemical energy storage systems have been extensively explored due to their high efficiency, tunability, and integration into next-generation sustainable energy technologies, such as, electric vehicles. Electrochemical energy storage systems typically converts and store energy through redox reactions but based on the energy storage mechanism it is categorized into three types i.e., fuel cells, batteries, and supercapacitors as shown in Fig. 6.4 [22]. The working principles, advantages, and the integration of LFPs within these electrochemical energy storage systems are discussed in detail within the following subsequent sections.

6.4.1.1 Fuel Cells

Fuel cells (FCs) convert chemical energy within the material into electricity via redox reactions by oxidizing a fuel (commonly H_2) at the anode while reducing

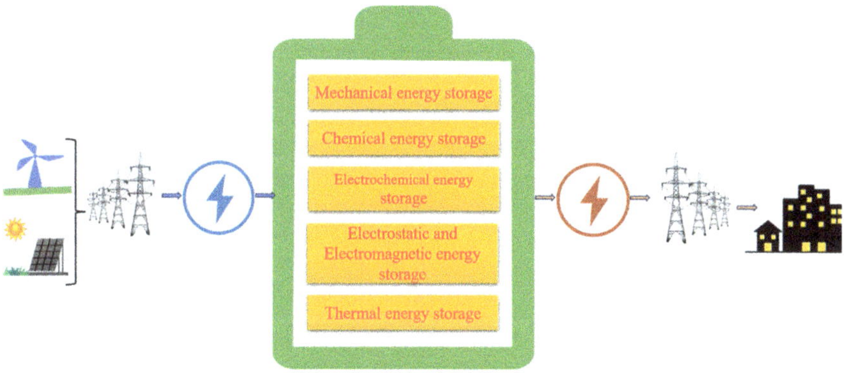

Fig. 6.3 Classification based on energy storage forms [26]. Copyright 2024. Reproduced with permission from American Chemical Society

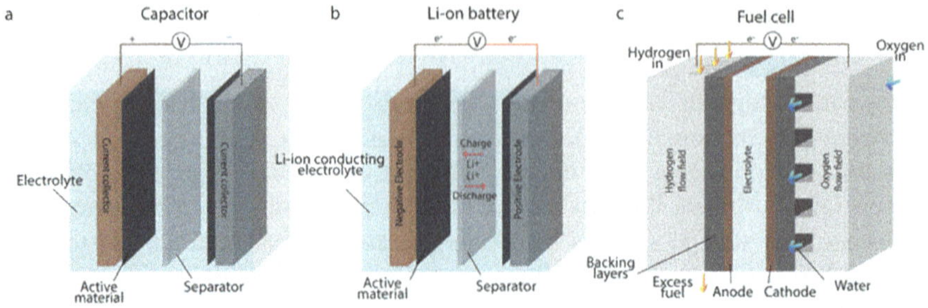

Fig. 6.4 Types of electrochemical energy storage devices [22]. Copyright 2022. Reproduced with permission from Wiley

an oxidant (usually O_2) on the cathode side, isolated by an electrolyte membrane [27]. They can be further categorized depending upon the type of electrolyte used as: solid oxide (SOFCs), molten carbonate (MCFCs), proton exchange membrane (PEMFCs), phosphoric acid fuel cells (PAFCs), and alkaline fuel cells (AFCs); among which PEMFCs and SOFCs are the most actively explored for both stationary and portable applications, owing to their high energy conversion efficiency and minimal environmental impact [28]. In this regard, LFPs, particularly, oxides, have demonstrated considerable promise in enhancing the performance of fuel cells due to their tunable redox activity, oxygen ion conductivity, and structural stability at elevated temperatures as already discussed above. For instance, $PrBaMn_{0.5}Fe_{1.5}O_{5+\delta}$ (PBMF), a double perovskite, was explored as a symmetric electrode for SOFCs by Li et al. and enhanced its catalytic activity and durability via the surface engineering approach by decorating the surface of PBMF using the atomic layer deposition method (ALD) with CeO_2 nanoparticles doped with iron, delivering a maximum power density of 573 mW/cm at 750 °C. This significant enhancement involved in both fuel oxidation and oxygen reduction mechanisms was due to the uniform surface modification, mixed ionic-electronic conductivity and structural stability of the double perovskite structure of PBMF, thus underlining the effective role of doped perovskites and ALD surface modification in boosting electrochemical performance (W. [35]). Further, Jeong et al. synthesized a double perovskite oxide $PrBa_{0.5}Sr_{0.5}Co_{2-x}FexO_{5+\delta}$ (PBSCF), and used as cathode material for the energy storage applications as fuel cells in which its catalytic and electrochemical performance was further optimized through plasma-enhanced ALD of cobalt oxide (CoO_x) nanoparticles on the surface of PBSCF material, improving its performance by approximately 36% at an operating temperature of 650 °C, without compromising the long-term stability. This study highlighted that how the strategic surface modification of double perovskite by uniformly decorating CoO_x nanoparticle significantly improved both the polarization resistance and ohmic losses attributed to the synergistic catalytic interaction between the PBSCF matrix and the CoO_x surface species, and self-catalytic effects of CoO_x, thus contributing to improved surface reactions and increased active sites at the electrode/electrolyte interface and hence the electrochemical output of PEMFCs while maintaining structural and operational robustness [23]. In addition to this, composite strategies, involving integration with materials like carbon nanotubes (CNTs), reduced graphene oxide (rGO), or MXene, etc., have also been employed in several studies to enhance conductivity, increase surface area, and provide structural stability to LFPs and develop high performance fuel cells. Further, polymer-perovskite composites are also being investigated in order to combine the mechanical flexibility of polymers with the electrochemical performance of perovskites, thereby making the practical viability of LFPs in flexible and wearable fuel cells.

6.4.1.2 Batteries

Batteries, the most commonly used electrochemical energy storage device utilizes the mechanism of storing the energy in chemical form and releasing it as electricity via the occurrence of a number of electrochemical redox reactions [40]. Typically, a battery includes a cathode, an anode, an electrolyte and a separator which when connected to an external circuit with load initiates the electrochemical reactions at the respective electrodes, causing the ions to migrate through the electrolyte to the electrodes and electrons to flow through the external circuit, thus producing electric current [27]. Thus, the fundamental characteristics and performance of a battery is majorly determined by the choice and morphology of the electrode material and type of electrolyte used. In certain battery systems, like lead-acid and alkaline batteries, electrodes undergo chemical transformation and the active materials are consumed during the electrochemical reaction, while in others, such as Li-ion batteries, intercalation/deintercalation processes dominate, allowing ion movement without significant degradation of the electrode materials.

Ongoing research efforts are focused on investigating and optimising the various compositions and structures of LFPs as electrode materials in a variety of rechargeable battery systems, including Li-ion, Na-ion, and Zn-ion batteries to achieve high-performance, durable, and sustainable energy storage systems. As outlined in the previous sections that diversity in structural types, dimensionality and morphologies offers significant improvement in electrochemical behaviour of LFPs. For instance, in a study by Bharathi et al., an advanced lead-free $Na_{0.5}Bi_{0.5}TiO_3$ based perovskite material was synthesized and used as electrode material for sodium-ion batteries delivering a reversible capacity of 470 mAh/g at 100 mA/g, attributed to its perovskite crystal structure which offered open diffusion channels and stable three-dimensional ion pathways, facilitating efficient and reversible Na^+ ion insertion and extraction [10]. The underlying charge storage mechanism in $Na_{0.5}Bi_{0.5}TiO_3$ based sodium-ion battery involved reversible intercalation of Na^+ ions into the anode material while charging, followed by deintercalation during discharging which was easily feasible due to its robust framework and mixed-cation (Na^+/Bi^{3+}) occupancy at the A-site, thereby enhancing its structural flexibility and redox activity via dual mechanism involving intercalation and alloying-type reactions and thus making it a compelling anode material for sodium-ion energy storage systems. Further, Wu et al. synthesized all-inorganic lead-free $Cs_2NaBiCl_6$ serving as the anode component in lithium-ion batteries using a simple grinding method and achieved ultrahigh Li^+ doping concentrations in the material delivering a stable specific capacity of around 300 mAh/g with Coulombic efficiencies of about 99% owing to its double-perovskite crystal structure which provided flexible lattice frameworks and abundant Li^+ insertion sites, enabling efficient ion transport favourable for Li^+ ion energy storage. Moreover, the crystal field generated by the Bi^{3+} and Na^+ cations in the $Cs_2NaBiCl_6$ lattice promoted stable Li^+ ion coordination, thereby reducing the diffusion barriers and enabling fast transport of ions; resulting in enhanced energy storage capacity [71]. Additionally, in a study by Chesta et al., caesium bismuth iodide ($CsBi_3I_{10}$), a lead-free inorganic perovskite was synthesized and used as cathode material in rechargeable aqueous zinc-ion batteries, delivering an impressive specific capacity of 220 mAh/g at 1000 mA/g, and retained around 50 mAh/g specific capacity after 20,000 cycles at a remarkable high current density of 20 A/g, which is accredited not only to its robust layered crystal structure (supporting rapid Zn^{2+} diffusion/intercalation.de-intercalation) structure but also due to the multistep charge storage mechanism through three sequential stages i.e., intercalation, conversion, and alloying processes [13].

In addition to acting as cathodes and anodes in sodium, lithium and zinc-based batteries, LFPs are also gaining attention in metal-air batteries, especially Li-air systems due to their outstanding bifunctional electrocatalytic performance toward ORR and OER, which is essential for the reversible operation of air batteries and are largely governed by the electronic structure and oxygen vacancy concentration of the material, both of which can be effectively tuned through A-site or B-site doping within the LFPs framework. For instance, Liu et al. demonstrated the role of lattice oxygen in enhancing OER activity in $LaNiO_3$ thin films, a lead-free metal oxide perovskite structure and concluded that strategic A-site doping helped in modulating its electronic structure and surface oxygen adsorption properties while retaining structural stability, thereby underscoring the potential of tailored perovskite compositions as high-performance electrode materials for metal-air batteries [39].

6.4.1.3 Supercapacitors

Supercapacitors are also an emerging clean electrochemical energy storage device with rapid charge/discharge ability, high energy and power density and a long cycle life than conventional batteries. Unlike batteries, supercapacitors store energy either through the charge accumulation and formation of double layer owing to electric double-layer formation at the electrode–electrolyte boundary or via faradaic processes involving fast and reversible electrochemical redox reactions, often termed pseudocapacitance. In addition to this, hybrid supercapacitors integrate the benefits of both the EDLC and pseudocapacitor, thus storing .charge both electrostatically as well as faradaically to achieve outstanding electrochemical properties for energy storage, making them well-suited for use in practical applications, such as, regenerative braking in hybrid and electric vehicles. Among several available materials,

LFPs have attained a huge interest as electrode material for supercapacitor applications owing to its unique properties as already discussed in previous sections. For instance, Chatterjee et al. synthesized all-inorganic $Cs_3Bi_2(Br/Cl)_9$ LFPs with hexagonal P3m1 phase via a low-cost, solution-based method and exhibited pseudocapacitive behaviour due to reversible redox transitions between Bi^{3+}/Bi^{5+} and intercalation/deintercalation of OH^- ions within the electrode framework with $Cs_3Bi_2Cl_9$ and $Cs_3Bi_2Br_9$ delivering a specific capacity of 1617 F/g and 613 F/g, respectively, at 2 mV/s scan rate with ~ 80% capacitance retention over 5000 cycles. This superior electrochemical performance was attributed to its optimized morphology and homogeneous microstructure with large electrochemical active surface area, facilitating fast ion diffusion, where the mixed halide composition (Br^-/Cl^-) modulated the band structure characteristics of the perovskite and provided multiple oxidation states enhancing the charge transfer kinetics and hence specific capacity [12]. In another example, a 0-D hybrid perovskite, methylammonium bismuth iodide ($(CH_3NH_3)_3Bi_2I_9$), was used as the electrochemically active material in a symmetrical electrochemical double-layer capacitor (EDLC), demonstrating an capacitance of 5.5 mF/cm^2 and exhibited excellent cycling stability, retaining 84.8% capacitance after 10,000 charge/discharge cycles which is attributed to the high surface area and porous nature of the $(CH_3NH_3)_3Bi_2I_9$ active layer, facilitating effective ion transport, while maintaining a low charge transfer resistance across electrode/electrolyte interface as confirmed by impedance spectroscopy; highlighting the crucial role of dimensional engineering in fabricating stable and high-performance EDLC supercapacitor using lead-free hybrid perovskite materials [53]. Additionally, in a study by Sasmal et al., $H_2W_2O_7$, a tungsten-based Ruddlesden-Popper (R–P) hybrid organic-inorganic perovskite was developed as a lead-free and ultrastable pseudocapacitive electrode for supercapacitor application, wherein, using the local structural engineering approach helped in achieving a high specific capacitance of 622 F/g (155.5 mAh/g at 64 C) and remarkable aqueous stability with ~100% capacity retention over thousands of cycles, thereby underscoring the key role of perovskite frameworks in enabling stable and high-performance pseudocapacitive electrodes for aqueous energy storage systems [58].

6.4.2 Mechanical Energy Storage

With the ever-increasing demand for compact, durable, and sustainable power sources for modern electronic devices, mechanical energy harvesting technologies have emerged as a reliable alternative to conventional batteries as devices such as wireless sensors, wearable health monitors, and implantable biomedical systems often require continuous and maintenance-free power, which traditional batteries struggle to provide due to their limited life span, bulkiness, and replacement challenges [61].

Based on various transduction energy conversion mechanisms explored, piezoelectric and triboelectric mechanical energy storage systems have attracted considerable interest owing to their excellent energy conversion efficiency and suitability for micro/nanoscale integration, discussed as below.

6.4.2.1 Piezoelectric Devices

Piezoelectric energy harvesting devices works on the principle of piezoelectricity, where certain materials possessing high piezoelectric coefficient and dielectric properties are placed under mechanical strain and upon deformation results in the displacement of positive and negative charges on their surfaces and generation of internal electric field, and thus creating a surface potential difference, which can be then harvested as electric energy. In this context, LFPs materials such as barium titanate ($BaTiO_3$), and bismuth ferrite ($BiFeO_3$), potassium sodium niobate (KNN), bismuth sodium titanate ($Bi_{0.5}Na_{0.5}TiO_3$, BNT), etc. have shown remarkable piezoelectric properties with strong piezoelectric coefficients and good mechanical flexibility and thus used as electrode material for flexible and wearable piezoelectric energy harvesting devices [59]. For instance, in a study by Khadtare et al., $Bi_{1-x}Sr_xFeO_3$ (x = 0.0–0.35), a lead-free piezoelectric material was synthesized using the sol-gel method and doped with strontium (Sr) in order to further enhanced its piezoelectric performance via the introduction of oxygen vacancies into the material in a controlled way and the composite with 25% Sr doping was found to exhibit highest remnant polarization (4.20 μC/cm^2), and maximal polarization (6.74 μC/cm^2), leading to superior piezoelectric behaviour; while the corresponding piezoelectric nanogenerator achieved an output voltage of 6.35 V and current of 0.64 μA, demonstrating excellent mechanical robustness and long-term stability over 1500 s of continuous operation. The fabricated piezoelectric device was able to power commercially available LEDs and charge a capacitor with 0.47 μF capacity while storing 7.29 J of energy within 500 s, highlighting the potential of Sr-doped $BiFeO_3$ as a promising lead-free piezoelectric material for durable and efficient energy harvesting applications [31]. Further, in another study by Mondal et al., a flexible piezoelectric nanogenerator (PENG) was fabricated using a composite of PVDF and rubidium copper chloride (Rb_2CuCl_3), a lead-free perovskite material, exhibiting strong interfacial interaction and enhanced β-phase content in PVDF to ~ 92.4%, crucial for high performing PENG. Specifically, the PENG device based on 2.5 wt% Rb_2CuCl_3 delivered an open-circuit voltage around 51.7 V with a short-circuit current of 4.6 μA and excellent charge output (~ 126 nC) while being stable over more than 10,000 mechanical cycles without voltage degradation, which was far superior as compared to pristine PVDF device. Additionally, due to the photoreactive properties of Rb_2CuCl_3, the composite showed piezo-phototronic effects, making it suitable for fast-response optoelectronic sensors, thus highlighting the multifunctionality of LFPs-based polymer composites in next-generation energy harvesters and wearable electronics as shown in Fig. 6.5 [48]. However, still improvements are being carried out focusing on different strategies,

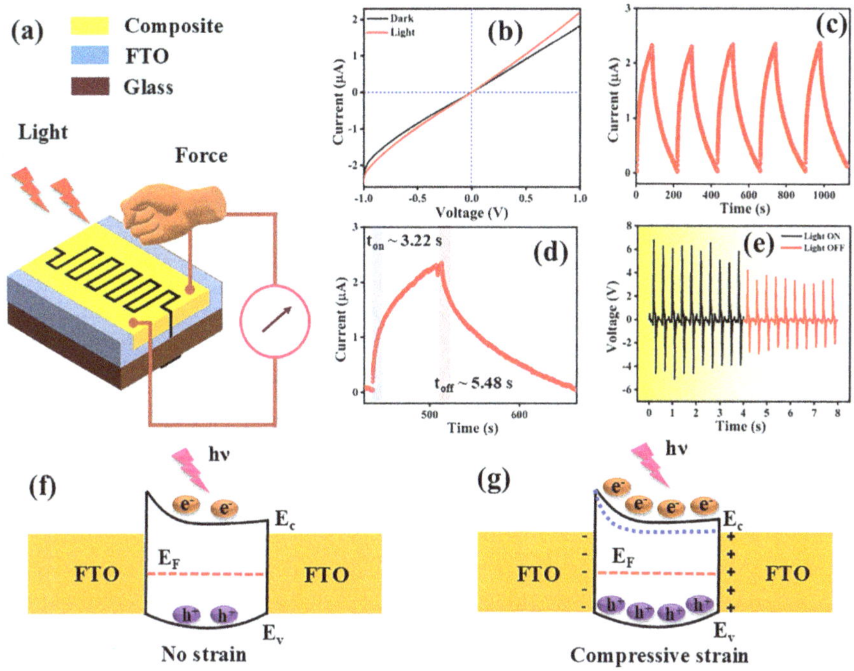

Fig. 6.5 Illustration of lead-free perovskite materials exhibiting piezo-phototronic effects for self-powered mechanical energy harvesting [48]. Copyright 2025. Reproduced with permission from American Chemical Society

such, elemental doping, surface modification, and composite formation with piezoelectric polymers to enhance charge generation, improve piezoelectric response, and boost energy conversion efficiency of LFPs-based piezoelectric devices.

6.4.2.2 Triboelectric Devices

Triboelectric energy harvesting devices, generally called triboelectric nanogenerators (TENGs), works on the principle of coupling of surface triboelectrification followed by charge redistribution, where mechanical interactions such as sliding, pressing, or vibrations between two materials with different electronegativity leads to the generation of an alternating current; which ultimately is harvested and stored as electrical energy [24]. In this context, LFPs have gained recognition as efficient electrodes in TENGs owing to their intrinsic ferroelectric properties, dielectric polarization, and chemical tunability, which can significantly enhance the surface charge density as well as charge retention, thus contributing to high-performing energy harvesting devices. The TENG performance is highly dependent upon the properties of the materials and the strategies used to synthesize them such as, compositional doping, surface engineering, composite formation, etc., which would help in boosting the output of the TENG by enhancing surface charge transfer, contact area, and effective dielectric response [24]. In this context, LFPs s with all the favourable characteristics required for a high performing TENG is widely explored as electrode material for next-generation self-powered electronics, particularly in wearable, biomedical, and environmental sensing applications. Recent studies have demonstrated the potential of lead-free halide perovskites, such as $CsBi_3I_{10}$, as an efficient triboelectric material which when integrated into the PVDF matrix via a one-step electrospinning method offered a significant enhancement in the output voltage of TENG, delivering a peak instantaneous power density of 7.11 W/m^2 owing to the improved dielectric properties and surface charge density, as confirmed by dielectric constant measurements and Kelvin probe force microscopy. Further, the fabricated device demonstrated effective biomechanical energy harvesting when worn on different parts of the body and proved its practical utility by functioning as a sensor in a wearable fall detection system for individuals with physical impairments [62]. In another study by Wei et al., $CsBi_3I_{10}$ based thin films were used in constructing photo-enhanced TENG integrating the material's semiconducting behaviour and electrical conductivity to have the synergistic effect due to triboelectric and photoelectric mechanisms. It was found to achieve a voltage of 158 V with current density of 45.5 mA/m^2 under dark conditions, which was further enhanced by 21.7% in voltage and 15.8% in current under light exposure with a power density of ~ 4.95 W/m^2, demonstrating the potential of hybrid energy harvesting from both mechanical and optical sources. Thus, this hybrid energy harvesting approach highlights that how lead-free perovskites can not only be used in mechanical energy conversion but can also in designing the next-generation multifunctional nanogenerators capable of harvesting both mechanical as well as solar energy [70].

6.4.3 Electrical Storage Systems

Electrical energy storage systems work on the principle of temporarily storing the electrical charge and release it whenever required enabling stable power delivery, and rapid energy discharge, playing a crucial role in modern electronics especially where fast response times and high-power densities are required [55]. Among them, electrostatic capacitors and electromagnetic storage devices represent two important categories, discussed in detail as follows.

6.4.3.1 Electrostatic Capacitors

Electrostatic capacitors, store energy in the form of electric field generated via the separation of positive and negative charges across a dielectric material when it is subjected to an external voltage and involves non-faradaic mechanisms [46]. The performance of these capacitors is largely governed by the dielectric constant, breakdown voltage, and polarization behaviour of the dielectric material. Since batteries offer high energy density, however, its limited power density restricts its usage in high-performance applications; on the other hand, supercapacitors resolve this issue by delivering high power density with nominal energy density and fast response; making them ideal for use in modern electronics. Nevertheless, its limited potential window, considerable leakage currents in the milliampere range, and a high economic cost, estimated around 9500 USD/kWh is still a bottleneck for its adoption at large-scale. Thus, in contrast, places where fast response and consistent performance is required, electrostatic capacitors are found to best suited owing to their high-voltage window, large-scale, and cost-effective fabrication. In this context, LFPs, particularly the materials possessing ferroelectric and antiferroelectric properties have shown promising dielectric properties, and robust temperature stability; crucial parameters for energy storage in electrostatic capacitors. For example, in recent research by Soheli et al., electrostatic energy storage capacitor was developed using an anti-ferroelectric LFPs material i.e., sodium niobate ($NaNbO_3$) owing to its desirable properties, including high dielectric breakdown strength and prominent recoverable energy density. Moreover, its crystal structure and phase transition behaviour played a crucial role in inducing relaxor-like behaviour resulting in a slim, non-hysteretic P-E loop, thereby maximizing the recoverable energy density, exceeding 18 J/cm^3, outperforming many conventional lead-based dielectric materials [60].

6.4.3.2 Electromagnetic Devices

Electromagnetic energy storage devices operate based on the principle of converting the magnetic energy to electrical energy using the principles of electromagnetic induction via the use of different components including inductors, transformers and superconducting magnetic energy storage (SMES) units (R. A. [20]). Till date, metallic and ceramic magnetic materials, the traditionally used materials have been widely explored in many of the electromagnetic devices, like, inductors in power electronics, ferrite-core transformers in high-frequency circuit, etc., however, the emergence of LFPs has garnered huge interest in this domain due to their inherent ferroelectric-ferromagnetic coupling (multiferroicity) and high electrical insulation properties. For instance, in a study by Wang et al., a modified lead-free perovskite composition of $Ba_{0.85}Ca_{0.15}Ti_{0.90}Zr_{0.10}O_3$ combined with 0.3 wt% CuO was synthesized via a high-temperature sintering route at 1450 °C to optimize its crystallinity and functional properties which was further used for magnetic field energy harvesting. Upon exposure to a weak magnetic of 250 μT only, the fabricated device was observed to deliver a significant output generating a peak power of ~ 8.2 mW and an energy density of 107.90 mW/cm^3, which is attributed to the enhancement in the piezoelectric energy conversion constant ($d_{33} \times g_{33}$) of the as synthesized composite [69]. Further, the fabricated device was successfully utilized to charge a capacitor and power a wireless sensor network, demonstrating its ability to support temperature sensing and radio frequency transmission in low-intensity magnetic environments. Another remarkable example of lead-free magnetoelectric energy harvesting is a composite of PVDF and $Ba_{0.85}Ca_{0.15}Ti_{0.9}Zr_{0.1}O_3$ integrated with magnetostrictive Metglas layer, the synergy between whom led to efficient conversion of low-frequency alternating magnetic fields into electrical energy. The fabricated device delivered a voltage of 14.8 V achieving a power density of ~ 4.7 $\mu W/cm^3$ even at a low-intensity AC magnetic field of ~ 10 Oe at 50 Hz frequency attributed to the strong interfacial coupling between the $Ba_{0.85}Ca_{0.15}Ti_{0.9}Zr_{0.1}O_3$ particles and the PVDF matrix, along with the high magnetic permeability of magnetostrictive Metglas layer; demonstrating its potential in wireless sensor networks, particularly for real-time magnetic field detection and energy harvesting from ambient magnetic noise sources, such as those emitted by high-temperature industrial equipment [25].

6.4.4 Thermal Energy Storage Systems

Thermal energy storage systems work on the principle of harnessing the thermal energy by absorbing or releasing heat via processes such as, sensible heat storage, latent heat associated with phase transitions, or reversible thermochemical reactions [11]. While the conventional thermal energy storage systems, such as, employing molten salts or organic phase changing materials, are widely used, however, the emerging studies are now focusing on exploiting the electrocaloric effects and field-induced phase transitions to achieve temperature regulation using the LFPs for solid-state thermal energy storage owing to their promising field-responsive thermal characteristics. For instance, Khardazi et al. synthesized $Ba_{0.9}Sr_{0.1}Ti_{0.9}Sn_{0.1}O_3$, a lead-free ferroelectric perovskite ceramic prepared through the sol-gel process, demonstrating a recoverable energy density of of 58.08 mJ/cm^3 with energy efficiency reaching 84.36% at room temperature and moderate electric field (20 kV/cm). Moreover, it exhibited a significant electrocaloric temperature change (~ 0.63 K) and a high EC responsivity of 0.31 K mm/kV owing to its high electrocaloric responsiveness, strong ferroelectric behaviour, and efficient energy storage capabilities thus, making it a strong candidate for solid-state thermal energy storage and cooling applications [32]. Further, in another study by Bai et al., electrocaloric strength of $BaTiO_3$ was found to be exceptionally high, reaching around 0.48 K cm/kV, making it reliable for solid-state thermal energy storage with both high performance and durability [9].

6.4.5 Chemical Energy Storage Systems

Chemical energy storage relies on the principle of capturing energy within chemical bonds chemical bonds of a molecule and which can be released later via controlled chemical reactions to generate power, playing a pivotal role in long-duration and large-scale storage systems. LFPs have been gaining significant attention in this domain due to their tunable catalytic properties, environment friendly nature, and structural flexibility [21].

6.4.5.1 Hydrogen Energy Storage

One of the most advanced and widely explored technologies in chemical energy storage is hydrogen-based energy conversion and storage, where energy is conserved through hydrogen gas produced via the process involving water electrolysis for hydrogen production, followed by electricity generation via fuel cells [21]. These systems are highly efficient and are very environment friendly as they emit only water as the by-product. A promising example of LFPs for hydrogen-based chemical energy storage systems includes $RbXH_3$ (X = Sc, Y) perovskite hydrides investigated using density functional theory (DFT) based methods by Masood et al., each of which showcased stable and high hydrogen storage potential with $RbScH_3$ demonstrating superior hydrogen storage capacity of 2.27 wt% and a desorption temperature of 374 K [45]. Further, in another study by Rehman et al., solid-state hydrogen storage potential of $NaMTH_3$ (MT = Sc, Ti, V) based hydride perovskites have been evaluated using DFT method and these materials were found to exhibit promising gravimetric hydrogen storage capacities of 4.26 wt% ($NaScH_3$), 4.09 wt% ($NaVH_3$), and 3.83 wt% ($NaTiH_3$) owing to their good mechanical stability, high metallic conductivity, dynamic phonon stability, and favourable thermodynamic behaviour [65].

6.4.5.2 Synthetic Natural Gas Storage

Another important chemical energy storage systems includes Synthetic Natural Gas Storage systems that convert the renewable hydrogen and captured CO_2 from atmosphere into methane via the Sabatier reaction ($CO_2 + 4H_2 \rightarrow CH_4 + 2H_2O$). This process not only stores chemical energy as fuel but also addresses CO_2 emissions (major greenhouse gases responsible for global warming) by converting it into methane that can be further stored and used as a substitute for fossil fuels for various applications [57]. One promising example of LFPs in chemical energy storage involves $LaNi_{0.9}Ru_{0.1}O_3$, a Ru-substituted lanthanum nickelate catalyst used for CO_2 methanation reaction supported on Al_2O_3 (30 wt%), achieving a 77% CO_2 conversion and > 90% CH_4 selectivity at just 400 °C, which was 50 °C lower than the bulk form; thus, demonstrating high efficiency and stability due to enhanced Ni dispersion and surface area [4]. Another promising approach employs Ni-doped rare-earth perovskite oxides of the form Ni/$AMnO_3$ (A = Sm, La, Nd, Ce, Pr), synthesized via an auto-combustion method for CO_2 methanation catalysis and among them Ni/$CeMnO_3$ exhibited superior with a CO_2 conversion rate of 6.6×10^{-5} mol CO_2 g/cat/s and high CH_4 selectivity due to the robust metal-support interaction, redox-active surface sites, high surface area, and CO_2 activation capability through associative and dissociative mechanisms, thus, highlighting its potential in sustainable CO_2-to-methane conversion [56].

6.5 Conclusion and Outlook

Since lead-free perovskites (LFPs) possess a number of unique and excellent properties which can be structurally controlled, making them promising renewable solutions to most energy storage applications. Despite this, there are still some issues associated with LFPs, which could be resolved using the modern advanced techniques, like, composite engineering, nano-structuring, and dopant modulation. This chapter has broadly discussed their structure and chemistry, methods of preparation, and their categorization in different energy storage systems.

Below listed are some of the still existing challenges faced by LFPs (also shown in Fig. 6.6):

1. Structural and phase stability issues: They are very reactive and undergo degradation upon exposure to humid or oxidative environments, therefore, limiting its long-term stability in some energy storage such as lithium-ion, zinc-ion, aluminium-ion batteries.
2. Limited awareness of structure-performance relationships: Higher characterization methods are necessary to tune the electronic structure and achieve the highly efficient LFPs as there is current lack in knowledge on how different dopants, surface terminations, crystal orientations, and lattice distortions would affect the performance of LFPs in all the energy storage technologies.
3. Flexibility and solid-state system compatibility: LFPs-based materials occur as rigid ceramic compounds that do not have the mechanical flexibility, hence restricting their applications to be parts of the future high performance flexible and wearable energy devices.

Thus, based on the aforementioned challenges, following future perspectives should be adopted for the development of LFPs in advanced energy storage devices (also shown in Fig. 6.6):

1. Use advanced doping and defect engineering techniques: The use of advanced doping and defect engineering techniques, such as, multivalent cation doping, A-site/B-site substitution, and controlled oxygen/non-oxygen vacancy formation should be adopted which would help ultimately help in improving the Charge storage and transfer performance.
2. Exploring novel hybrid composites using theoretical studies: Various theoretical studies should be explored which would in turn help in predicting the optimal combination of materials, understand interfacial interactions, and aid in engineering materials with the desired tailored properties for improved performance in diverse energy storage systems.
3. Use of Machine learning (ML) and artificial intelligence (AI) to synthesize lead-free materials: Since AI and ML are the emerging powerful tools in today's world, thus these predictive models would reduce the experimental workload and help in synthesizing the desired material with required properties, thereby accelerating the development of high-performance LFPs in a wide range of energy storage devices.

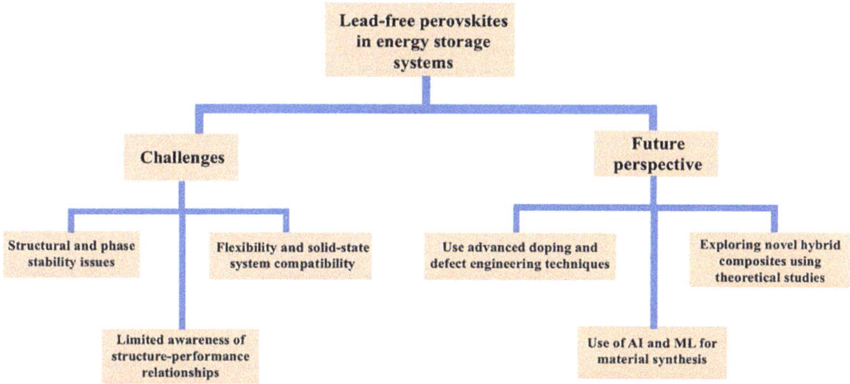

Fig. 6.6 Challenges and future perspectives of lead-free perovskites in energy storage applications

References

1. T.R. Abraham, K. Sunil, M. Shah, N. Ashok, S. Thomas, Energy storage devices: batteries and supercapacitors, in *Nanobiohybrids for Advanced Wastewater Treatment and Energy Recovery*, (IWA Publishing, 2023), pp. 61–84. https://doi.org/10.2166/9781789063592_0061
2. K. Aishwarya, I. Hannah Jeniffer, S. Maruthasalamoorthy, R. Nirmala, N. Punithavelan, R. Navamathavan, Review—state of the art of the multifunctional bismuth ferrite: synthesis method and applications. ECS J. Solid State Sci. Technol. **11**(4), 043010 (2022). https://doi.org/10.1149/2162-8777/ac627a
3. T. Alammar, I. Hamm, V. Grasmik, M. Wark, A.-V. Mudring, Microwave-assisted synthesis of perovskite $SrSnO_3$ nanocrystals in ionic liquids for photocatalytic applications. Inorg. Chem. **56**(12), 6920–6932 (2017). https://doi.org/10.1021/acs.inorgchem.7b00279
4. A.A. Alkhoori, E.P. Komarala, A.A. Dabbawala, A.G.S. Hussien, D.H. Anjum, S. Mao, K. Polychronopoulou, Exsolved $LaNiRuO_3$ perovskite-based catalysts for CO_2 methanation reaction. RSC Sustain. **2**(12), 3866–3878 (2024). https://doi.org/10.1039/D4SU00410H
5. A.A. Ansari, N. Ahmad, M. Alam, S.F. Adil, M.E. Assal, A. Albadri, A.M. Al-Enizi, M. Khan, Optimization of redox and catalytic performance of $LaFeO_3$ perovskites: synthesis and physicochemical properties. J. Electron. Mater. **48**(7), 4351–4361 (2019). https://doi.org/10.1007/s11664-019-07216-4
6. T. Ao, A.T. Jafry, N. Abbas, Sustainability and scalability of photoelectrochemical and photocatalytic water splitting by using perovskite materials for hydrogen production. Electrochem. Commun. **177**, 107948 (2025). https://doi.org/10.1016/j.elecom.2025.107948
7. A. Arabkoohsar, 1—Classification of energy storage systems, in *Future Grid-Scale Energy Storage Solutions*, ed. by A. Arabkoohsar, (Academic, 2023), pp. 1–30. https://doi.org/10.1016/B978-0-323-90786-6.00011-X
8. E.A.R. Assirey, Perovskite synthesis, properties and their related biochemical and industrial application. Saudi Pharm. J. **27**(6), 817–829 (2019). https://doi.org/10.1016/j.jsps.2019.05.003
9. Y. Bai, X. Han, X.-C. Zheng, L. Qiao, Both high reliability and giant electrocaloric strength in $BaTiO_3$ ceramics. Sci. Rep. **3**(1), 2895 (2013). https://doi.org/10.1038/srep02895
10. K.K. Bharathi, B. Moorthy, H.K. Dara, L. Durai, D.K. Kim, Electrochemical properties of $Na_{0.5}Bi_{0.5}TiO_3$ perovskite as an anode material for sodium ion batteries. J. Mater. Sci. **54**(20), 13236–13246 (2019). https://doi.org/10.1007/s10853-019-03834-9
11. V. Busico, C. Carfagna, V. Salerno, M. Vacatello, F. Fittipaldi, The layer perovskites as thermal energy storage systems. Sol. Energy **24**(6), 575–579 (1980). https://doi.org/10.1016/0038-092X(80)90356-4

12. A. Chatterjee, S. Das, P. Pal, S. Maiti, T. Paul, M. Ghosh, K.K. Chattopadhyay, Unveiling the electrochemical supercapacitor dynamics in lead-free $Cs_3Bi_2X_9$ (Cl/Br) perovskite: impact of halogen alteration. J. Energy Storage **119**, 116299 (2025). https://doi.org/10.1016/j.est.2025.116299
13. C. Chesta, J. Subbiah, S. Srinivasan, D.J. Jones, Unravelling the performance of lead-free perovskite cathodes for rechargeable aqueous zinc-ion batteries. J. Power Sources **624**, 235591 (2024). https://doi.org/10.1016/j.jpowsour.2024.235591
14. A.E. Danks, S.R. Hall, Z. Schnepp, The evolution of 'sol–gel' chemistry as a technique for materials synthesis. Mater. Horiz. **3**(2), 91–112 (2016). https://doi.org/10.1039/C5MH00260E
15. N.A. Fauziyah, A.W. Permatasari, P.C. Wardhani, Musyarofah, Z. Mohamed, W. Kartini, Potential applications of the hydrothermal process for the development of perovskite-based PCM materials. IOP Conf. Ser.: Earth Environ. Sci **1454**(1), 012008 (2025). https://doi.org/10.1088/1755-1315/1454/1/012008
16. Y. Gao, Y. Pan, F. Zhou, G. Niu, C. Yan, Lead-free halide perovskites: a review of the structure–property relationship and applications in light emitting devices and radiation detectors. J. Mater. Chem. A **9**(20), 11931–11943 (2021). https://doi.org/10.1039/D1TA01737C
17. G. Getachew, A. Wibrianto, A.S. Rasal, S. Kizhepat, W.B. Dirersa, V. Gurav, J.-Y. Chang, Lead-free metal halide perovskites as the rising star in photocatalysis: the past, present, and prospective. Prog. Mater. Sci. **140**, 101192 (2023). https://doi.org/10.1016/j.pmatsci.2023.101192
18. P. Goel, S. Sundriyal, V. Shrivastav, S. Mishra, D.P. Dubal, K.-H. Kim, A. Deep, Perovskite materials as superior and powerful platforms for energy conversion and storage applications. Nano Energy **80**, 105552 (2021). https://doi.org/10.1016/j.nanoen.2020.105552
19. L. He, P.-P. Shi, M.-M. Zhao, C.-M. Liu, W. Zhang, Q. Ye, Emergent chirality and nonlinear optical switching in a ferroelastic molecular perovskite solid solution. Chem. Mater. **33**(2), 799–805 (2021). https://doi.org/10.1021/acs.chemmater.0c04735
20. R.A. Huggins, Electromagnetic energy storage, in *Energy Storage: Fundamentals, Materials and Applications*, ed. by R. Huggins, (Springer International Publishing, 2016), pp. 69–93. https://doi.org/10.1007/978-3-319-21239-5_7
21. H.L. Huynh, W.M. Tucho, X. Yu, Z. Yu, Synthetic natural gas production from CO_2 and renewable H_2: towards large-scale production of Ni–Fe alloy catalysts for commercialization. J. Clean. Prod. **264**, 121720 (2020). https://doi.org/10.1016/j.jclepro.2020.121720
22. M.R. Islam, S. Afroj, K.S. Novoselov, N. Karim, Smart electronic textile-based wearable supercapacitors. Adv. Sci. **9**(31), 2203856 (2022). https://doi.org/10.1002/advs.202203856
23. H.J. Jeong, H.S. Park, K.H. Kim, W. Chang, Y.S. Kim, Y.S. Choi, J.H. Shim, Performance improvement of proton ceramic fuel cells through surface treatment of cobalt oxide nanoparticles on perovskite oxide. ECS Trans. **111**(6), 2155 (2023). https://doi.org/10.1149/11106.2155ecst
24. Z. Jin, H. Xing, Y. Hu, Y. Cho, Perovskite-based triboelectric nanogenerator and its application towards self-powered devices. Nanoenergy Adv. **5**(1), 2 (2025). https://doi.org/10.3390/nanoenergyadv5010002
25. J. Kaarthik, N. Ram, R. Aepuru, S.G. Reddy, D.P. Pabba, A. Venkateswarlu, Robust magnetic energy harvesting with flexible lead-free poly(vinylidene fluoride)-$Ba_{0.85}Ca_{0.15}Ti_{0.9}Zr_{0.1}O_3$ fibers and Metglas-based magnetoelectric composites. Sustain. Energy Fuel **8**(12), 2583–2592 (2024). https://doi.org/10.1039/D3SE01517C
26. M. Kandhasamy, B.K. Duvaragan, S. Kamaraj, G. Shanmugam, An overview on classification of energy storage systems, in *Materials for Boosting Energy Storage. Volume 2: Advances in Sustainable Energy Technologies*, vol. 1484, (American Chemical Society, 2024), pp. 1–25. https://doi.org/10.1021/bk-2024-1484.ch001
27. V. Karol, P. Sharma, A. Singh, D. Goel, S. Kaur, Review of energy storage devices: fuel cells, hydrogen storage fuel cells, rechargeable batteries, PV solar cells, 1–15 (2024a). https://doi.org/10.1021/bk-2024-1488.ch001
28. V. Karol, P. Sharma, A. Singh, D. Goel, S. Kaur, Review of energy storage devices: Fuel cells, hydrogen storage fuel cells, rechargeable batteries, PV solar cells, in *Materials for Boosting*

Energy Storage. Volume 3: Advances in Sustainable Energy Technologies, vol. 1488, (American Chemical Society, 2024b), pp. 1–15. https://doi.org/10.1021/bk-2024-1488.ch001
29. E.A. Katz, Perovskite: name puzzle and German-Russian odyssey of discovery. Helv. Chim. Acta **103**(6), e2000061 (2020). https://doi.org/10.1002/hlca.202000061
30. V.G. Kessler, G.I. Spijksma, G.A. Seisenbaeva, S. Håkansson, D.H.A. Blank, H.J.M. Bouwmeester, New insight in the role of modifying ligands in the sol-gel processing of metal alkoxide precursors: a possibility to approach new classes of materials. J. Sol-Gel Sci. Technol. **40**(2), 163–179 (2006). https://doi.org/10.1007/s10971-006-9209-6
31. S. Khadtare, A. Kumar, U. Jung, C. Sang, J. Park, Piezoelectric nanogenerator based on lead-free $BiFeO_3$:Sr perovskite. J. Alloys Compd. **926**, 166738 (2022). https://doi.org/10.1016/j.jallcom.2022.166738
32. S. Khardazi, H. Zaitouni, S. Belkhadir, D. Mezzane, M. Amjoud, Y. Gagou, B. Asbani, I. Lukyanchuk, S. Terenchuk, Improvement of the electrocaloric effect and energy storage performances in Pb-free ferroelectric $Ba_{0.9}Sr_{0.1}Ti_{0.9}Sn_{0.1}O_3$ ceramic near room temperature. J. Solid State Chem. **311**, 123112 (2022). https://doi.org/10.1016/j.jssc.2022.123112
33. R. Kumar, D. Lee, Ü. Ağbulut, S. Kumar, S. Thapa, A. Thakur, R.D. Jilte, C.A. Saleel, S. Shaik, Different energy storage techniques: recent advancements, applications, limitations, and efficient utilization of sustainable energy. J. Therm. Anal. Calorim. **149**(5), 1895–1933 (2024). https://doi.org/10.1007/s10973-023-12831-9
34. A. Larrañaga, A. Ecija, K. Vidal, L. Ortega-San-Martín, M.I. Arriortua, Synthetic methods for perovskite materials; structure and morphology, in *Advances in Crystallization Processes*, ed. by Y. Mastai, (IntechOpen, 2012). https://doi.org/10.5772/36540
35. W. Li, X. Li, C. Wang, Y. Gong, R. Wang, H. Wang, J. Jin, L. Zhao, B. He, Enhancing the electrocatalytic activity of perovskite electrodes by atomic layer-deposited doped CeO_2 for symmetrical solid oxide fuel cells. Sep. Purif. Technol. **302**, 122135 (2022). https://doi.org/10.1016/j.seppur.2022.122135
36. X. Li, S. Aftab, S. Hussain, F. Kabir, A.M.A. Henaish, A.G. Al-Sehemi, M.R. Pallavolu, G. Koyyada, Dimensional diversity (0D, 1D, 2D, and 3D) in perovskite solar cells: exploring the potential of mixed-dimensional integrations. J. Mater. Chem. A **12**(8), 4421–4440 (2024). https://doi.org/10.1039/D3TA06953B
37. A.R. Lim, Exploring the potential applications of lead-free organic–inorganic perovskite type $[NH_3(CH_2)_nNH_3]MCl_4$ (n = 2, 3, 4, 5, and 6; M = Mn, Co, Cu, Zn, and Cd) crystals. Sci. Rep. **14**(1), 11808 (2024). https://doi.org/10.1038/s41598-024-62705-9
38. A. Liu, H. Zhu, Y. Reo, M.-G. Kim, H.Y. Chu, J.H. Lim, H.-J. Kim, W. Ning, S. Bai, Y.-Y. Noh, Modulation of vacancy-ordered double perovskite Cs_2SnI_6 for air-stable thin-film transistors. Cell Rep. Phys. Sci. **3**(4), 100812 (2022). https://doi.org/10.1016/j.xcrp.2022.100812
39. J. Liu, E. Jia, K.A. Stoerzinger, L. Wang, Y. Wang, Z. Yang, D. Shen, M.H. Engelhard, M.E. Bowden, Z. Zhu, S.A. Chambers, Y. Du, Dynamic lattice oxygen participation on perovskite $LaNiO_3$ during oxygen evolution reaction. J. Phys. Chem. C **124**(28), 15386–15390 (2020). https://doi.org/10.1021/acs.jpcc.0c04808
40. L. Liu, X. Zhang, Y. Liu, X. Gong, Electrochemical energy storage devices-batteries, supercapacitors, and battery–supercapacitor hybrid devices. ACS Appl. Electron. Mater. **7**(6), 2233–2270 (2025). https://doi.org/10.1021/acsaelm.5c00069
41. C.-H. Lu, N. Naresh, P.S. Kumar, S. Som, Microwave-assisted solvothermal synthesis and electrochemical characterizations of ternary perovskite $NiTiO_3$ anode materials for lithium-ion batteries. Ceram. Int. **45**(15), 19517–19521 (2019). https://doi.org/10.1016/j.ceramint.2019.06.057
42. M. Lyu, J. Yun, P. Chen, M. Hao, L. Wang, Addressing toxicity of lead: progress and applications of low-toxic metal halide perovskites and their derivatives. Adv. Energy Mater. **7**(15) (2017). https://doi.org/10.1002/aenm.201602512
43. K.V. Madhale, M.M. Faras, A.A. Mohite, U.V. Shembade, S.R. Gurav, A.P. Torane, S.B. Kulkarni, Facile combustion synthesis of granular-like $La_{1-x}Sr_xMnO_3$ perovskites as electrode material for supercapacitor application. J. Energy Storage **101**, 113841 (2024). https://doi.org/10.1016/j.est.2024.113841

44. B. Malič, J. Koruza, J. Hreščak, J. Bernard, K. Wang, J. Fisher, A. Benčan, Sintering of lead-free piezoelectric sodium potassium niobate ceramics. Materials **8**(12), 8117–8146 (2015). https://doi.org/10.3390/ma8125449
45. M.K. Masood, W. Khan, S. Bibi, M. Urooj, A. Ali, J. Rehman, A.A.A. Bahajjaj, Exploring the hydrogen storage potential of lead-free rubidium based perovskite hydrides: a first-principles study. Int. J. Hydrog. Energy **147**, 150036 (2025). https://doi.org/10.1016/j.ijhydene.2025.150036
46. J.R. Miller, CAPACITORS | Overview, in *Encyclopedia of Electrochemical Power Sources*, ed. by J. Garche, (Elsevier, 2009), pp. 587–599. https://doi.org/10.1016/B978-044452745-5.00350-6
47. S. Mondal, S. Poddar, S. Bhattacharjee, S. Maiti, A. Banerjee, K.K. Chattopadhyay, Lead free halide perovskite embedded PVDF based efficient mechanical energy harvester: self-driven respiratory sensor. Nano Energy **115**, 108689 (2023). https://doi.org/10.1016/j.nanoen.2023.108689
48. S. Mondal, M. Salesh, U. Latief, A. Aishwarya, A. Yella, A.R. Bhattacharyya, Multifunctional lead-free halide perovskite based poly(vinylidene fluoride) composites for biomechanical energy harvesting and self-powered piezo-optoelectronic applications. ACS Appl. Mater. Interfaces **17**(12), 18813–18830 (2025). https://doi.org/10.1021/acsami.4c21089
49. J. Mu, K. Xu, L. He, T. Yin, Y. Xu, J. Men, Q. Ye, Lead-free semiconducting organic–inorganic hybrid perovskite with mixed organic cations. Cryst. Growth Des. **23**(4), 2042–2047 (2023). https://doi.org/10.1021/acs.cgd.2c00931
50. M. Mukhtar, B.S. Goud, Z. Ali, M.W. Shaid, B. Naz, Q.U. Ain, M.A. Assiri, S. Sonmezoglu, A.H. Rajpar, J.H. Kim, S. Aftab, Eco-friendly lead-free metal halide perovskites: progress and prospects in multifunctional applications. J. Alloys Compd. **1026**, 180365 (2025). https://doi.org/10.1016/j.jallcom.2025.180365
51. C.N. Nwagu, C.O. Ujah, D.V.V. Kallon, V.S. Aigbodion, Integrating solar and wind energy into the electricity grid for improved power accessibility. Unconventional Resour. **5**, 100129 (2025). https://doi.org/10.1016/j.uncres.2024.100129
52. A.G. Olabi, K. Elsaid, K. Obaideen, M.A. Abdelkareem, H. Rezk, T. Wilberforce, H.M. Maghrabie, E.T. Sayed, Renewable energy systems: comparisons, challenges and barriers, sustainability indicators, and the contribution to UN sustainable development goals. Int. J. Thermofluids **20**, 100498 (2023). https://doi.org/10.1016/j.ijft.2023.100498
53. J.K. Pious, M.L. Lekshmi, C. Muthu, R.B. Rakhi, C. Vijayakumar, Zero-dimensional methylammonium bismuth iodide-based Lead-free perovskite capacitor. ACS Omega **2**(9), 5798–5802 (2017). https://doi.org/10.1021/acsomega.7b00973
54. P. Priyadarshini, S. Senapati, R. Naik, Lead-free organic inorganic hybrid halide perovskites: an emerging candidate for bifunctional applications. Renew. Sust. Energ. Rev. **186**, 113649 (2023). https://doi.org/10.1016/j.rser.2023.113649
55. M.M. Rana, M. Uddin, M.R. Sarkar, S.T. Meraj, G.M. Shafiullah, S.M. Muyeen, M.A. Islam, T. Jamal, Applications of energy storage systems in power grids with and without renewable energy integration—a comprehensive review. J. Energy Storage **68**, 107811 (2023). https://doi.org/10.1016/j.est.2023.107811
56. M. Safdar, N. Shezad, F. Akhtar, H. Arellano-García, Development of Ni-doped A-site lanthanides-based perovskite-type oxide catalysts for CO_2 methanation by auto-combustion method. RSC Adv. **14**(28), 20240–20253 (2024). https://doi.org/10.1039/D4RA02106A
57. S. Sami, M. Gholizadeh, M. Deymi-Dashtebayaz, A comprehensive 5E analysis of synthetic natural gas production through direct air capture and renewable hydrogen: based on a specified-scale residential application. Renew. Sust. Energ. Rev. **212**, 115376 (2025). https://doi.org/10.1016/j.rser.2025.115376
58. S. Sasmal, S. Roy, P.K. Gupta, K. Ramkumar, P. Biswas, B. Ghorui, S.K. Saju, P.M. Ajayan, S. Valiyaveettil, R.G.S. Pala, S. Sivakumar, Lead-free, ultrastable, tungsten-based Ruddlesden–Popper and hybrid perovskite frameworks for aqueous pseudocapacitive charge storage. Energ. Technol., 2400531 (2025). https://doi.org/10.1002/ente.202400531

59. N. Sezer, M. Koç, A comprehensive review on the state-of-the-art of piezoelectric energy harvesting. Nano Energy **80**, 105567 (2021). https://doi.org/10.1016/j.nanoen.2020.105567
60. S.N. Soheli, Z. Lu, D. Sun, I. Shyha, Lead-free $NaNbO_3$-based ceramics for electrostatic energy storage capacitors. Ceramics **7**(2), 712–734 (2024). https://doi.org/10.3390/ceramics7020047
61. W.-D. Steinmann, Overview of the section on mechanical energy storage, in *Encyclopedia of Energy Storage*, (Elsevier, 2022), pp. 1–4. https://doi.org/10.1016/B978-0-12-819723-3.00152-9
62. H. Su, J. Zheng, L. Tang, F. Liu, M. Lu, F. Wang, L. Wang, H. Zheng, B. Cai, W.W. Yu, One-step electrospinning synthesis of lead-free perovskite nanofiber-based triboelectric nanogenerator for wearable intelligent fall alarm. J. Colloid Interface Sci. **696**, 137836 (2025). https://doi.org/10.1016/j.jcis.2025.137836
63. T. Tabari, D. Singh, A. Calisan, M. Ebadi, H. Tavakkoli, B. Caglar, Microwave assisted synthesis of $La_{1-x}Ca_xMnO_3$ (x = 0, 0.2 and 0.4): structural and capacitance properties. Ceram. Int. **43**(17), 15970–15977 (2017). https://doi.org/10.1016/j.ceramint.2017.08.182
64. N. Tewari, S.B. Shivarudraiah, J.E. Halpert, Photorechargeable lead-free perovskite lithium-ion batteries using hexagonal $Cs_3Bi_2I_9$ nanosheets. Nano Lett. **21**(13), 5578–5585 (2021). https://doi.org/10.1021/acs.nanolett.1c01000
65. Z. ur Rehman, M.A. Rehman, S.Y. Alomar, B. Rehman, M. Awais, M. Amjad, S. Sikiru, E.M. Ali, A. Hamad, Hydrogen storage capacity of lead-free perovskite NaM_TH_3 (M_T=Sc, Ti, V): A DFT study. Int. J. Energy Res. **2024**(1), 4009198 (2024). https://doi.org/10.1155/2024/4009198
66. M. Varlam, M. Dragan, S. Enache, K. Petrov, Perovskite-based materials for energy applications, in *Perovskite Materials, Devices and Integration*, ed. by H. Tian, (IntechOpen, 2020). https://doi.org/10.5772/intechopen.91271
67. A. Varma, A.S. Mukasyan, A.S. Rogachev, K.V. Manukyan, Solution combustion synthesis of nanoscale materials. Chem. Rev. **116**(23), 14493–14586 (2016). https://doi.org/10.1021/acs.chemrev.6b00279
68. L. Wang, H. Liu, Y. Zhang, O.F. Mohammed, Photoluminescence origin of zero-dimensional Cs_4PbBr_6 perovskite. ACS Energy Lett. **5**(1), 87–99 (2020). https://doi.org/10.1021/acsenergylett.9b02275
69. Q. Wang, K.-B. Kim, S.B. Woo, T.H. Sung, Magnetic field energy harvesting with a lead-free piezoelectric high energy conversion material. Energies **14**(5), 1346 (2021). https://doi.org/10.3390/en14051346
70. Z. Wei, L. Ding, N. Sun, L.-Y. Dang, H.-R. Sun, J.-C. Han, J.-Q. Zhu, G.-G. Wang, Lead-free $CsBi_3I_{10}$ perovskite based photo-enhanced triboelectric nanogenerator. Nano Energy **108**, 108209 (2023). https://doi.org/10.1016/j.nanoen.2023.108209
71. H. Wu, J. Pi, Q. Liu, Q. Liang, J. Qiu, J. Guo, Z. Long, D. Zhou, Q. Wang, All-inorganic lead free double perovskite Li-battery anode material hosting high Li^+ ion concentrations. J. Phys. Chem. Lett. **12**(17), 4125–4129 (2021). https://doi.org/10.1021/acs.jpclett.1c00041
72. A. Yadav, A. Kumar, M. Bag, Enhancing device performance and stability of lead-free quasi-2D halide perovskite supercapacitor through Ag^+/Bi^{3+} cation interaction. FlatChem **47**, 100717 (2024). https://doi.org/10.1016/j.flatc.2024.100717
73. C. Yadav, S. Kumar, Review on perovskite solar cells via vacuum and non-vacuum solution based methods. Results Surf. Interfaces **14**, 100210 (2024). https://doi.org/10.1016/j.rsurfi.2024.100210
74. R. Zhang, Z. Chen, J. Ma, P. Zhang, M. Liu, X. Li, R. Zhao, J. Tang, Z. Ren, S. Li, An all-solid-state photo-rechargeable battery based on $Cs_3Bi_2I_9$. Chem. Commun. **59**(20), 2911–2914 (2023). https://doi.org/10.1039/D2CC06662A
75. C. Zhao, X. Meng, W. Wang, Y. Zhou, Energy storage performance of (K, Na)NbO_3 ferroelectric thin films with Mn-Ta and Mn-Ti co-doping. Ceram. Int. **45**(11), 13772–13779 (2019). https://doi.org/10.1016/j.ceramint.2019.04.073
76. L. Zhou, J.-F. Liao, Z.-G. Huang, X.-D. Wang, Y.-F. Xu, H.-Y. Chen, D.-B. Kuang, C.-Y. Su, All-inorganic lead-free Cs_2PdX_6 (X = Br, I) perovskite nanocrystals with single unit cell thickness and high stability. ACS Energy Lett. **3**(10), 2613–2619 (2018). https://doi.org/10.1021/acsenergylett.8b01770

Chapter 7
Lead-Free Perovskites for Thin-Film Transistors

Taame Abraha Berhe, Chia-Hsun Nieh, Chung-Yen Yang, Amare Aregahegn Dubale, and Chu-Chen Chueh

Abstract Sustainability is a key issue for the next-generation optoelectronic semiconductor devices. Currently, devices based on lead halide perovskite semiconductors are revolutionizing the optoelectronics field beyond photovoltaic. Despite breakthroughs in a wide range of applications, including solar cells, light-emitting diodes, lasers, sensors, transistors, X-ray detection, sustainability issues have emerged as a bottleneck challenge for further breakthroughs at the industrial level. To overcome this challenge, the development of lead-free perovskite semiconductors is a top priority of the next-generation semiconductor devices. Lead-free perovskite thin-film transistors (TFTs) are one such example. Currently, progress in this technology is on-going and holds great promise for achieving high performance in displays and other invisible electronic applications. Therefore, this chapter aims to introduce the main discussions and key aspects of lead-free perovskite transistors. These discussions include, but are not limited to, the basic principles of TFTs and their manufacturing techniques, key performance parameters, key components, operating principles, transition requirements for lead-free perovskite transistors, understanding charge transport, high-mobility lead-free perovskites transistors, key dimensions, and primary device architecture layouts, as well as the future development directions of these transistors.

Keywords: Lead-free · Perovskite · Mobility · Dimension · Transistors · Stability

---

T. A. Berhe (✉) · C. Nieh · C.-Y. Yang · C.-C. Chueh (✉)
Department of Chemical Engineering, National Taiwan University, Taipei, Taiwan
e-mail: taame.abraha@adu.edu.et

C.-C. Chueh
e-mail: cchueh@ntu.edu.tw

T. A. Berhe
Department of Chemistry, College of Natural and Computational Sciences, Adigrat University, Adigrat, Ethiopia

A. A. Dubale
Center for Material Science and Engineering, College of Natural and Computational Sciences, Addis Ababa University, Addis Ababa, Ethiopia

V. Chauhan et al. (eds.), *Sustainable Lead-Free Perovskite Materials*, Topics in Applied Physics 157, https://doi.org/10.1007/978-3-032-24873-2_7

7.1 Introduction

In modern society, power electronics devices have become extremely important to human life. These devices combine principles from control engineering, electronics, and electrical engineering to understand how to manage, control, and transform electrical energy. The ultimate goal of power electronics is to provide economical, compact, and efficient devices to achieve the objectives of controlling and transforming electrical energy. Power electronics devices use semiconductor electronics to achieve the primary objectives of controlling and transforming electrical energy. These semiconductor devices have received considerable attention due to their versatility. These devices include three types that play a critical role in controlling and regulating high power in electronic circuits: thyristors, transistors, and diodes. Transistors are essential semiconductor devices for power and electronic signals, acting as switches and amplifiers. Modern electronics use transistors as basic structural elements to monitor the response between different terminals to an applied voltage. These semiconductor devices are widely used in various applications because they can monitor whether the current flowing between different terminals is proportional to the applied voltage. In essence, these semiconductor devices have become the backbone of microelectronics with the ultimate goal of providing fast electronic operations for the modern world.

Furthermore, transistors come in different types. These include field-effect transistors (FETs), bipolar junction transistors (BJTs), metal–oxide–semiconductor FETs (MOSFETs), and digital transistors. A FET is a 3-terminal device with a drain, gate, and source that can control the direction of current flow by applying an electric field to the semiconductor. The fast switching speed and high input impedance of FETs have made them highly desirable in various semiconductor power electronics devices such as digital, analog, and display applications, especially liquid crystal displays. Therefore, FETs are classified into junction FETs (JFETs) and MOSFETs and metal–semiconductor FETs (MSFETs), as well as high-electron-mobility FETs (HEMTs). Among these, MOSFETs have the widest range of applications. Compared to other types of FETs, JFETs have a reduction mode, low frequency drift, low input resistance, high drain resistance, and high input capacitance, which allows both n-type and p-type channels. The high input capacitance extends the operating time, while the low frequency makes them suitable devices for real-time circuits and oscillators. HEMTs are also designed to provide high power outputs. This is due to the low manufacturing costs and high mobility required for modern and flexible integrated systems and electronic devices. Therefore, FETs can provide efficient power distribution at high frequencies using high electron mobility heterostructure devices and epitaxial layers. As a result, this type of FET requires semiconductors with high electron mobility to function properly. Therefore, these types of thin film transistors (TFTs) are semiconductor devices with advantages such as low power consumption, better contrast, better color saturation, low cost, high response speed, and fast response times. In order to ensure a bright future and reduce greenhouse gas emissions, the society of this century must prioritize energy saving. As a result, the demand for TFTs that can display dynamic images is growing. It should be noted that poor resolution, depth of field, and color accuracy cannot achieve good visual quality, and great attention should be paid to completely changing the market. Therefore, it is necessary to explore new high-performance semiconductor devices with high electron mobility transistors.

For this purpose, lead halide perovskites have been introduced into FETs. However, due to the toxicity of lead metal and the resulting sustainability issues, FETs applications urgently require lead-free perovskite semiconductors. The first lead-free perovskite FET was fabricated using a tin halide perovskite, PEA_2Sn_4, with a hole mobility of 0.6 $cm^2 \cdot V^{-1} \cdot s^{-1}$ [19]. Due to the strong quantum confinement and low mobility of organic spacers, performance improvements have been slow [46]. Furthermore, the mobility of $CsSnI_3$ and $CH_3NH_3SnI_3$ was measured to be 580 and 2000 $cm^2 \cdot V^{-1} \cdot s^{-1}$ using other measurement methods. However, there are still many bottleneck challenges, such as material degradation or stability issues, ion migration under bias, poor charge carrier injection, and bulk and interface defects [11]. Although tin halide perovskites have achieved mobility of over 10 $cm^2 \cdot V^{-1} \cdot s^{-1}$, their intrinsic instability makes them unsuitable for FETss [30]. This stability challenge has been addressed by introducing the conjugate ligand, 2-(3‴, 4′-dimethyl-[2, 2′,5′,2′,5″,2‴-quaterthiophen]-5-yl)ethan-1-ammonium, thereby increasing the mobility of $(4Tm)_2SnI_4$ by a factor of 10 [11]. This demonstrates the important role of lead-free lenses in the development of FETs. Thus, this chapter will focus on the application of lead-free perovskite high-mobility FETs as the primary objective to deepen understanding of this research field.

7.2 Fundamentals of Perovskite TFTs

Transistors are a modern technological advancement used in a wide range of electronic applications that require the amplification or switching of electronic signals. These technologies are based on semiconductor materials and can be broadly classified into insulated-gate bipolar transistors (IGBTs), FETs, and BJTs. An IGBT is a semiconductor-based transistor that combines a MOSFET with a BJT to handle high current loads. IGBTs are essential in applications that require high current, high voltage, and fast switching speeds, such as renewable energy systems, industrial drives, and electric vehicle functions. BJTs, on the other hand, use charge carriers (such as holes and electrons) and are therefore divided into two types: NPN and PNP transistors. These transistor devices are 3-terminal semiconductor devices that can modulate or amplify electronic signals. Oscillators, switches, and on/off switches are all made of BJTs, which are current-sensing devices and are therefore very attractive in both digital and analog circuits. The third type of transistor is the FET, which is controlled by applying a voltage to induce an electric field across the gate. The gate, drain, and source are connected to a single semiconductor material and are considered a single semiconductor device with three terminals. This device does not have a p–n junction in the main circuit, which allows for the creation of p-channel and n-channel FETs. In addition, JFETs, TFTs, and MOSFETs are all types of FETs, as listed in Table 7.1.

Due to its inherent limitations, TFT has become extremely important in various applications, as described in Table 7.1. This type of transistor requires high-mobility semiconductors, such as halide perovskites [38, 59, 90, 92, 93].

7.3 The Appearance of Perovskite TFTs

Due to their solution processing and tunable semiconductor properties, perovskite TFTs (PetFTs) are in demand for next-generation semiconductor electronic devices such as optoelectronics, amplifiers, switches, sensors, and displays. The first 3D perovskite ($CH_3NH_3PbI_{3-x}Cl_x$) transistors based on lead-hydride perovskites had an electron mobility of 1.0 cm^2V^{-1} s^{-1} and hole mobility of 1.3 cm^2V^{-1} s^{-1} as reported as ambipolar carrier in 2015 [45], but the field effect mobility for the first 2D perovskite ($(C_6H_5C_2H_4NH_3)_2SnI_4$) is 0.6 cm^2V^{-1} s^{-1} in 1999 [19]. The hall mobility of the first $CH_3NH_3SnI_3$ was also reported as 50 cm^2 V^{-1} s^{-1} in 1994 [47]. These transistors use perovskite materials, which are known for their excellent charge transport properties, as the semiconductor channel layer. The advantages of PetFTs, such as solution processing, tunable band gaps, and high mobility, make them very attractive for applications in displays, sensors, and flexible electronics. Despite their promising prospects, PETFs still face challenges such as stability, high-quality film orientation, hole density control, and interface chemistry limitations. These challenges require improved processing techniques to produce high-quality films with consistent interconnecting morphology and optimize key performance parameters.

7.3.1 Fabrication Techniques

The fabrication of perovskite semiconductors involves three important considerations, such as selecting cost-effective precursor preparation standards, selecting and designing additives, and deposition methods [83].

- **Cost-effectiveness and high-quality precursor preparation selection criteria**: Cost and quality are important considerations in the preparation of thin-film perovskites. Cost makes thin-film perovskites readily available, while the quality of the raw materials ensures their attractiveness to end users. This criterion depends on the choice of inorganic halide salts, such as lead-free metal halides (CsX, where X = I, Br, or Cl) and organic salts, such as formamidinium halides (FAX) and methylammonium halides (MAX). The quality of the precursors depends on their purity. Impure precursor solutions produce low-quality precursors with numerous sequence defects, while high-purity precursor solutions

Table 7.1 Various types of Transistor and associated features

Feature	TFT	JFET	MOSFETs
Threshold voltage	Depends on fabrication process and the material	0.1 and 0.3 V	0.2–0.4 V
Materials	Perovskites, organic, metal oxides, polycrystalline silicon, or amorphous silicon	Usually bulk semiconductor	Gallium Nitride (GaN), Silicon Germanium (SiGe), Silicon on Insulator (SOI), but usually Silicon (Si)
Power handling	Lower than MOSFETs	Lower power handling than TFT and MOSFET	Higher power than TFTs and JFETs
Mobility	Sufficient for display, though lower mobility than MOSFETs	Lower mobility	Normally higher mobility
Input impedance	Lower than both JFETs and MOSFETs	High, but lower than MOSFET	Highest
Gate current	Ideally zero, very low	Small gate current	Ideally approaching zero or very low
Noise	Low-noise performance	Low-noise	Higher 1/f noise
Substrate	Flexible or transparent glass or plastic	Si, SiC and GaN	Rigid silicon wafers
Type of charge carrier	Frequently n-type, but ambipolar TFTs and p-type exist	Either p-channel or n-channel	Both p-type and n-type
Transparency	Transparent	Non-transparent	Opaque
Transconductance	Variable	Constant	Lower than JFETs
Size	Smaller size	Larger size	Normally smaller than TFTs and JFETs
Suitability	Low-power, flexible displays and large area	Excel in high-input impedance and low-noise	Digital and high-speed switching
Stability	Face challenges	Less stable than MOSFET	Better stability
Construction	Deposited on a flexible or glass materials	Gallium nitride (GaN) and silicon (Si) and	Deposited on silicon wafers
Scalability	Scalable for large areas	Limited scalability	More scalable
Cost	More cost-effective	More expensive	Most cost-effective

(continued)

Table 7.1 (continued)

Feature	TFT	JFET	MOSFETs
Manufacturing	Simpler, bottom-up processes	Simpler than MOSFETs and applies photolithography top-down approach	Top-down approach more complex
Applications	Flexible electronics, sensors, displays (OLED, LCD)	High-frequency, analog circuits, applications	Memory, microprocessors, digital circuits, high-performance computing

provide high-quality perovskite crystal structures, enabling high efficiency and stability in photovoltaic cells, field-effect transistors (FETs), light-emitting diodes (LEDs), and other optical devices. The stoichiometric ratio is also considered a key factor influencing the composition, shape, and crystallinity of perovskite crystals. An imbalance in the ratio of metal halide to organic halide can lead to the formation of non-perovskite phases, ultimately reducing device performance. Therefore, the purity of the precursor solution should be above 99.99% [86, 94].

- **Additive selection and engineering design**: The selection and engineering of additives are not only important steps but also key criteria for obtaining high-quality perovskite films. Additives help improve the quality and crystallinity of the films, while engineering helps achieve well-aligned interfacial contacts and simplified devices, enabling cost-effective, reproducible, and scalable large-area manufacturing, thereby improving device stability and efficiency. Various types of additives, solvents, and dopant molecules can be used to facilitate solution-based perovskite film preparation [83]. This solution processing approach involves dissolving the perovskite precursor material in additives, solvents, and functional additives before depositing it onto a substrate.
- **Composition engineering**: This engineering technique helps design additives, solvents, and perovskite materials into different stoichiometric ratios to improve stability and efficiency by monitoring film morphology, crystallinity, grain size, and grain boundaries.
- **Deposition methods**: The most common deposition methods are solution-based processes, such as electrochemical methods, sequential deposition methods, and physical vapour deposition (PVD), such as sputtering and thermal evaporation as shown in Fig. 7.1. PVD is the most promising method for producing high-quality thin films with uniformity and desired morphology [58], while solution-based methods such as blade coating and spin-coating are favored for their high throughput and cost-effectiveness, primarily used in optical and electronic applications [43, 93]. In solution-based PeTFT deposition methods, the formation of defect-rich deposits and limited control over crystallization remains bottleneck challenges, reducing reproducibility and performance [92]. Despite these

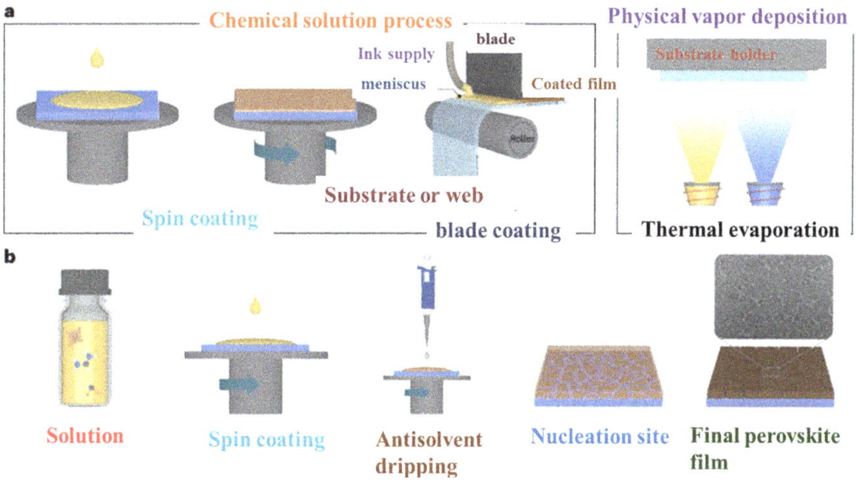

Fig. 7.1 Typical approach for depositing tin-halide perovskite thin films; (**a**) examples of film deposition methods and (**b**) spin-coating deposition of Sn^{2+}-perovskite film [92]. Copyright 2025 reproduced with permission from Nature Publishing Group

challenges, solution-based PeTFT deposition methods still offer the following advantages:

- **Low-temperature process**: The solution process can deposit perovskite thin films at relatively low temperatures of ~ 100 °C, which is advantageous for integration with temperature-sensitive plastic substrate to manufacture flexible devices.
- **Versatility in composition tuning**: Solution processing offers greater flexibility and higher efficiency in selecting materials and additives for perovskite film deposition. This enables easy tuning of perovskite composition across a wide range of organic, inorganic, and hybrid materials, thereby rapidly screening out the optimal additives and compositions.
- **Cost-effectiveness**: Compared to vacuum-based physical deposition methods, solution processing requires less expensive equipment and consumables. This convenience makes it particularly suitable for research laboratories or small-scale production facilities with limited budgets.

7.3.2 Key Performance Parameters

During the operation of a PeTFT device, the application of source voltage and gate voltage causes charge carriers to accumulate at the interface between the perovskite layer and the insulating layer. When a S-D voltage is applied, current flows from the S

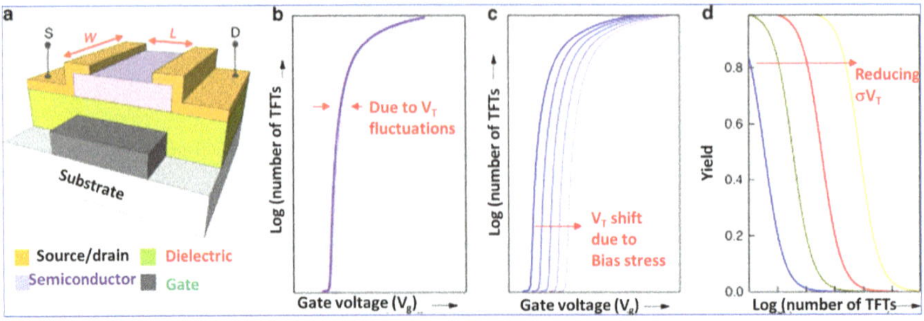

Fig. 7.2 TFT architecture and important device metrics [51]; Copyright 2018 reproduced with permission from Nature Publishing Group (OA)

electrode to the D electrode through the channel. Unipolar devices utilize electrons (n-channel) or holes (p-channel) as the major carrier type. The bottom-gate top-contact (BG-TC) geometry made from silicon/silicon oxide (Si/SiO_2) is the most commonly used device architecture for perovskite to simplify the fabrication process. Several key parameters determine the operational characteristics of the transistor, such as sub-threshold swing (*SS*), on/off current ratio (I_{ON}/I_{OFF}), threshold voltage (V_{Th}), and field-effect mobility (μ). These electrical parameters are considerably influenced by several factors in PeFET devices, such as perovskite film morphology, perovskite properties, and device geometry. Generally, a small *SS*, large I_{ON}/I_{OFF}, low V_{Th}, and high μ values are essential for the health operation of a transistor [85]. The mobility of a transistor is the first important parameter determining the efficiency of charge carriers transport along the conducting channel [71].

7.3.2.1 Mobility

Since the discovery of the transistor in 1984, the speed at which charge charrier move through the channel has determined the transistor's mobility, thereby influencing its switching speed [4]. As shown in Fig. 7.2a, the device architecture can drive individual picture elements (pixels) in the display industry, making it a current and future choice. Mobility is influenced by the extrinsic properties of specific semiconductors. The mobility principle is that the higher the carrier mobility, the better the transistor and the more suitable it is for application. Table 7.2 shows properties TFTs and corresponding semiconductors.

7.3.2.2 On/Off Ratio

This parameter is a critical parameter for switching applications, used to describe the ratio of the transistor current during the on- and off-states, reflecting the quality of the switching behavior. A high switching ratio indicates that the TFT device can

Table 7.2 TFTs properties and semiconductor types

TFTs properties	Semiconductor type				
	Organic	α-Si	Poly-Si	Metal oxides	Perovskite
Mobility/ cm^2/Vs	1–20	0.1–1	Up to 100	Up to 100	10^{-7}–10
Lifetime stability	Low	Low	High	High	Low
Uniformity	Low	High	Low	High	Good
Commercial applications	–	LCD display	Smart phone displays	Laptop displays	Still under development, with great promise to surpassing others
Costs	Cheap	Cheap	Expensive	Cheap	More cost-effective
Potential for large area application	Excellent	Poor	Poor	Excellent	More suitable
Threshold voltage, V	< 1	1–5	− 1 to 3	0.5–5	− 1 to 2
On/Off Ratio range	~ 10^2 to 10^4	~ 10^5 to 10^6	~ 10^6 to 10^7	~ 10^4 to 10^7	~ 10^8

efficiently switch between on- and off-states, which is a major demand for switching technology. In general, a high I_{ON}/I_{OFF} is desired because it indicates a high contrast between on- and off-states, thereby preventing water leakage and improving the switching efficiency [22]. Material properties, device fabrication process, V_{Th}, gate voltage, and hysteresis are the main factors affecting the I_{ON}/I_{OFF} of PETFTs. By implementing anion engineering, perovskite transistors have achieved hysteresis-free operation, mobility of 20 cm^2V^{-1} s^{-1}, and current ion/Iof ratios of more than 10^7 [91].

7.3.2.3 Threshold Voltage (V_{Th})

V_{Th} is the voltage applied between the gate and source terminals of a PETF that controls the current flow between the source and drain terminals, which causes the channel to switch to a single state, and thus the power consumption and device performance. Since low-power applications prefer low V_{Th} values, PeTFTs exhibit a wide range of V_{Th} values, especially for low-power applications that require p-channel devices. This can be achieved through anion engineering using methylammonium tin iodide ($MASnI_3$) perovskite films, which reduces vacancy defects and improves crystallinity, thereby enabling low V_{Th} values for high-performance PeTFTs [22, 59, 91]. As is known, V_{Th} is of crucial importance in digital circuits, which makes it necessary to use TFT with standard deviation. Compared with other important parameters, σV_{Th} is a key indicator [51, 67].

Furthermore, for a large set of identical TFTs, the variation of V_{Th} can cause the transfer curve (shaded area) to deviate from its true characteristics (solid line) as shown in Fig. 7.2b. The deviation may be due to the semiconductor used, the manufacturing process (Fig. 7.2a), or the operational stability of the device. For example, a-Si is known for its small σV_{Th} and is also characterized by low carrier density and operational stability (Fig. 7.2c) even in large-scale manufacturing plants (Table 7.2). Alternatively, poly-Si TFTs are known for their operational stability, but due to σV_{Th}, they are also susceptible to process variations. Therefore, complex integrated circuits make σV_{Th} more critical [68]. Figure 7.2d illustrates the impact of σV_T on the number of concatenated unipolar logic NOT gates. Strong integration has been employed to overcome this logical complementarity circuit, but at the expense of increased production cost and complexity [12].

7.3.2.4 Output Impedance

The resistance of a transistor when it is in the "on" state is called the output impedance. The output impedance of a PeFET is the resistance to current flow when a voltage is applied to the output terminal (the drain of the TFT). In general, transistors with high output impedance are considered to be better transistors. To calculate the carrier mobility of the semiconductor layer in a PeFET, the transfer characteristic curve is based on two important assumptions: (i) the transverse gate electric field is much larger than the applied source and drain fields; (ii) the mobility is independent of the carrier density. The values of the IDs at low applied voltages, V_{ds}, are given by the transfer characteristic curves, where $V_g - V_{Th} \gg V_{ds}$. The drift velocities of charge carriers in the perovskite channel in this region are related to the linear mobility μlin, which is given by Eq. 7.1 [82].

$$I_{ds,lin} = \frac{W}{L} C_i \mu_{lin} \left(\left(V_g - V_T \right) V_{ds} - \frac{1}{2} V_{ds}^2 \right) \tag{7.1}$$

In the linear regime with V_{ds} and V_g, $I_{ds,lin} = \frac{W}{L} C_i \mu_{lin} \left(\left(V_g - V_{Th} \right) V_{ds} \right)$.

Where μ_{lin} in linear is defined as in Eq. 7.2. The drain current is directly proportional to V_g, and μ_{lin} is calculated from V_g at constant V_{ds} versus the gradient of I_d.

$$\mu_{lin} = \left(\frac{dI_{ds,lin}}{dV_g} \right) \frac{L}{C_i V_{ds} W} \tag{7.2}$$

where L and W represent the channel length and width, respectively, and C_i is the gate capacitance per unit area, $C_i = \frac{\varepsilon_0 \varepsilon_{SiO_2}}{t_{SiO_2}}$. I_{ds} is the drain-source current when the gate-source electric field, V_g, is applied and $\frac{dI_{ds,lin}}{dV_g}$ can be obtained from Fig. 7.3a.

When $V_{ds} = V_g - V_{Th}$, the channel is pinched off as shown in Fig. 7.3b. Therefore, current remains constant and then saturates ($I_{ds,sat}$) as shown in Fig. 7.3a. Thus,

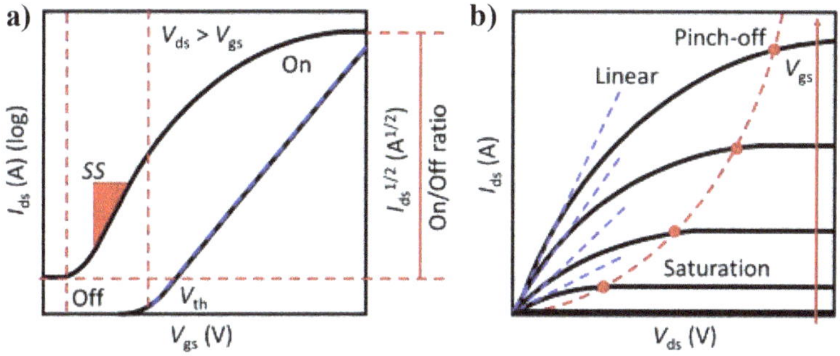

Fig. 7.3 Transfer (**a**) and output (**b**) characteristic curves of a PeFET with linear and saturated regimes [32]. Copyright 2022 reproduced with permission from Cell Press

substituting V_{ds} with $V_g - V_{Th}$ in Eq. 7.1 provides Eq. 7.3.

$$I_{ds,sat} = \frac{W}{2L}\mu_{sat}C_i\left(V_g - V_{Th}\right)^2 \tag{7.3}$$

In the saturation regime of PeFET, the saturation mobility is calculated using Eq. 7.4:

$$\mu_{sat}\left(V_g\right) = \left(\frac{dI_{ds,sat}^{1/2}}{dV_{gs}}\right)\frac{2L}{wC_i}\cdot\frac{1}{\left(V_g - V_{Th}\right)} \tag{7.4}$$

where g_d is output conductance and C_i is the gate capacitance per unit area. $\frac{dI_{ds}^{1/2}}{dV_{gs}}$ can be obtained from Fig. 7.3a. At the source contact where $V_g > V_{Th,\text{ the}}$ mobile charges, Q_{mob} per unit area is given by, $Q_{mob} = C_i(V_g - V_{Th})$. In the absence of diffusion, charge carrier induced current at the source drain, $I_d = W\mu C_i Q_{mob} E_x$, where μ is mobility, W is channel width, and E_x is the electric field at x. The $V_d \le V_g - V_{Th}$ belongs to the linear mobility while $V_d \ge V_g - V_{Th}$ belongs to the saturation mobility.

The mobility of a FET is given by Eq. 7.5 [93].

$$\mu_{FE} = \frac{g_m L}{WC_i V_d} \tag{7.5}$$

where V_d is the drain voltage and V_g is the gate voltage. The effective mobility is given by Eq. 7.6.

$$\mu_{eff} = \frac{g_d L}{WC_i(V_g - V_{Th})} \tag{7.6}$$

V_{Th} is defined as the gate bias at which a certain amount of charge accumulates at the interface. g_m of a TFTs is the ratio of the change in I_d to the change in V_{gs} at a constant drain-source voltage and can be defined by Eq. 7.7 [20, 93].

$$g_m = \frac{dI_D}{dV_{gs}} = \frac{\Delta I_d}{\Delta V_{gs}} \tag{7.7}$$

The signal amplification of the transistor is controlled by transconductance, which is considered as a main parameter. The frequency dependence of transconductance can be calculated using Eq. 7.8 [20].

$$\frac{dI_d}{dt} = \frac{-I_d}{\tau_e} \tag{7.8}$$

where τ_e is electron transit time, where hole is not the limiting factor. The ionic charge injected to the channel is calculated by Eq. 7.9:

$$Q = \int I_d \mathrm{d}t \tag{7.9}$$

where, $I_g = \frac{V_g}{|Z|}$.

Furthermore, in the subthreshold range, the interface trap density (N_{it}) can be calculated from the subthreshold swing S (Eq. 7.10) using Eq. 7.11 [79].

$$S = \frac{dV_{gs}}{d(\log I_{ds})} \tag{7.10}$$

$$N_{it} = \left(\frac{Sq}{k_B T In 10} - 1\right)\frac{C_i}{q} \tag{7.11}$$

where, T is temperature, k_B is the Boltzmann constant, and q is the elementary charge. The smaller the N_{it} value, the better the dielectric-semiconductor interface quality. The photoconductive gain, G, can be calculated using Eq. 7.12 [31].

$$G = \frac{\tau_{life}}{\tau_{tran}} = \frac{\tau_{life}}{L^2/\mu.V_{ds}} \tag{7.12}$$

where, τ_{tran} is the carrier transit time, τ_{life} is carrier recombination time. Alternatively, it can be calculated using Eq. 7.13 [24].

$$G_{th} = \tau_{lifetime} \times QE \times \mu \times \frac{V_{ds}}{A} \tag{7.13}$$

where, QE is quantum efficiency, A is area of semiconductor flake.

7.3.3 Description of Key Device Components

Figure 7.4a shows the first reported perovskite transistor, and Fig. 7.4b shows a typical perovskite structure used in TFTs. PETFTs use perovskite semiconductor materials as the channel layer, typically in the form of a thin film. These transistors include field effect transistors (Fig. 7.4c), phototransistors (Fig. 7.4d), and light-emitting transistors (Fig. 7.4e), each with three terminals: source, drain, and gate, similar to transistors such as MOSFETs and JFETs. The structure generally includes source/drain electrodes, a perovskite channel layer, a gate dielectric layer, and a gate electrode.

- **Substrate**: TFTs use glass, silicon wafers, flexible substrates, such as colorless polyimide (CPI), polyestersulfone (PES), cellulose paper, and thermoplastic polyurethane (TPU), alumina, etc. This substrate is the foundation of the device, and the reaming layers are sequentially deposited on this substrate.
- **Source/Drain (S/D) Electrodes**: Allows current to flow in and out. Silver (Ag), gold (Au), platinum (Pt), palladium (Pd) and nickel (Ni) are the most common S/D electrodes in TFTs, allowing establishing electrical contact with the active channel to allow current to flow in and out.
- **Active Channel**: Semiconductors, such as perovskites, organics, silicon (polycrystalline and amorphous), and metal oxides, have been used as active channel for n-type, p-type, or ambipolar active channels, with current flowing from these channels.

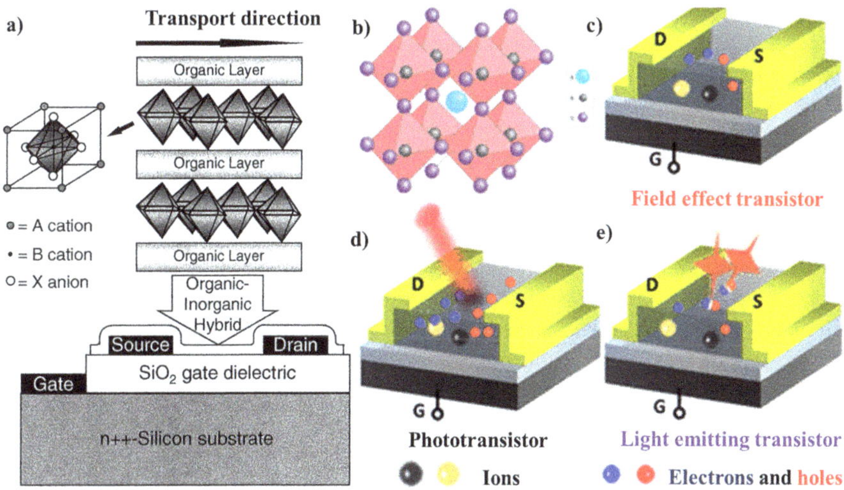

Fig. 7.4 Various device structures reported: (**a**) the first FET made from 2D $(C_6H_5C_2H_4NH_3)_2SnI_4$ [19]. Copyright 1999 reproduced with permission from AAAS. **Perovskite FETs with different functionalities**: reproduced with permission [33]; Copyright 2022 American Chemical Society; (**b**) representative perovskite structure; (**c**) schematic of field effect transistor; (**d**) schematics of phototransistor; and (**e**) schematics of light emitting transistor

- **Gate Insulator (GI)**: Separates the gate electrode from the active channel to act as a capacitor. Tantalum oxide (Ta_2O_5), zirconium oxide (ZrO_2), hafnium oxide (HfO_2), aluminium oxide (Al_2O_3), and silicon dioxide (SiO_2) are the primary gate insulators for TFTs, used to separate the active channel from the gate electrode.

- **Gate Electrode (GE)**: Control the current in the channel. Indium tin oxide (ITO) is primarily used as the top gate, while other metal electrodes such as silicon (Si), chromium (Cr), copper (Cu), molybdenum (Mo), titanium (Ti), and aluminium (Al) are used as gate electrodes to monitor the current flow in the channel.
- **The PeTFT device structure** includes a channel made of a perovskite thin film, with the drain and source electrodes connected to the two ends, and the gate electrode serving as the third electrode to control the current between the sources and drain electrodes. Applying a voltage to the gate helps control the number of charge carriers. Thus, depending on the voltage applied to the gate, PeTFTs can be classified into enhancement mode and depletion mode.

7.3.4 Working Principle

When a voltage is applied to the gate electrode, an electric field is generated on the insulating layer, which in turn forms a depletion region in the channel. The depletion region reduces the number of free charge carriers in the channel, thereby decreasing the channel's conductivity. This effect is known as the field effect and forms the basis of FET operation. For an n-type FET, a negative voltage applied to the gate electrode creates a depletion region in the channel, reducing the flow of electrons from the source to the drain. Conversely, a positive voltage applied to the gate electrode of a p-type FET creates a depletion region reducing the flow of holes from the source to the drain. Thus, by changing the voltage applied to the gate electrode, the conductivity of the channel can be controlled, thereby regulating the current flowing through the FET. To explain the operating principle of a PeTFT, its operational advantage lies in monitoring the current between the drain and source terminals. The operating principle of a PeTFT involves applying a voltage to the gate electrode to control the current between the drain and source terminals. The primary difference between PeTFT and other TFTs lies in the use of perovskite thin films as the semiconductor channel, typically lead-free metal halide perovskites, to achieve sustainability. The electric field generated by the gate voltage can be used to modulate the channel's conductivity, allowing the PeTFT to switch between non-conductive (off) and conductive (on) states. Further details are provided below:

- **Perovskite material**: Perovskite materials with tunable optoelectronic properties are used as active layers in TFT channels. These materials can be p-type, n-type, or even ambipolar, meaning they can conduct both positive and negative charge carriers.

- **Gate control**: The gate electrode is positioned near the perovskite channel and separated by an insulating layer. Applying a voltage to the gate electrode generates an electric field that attracts or repels charge carriers within the perovskite channel.
- **Channel modulation**: The gate voltage controls the number of charge carriers (electrons or holes) available in the channel, thereby effectively altering its conductivity. Increasing the gate voltage increases the number of charge carriers, reduces the channel resistance, and allows more current to flow between the source and drain (on state). Conversely, decreasing the gate voltage reduces the number of charge carriers, increases the channel resistance, and blocks current flow (off state).
- **Ambipolar transport**: Some perovskite materials exhibit ambipolar transport properties, meaning that they can conduct both electrons and holes. This property enables unique transistor designs and applications, such as light-emitting TFTs, where both types of charge carriers are injected and recombine to produce light.

In general, material properties such as doping and grain boundaries are affecting the interactions with phonons due to the presence of defects, ion migration, and polaron scattering in PeTFTs. In addition, the temperature-dependent dielectric action affects the mobility, while strong polaron-acoustic phonon interactions occur due to the "softness" of perovskite. The most influential factor is ionic migration, but surface passivation and mixed-metal compositions can overcome this. Furthermore, it is important to develop charge transport models to better understand the factors affecting the mobility in lead-free perovskites. In nature, lead-free perovskites are polycrystalline, which combines both thermally driven hopping and band-like transport models. Such semiconducting materials exhibit different ion migration behaviors and complex defects compared to lead-based hybrid perovskites. This requires improved transport models. Additionally, lead-free hybrid perovskites exhibit lattice softness, which is influenced by the Frohlich scattering owing to the presence of phonon interaction. Since lead-free hybrid perovskites are highly defect-rich, a large amount of charge carriers need thermal energy to jump into the local states. In this case, the charge density increases with temperature. Therefore, interface engineering, defect engineering and functional bridges, modifiers and stabilizers such as fillers, spacer ions, additives, inter layers, buffer layers and self-assembling molecules are necessary to reduce the defect level, ion migration and trapping conditions considered as experimental models [63, 81, 89]. To further understand the behavior, mechanism and underlying physics of charge transport, the application of theoretical models such as first-principles calculations, drift–diffusion models and Boltzmann transport theory are useful.

7.4 Demand for Lead-Free Perovskites in TFTs

Their environmental friendliness and promising electronic properties make emerging lead-free perovskite materials attractive candidates for next-generation FETs. Therefore, understanding and optimizing charge transport is crucial to achieve high-performance lead-free PeFETs. The transition to such FETs is achieved by adopting or improving the PeFET device to overcome the problems associated with lead-based perovskites. Lead-free perovskites are environmentally friendly alternatives, and their optimized optical properties and high carrier mobility make them suitable for FETs. Furthermore, toxicity concerns, environmental regulations, improved stability, tunable properties, and high performance are key parameters driving the transition from lead to lead-free FETs and other optoelectronic applications. These new perovskite families are mainly classified into Sn-based perovskites ($MASnI_3$, $CsSnI_3$, $(PEA)_2SnI_4$), bismuth-based perovskites ($Cs_2AgBiBr_6$), antimony (Sb) and germanium (Ge) based perovskites. Lead-free perovskites offer significant advantages but also face challenges such as Sn oxidation, crystallization, and defect control, stability under ambient conditions, cost, and scalability, as well as strong quantum confinement effects and insulating organic spacers. These issues require careful attention and further improvement to enable industrial-scale production and fulfil their roles in meeting societal needs. Thus, anion engineering, composition engineering, additive engineering, gran engineering, and π-conjugated ligand are several important improvement strategies for enhancing the performance and stability of lead-free PeTFTs [11, 35, 90–92].

7.4.1 *Understanding Charge Transport in Lead-Free PeFETs*

Although it is challenging to reliably produce high-quality devices, the high carrier mobility of lead-free PETFs makes them promising candidates for high-performance TFTs. In addition, less toxic elements have been able to replace a lead atom, which has led to research into lead-free perovskite-based devices [5, 37, 95]. Among the possible alternative candidates, Sn-based perovskites are the most popular, and compared to their Pb-based counterparts, Sn-based perovskites exhibit comparable or even better properties. For example, CsSnI3 thin films and $MASnI_3$ (MA: methylammonium) single crystals have been reported to have mobility values of 585 and 2000 $cm^2\ V^{-1}\ s^{-1}$, respectively [8, 66]. Pb-based perovskites, especially MAPbI3 crystals, have been reported to have mobility values of 500 to 800 $cm^2\ V^{-1}\ s^{-1}$ [16, 18, 69] Mitzi et al. first reported the first PeFET in 1999, whose two-dimensional (2D) system, $(C_6H_5C_2H_4NH_3)_2SnI_4$ ($(PEA)_2SnI_4$), exhibited p-channel behavior with a hole mobility (μ_h) of 0.62 $cm^2\ V^{-1}\ s^{-1}$ [19].

Moreover, as shown in Fig. 7.5, the charge transport properties of lead-free halide perovskites depend on two vital parameters: composition and dimensionality [40]. Due to their ideal isotropic charge transport properties, high-mobility TFT

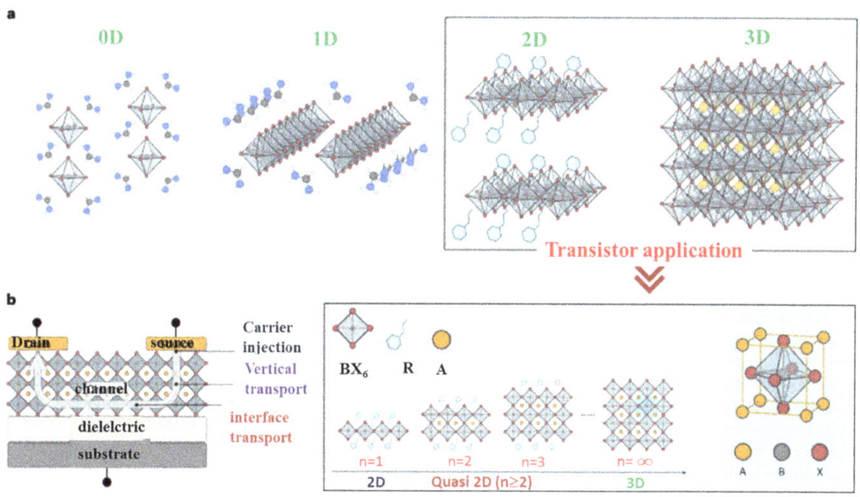

Fig. 7.5 Halide perovskites for high-performance TFT applications [92]. Copyright 2025 reproduced with permission from Nature Publishing Group

channel applications prefer 3D perovskites (Fig. 7.5a). Due to their weaker ion migration, more dispersed valence band edge, and the presence of small polarons, Sn^{2+}-perovskites outperform their Pb counterparts, making them the optimal choice [26, 40]. In a 2D layered structure ($(PEA)_2SnI_4$), a hole mobility (μ_h) of ~ 10 $cm^2\ V^{-1}\ s^{-1}$ was observed for the first time, along with improved stability [35, 44]. Nevertheless, the insulating properties of bulky organic spacers in 2D perovskites may limit further improvements in device performance (Fig. 7.5b) [92].

7.4.2 Key Aspects of Charge Transport in Lead-Free PeFETs

The origin of hysteresis, the role of ion migration, the electronic structure of grain boundaries, the mechanisms of carrier scattering and trapping, material selection, crystallization, charge carrier mobility, temperature dependence, and contact resistance, among others, are all critical aspects of charge transport in lead-free PeFETs devices. Therefore, lead-free PeFETs provide a platform for accurately characterizing the charge density and charge transport properties and mechanisms of semiconductors in a controlled manner.

7.4.2.1 Material Selection

For successful FETs, the first consideration is material selection. While there are many parameters for success, the durability, optical, electrical, and stability of the material are critical factors to consider. For optoelectronic devices (e.g., solar cells, transistors, and light-emitting devices), lead-free double perovskites are preferred over lead-based perovskite semiconductor materials. To this end, strategies involving partial replacement of lead atoms and complete replacement of lead atoms have been pursued to develop lead-free perovskites, addressing health concerns. Lead-free perovskites, such as mixed Sn–Pb, and $Cs_2AgBiBr_6$ perovskites, have been developed as durable lead-free perovskites for current and future applications. First, $(PEA)_2Sn_xP$ was modified to explore the possibility of replacing lead atoms. Although the Sn/Pb ratio is relatively low, $Cs_2AgBiBr_6$ has a nonlinear band gap, its longer charge cycle life and stability offer great promise for further development. Moreover, other alternatives such as tin (Sn^{2+}) halide perovskites, including formamidinium tin triiodide ($FASnI_3$) [63, 90], methylammonium tin triiodide ($MASnI_3$) [77], and cesium tin triiodide ($CsSnI_3$) [10, 62], have been also been reported to aid in expanding the applications of high-performance p-channel TFTs [6, 35, 57, 70].

7.4.2.2 Controlling Crystal Growth

The rapid crystallization characteristics of lead-free perovskites present a significant challenge in the development of PeTFTs. For that reason, monitoring the crystallization properties and optimizing the production process are crucial steps in the preparation of PETFTs to reduce grain boundaries and produce crystals with minimal defects and uniform shapes. Well-controlled crystallization improves charge transport, a key property of TFTs. Therefore, spin coating, doctor-blading, hot-casting, one-step deposition, two-step sequencing, and physical vapor deposition methods are all important for preparing PETFTs and require careful optimization to control the crystallization process.

7.4.2.3 Charge Carrier Mobility

TFTs are important semiconductor devices for studying the charge transport properties and carrier mobility of certain semiconductor materials. This charge mobility is a key parameter for measuring the performance of high-efficiency TFTs, as it indicates the rate at which charge carriers move through the semiconductor device. Factors such as the presence of defects, temperature, and grain size affect this charge mobility. These factors can be optimized during the thin film preparation process, especially at the optimum temperature for successful perovskite film preparation.

7.4.2.4 Temperature Dependence

Almost all optoelectronic devices require an optimal temperature range to operate properly. Finding the optimal temperature range for optoelectronic devices is a challenging task, but it must be optimized regardless of the cost. For TFTs, the charge mobility depends on temperature. The temperature dependence of the charge mobility is important to distinguish between various transport-limiting factors and scattering mechanisms. It is clear that perovskite materials are thermoelectric materials, and therefore their mobility behavior is highly dependent on temperature.

7.4.2.5 Contact Resistance

Another key aspect of charge transport is the contact resistance between the perovskite semiconductor layer and the electrodes, which can have a positive or negative effect on charge mobility. Therefore, improving the crystallization process and the quality of the electrodes can reduce the negative impact of contact resistance.

7.5 High Mobility Lead-Free PeFETs

The carrier mobility of metal halide perovskites can reach 100 $cm^2\ V^{-1}\ s^{-1}$, as measured by terahertz techniques, making them highly promising for sue as TFTs in modern electronic technology [56]. A wide variety of lead-free perovskites are suitable applicable for TFTs. The following discussion focuses on the main lead-free perovskites as representative examples of perovskites, though they are not the only series available. These perovskite series are not only applicable to TFTs but also to lasers and LEDs [42].

7.5.1 SN-Based PeTFTs

The Sn-based perovskite series for TFT applications include MASnI3 and $CsSnI_3$, as well as $(PEA)_2SnI_4$. The charge mobility of p-type Sn-based PETFs is 20 $cm^2\ V^{-1}\ s^{-1}$. In addition, 3D PeTFTs incorporating 2-thiopheneethylamine thiocyanate show an average μ_h of over 63 $cm^2\ V^{-1}\ s^{-1}$ [74]. Furthermore, the excess carrier concentration in $CsFASnI_3$ films reaches $10^{19}\ cm^{-3}$, which increases the mobility of TEASCN-3-FASn s^{-1}, and the mobility of TEAI-$CsFASnI_3$ perovskite is 12 $cm^2\ V^{-1}\ s^{-1}$ with the existing I_{ON}/I_{OF}F of 1.6×10^7 and V_{Th} of 12 V [74]. Based on the triplet A-site cations (caesium–formamidinium–phenethylammonium mobility). ~ 70 $cm^2\ V^{-1}\ s^{-1}$ and I_{ON}/I_{OFF} above 10^8 [93]. The focus of this study is on improving the performance of Sn-based PeTFTs by improving the starting compositions and deposition methods.

7.5.2 $Cs_2AgBiBr_6$ FETs

In addition to Sn-based PeFETs, the highest mobility reported of the $Cs_2AgBiBr_6$ FET is 0.29 (± 00.07) $cm^2\ V^{-1}\ s^{-1}$ (with 30 μm channel), which is attributed to the ability to synthesize larger grain sizes in $Cs_2AgBiBr_6$ thin-film based transistor [1]. Although their efficiency is relatively low, these perovskites exhibit excellent stability, making them highly promising for use in TFTs.

7.5.3 Low-Voltage PeFETs

Lead-free perovskites exhibit a wide voltage range from low to high. This makes them suitable for low-power electronic devices, such as ultra-low voltage TFTs. Moreover, photosensitivity, non-volatile memory effects, and high carrier mobility are special properties of low-voltage PeFETs, making them attractive for various potential applications. The possibility of fabricating them via low-temperature solution processes opens up significant potential for printed and flexible electronics, and also provides a favorable expansion of the development of TFTs in the future.

7.5.4 Transparent PeFETs

Currently, transparent electronics, which are difficult to detect, are attracting attention for various invisible electronic applications, such as invisible displays and electronic transistors. To further develop these electronic applications, semiconductor materials with high mobility and durability, such as lead-free perovskites, are needed to advance the development of this field. For example, $MAPbCl_3$ perovskite has been used for this purpose and holds potential applications in this field [75]. $CsPbBr_3$ QDs have also shown great promise for transparent electronics [87]. We need to develop new perovskites that are free from lead due to the sustainability and health concerns associated with metallic lead atoms. The low-temperature solution process of lead-free perovskites makes them suitable for large-area, low-cost transparent electronics. This transparency makes these materials very suitable for advanced display and invisible electronic technologies that are in high demand in modern society. It goes without saying that successfully developing high-performance transistors and transparent transistors is not easy. For transparent transistors and high-performance transistors, lead-free perovskites are being developed.

7.5.5 2D Lead-Free PeFETs for Image Detection

Image recognition technology uses high-performance switches to track the electronic signal flows from individual pixels. 2D PeFETs have been demonstrated as phototransistors for image acquisition. Such technological advances require high-performance TFTs. Therefore, lead-free perovskites hold great promise due to their durability and high charge carrier mobility properties.

7.5.6 Dimensions of Perovskites and Their Impacts on Transistor Applications

The size of lead-free perovskite materials is a critical parameter that affects the charge carrier mobility in TFTs. Therefore, it is important to consider the dimensions of lead-free perovskites to understand the charge carrier mobility of each perovskite material and to identify their applications. Lead-free perovskites are available in four dimensions: 0D, 1D, 2D and 3D perovskites. Each dimension exhibits different charge mobility. Dimensional changes lead to changes in scattering effects, quantum confinement, density of states, and electronic band structure, which are parameters that characterize charge transport between layers in a transistor. These different dimensions provide unique properties that influence the performance of PETFTs [21].

7.5.7 3D Lead-Free Perovskites

Perovskite solar cells have emerged since the discovery of 3D $CH_3NH_3PbI_3$ photovoltaics in 2009. Due to its remarkable properties, this perovskite has found many applications in a variety of optical devices beyond photovoltaic. However, the presence of lead atoms makes it toxic, raising sustainability concerns. Therefore, lead-free 3D perovskites, such as 3D Sn perovskites, have emerged. High-performance TFTs require high carrier mobility, which requires 3D lead-free perovskites. However, these materials are susceptible to moisture and oxygen, which can compromise their stability. Not only Sn-based, but also bismuth-based lead-free perovskites have been prepared, such as 3D $Cs_2AgBiBr_6$ (Fig. 7.5a) with great promise.

7.5.8 2D Lead-Free Perovskites

As shown in Fig. 7.5a, 2D lead-free perovskites are more stable than 3D lead-free perovskites because the organic layer forms a layered structure that acts as a protective layer while acknowledging the weakness. Due to the need for stable materials, these 2D lead-free perovskites are promising candidates for use in PETFs, with low hysteresis and high I_{ON}/I_{OFFs} [7].

7.5.9 Quasi-2D Lead-Free Perovskites

Quasi-2D lead-free perovskites have achieved the balanced mobility of 3D perovskites and the stability of 2D perovskites, thereby enhancing the durability, performance, and stability required for PETFT. The presence of both organic ligand layers and inorganic perovskite layers in the same material offers great opportunities to design specific phase distributions and quantum-well structures, thereby influencing their optoelectronic properties. Quasi-2D nanoplatelets exhibit strong anisotropic light emission, enhanced optical gain [65], enhanced energy transfer efficiency [60], high absorption cross-sections [2, 78], and a more "compact" configuration [39] due to their anisotropic shape and charge carrier mobility enhancement induced by nanoparticle formation [17, 48].

7.5.10 Nanoplatelet-Based Lead-Free Perovskites

Nanoplatelet materials exhibit unique quantum confinement effects, which is a highly desirable property for TFTs. Therefore, the preparation of lead-free nanoplatelet perovskite materials would be very suitable for TFT devices requiring quantum confinement. In particular, $Cs_3Bi_2I_9$, $Cs_3Sb_2(Br_xI_{1-x})_9$ ($0 \leq x \leq 1$) and Cs_2AgBiX_6 ($X = Cl, Br, I$) have been reported in the literature [41, 54, 73]. If the nanoplatelet size is considered to be the appropriate semiconductor size for TFTs, these three reported perovskites are suitable semiconductors for creating unique quantum confinement effects for high-performance PETFTs at nanoplatelet sizes. Moreover, thickness-dependent optical properties have attracted much attention for quasi-2D or 2D nanoplatelets [50]. The quantum confinement effect creates a numerically ordered set of energy levels for electrons, with a high number of states in specific energy levels. The possibility of creating coherent energy levels ultimately leads to a high carrier density in the coherent energy levels. This quantum confinement effect leads to saturated mobility, as observed with increasing In_2O_3 layer thickness, where the saturation mobility increased from 12.76 to 97.37 $cm^2\ V^{-1}\ s^{-1}$ [61]. Consequently, engineering ultra-thin channel transistors best exploit the quantum confinement effect by reducing energy barriers at interfaces to overcome current density decay [34]. This ultra-thin channel transistor allows electrons to easily tunnel for fast transfer rate across the interfaces [72]. It has also been observed an enhanced morphological and electrical interconnection-an indication of high quality thin film, resulting in high performance TFTs [28]. These interconnected morphological characteristics reveal uniform, ultra-thin, and highly ordered structural growth, with nearly aligned quantized energy levels that are close to each other, thereby enhancing carrier mobility [25, 55]. The presence of disorder, defects, and long distances between quantized energy levels can all lead to reduced mobility. Hence careful analysis of these unfavorable factors will result in higher device performance [76].

7.6 Factors Affecting the Performance of Lead-Free PeTFTs

Each type of electronic semiconductor device has its own basic parameters and factors that contribute to its optimal performance. Like other semiconductor devices, TFTs also have factors that affect their performance. The performance of TFTs depends not only on the charge carrier mobility, but also on the channel width and length, dielectric layers, interconnect materials, crystallinity, defects, and environmental conditions (such as humidity, light, air, and temperature).

7.6.1 Channel Width and Length

Lead-free PeFETs have n-type and p-type channels. These channels have specific length and width requirements to achieve optimal performance. The channel width is the dimension of the channel perpendicular to its length, while the channel length is defined as the distance from the drain to the source. The primary effects of channel width and length are on the short-channel effect, V_{Th}, and current–voltage characteristics, which are critical for determining the switching speed and current of the transistor. High-performance perovskite phototransistors require a reduced channel length [31]. These two critical parameters are key to enhancing performance and must be optimized according to the specific requirements of the TFTs.

7.6.2 Dielectric Layers

The dielectric layer is the protective layer between the semiconductor channel and the gate electrode. This layer provides the applied voltage to the gate electrode but does not affect the short-circuit current. This layer affects the electrical properties of the transistor, such as V_{Th} and mobility. The thickness and capacitance are two important properties of the dielectric layer, which greatly affect the performance of TFTs. Therefore, to achieve stable operation and high performance of the lead-free PeFET, the dielectric layer requires proper interface compatibility, low leakage current, high breakdown strength, and high dielectric constant.

7.6.3 Contact Materials

The selection of contact material greatly affects the performance and stability of the devices. To achieve high efficiency, lead-free PeFETs require low-resistance contact materials. The choice of contact material depends on its ability to form low-resistance contacts with the perovskite active layer. This greatly affects the charge transport during operation of the TFT. For example, aluminum (Al), platinum (Pt), gold (Au), molybdenum (Mo), titanium (Ti), and silver (Ag) are commonly used as materials for the drain and source electrodes due to their low-resistance contact properties. The advantage of lead-free PeFETs lies in their ability to create high-performance TFTs using low-resistance contact materials.

7.6.4 Defects and Crystallinity

Low-purity inorganic materials, organic salts, and solvents can degrade the quality of lead-free perovskite thin films, leading to morphological defects that affect the stability and performance of TFTs during application. In addition, the crystallization and crystal growth processes play a critical role in reducing defects to improve device performance. Controlling the crystallization and growth of crystals, such as single crystals, is essential for obtaining high-quality perovskite films to enhance TFT performance. The advantages are that high-quality films exhibit high carrier mobility, which is essential for high-performance TFTs.

7.6.5 Environmental Factors

Not only defects, ion migration, oxidation of Sn^{2+} to Sn^{4+} and crystallinity in the film, but also environmental factors such as UV exposure, temperature, and humidity affect the performance and stability of PeFETs. This is because perovskite materials, especially Sn-based perovskites, are very sensitive to environmental conditions. Therefore, optimizing these environmental conditions requires great attention. Looking other alternative lead free provkites that can resist such environmental factors can also advance the field.

7.7 Key Device Architecture of Lead-Free PeFETs

Perovskite solar cells exhibit a variety of device architectures, such as planar, inverted, tandem, and singly junction structures, which offer great potential for various photovoltaic applications. Similarly, PeFETs exhibit a variety of device architectures, such as top-gate/bottom-contact (TCBC), top-gate/top-contact (TCTC), bottom-gate/bottom-contact (BGBC), and bottom-gate/top-contact (BGTC) [29], using planar and vertical configuration. Lead-free PEFET device architectures have attracted more attention, primarily due to their improved current, gate control, charge carrier injection, as well as simple device fabrication, low V_{Th} values, and high current-on time, which are required for high-performance transistors with sustained power consumption, fast switching speed, and high charge carrier mobility. On the other hand, understanding device architecture allows us to explore the potential of transparent, flexible, and invisible electronic products and the resilience of flexible transistors under the stress of applications. Based on this understanding, the following discussion explores the architecture of several devices.

7.7.1 *Bottom-Gate/Top-Contact (BGTC)*

In this well-known configuration, known as bottom gate/top-contact (BGTC), the source/drain electrodes are located on top, with the perovskite channel layer above the gate electrode. The fabrication process for this configuration involves substrate preparation, gate electrode deposition, dielectric layer, then perovskite semiconductor layer, and finally drain/source contacts (Fig. 7.6a). With proper compatibility, high performance potential, and simple fabrication process, it can be applied to a wide range of applications. However, it also has limitations (compared to single crystal PFATs), such as degradation, leading to surface defects, ion migration, and poor perovskite morphology leading to high contact resistance and low mobility [79].

7.7.2 *Top-Gate/Bottom-Contact (TGBC)*

The top-gate/bottom-contact (TGBC) transistor device architecture is a modified version of the BGTC device architecture. The fabrication sequence involves preparing the substrate, depositing drain/source electrodes, depositing a perovskite layer, then a dielectric layer, and finally a gate electrode (Fig. 7.6b). Compared to the BGTC device architecture, this architecture offers advantages such as improved carrier availability, integration density, and simplified designs. However, it has limitations, including higher contact resistance, fabrication difficulties, and stability issues compared to the BGTC device architecture.

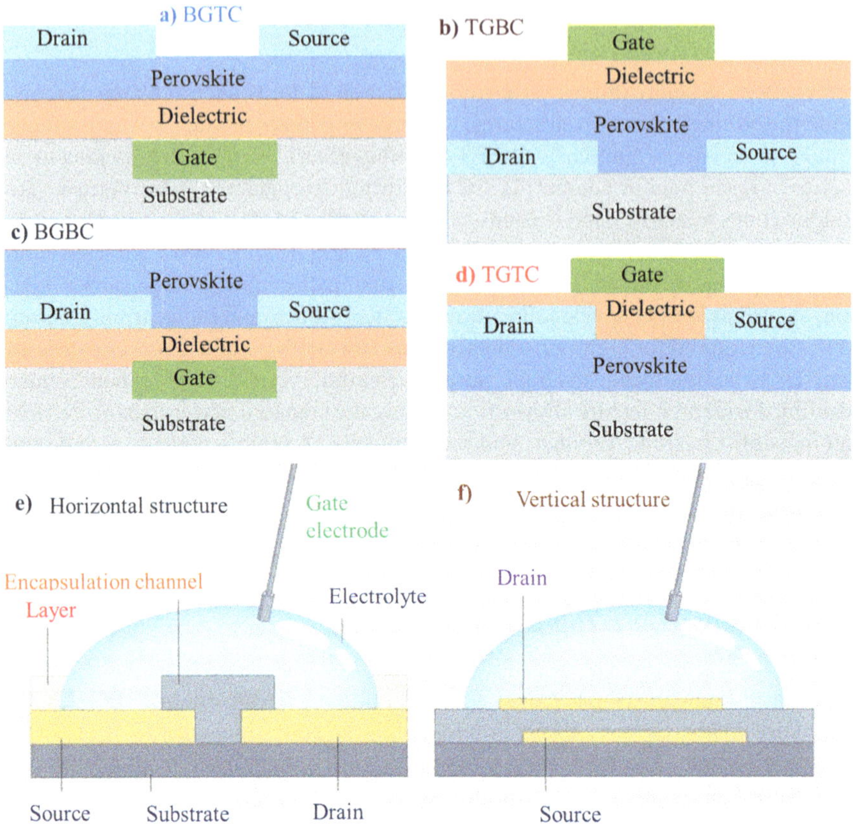

Fig. 7.6 PeFET device structures with planar and vertical layouts [9]. Copyright 2023 reproduced with permission from Nature Publishing Group

7.7.3 *Bottom-Gate/Bottom-Contact (BGBC)*

For the bottom gate/bottom-contact (BGBC) configuration, first prepare a gate-coated substrate and then deposit a dielectric layer. Next, deposit the source and drain electrodes on top of the dielectric layer. Finally, deposit the perovskite layer on the dielectric layer, placing the perovskite layer between the source and drain electrodes and covering it (Fig. 7.6c).

7.7.4 *Top-Gate/top Contact (TGTC)*

In a top-gate/top-contact PeFET, the fabrication sequence is as follows: substrate preparation, followed by deposition of perovskite, drain and source electrodes, dielectric layer, and finally gate electrode (Fig. 7.6d). Although a challenging fabrication process, this results in better stability and higher I_{ON}/I_{OFFs} and carrier mobility, thereby enabling high-performance transistors.

7.7.5 *Planar Layout*

This configuration is favored for its high performance and simple fabrication process. All components must be deposited on a flat substrate to provide a simple structure (Fig. 7.6e). When a perovskite semiconductor is deposited between the drain and source terminals monitored by the gate of a planar structure, the device is referred to as a planar PeFET. Scalability, hysteresis, and stability issues are the primary challenges associated with this device structure.

7.7.6 *Vertical Layout*

Vertically structured PeFETs have a stacked layer configuration, with the channel made of a perovskite semiconductor. The advantages of this structure include miniaturization, high current density, and short path lengths. As shown in Fig. 7.6f, the unique arrangement of this structure involves the vertical alignment of the gate, drain, and source electrodes, with a perovskite layer sandwiched between them.

7.8 Future Directions

Lead-free PeFETs offer high carrier mobility, tunable optoelectronic properties, and low-cost fabrication capabilities for large areas, which open up significant opportunities for future electronic devices. Future research directions include improving device stability, exploring new perovskite compositions, optimizing interface engineering, and developing perovskite-based complementary circuits and neuromorphic computing applications. Further research is needed to optimize the performance and stability of PeFETs by exploring different perovskite compositions, dimensions, and device architectures. Developing high-quality perovskite films and reliable device techniques are crucial to fully realize the potential of PeFETs for future electronic applications. The following outlines current research directions and future prospects to stimulate further research and discoveries on PeFETs.

7.8.1 Stability Enhancement

High-performance lead-free PeFETs (e.g., Sn-based lead-free perovskites) face several challenges that limit long-term device performance. These challenges arise from intrinsic properties such as defects and ion migration, as well as extrinsic conditions such as UV light, air, humidity, and temperature. A major goal of transistor development is to identify new lead-free perovskites such as tin, antimony and bismuth-based perovskites or to conduct in-depth research to improve the intrinsic properties of these unstable perovskites and their ability to withstand external environmental conditions. Therefore, moisture-resistant superhydrophobic lead-free perovskites are essential, and future research efforts should focus in this direction to better develop long-lasting PeFETs.

7.8.2 Material Innovation

In addition to stability improvement strategies, new innovations in lead-free perovskites are attracting great attention. Such innovations should be based on cation and anion engineering, spacer engineering, dimer engineering, cluster engineering, or other novel perovskite structures. This strategy should be applied to improved and novel lead-free PeFETs, as well as other optoelectronic device applications such as displays and non-visible electronics.

7.8.3 Interfacial Engineering

Due to the interface in transistors, challenges related to defects, ion migration, and the formation of new states are particularly evident at buried junctions. Understanding this interface and exploring possible solutions through interface engineering is a crucial step towards developing lead-free PeFETs. The application of self-assembled molecular monolayers offers a great opportunity to address the interface chemistry limitations that hinder high-performance PeFETs. In lead-free PeFETs, the use of self-assembled molecules offers two key advantages [59]:

- Interfacial modification of perovskite active channels and dielectric layers using self-assembled molecule enables efficient charge carrier transport.
- TFTs with interface engineering demonstrate overall performance improvements in operational stability, subthreshold swing, dual-scan hysteresis, V_{Th}, I_{ON}/I_{OFF}, and carrier mobility. These primary benefits are achieved by aligning the two surfaces of the PEA2SnI4 channel and SiO_2 gate with self-assembled molecules, reducing non-radiative recombination and carrier scattering. This creates better energy alignment and defects to realize the desired performance.

7.8.4 Large-Area Fabrication and Integration

The industrial-scale production capacity of lead-free PeFETs represents significant on-going research advancement. The development of large-area PeFET arrays requires cost-effective manufacturing methods that can be scaled up for mass production to enhance practical applications. Invisible electronics, sensors and displays are awaiting innovative manufacturing methods for lead-free PeFETs.

7.8.5 Understanding Charge Transport and Ion Migration

High-performance lead-free PeFETs are characterized by high charge carrier mobility and low ion migrations. Thus, understanding the avoidable ion migration behaviour, charge carrier mobility, and transport mechanisms in perovskites should be the focus of further research on lead-free PeFETs [52]. The emphasis on the effects of ion migration stem from its screening of gate voltage, which influences the shape of transistor electrical properties, the density of mobile charge carriers, and channel current [16]. Consequently, there are conflicting reports on the role of ion migration: the first report indicates that ion migration is suppressed by grain boundaries [53], while another report suggests that ion migration is promoted by grain boundaries [64, 80]. These reports highlights that ion migration has become a significant concern, and thus the intrinsic ion migration-induced device failures and their detrimental effects require further in-depth investigation [3]. In addition to the influence of manufacturing environments, metal lead atoms can also cause ion migration, impairing device performance [23]. This challenge necessitates the removal of metallic lead, and lead-free perovskites must be developed rapidly. Consequently, research has followed methods reported elsewhere to suppress ion migration, such as chemical fixation, adding buffer layers, and improving the quality of perovskite films [3, 13–15, 27, 88]. Understanding the complex properties of perovskites is an urgent research issue not only in transistors but also in all optoelectronic applications beyond photovoltaic.

7.8.6 Focus on Low-Voltage Operation

Current technological advancements require low-power operation to address sustainability challenges. Therefore, lead-free PeFETs are the optimal choice for achieving this goal. Consequently, future research should focus on innovating or discovering lead-free PeFETs capable of operating at low voltages. This represents a major objective in the development of low-power electronics, such as wearable electronics and portable devices.

7.8.7 Focus on the Development of Ambipolar Transistors

Future new features and simplified circuit designs stem from the application development of new ambipolar PeFETs. These advantages have substantial implications because ambipolar PeFETs can simultaneously provide both hole and electron conduction. New lead-free perovskites will provide the ambipolar, p-type, and/or n-type perovskites required for ambipolar, p-type, and/or n-type PeFETs. In this development process, ambipolar perovskites can simultaneously provide electrons and holes, thereby gaining additional advantages in the development of better transistors and other optoelectronic devices. The ambipolarity of lead-based perovskites such as $CH_3NH_3PbI_{3-x}Cl_x$ and $CH_3NH_3PbI_3$ has been reported in phototransistor devices [31], but lead-free ambipolar perovskites have not been reported yet [84]. Tin- and bismuth-based perovskites are lead-free hybrid perovskites, and they do not exhibit ambipolar transport conductivity, but are usually N-type or P-type. Further research is needed to exploit the unique amipolar properties of TFTs from lead-free hybrid perovskites. Note that ambipolar transistor operation occurs when both holes (h) and electrons (e) are present in the channel. This occurs when V_g is greater than $V_{Th,e}$ and between V_s and V_d, with $V_g - V_d < V_{Th,h}$, indicating that both the drain and source are injected with the corresponding charge carriers. In the unipolar state, the channel contains only one charge carrier, depending on its p-type or n-type nature.

7.8.8 Focus on Photodetection Development

For semiconductor devices, light-based sensing and imaging applications require fast-response and highly sensitive charge transport devices such as TFTs. Lead-free perovskites are particularly suitable for this purpose due to their high charge carrier mobility. Future development efforts will focus on improving the response speed and sensitivity of lead-free PeFETs. This will enable photodetection systems to achieve consistent performance and safety in data transmission and security applications.

7.8.9 Non-volatile Memory

Not only photodetection enhancement strategies, but also the development of non-volatile memory devices based on lead-free PeFETs is an on-going research direction, aiming to demonstrate the non-volatile memory effects required for future portable memory applications [49].

7.8.10 Nanospike Electrodes: New Nanoscale Electrode Design

In the design of devices with issues related to short channels and contact resistance effects, designing devices at the nanoscale is a very useful strategy. The nanospike strategy has many advantages [36]:

(i) Eliminate the resistance reduction induced by the drain.
(ii) The formation of charge nanoribbons increase the Fermi level, thereby improving charge injection and further enhancing local carrier densities realizing in higher speed and mobility during charge transport.
(iii) Due to the reduction in overlapping capacitance (between the gate-source and gate-drain), larger alignment errors and mismatches can be tolerated during fabrication, thereby improving speed.
(iv) Increasing the thickness of the dielectric can reduce capacitive coupling, thereby improving speed.
(v) Using a lower V_{Th} allows operation at lower power and lower voltages.
(vi) The arrangement of thick gate insulators and short channel lengths enhance printability/flexibility/electrical properties.
(vii) Healthier device characteristics and improved contact points.
(viii) Eliminating drain-induced resistance drop and the formation of charge nanoribbons are the reasons for maintaining good turn-off performance, improving subthreshold characteristics, and reducing contact resistance per unit area.

Figure 7.7 shows the nanospike electrode design engineering, which can provide TFTs with improved performance and holds promise for the development of novel TFTs with significant potential. Figure 7.7a illustrates the side-guard electrodes, while Fig. 7.7b shows the cross-sectional view of the electrode layer thicknesses and materials. The device design features a nanospike array, as depicted in Fig. 7.7c, and the dimension of the nanospike array electrode are shown in Fig. 7.7d. Generally, future electronic developments require lead-free perovskites, which hold significant potential for a wide range of electronic applications, including transistors, solar cells, LEDs, and other semiconductor devices. Thus, elucidating the potential of lead-free PeFETs and addressing their persistent challenges are two major research directions for the future.

7.9 Conclusion and Outlook

In summary, the main goal of lead-free PeFETs is to understand the requirements of future-generation transistors. This challenge involves developing the ability to create sustainable transistors with an unquestionable wide range of electronic applications. When interpreting sustainability issues for next-generation transistors and

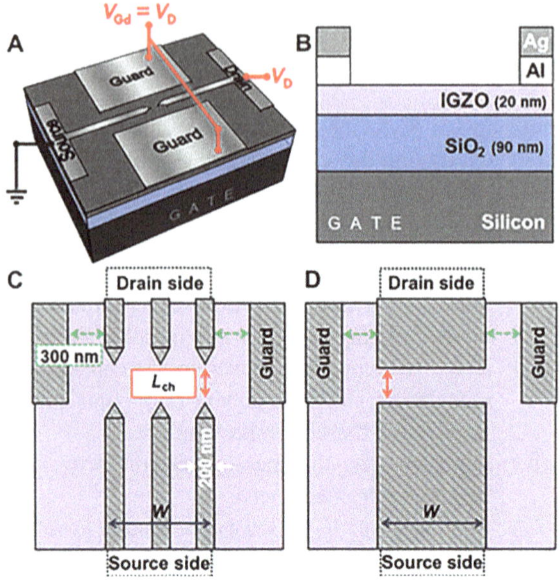

Fig. 7.7 Nanospike electrode TFT characteristics and design engineering [36]. Copyright 2022 reproduced with permission from AAAS

other electronic applications, the primary focus is on the environmental impact of mass production and the long-term operation or lifetime of the devices. However, this is not limited to addressing issues such as energy consumption, emissions, material abundance or resource depletion, but also aims to meet the requirements of long-term viability and circular economy. For lead-based PeFETs, the main sustainability challenges focus on the environmental impact of the manufacturing process and the long-term viability, which is highly promising given the high carrier mobility and device efficiency. Therefore, the current demand for high-mobility lead-free PeFETs is primarily driven by the need to reduce the environmental impact of large-scale production or manufacturing processes, while simultaneously promoting new research and development of long-lasting high-performance lead-free PeFET devices. This is the problem addressed in this chapter. Understanding the basic concepts, design processes, key performance parameters (such as mobility, I_{ON}/I_{OFF}, V_{Th}, and output impedance), and operating principles are essential foundations in this field to achieve the ultimate goal of lead-free PeFETs. Furthermore, creating and maintaining a clear transition path to improve the direction of lead-free PeFETs requires significant attention to bring new perspectives, opportunities, and continuous success. Therefore, understanding the charge carrier transport pathways, developing high-mobility lead-free perovskites, identifying the key effects of the dimensionality of the invention for easy carrier mobility (such as 1D, 2D, quasi-2D and 3D) (such as channel width and length, dielectric layer, contact materials with minimal contact resistance, oxygen deficiency, light conditions, temperature conditions such as crystalline), re-exploring device architectures and layouts (such as BGTC, TGBC, TGTC and BGBC device structures in planar or vertical orientation), are key research topics that are expected to bring new discoveries on the development roadmap of lead-free

PFETs. Furthermore, future developments of such transistors will focus primarily on improved stability, perovskite innovations, interface engineering, large-area fabrication and potential integration, including further understanding of charge transport pathways and ion migration, low-voltage operation, the discovery of amphoteric lead-free PeFETs, and the enhancement and enhancement of photo-voltage systems.

References

1. G. Abiram, F.H. Gourji, S. Pitchaiya, P. Ravirajan, T. Murugathas, D. Velauthapillai, Air processed $Cs_2AgBiBr_6$ lead-free double perovskite high-mobility thin-film field-effect transistors. Sci. Rep. **12**(1), 2455 (2022). https://doi.org/10.1038/s41598-022-06319-z
2. A.W. Achtstein, A. Antanovich, A. Prudnikau, R. Scott, U. Woggon, M. Artemyev, Linear absorption in CdSe nanoplates: thickness and lateral size dependency of the intrinsic absorption. J. Phys. Chem. C **119**(34), 20156–20161 (2015). https://doi.org/10.1021/acs.jpcc.5b06208
3. E. Bi, Z. Song, C. Li, Z. Wu, Y. Yan, Mitigating ion migration in perovskite solar cells. Trends Chem. **3**(7), 575–588 (2021). https://doi.org/10.1016/j.trechm.2021.04.004
4. T.P. Brody, The thin film transistor—a late flowering bloom. IEEE Trans. Electron Devices **31**(11), 1614–1628 (1984). https://doi.org/10.1109/T-ED.1984.21762
5. S. Chakraborty, W. Xie, N. Mathews, M. Sherburne, R. Ahuja, M. Asta, S.G. Mhaisalkar, Rational design: a high-throughput computational screening and experimental validation methodology for lead-free and emergent hybrid perovskites. ACS Energy Lett. **2**(4), 837–845 (2017). https://doi.org/10.1021/acsenergylett.7b00035
6. I.-H. Chao, Y.-T. Yang, M.-H. Yu, C.-H. Chen, C.-H. Liao, B.-H. Lin et al., Performance enhancement of lead-free 2D Tin Halide perovskite transistors by surface passivation and its impact on non-volatile photomemory characteristics. Small **19**(20), 2207734 (2023). https://doi.org/10.1002/smll.202207734
7. J. Chen, M.-Y. Sun, Z.-H. Wang, Z. Zhang, K. Zhang, S. Wang et al., Performance limits and advancements in single 2D transition metal dichalcogenide transistor. Nano Micro Lett. **16**(1), 264 (2024). https://doi.org/10.1007/s40820-024-01461-x
8. I. Chung, J.-H. Song, J. Im, J. Androulakis, C.D. Malliakas, H. Li et al., $CsSnI_3$: semiconductor or metal? High electrical conductivity and strong near-infrared photoluminescence from a single material. High hole mobility and phase-transitions. J. Am. Chem. Soc. **134**(20), 8579–8587 (2012). https://doi.org/10.1021/ja301539s
9. C. Cunin, A. Gumyusenge, Vertical architecture improves performance of transistor family. Nature **613**, 444–445 (2023). https://doi.org/10.1038/d41586-023-00026-z
10. F. De Angelis, The revival of metal-halide perovskites transistors. ACS Energy Lett. **7**(4), 1490–1491 (2022). https://doi.org/10.1021/acsenergylett.2c00677
11. Y. Gao, Z. Wei, P. Yoo, E. Shi, M. Zeller, C. Zhu et al., Highly stable lead-free perovskite field-effect transistors incorporating linear π-conjugated organic ligands. J. Am. Chem. Soc. **141**(39), 15577–15585 (2019). https://doi.org/10.1021/jacs.9b06276
12. G. Gelinck, P. Heremans, K. Nomoto, T.D. Anthopoulos, Organic transistors in optical displays and microelectronic applications. Adv. Mater. **22**(34), 3778–3798 (2010). https://doi.org/10.1002/adma.200903559
13. H. Guo, Y. Fang, Y. Lei, J. Wu, M. Li, X. Li et al., Mitigating ion migration with an ultrathin self-assembled ionic insulating layer affords efficient and stable wide-bandgap inverted perovskite solar cells. Small **19**(38), 2302021 (2023). https://doi.org/10.1002/smll.202302021
14. D. He, D. Ma, J. Zhang, Y. Yang, J. Ding, C. Liu et al., Universal ion migration suppression strategy based on supramolecular host-guest interaction for high-performance perovskite solar cells. Adv. Mater. **37**, 2505115 (2025). https://doi.org/10.1002/adma.202505115

15. D. He, Y. Yu, X. Liu, X. Shai, J. Chen, Methods of suppressing ion migration in n-i-p perovskite solar cells. iEnergy **3**(4), 242–251 (2024). https://doi.org/10.23919/IEN.2024.0029
16. L.M. Herz, Charge-carrier mobilities in metal halide perovskites: fundamental mechanisms and limits. ACS Energy Lett. **2**(7), 1539–1548 (2017)
17. B. John, *Role of Nanoparticles in Enhancing Semiconductor Performance* (2025)
18. M.B. Johnston, L.M. Herz, Hybrid perovskites for photovoltaics: charge-carrier recombination, diffusion, and radiative efficiencies. Acc. Chem. Res. **49**(1), 146–154 (2016)
19. C.R. Kagan, D.B. Mitzi, C.D. Dimitrakopoulos, Organic-inorganic hybrid materials as semiconducting channels in thin-film field-effect transistors. Science **286**(5441), 945–947 (1999). https://doi.org/10.1126/science.286.5441.945
20. D. Khodagholy, J. Rivnay, M. Sessolo, M. Gurfinkel, P. Leleux, L.H. Jimison et al., High transconductance organic electrochemical transistors. Nat. Commun. **4**(1), 2133 (2013). https://doi.org/10.1038/ncomms3133
21. H. Kim, Comparative review of field-effect transistors based on three-dimensional, two-dimensional, and double halide perovskites. Electron. Mater. Lett. (2025). https://doi.org/10.1007/s13391-025-00588-2
22. J. Kim, Y.S. Shiah, K. Sim, S. Iimura, K. Abe, M. Tsuji et al., High-performance P-channel tin halide perovskite thin film transistor utilizing a 2D–3D core-shell structure. Adv. Sci. (Weinh) **9**(5), e2104993 (2022). https://doi.org/10.1002/advs.202104993
23. T. Kim, S. Park, V. Iyer, B. Shaheen, U. Choudhry, Q. Jiang et al., Mapping the pathways of photo-induced ion migration in organic-inorganic hybrid halide perovskites. Nat. Commun. **14**(1), 1846 (2023). https://doi.org/10.1038/s41467-023-37486-w
24. G. Konstantatos, M. Badioli, L. Gaudreau, J. Osmond, M. Bernechea, F.P.G. de Arquer et al., Hybrid graphene–quantum dot phototransistors with ultrahigh gain. Nat. Nanotechnol. **7**(6), 363–368 (2012). https://doi.org/10.1038/nnano.2012.60
25. J.G. Labram, Y.-H. Lin, K. Zhao, R. Li, S.R. Thomas, J. Semple et al., Signatures of quantized energy states in solution-processed ultrathin layers of metal-oxide semiconductors and their devices. Adv. Func. Mater. **25**(11), 1727–1736 (2015). https://doi.org/10.1002/adfm.201403862
26. Z. Le, A. Liu, Y. Reo, S. Bai, Y.-Y. Noh, H. Zhu, Ion migration in tin-halide perovskites. ACS Energy Lett. **9**(4), 1639–1644 (2024). https://doi.org/10.1021/acsenergylett.4c00198
27. J.-W. Lee, S.-G. Kim, J.-M. Yang, Y. Yang, N.-G. Park, Verification and mitigation of ion migration in perovskite solar cells. APL Mater. **7**(4), 041111 (2019). https://doi.org/10.1063/1.5085643
28. J. Lee, J. Jeong, Correlation between quantum confinement effect and characteristics of thin-film transistors in solution-processed oxide-based thin-films. IEEE J. Electron Devices Soc. **12**, 912–918 (2024). https://doi.org/10.1109/JEDS.2024.3468300
29. W.-Y. Lee, J. Mei, Z. Bao, *OFETs: Basic Concepts and Material Designs; The WSPC Reference on Organic Electronics: Organic Semiconductors* (World Scientific Publishing Co Pte Ltd., 2016), pp. 19–83
30. Y.M. Lee, J. Park, B.D. Yu, S. Hong, M.-C. Jung, M. Nakamura, Surface instability of Sn-based hybrid perovskite thin film, $CH_3NH_3SnI_3$: the origin of its material instability. J. Phys. Chem. Lett. **9**(9), 2293–2297 (2018). https://doi.org/10.1021/acs.jpclett.8b00494
31. F. Li, C. Ma, H. Wang, W. Hu, W. Yu, A.D. Sheikh, T. Wu, Ambipolar solution-processed hybrid perovskite phototransistors. Nat. Commun. **6**(1), 8238 (2015). https://doi.org/10.1038/ncomms9238
32. F. Li, T. Yang, R. Zheng, Solution-processed perovskite crystals for electronics: moving forward. Matter **5**(6), 1700–1733 (2022). https://doi.org/10.1016/j.matt.2022.04.022
33. F. Li, W. Yu, X. Guan, T. Wu, Emerging transistor applications enabled by halide perovskites. Acc. Mater. Res. **3**(1), 8–20 (2022). https://doi.org/10.1021/accountsmr.1c00103
34. X. Li, S. Huang, J. Wang, L. Wang, L. Li, Fundamental understanding of quantum confinement effect on gate oxide reliability for gate-all around field-effect transistor. J. Appl. Phys. **136**(23), 235701 (2024). https://doi.org/10.1063/5.0229625

35. A. Liang, Y. Gao, R. Asadpour, Z. Wei, B.P. Finkenauer, L. Jin et al., Ligand-driven grain engineering of high mobility two-dimensional perovskite thin-film transistors. J. Am. Chem. Soc. **143**(37), 15215–15223 (2021). https://doi.org/10.1021/jacs.1c06337
36. K. Liang, X. Xu, Y. Zhou, X. Wang, C.M. McCulley, L. Wang et al., Nanospike electrodes and charge nanoribbons: a new design for nanoscale thin-film transistors. Sci. Adv. **8**(4), eabm1154 (2022). https://doi.org/10.1126/sciadv.abm1154
37. L. Liang, P. Gao, Lead-free hybrid perovskite absorbers for viable application: can we eat the cake and have it too? Adv. Sci. **5**(2), 1700331 (2018). https://doi.org/10.1002/advs.201700331
38. J. Lim, M. Kober-Czerny, Y.-H. Lin, J.M. Ball, N. Sakai, E.A. Duijnstee et al., Long-range charge carrier mobility in metal halide perovskite thin-films and single crystals via transient photo-conductivity. Nat. Commun. **13**(1), 4201 (2022). https://doi.org/10.1038/s41467-022-31569-w
39. Y. Lin, X. Ling, L. Yu, S. Huang, A.L. Hsu, Y.-H. Lee et al., Dielectric Screening of excitons and trions in single-layer MoS_2. Nano Lett. **14**(10), 5569–5576 (2014). https://doi.org/10.1021/nl501988y
40. A. Liu, H. Zhu, S. Bai, Y. Reo, M. Caironi, A. Petrozza et al., High-performance metal halide perovskite transistors. Nat. Electron. **6**(8), 559–571 (2023). https://doi.org/10.1038/s41928-023-01001-2
41. Z. Liu, H. Yang, J. Wang, Y. Yuan, K. Hills-Kimball, T. Cai et al., Synthesis of lead-free Cs_2AgBiX_6 (X = Cl, Br, I) double perovskite nanoplatelets and their application in CO_2 photocatalytic reduction. Nano Lett. **21**(4), 1620–1627 (2021). https://doi.org/10.1021/acs.nanolett.0c04148
42. I. López-Fernández, D. Valli, C.-Y. Wang, S. Samanta, T. Okamoto, Y.-T. Huang et al., Lead-free halide perovskite materials and optoelectronic devices: progress and prospective. Adv. Func. Mater. **34**(6), 2307896 (2024). https://doi.org/10.1002/adfm.202307896
43. J.T. Matondo, H. Hu, Y. Ding, G. Cheng, J. Ding, Scalable solution-phase processing of halide perovskite films: solvent removal, quenching methods, and film drying. Nano Energy **132**, 110402 (2024). https://doi.org/10.1016/j.nanoen.2024.110402
44. T. Matsushima, S. Hwang, A.S.D. Sandanayaka, C. Qin, S. Terakawa, T. Fujihara et al., Solution-processed organic-inorganic perovskite field-effect transistors with high hole mobilities. Adv. Mater. **28**(46), 10275–10281 (2016). https://doi.org/10.1002/adma.201603126
45. Y. Mei, C. Zhang, Z.V. Vardeny, O.D. Jurchescu, Electrostatic gating of hybrid halide perovskite field-effect transistors: balanced ambipolar transport at room-temperature. MRS Commun. **5**(2), 297–301 (2015). https://doi.org/10.1557/mrc.2015.21
46. R.L. Milot, R.J. Sutton, G.E. Eperon, A.A. Haghighirad, J. Martinez Hardigree, L. Miranda et al., Charge-carrier dynamics in 2D hybrid metal-halide perovskites. Nano Lett. **16**(11), 7001–7007 (2016). https://doi.org/10.1021/acs.nanolett.6b03114
47. D.B. Mitzi, C.A. Feild, W.T.A. Harrison, A.M. Guloy, Conducting tin halides with a layered organic-based perovskite structure. Nature **369**(6480), 467–469 (1994). https://doi.org/10.1038/369467a0
48. R. Momper, H. Zhang, S. Chen, H. Halim, E. Johannes, S. Yordanov et al., Kinetic control over self-assembly of semiconductor nanoplatelets. Nano Lett. **20**(6), 4102–4110 (2020). https://doi.org/10.1021/acs.nanolett.9b05270
49. C.-H. Nieh, Y.-C. Kao, Y.-H. Sun, M.-H. Yu, I.-H. Chao, B.-H. Lin et al., Enhancing charge transport and fast non-volatile memory in 2D tin-based perovskite transistors through porphyrin-like additives. Adv. Funct. Mater. **36**(4), e10256 (2026). https://doi.org/10.1002/adfm.202510256
50. C. Otero-Martínez, J. Ye, J. Sung, I. Pastoriza-Santos, J. Pérez-Juste, Z. Xia et al., Colloidal metal-halide perovskite nanoplatelets: thickness-controlled synthesis, properties, and application in light-emitting diodes. Adv. Mater. **34**(10), 2107105 (2022). https://doi.org/10.1002/adma.202107105
51. A.F. Paterson, T.D. Anthopoulos, Enabling thin-film transistor technologies and the device metrics that matter. Nat. Commun. **9**(1), 5264 (2018). https://doi.org/10.1038/s41467-018-07424-2

52. F. Paulus, C. Tyznik, O.D. Jurchescu, Y. Vaynzof, Switched-on: progress, challenges, and opportunities in metal halide perovskite transistors. Adv. Func. Mater. **31**(29), 2101029 (2021). https://doi.org/10.1002/adfm.202101029
53. N. Phung, A. Al-Ashouri, S. Meloni, A. Mattoni, S. Albrecht, E.L. Unger et al., The role of grain boundaries on ionic defect migration in metal halide perovskites. Adv. Energy Mater. **10**(20), 1903735 (2020)
54. Z. Qi, X. Fu, T. Yang, D. Li, P. Fan, H. Li et al., Highly stable lead-free $Cs_3Bi_2I_9$ perovskite nanoplates for photodetection applications. Nano Res. **12**(8), 1894–1899 (2019). https://doi.org/10.1007/s12274-019-2454-0
55. C. Qian, J. Sun, Y. Gao, Transport of charge carriers and optoelectronic applications of highly ordered metal phthalocyanine heterojunction thin films. Phys. Chem. Chem. Phys. **23**(16), 9631–9642 (2021). https://doi.org/10.1039/D1CP00889G
56. X. Qiu, Y. Liu, J. Xia, J. Guo, P.-A. Chen, H. Wei et al., Room temperature two-dimensional lead halide perovskite thin-film transistors with high stability. Cell Rep. Phys. Sci. **4**(1), 101217 (2023). https://doi.org/10.1016/j.xcrp.2022.101217
57. X. Qiu, J. Xia, Y. Liu, P.-A. Chen, L. Huang, H. Wei et al., Ambient-stable 2D Dion-Jacobson phase tin halide perovskite field-effect transistors with mobility over 1.6 $Cm^2\ V^{-1}\ s^{-1}$. Adv. Mater. **35**(44), 2305648 (2023). https://doi.org/10.1002/adma.202305648
58. Y. Reo, T. Zou, T. Choi, S. Kim, J.-Y. Go, T. Roh et al., Vapour-deposited high-performance tin perovskite transistors. Nat. Electron. **8**(5), 403–410 (2025). https://doi.org/10.1038/s41928-025-01380-8
59. Y. Rong, D. Yu, X. Zhang, T. Wang, J. Wang, Y. Li et al., Perovskite thin-film transistors for ultra-low-voltage neuromorphic visions. Adv. Sci. **11**(48), 2410015 (2024). https://doi.org/10.1002/advs.202410015
60. C.E. Rowland, I. Fedin, H. Zhang, S.K. Gray, A.O. Govorov, D.V. Talapin, R.D. Schaller, Picosecond energy transfer and multiexciton transfer outpaces Auger recombination in binary CdSe nanoplatelet solids. Nat. Mater. **14**(5), 484–489 (2015). https://doi.org/10.1038/nmat4231
61. J.K. Saha, J. Jang, Saturation mobility of 100 $cm^2\ V^{-1}\ s^{-1}$ in ZnO thin-film transistors through quantum confinement by a nanoscale In_2O_3 interlayer using spray pyrolysis. ACS Nano **18**(44), 30484–30496 (2024). https://doi.org/10.1021/acsnano.4c08644
62. S.P. Senanayak, K. Dey, R. Shivanna, W. Li, D. Ghosh, Y. Zhang et al., Charge transport in mixed metal halide perovskite semiconductors. Nat. Mater. **22**(2), 216–224 (2023). https://doi.org/10.1038/s41563-022-01448-2
63. S. Shao, W. Talsma, M. Pitaro, J. Dong, S. Kahmann, A.J. Rommens et al., Field-effect transistors based on formamidinium tin triiodide perovskite. Adv. Func. Mater. **31**(11), 2008478 (2021). https://doi.org/10.1002/adfm.202008478
64. Y. Shao, Y. Fang, T. Li, Q. Wang, Q. Dong, Y. Deng et al., Grain boundary dominated ion migration in polycrystalline organic–inorganic halide perovskite films. Energy Environ. Sci. **9**(5), 1752–1759 (2016)
65. C. She, I. Fedin, D.S. Dolzhnikov, A. Demortière, R.D. Schaller, M. Pelton, D.V. Talapin, Low-threshold stimulated emission using colloidal quantum wells. Nano Lett. **14**(5), 2772–2777 (2014). https://doi.org/10.1021/nl500775p
66. C.C. Stoumpos, C.D. Malliakas, M.G. Kanatzidis, Semiconducting tin and lead iodide perovskites with organic cations: phase transitions, high mobilities, and near-infrared photoluminescent properties. Inorg. Chem. **52**(15), 9019–9038 (2013). https://doi.org/10.1021/ic401215x
67. R.A. Street, Thin-film transistors. Adv. Mater. **21**(20), 2007–2022 (2009). https://doi.org/10.1002/adma.200803211
68. K. Terada, T. Mogami, Measurement of standard deviation for threshold voltage using parallel-connected MOSFETs. Paper presented at the ESSDERC '96: proceedings of the 26th European solid state device research conference (1996)
69. D.A. Valverde-Chávez, C.S. Ponseca, C.C. Stoumpos, A. Yartsev, M.G. Kanatzidis, V. Sundström, D.G. Cooke, Intrinsic femtosecond charge generation dynamics in single crystal CH_3 NH_3 PbI_3. Energy Environ. Sci. **8**(12), 3700–3707 (2015)

70. S. Wang, S. Kalyanasundaram, L. Gao, Z. Ling, Z. Zhou, M. Bonn et al., Unveiling the role of linear alkyl organic cations in 2D layered tin halide perovskite field-effect transistors. Mater. Horiz. **11**(5), 1177–1187 (2024). https://doi.org/10.1039/d3mh01883k
71. S. Wang, H. Lian, Y. Yang, Z. Wu, Y. Li, H. Ling et al., Morphological engineering for high-performance perovskite field-effect transistors. FlexMat **2**(1), 82–106 (2025). https://doi.org/10.1002/flm2.39
72. Z. Wang, L. Wang, Y. En, X.-W. Jiang, Quantum confinement modulation on the performance of nanometer thin body GaSb/InAs tunnel field-effect transistors. J. Appl. Phys. **121**(22), 224503 (2017). https://doi.org/10.1063/1.4985610
73. D. Wu, B. Huo, Y. Huang, X. Zhao, J. Yang, K. Hu et al., Synthesis of stable lead-free $Cs_3Sb_2(Br_xI_{1-x})9$ $(0 \leq x \leq 1)$ perovskite nanoplatelets and their application in CO_2 photocatalytic reduction. Small **18**(12), 2106001 (2022). https://doi.org/10.1002/smll.202106001
74. Y. Wu, F. Yuan, S. Yang, E. Li, W. Wang, Y. Liu et al., Tailoring tin-based perovskite crystallization via large cations and pseudo-halide anions for high mobility and high stable transistors. Sci. Adv. **11**(31), eadv4138 (2025). https://doi.org/10.1126/sciadv.adv4138
75. J. Xia, X. Qiu, Y. Liu, P.-A. Chen, J. Guo, H. Wei et al., Ferroelectric wide-bandgap metal halide perovskite field-effect transistors: toward transparent electronics. Adv. Sci. **10**(10), 2300133 (2023). https://doi.org/10.1002/advs.202300133
76. Y. Xing, N. Yazdani, W.M.M. Lin, M. Yarema, R. Zahn, V. Wood, Effect of positional disorders on charge transport in nanocrystal quantum dot thin films. ACS Appl. Electron. Mater. **4**(2), 631–642 (2022). https://doi.org/10.1021/acsaelm.1c01011
77. S. Yang, J. Wen, Y. Wu, H. Zhu, A. Liu, Y. Hu et al., Unlocking the potential of tin-based perovskites: properties, progress, and applications in new-era electronics. Small **20**(1), e2304626 (2024). https://doi.org/10.1002/smll.202304626
78. A. Yeltik, S. Delikanli, M. Olutas, Y. Kelestemur, B. Guzelturk, H.V. Demir, Experimental determination of the absorption cross-section and molar extinction coefficient of colloidal CdSe nanoplatelets. J. Phys. Chem. C **119**(47), 26768–26775 (2015). https://doi.org/10.1021/acs.jpcc.5b09275
79. W. Yu, F. Li, L. Yu, M.R. Niazi, Y. Zou, D. Corzo et al., Single crystal hybrid perovskite field-effect transistors. Nat. Commun. **9**(1), 5354 (2018). https://doi.org/10.1038/s41467-018-07706-9
80. J.S. Yun, J. Seidel, J. Kim, A.M. Soufiani, S. Huang, J. Lau et al., Critical role of grain boundaries for ion migration in formamidinium and methylammonium lead halide perovskite solar cells. Adv. Energy Mater. **6**(13), 1600330 (2016)
81. X. Zai, H. Dong, Z. Shen, D. Ji, X. Dong, C. Ran, Z. Wu, Defect engineering toward high-performance tin-based perovskite field-effect transistors. Adv. Mater. **37**(42), e04087 (2025). https://doi.org/10.1002/adma.202504087
82. J. Zaumseil, H. Sirringhaus, Electron and ambipolar transport in organic field-effect transistors. Chem. Rev. **107**(4), 1296–1323 (2007). https://doi.org/10.1021/cr0501543
83. C. Zhang, N.-G. Park, Materials and methods for cost-effective fabrication of perovskite photovoltaic devices. Commun. Mater. **5**(1), 194 (2024). https://doi.org/10.1038/s43246-024-00636-8
84. F. Zhang, Q. Zhang, X. Liu, L. Qin, Y. Hu, Z. Lou et al., Ambipolar transport in two-dimensional Sn-based perovskite field-effect transistors using an aliphatic polymer-assisted method. J. Mater. Chem. A **9**(40), 22842–22853 (2021). https://doi.org/10.1039/D1TA06277H
85. Q. Zhang, Y. Zhang, Y. Luo, H. Yin, New structure transistors for advanced technology node CMOS ICs. Natl. Sci. Rev. **11**(3), nwae008 (2024). https://doi.org/10.1093/nsr/nwae008
86. Y. Zhang, S. Seo, S.Y. Lim, Y. Kim, S.-G. Kim, D.-K. Lee et al., Achieving reproducible and high-efficiency (>21%) perovskite solar cells with a presynthesized $FAPbI_3$ powder. ACS Energy Lett. **5**(2), 360–366 (2020). https://doi.org/10.1021/acsenergylett.9b02348
87. C. Zhao, Y. Liu, L. Chen, J. Li, H.Y. Fu, S. Zhao et al., Transparent $CsPbBr_3$ quantum dot photodetector with a vertical transistor structure. ACS Appl. Electron. Mater. **3**(1), 337–343 (2021). https://doi.org/10.1021/acsaelm.0c00877

88. Y. Zhao, W. Zhou, Z. Han, D. Yu, Q. Zhao, Effects of ion migration and improvement strategies for the operational stability of perovskite solar cells. Phys. Chem. Chem. Phys. **23**(1), 94–106 (2021). https://doi.org/10.1039/D0CP04418K
89. S. Zhou, G. Zhou, Y. Li, X. Xu, Y.-J. Hsu, J. Xu et al., Understanding charge transport in all-inorganic halide perovskite nanocrystal thin-film field effect transistors. ACS Energy Lett. **5**(8), 2614–2623 (2020). https://doi.org/10.1021/acsenergylett.0c01295
90. Z. Zhou, Q. Li, M. Chen, X. Zheng, X. Wu, X. Lu et al., High-Mobility and bias-stable field-effect transistors based on lead-free formamidinium tin iodide perovskites. ACS Energy Lett. **8**(10), 4496–4505 (2023). https://doi.org/10.1021/acsenergylett.3c01400
91. H. Zhu, A. Liu, K.I. Shim, H. Jung, T. Zou, Y. Reo et al., High-performance hysteresis-free perovskite transistors through anion engineering. Nat. Commun. **13**(1), 1741 (2022). https://doi.org/10.1038/s41467-022-29434-x
92. H. Zhu, Y. Reo, G. Park, W. Yang, A. Liu, Y.-Y. Noh, Fabrication of high-performance tin halide perovskite thin-film transistors via chemical solution-based composition engineering. Nat. Protoc. **20**(7), 1915–1929 (2025). https://doi.org/10.1038/s41596-024-01101-z
93. H. Zhu, W. Yang, Y. Reo, G. Zheng, S. Bai, A. Liu, Y.-Y. Noh, Tin perovskite transistors and complementary circuits based on A-site cation engineering. Nat. Electron. **6**(9), 650–657 (2023). https://doi.org/10.1038/s41928-023-01019-6
94. P. Zhu, D. Wang, Y. Zhang, Z. Liang, J. Li, J. Zeng et al., Aqueous synthesis of perovskite precursors for highly efficient perovskite solar cells. Science **383**(6682), 524–531 (2024)
95. I. Zimmermann, S. Aghazada, M.K. Nazeeruddin, Lead and HTM free stable two-dimensional tin perovskites with suitable band gap for solar cell applications. Angew. Chem. Int. Ed. **58**(4), 1072–1076 (2019). https://doi.org/10.1002/anie.201811497

Chapter 8
Lead-Free Perovskite Materials for Piezoelectric Applications

Supratim Mitra

Abstract Lead-free perovskite materials have gained significant attention for piezoelectric applications due to environmental concerns. In recent years, regulatory bodies like RoHS and WEEE have further boosted the demand for eco-friendly technologies. This chapter offers a concise overview of the design strategies and structure-property relationships of major lead-free systems, including (K,Na)NbO_3 (KNN), $(Bi_{0.5}Na_{0.5})TiO_3$-$BaTiO_3$ (BNT-BT), and Ba(Zr,Ti)O_3-(Ba,Ca)TiO_3 (BZT-BCT). Their applications are also highlighted, emphasizing these materials as viable alternatives to hazardous lead-based compounds such as Pb(Zr,Ti)O_3 (PZT). Key design approaches like compositional engineering, domain tuning, and advanced processing techniques, particularly spark plasma sintering and templated grain growth have been shown to greatly improve electromechanical performance. The origins of excellent electromechanical properties are further examined through intrinsic and extrinsic mechanisms. Special focus is given to morphotropic and polymorphic phase boundary effects, similar to those seen in PZT. The chapter ends with a discussion of how these materials are integrated into real-world applications, including sensors, actuators, energy harvesters, and biomedical devices. Lead-free piezoelectrics, thanks to their unique performance benefits, offer promising pathways toward next-generation smart, wearable, and environmentally sustainable technologies.

Keywords: Lead-free piezoelectrics · Perovskite structure · Electromechanical properties · Energy harvesting · Sensors · Actuators · Transducers

8.1 Introduction

Piezoelectric materials are best known for their ability to convert mechanical energy into electrical energy and vice versa. This functional property makes them an attractive material for various modern technologies, including electronics, healthcare,

S. Mitra (✉)
Department of Physical Sciences, Banasthali Vidyapith, Banasthali, Rajasthan, India
e-mail: supratimmitra2003@gmail.com; supratimmitra@banasthali.in

V. Chauhan et al. (eds.), *Sustainable Lead-Free Perovskite Materials*, Topics in Applied Physics 157, https://doi.org/10.1007/978-3-032-24873-2_8

energy, and defense. Ceramic-based piezoelectrics are of much interest due to their high sensitivity, rapid electromechanical response, and compatibility with device miniaturization. These advantages enable their use in various advanced applications, including precision actuators, medical ultrasound transducers, vibration sensors, sonar systems, energy harvesters, and microelectromechanical systems (MEMS) [1].

For several decades, lead-based piezoelectric ceramics have been regarded as the most demanding materials for commercial piezoelectric applications. In particular, lead zirconate titanate, $Pb(Zr,Ti)O_3$ (PZT), exhibits excellent electromechanical properties and consistent performance over a wide temperature window, positioning it as the predominant material for piezoelectric devices [1]. However, due to the increasing concerns over lead toxicity, statutory restrictions have been imposed on the manufacturing of lead-containing devices. These regulatory constraints are enforced through various global legislative directives, such as the Restriction of Hazardous Substances (RoHS) and Waste Electrical and Electronic Equipment (WEEE) [2, 3]. As a consequence, substantial research has focused on the development of lead-free piezoelectric ceramics as alternatives to lead-based PZT [4]. Nonetheless, achieving performance comparable to that of PZT remains extremely challenging, as lead-free piezoelectric ceramics must simultaneously demonstrate high piezoelectric coefficients, thermal stability, mechanical robustness, and scalability for commercial applications.

Among the various lead-free piezoelectrics that have been studied, including perovskite, tungsten-bronze, and bismuth-layered structures, perovskite-based systems have emerged as the most promising candidate for next-generation device applications [5]. The perovskite lattice provides flexibility to accommodate a variety of ions through substitutions. This facilitates precise control over lattice distortions, tuning phase transitions, and polarization mechanisms. Leveraging this structural flexibility, lead-free perovskite materials are often designed to exhibit morphotropic phase boundaries (MPBs) or polymorphic phase transitions (PPTs) [4–8]. The structural instabilities associated with these phase boundaries promote polarization rotation and enhance domain-wall mobility, resulting in piezoelectric responses that are comparable to those observed in PZT [9, 10].

Amid several lead-free piezoelectric classes, potassium sodium niobate (KNN)-based, bismuth sodium titanate (BNT)-based, and barium titanate (BT)-based solid solutions are of particular importance, as discussed later [4, 5, 11]. KNN-based ceramics are advantageous owing to their relatively high Curie temperatures exceeding 400 °C and strong piezoelectric properties. However, alkali volatilization during sintering, poor densification, and compositional sensitivity continue to hinder their reproducibility and large-scale applications. BNT-based systems exhibit strong ferroelectricity with high remnant polarization (38 $\mu C/cm^2$) and Curie temperatures around 320 °C, but their usability is constrained by high coercive fields (73 kV/cm), significant hysteresis losses, and modest piezoelectric coefficients [12]. $BaTiO_3$-based solid solutions constitute another important group of lead-free piezoelectric ceramics and have been widely explored. Compositional modification in $BaTiO_3$, particularly in the $Ba(Zr,Ti)O_3$-$(Ba,Ca)TiO_3$ (BZT–BCT) system, where the substitution of Zr^{4+} at the Ti site and Ca^{2+} at the Ba site dramatically improves the functional properties. This enhancement in piezoelectric properties is attributed to the presence of a MPB, where cubic, rhombohedral, and tetragonal phases coexist. Compositions close to this triple point exhibit exceptionally high piezoelectric charge coefficients (d_{33} > 500 pC/N) and electromechanical coupling factors that comparable to, or even superior, those of soft PZT ceramics [13]. The associated flattening of free-energy landscape near the MPB facilitates polarization rotation and enhances domain wall mobility. Furthermore, dense ceramics with tailored microstructures can be readily obtained using conventional solid-state sintering, sol–gel processing, hydrothermal synthesis, combustion methods, or advanced techniques such as spark plasma sintering. As a result, BZT–BCT has become a central focus in the development of lead-free piezoelectrics owing to its outstanding electromechanical response, structural tunability, and relatively ease of processing. However, the relatively low Curie temperature ~ 90–120 °C near MPB compositions limits their utilization in high-temperature environments. Therefore, there is a pressing need for continued compositional engineering, domain structure optimization, and integration of novel processing methods to fully realize their potential in commercial applications [14].

This chapter presents an overview of the key lead-free perovskite piezoelectric systems, emphasizing their structural and phase evolution, enhancements in the electromechanical properties through chemical and processing strategies, and their relevance to sustainable, next-generation devices. It also highlights the influence of structural instabilities, compositional modifications, and processing techniques to achieve high-performance lead-free piezoelectrics, while discussing the strengths and limitations of KNN-, BNT-, and BZT-BCT-based ceramics are discussed in contrast PZT. These insights offer perspectives on future directions for designing environmentally sustainable piezoelectric materials for next-generation electronic and electromechanical applications.

8.2 Piezoelectric Properties

Piezoelectric properties of materials provide the fundamental parameters and key figure of merit for their practical applications. These properties establish the correlation between electric fields and mechanical stress or strain through coupling of electrical and mechanical energies. The piezoelectric charge coefficient (d_{ij}), a pivotal parameter, defines the linear relationship between applied mechanical stress (σ_j) and the resulting electric displacement (D_i) or vice-versa. When a small mechanical stress is applied to a piezoelectric material, it generates an electric displacement proportional to the applied force, expressed as $D_i = d_{ij}\sigma_j$. Since this electric displacement is leads to the accumulation of charges at the electrodes, the coefficient is therefore referred to as piezoelectric charge coefficient (d_{ij} = [pC/N]). Conversely, upon application of an electric field (E_j), the material experiences an equivalent mechanical strain (S_i), described by $S_i = d_{ij}E_j$. This bidirectional electromechanical interaction is characterized by the same coefficient, which is defined as the piezoelectric strain coefficient (d_{ij} = [pm/V]) in the converse effect. In ferroelectric materials, when the applied electric field is small, the piezoelectric charge or strain coefficient (d_{ij}) are referred to as small signal d_{ij} parameters, reflecting the linear relationship between applied electric field and the induced strain. This behavior originates from a combination of both the intrinsic lattice responses and extrinsic contributions, primarily reversible domain-wall motion. However, when the applied electric field or mechanical stress exceeds a certain threshold, such as the local coercive field, irreversible domain wall motion and partial polarization switching start dominating. Under such conditions, the piezoelectric response becomes nonlinear, and the effective coefficient varies with the amplitude of the driving signal. This nonlinear regime is characterized by the large-signal piezoelectric coefficient ($d_{ij}{}^*$), defined as the ratio of the maximum strain (S_{max}) to the maximum applied unipolar electric field (E_{max}), i.e., $d_{ij}{}^* = S_{max}/E_{max}$.

Another important material parameter is the piezoelectric voltage coefficient (g_{ij}), which is obtained by dividing the piezoelectric charge coefficient (d_{ij}) by the dielectric permittivity (ε_{ij}). This coefficient is defined as the achievable voltage output for a given mechanical stress, and conversely the mechanical strain produced per unit electric displacement. It demonstrate the ability to convert mechanical energy into electrical energy, and vice versa, in terms of voltage response (g_{ij} = [Vm/N)], which is a key figure of merit in sensor and energy harvesting applications [14].

The next important parameter is the piezoelectric coupling coefficient, also known as the electromechanical coupling coefficient (k_{ij}), which quantifies the efficiency of energy conversion between electrical and mechanical forms. Specifically, it is defined as the ratio of the converted energy (mechanical to electrical or vice versa) to the total input energy. A higher coupling coefficient indicates more efficient energy transduction, making it a critical figure of merit for resonant devices such as ultrasonic transducers and actuators. However, the coupling factor depends on the geometry, boundary condition, polarization direction, and mode of vibration of the piezoelectric element. It is generally computed experimentally using resonance and anti-resonance technique by analyzing impedance spectrum of the piezoelectric element. For a given mode (e.g., thickness, radial, longitudinal), the coupling factor can be calculated from the measured resonance (f_r) and anti-resonance (f_a) frequency using $= \sqrt{(f_a{}^2 - f_r{}^2)/f_a{}^2}$. The mechanical quality factor (Q_m) is used to quantify the relationship between sharpness of resonance frequency and the reactance to resistance ratio of the series equivalent circuit that serves as the piezoelectric resonator. It is evaluated using, $Q_m = f_r/(f_1 - f_2)$, where, f_1 and f_2 are frequencies at -3db of the maximum admittance [1].

8.3 Perovskite Structure

The piezoelectric properties discussed in the previous section are intrinsically linked to their crystal structure. For real-world applications, especially for ceramics, polycrystalline ferroelectrics are preferred over single-crystal due to their ease of fabrication, cost-effectiveness, and scalability. However, because the grains in a polycrystalline ceramic are randomly oriented, piezoelectricity arises only when a net polarization is induced through application of an external DC electric field. This is called a poling process, which is essential to breaks the inversion symmetry required to bring out the piezoelectric response by aligning their crystallographic polar axis. Therefore, from crystallography point of view only non-centrosymmetric point groups are capable of exhibiting piezoelectricity. Out of 20 non-centrosymmetric point groups only 10 are polar and relevant to ceramics: triclinic (1), monoclinic (2, m), orthorhombic (2 mm), rhombohedral (3, 3 m), tetragonal (4, 4 mm), and hexagonal (6, 6 mm).

Among these, the perovskite structure (ABO_3) is the most widely adopted in piezoelectric ceramics due to its technological relevence. The ideal cubic form consists of the A-site cation that occupies the cube corners, the B-site cation sits at the center, and the oxygen anions are located at the face centers, creating a corner-sharing BO_6 octahedra network (Fig. 8.1a). The stability of this cubic perovskite structure is governed by the Goldschmidt tolerance factor [15]:

$$A = \frac{R_A + R_O}{\sqrt{2}(R_B + R_O)}$$

where R_A, R_B, and R_O are the ionic radii of the *A*-site, *B*-site, and oxygen ions, respectively. A perovskite structure is typically stable for $0.9 < t < 1.1$. Deviations outside this range often destabilize the perovskite phase. However, perovskites can undergo various distortions due to vacancies or substitutions at the sites:

- along [100] to form tetragonal cells,
- along [110] for orthorhombic symmetry,
- along [111] to yield rhombohedral structures,
- Or along general directions to produce monoclinic or triclinic phases (Fig. 8.1b).

These distortions, often accompanied by tilting of the oxygen octahedra [16], result in multiple domain variants: 6 for tetragonal, 12 for orthorhombic, 8 for rhombohedral, 24 for monoclinic, and 48 for triclinic structures (Fig. 8.1b). Domain reorientation under electric fields contributes significantly to the piezoelectric response. Therefore, the crystal structure demonstrates not only the intrinsic polarization behaviour but also the number and mobility of domain variants, influencing the extrinsic contributions to the piezoelectric response. The perovskite structure remains the most favourable due to its alterable chemistry, multiple distortion pathways, and compatibility with domain engineering strategies.

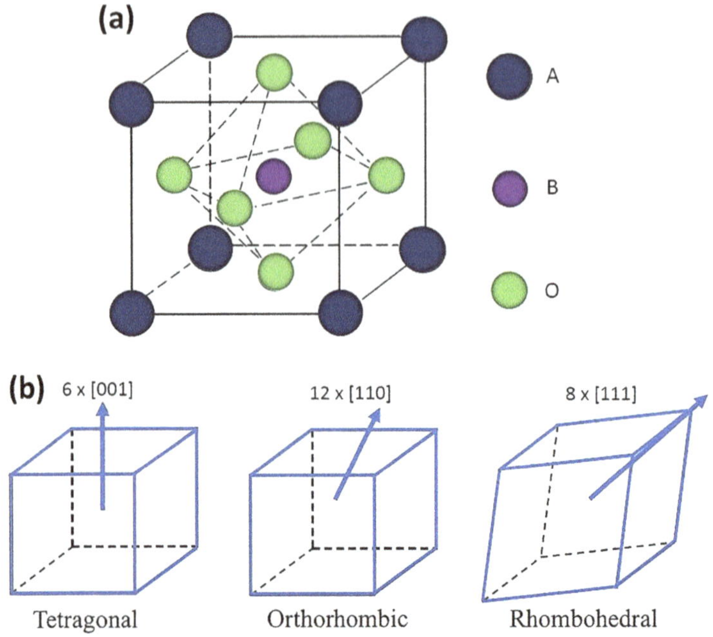

Fig. 8.1 (**a**) Schematic of perovskite structure (ABO_3) showing cation (A^{2+}, B^{4+}) and anion O^{2-} position. (**b**) Schematic for polarization rotation, where arrow represent the direction of polarization

8.4 Lead-Free Perovskite Systems

8.4.1 Potassium Sodium Niobate (K,Na)NbO₃ (KNN)-Based Systems

$K_{1-x}Na_xNbO_3$ (KNN) is one of the earliest lead-free piezoceramics explored, known for their relatively high Curie temperatures, low dielectric losses, and excellent piezoelectric coefficients [17]. However, a major breakthrough emerged in 2004, when researchers at the Toyota Central Research Laboratory demonstrated that KNN ceramics modified with Li^+, Ta^{5+}, and Sb^{5+} could achieve piezoelectric coefficients at room temperature comparable to those of conventional lead-based PZT materials [18]. It is crucial to remember that the industrial interest in Li- and Ta-modified KNN systems began earlier, even though this breakthrough is usually regarded as a turning point. Notably, Murata Manufacturing Co. had already filed patents on such compositions between 1998 and 1999 [9]. The preliminary challenges in KNN-based systems arises from their compositional complexity and the volatility of alkali components during high-temperature processing. Recent developments have therefore concentrated on optimizing the orthorhombic–tetragonal phase boundary to boost piezoelectric performance. Strategies such as targeted doping and grain orientation control via tape casting and texturing, have led to significant improvements in

d_{33} values and thermal stability. Nevertheless, issues such as achieving high densification and enhancing fatigue resistance in polycrystalline ceramics continue to pose limitations.

8.4.1.1 Structure and Phase Diagram of KNN

Perovskite KNN is a solid solution of the end members ferroelectric $KNbO_3$ (KN) and antiferroelectric $NaNbO_3$ (NN), both of which exhibit orthorhombic structure with *Amm2* symmetry at room temperature [19]. However, the perovskite-type ABO_3 subcell of KNN exhibits a monoclinic symmetry with lattice parameters $a_m = c_m > b_m$, where b_m is perpendicular to the $a_m c_m$ plane and the angle β is slightly greater than 90° (Fig. 8.2a). XRD measurements show no superlattice reflections, indicating no octahedral tilting. A projection along the b_m axis shows a_m and c_m appearing nearly perpendicular (Fig. 8.2b). To clarify geometry, an exaggerated β is used to illustrate four subcells, revealing that the diagonals form a rectangle, confirming that while the subcell is monoclinic, the overall unit cell has orthorhombic symmetry at room temperature (Fig. 8.2c) [20].

Tuning phase boundaries is an effective strategy for enhancing the electrical properties of piezoelectric materials. The nature and characteristics of these boundaries play a decisive role in improving piezoelectric activity, regardless of whether the system is lead-based or lead-free. Such enhancements are largely attributed to the high degree of dipole alignment. A widely recognized illustration of this concept is the establishment of a morphotropic phase boundaries (MPBs) in PZT, regarded is as a classic example in piezoelectric perovskite materials. At these boundaries, abrupt structural changes results in maximized piezoelectric responses at specific compositions [19]. An MPB represents a compositionally fixed boundary separating two ferroelectric phases of differing symmetry, most commonly orthorhombic and tetragonal, and ideally depends only on composition and exhibits minimal temperature dependence (Fig. 8.3a). When the MPB is nearly vertical in the temperature–composition phase diagram, the material remains close to this structural instability over a wide temperature range, thereby retaining high functional properties.

In PZT, the compositionally driven MPB appears at a Zr/Ti ratio of 52/48, where the enhanced piezoelectric responses arise from the transition between tetragonal and rhombohedral phases. The anisotropic flattening of Gibbs free-energy for theses phases near the MPB facilitates domain-wall mobility and enables easier polarization rotation [22]. This flattened Gibbs free-energy profile is also associated with monoclinic distortion and, in some cases, leads to the formation of an intermediate monoclinic phase [23, 24]. In contrast, the phase boundaries in KNN-based systems are more complex than those in PZT [4]. Transitions such as rhombohedral–orthorhombic (R-O) and orthorhombic-tetragonal (O-T) correspond to intrinsic polymorphic phase transitions (PPTs). Although the Gibbs free-energy profile near a PPT can resemble that of an MPB, these boundaries differ in that they are highly sensitive not only to composition but also to temperature (Fig. 8.3b).

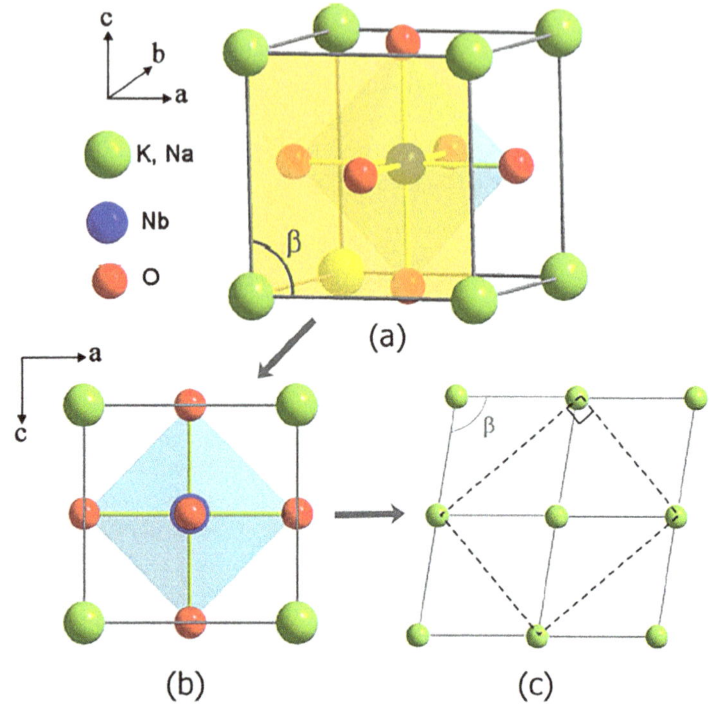

Fig. 8.2 The structure illustration for KNN of orthorhombic structure at room temperature (**a**) the perovskite type ABO_3 subcell; (**b**) the projection of the subcell along *b* axis; (**c**) four adjacent subcell projections in an exaggerated manner to show the geometrical relationship, while omitting Nb and O atoms. (Reprinted with permission © 2007, American Institute of Physics [20])

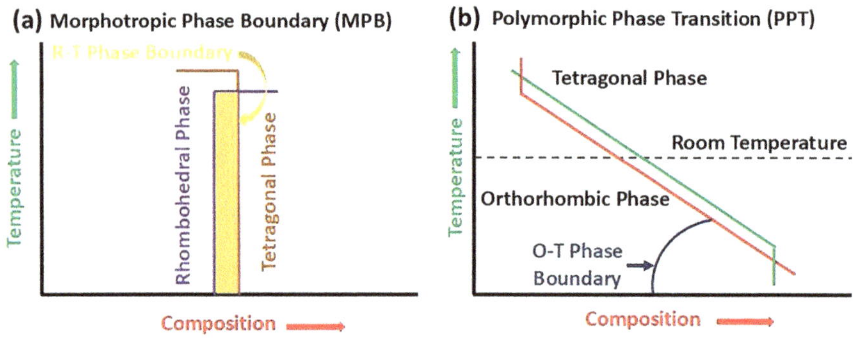

Fig. 8.3 Difference in **a** morphotropic phase boundary (MPB), **b** polymorphic phase transition (PPT) [9, 21]

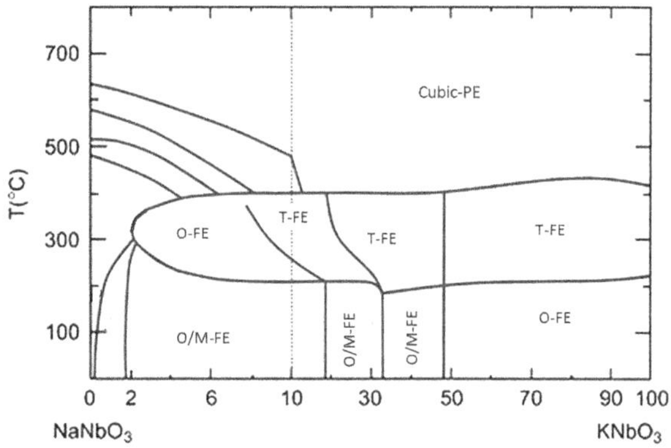

Fig. 8.4 Phase diagram of $K_{1-x}Na_xNbO_3$ with end members $NaNbO_3$ and $KNbO_3$. Regions labeled as: orthorhombic with monoclinic distortions ferroelectric: O/M-FE; orthorhombic ferroelectric: O-FE; tetragonal ferroelectric: T-FE; cubic paraelectric: cubic-PE [9]

The phase diagram of KNN exhibits several thermally (PPT) and compositionally (MPB) induced phase boundaries [25–27]. Undoped KNN undergoes a sequence of structural phase transitions with increasing temperature (Fig. 8.4). A low-temperature polymorphic phase transition (PPT) from the orthorhombic to the tetragonal ferroelectric phase occurs near 200 °C, and finally to a cubic paraelectric phase around 400 °C. Earlier studies on KNN co-doped with Li, Ta, and Sb interpreted the sharp property enhancement near room temperature as indicative of an MPB [18]. However, subsequent investigations have shown that this transition is predominantly temperature-driven rather than purely compositional. As a result, the boundary is more accurately described as a shifted PPT rather than a stable, vertical MPB [4]. In addition, crossing such a phase boundary alters the domain wall structure, which can essentially depolarize the material and reduce performance stability. To address this, compositions are designed either to lie close to an MPB or by shifting orthorhombic-tetragonal (O-T) transition (PPT) below room temperature. Nevertheless, MPBs are generally preferred, as their enhanced properties are not confined to a narrow temperature range.

Therefore, to investigate the presence of MPBs, several complex solid solutions, such as KNN-$Bi_{0.5}Na_{0.5}TiO_3$ (BNT), KNN-$BaTiO_3$ (BT), and KNN-$Bi_{0.5}K_{0.5}TiO_3$ (BKT) have been extensively studied [12, 27, 28]. In the phase diagrams of these systems, MPB-like features can be engineered, in which near-degenerate ferroelectric phases and monoclinic bridging regions facilitate polarization rotation mechanisms. This leads to enhanced piezoelectric coefficients (d_{33}), large electromechanical coupling, and improved thermal stability. Furthermore, the introduction of dopants such as $CaTiO_3$ shifts the orthorhombic-tetragonal (O–T) phase boundary below room temperature, thereby broadening the operational window and reducing susceptibility to depolarization [29, 30]. Although existence of a classical MPB analogous to that in lead zirconate titanate (PZT) is not observed in binary KNN, strategic chemical modifications and multi-component engineering facilitate an MPB-like behavior. These strategies have significantly improved the functional performance and thermal reliability of KNN-based lead-free piezoelectrics, making them increasingly suitable for a wide range of sensing, actuation, and transduction applications.

8.4.1.2 Domain Engineering and Processing Techniques

Domain engineering is a well-established and effective strategy for enhancing the piezoelectric performance of lead-free perovskite materials. By carefully controlling the orientation, density, and mobility of ferroelectric domains, the extrinsic contributions to the overall electromechanical response can be effectively optimized [31]. Conventional approaches, such as electric-field poling, grain orientation control, and the application of uniaxial pressure during sintering, are widely adopted to promote preferred crystallographic alignment, thereby enhancing piezoelectric coefficients and strain response.

In addition, the role of advanced processing techniques has become evident in attaining high-performance lead-free ceramics. For instance, spark plasma sintering (SPS) facilitates rapid densification at lower temperatures and shorter dwell times. This approach also assists maintaining phase purity and minimizing volatilization of elements like Na and K in KNN systems [27]. Similarly, tape casting and templated grain growth (TGG) are employed to produce textured ceramics, which enhance d_{33} values due to improved crystallographic anisotropy. Further, the introduction of sintering aids such as CuO, MnO_2, or rare-earth oxides can refine grain size and inhibit abnormal grain growth without compromising the desirable phase [27]. Alongside these approaches, chemical synthesis routes such as sol-gel and hydrothermal synthesis provide improved compositional homogeneity and phase purity, making them attractive for producing fine-grained powders conducive to achieving high sintered density. Recently, additive manufacturing and inkjet printing have demonstrated strong potential as scalable and sustainable techniques for fabricating patterned piezoelectric structures suitable for miniaturized and flexible device architectures. Altogether, the synergy between domain engineering and modern processing is pivotal to bridging the performance gap between lead-free perovskites and their lead-based counterparts.

8.4.1.3 Applications and Device Integration

Lead-free perovskite piezoelectrics are extensively utilized into a variety of applications including industrial, biomedical, and consumer electronics. In sensor applications, KNN-based ceramics are commonly integrated into vibration sensors, accelerometers, and pressure transducers. Energy harvesting is another emerging field where lead-free materials show tremendous potential. Lead-free KNN-based perovskite systems have been demonstrated to efficaciously convert ambient mechanical vibrations into electrical energy, suitable for powering wireless sensors and wearable electronics. Furthermore, flexible and printed piezoelectric devices are developed using composite forms of these materials, enabling their use in self-powered biomedical sensors and skin-mountable electronics.

8.4.2 *Bismuth Sodium Titanate ($(Bi_{0.5}Na_{0.5})TiO_3$ (BNT)-Based Systems*

$(Bi_{0.5}Na_{0.5})TiO_3$ (BNT) is another prominent Bi-based lead-free perovskite ferroelectric system, first reported by Smolenskii et al. in 1960 [32]. It has attracted considerable attention due to its strong ferroelectric properties, high Curie temperature (~320 °C), and relatively large remnant polarization (38 μC/cm^2) [33]. Although BNT exhibit strong intrinsic ferroelectric properties, its applicability in actuator applications under low driving fields is limited by its large coercive field (~73 kV/cm) and high hysteresis [34, 35]. Undoped BNT possess high electrical conductivity which make the poling process difficult. In addition, BNT require a high sintering temperature (> 1200 °C) for achieving dense body, which results in Bi-volatization. To address these issues, extensive efforts have focused on processing of ceramics and chemical modification through ionic substitutions, by forming solid solutions of BNT with other perovskite-type materials. Binary systems such as BNT–$BaTiO_3$ (BT), BNT–$SrTiO_3$ (ST), BNT–$(Bi_{0.5}K_{0.5})TiO_3$ (BKT), and BNT–(K, Na)NbO_3 (KNN), as well as ternary systems including BNT-BT-BKT and BNT-BT-KNN, have been developed near MPB compositions to enhance the functional properties and tailor the phase behavior [9, 12, 36].

8.4.2.1 Structure and Phase Diagram of BNT-Based Systems

$(Bi_{0.5}Na_{0.5})TiO_3$ (BNT) crystallizes in a perovskite ABO_3 structure and exhibits a rhombohedrally distorted *R3c* symmetry at room-temperature. Accordingly, extensive structural investigations have been conducted to elucidate its phase evaluation and phase transition behavior. With increasing temperature, BNT undergoes a diffuse phase transition characteristic of a relaxor ferroelectrics, transforming from a rhombohedral to a tetragonal phase in the temperature range of 200–320 °C. Subsequently, a further transition from tetragonal to cubic phase occurs approximately at 540 °C. These phase transitions were identified from neutron powder diffraction studies performed by Jones and Thomas [37, 38]. Subsequent in-*situ* temperature-dependent transmission electron microscopy (TEM) studies by Dorcet and Trolliard affirmed a more complex scenario. Their studies suggested that, within the said temperature range of 200–320 °C, a transition from a ferroelectric rhombohedral phase to an anti-ferroelectric orthorhombic phase occurs. This transition is modulated by an anti-ferroelectric phase consisting of orthorhombic platelets within a rhombohedral matrix [39, 40]. As a consequence, BNT is structurally heterogeneous material in which rhombohedral, tetragonal, orthorhombic, and monoclinic regions coexist and evolve with temperature. Therefore, BNT exhibits pronounced local disorder, nanoscale phase coexistence, and intermediate octahedral tilt systems across a wide temperature range. This distinctive structural complexity underpins many of its unusual functional properties.

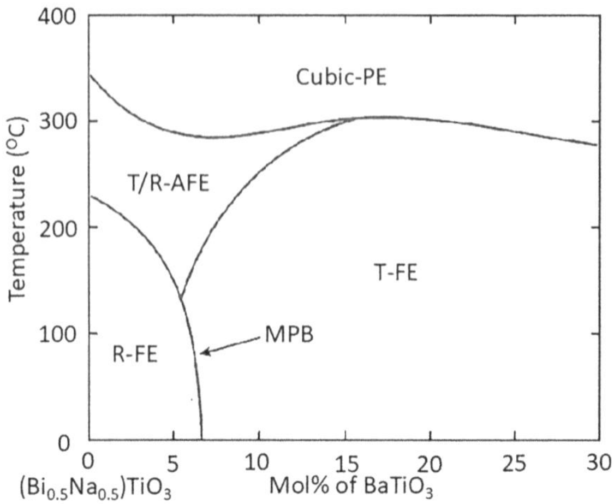

Fig. 8.5 Partial phase diagram of (1-x)($Bi_{0.5}Na_{0.5}$)TiO_3–$x$$BaTiO_3$ system. Regions labeled as, Rhombohedral ferroelectric: R-FE, tetragonal ferroelectric: T-FE, tetragonal + rhombohedral anti-ferroelectric: A/R-AFE [28]

In order to improve the functional properties of BNT, new BNT-based multi-component systems have been developed [12, 36]. Among the BNT-based solid solutions, the most widely investigated solid solution is (1-x)($Bi_{0.5}Na_{0.5}$)TiO_3–$x$$BaTiO_3$. The composition near $x = 0.06$ an morphotropic phase boundary (MPB) exists, which separates a rhombohedral ferroelectric phase from a tetragonal or pseudo-cubic phase (Fig. 8.5) [28]. Around the MPB, the piezoelectric constant ($d_{33} = 125$ pC/N) and coupling constant ($k_{33} = 55\%$) reach their maximum values. However, the Curie temperature (T_C ~ 288 °C) and depolarization temperature (T_d) are significantly lowered due to formation of an antiferroelectric phase. In contrast to the sharp MPB in lead-based PZT, the MPB in BNT–BT is strongly curved, which results in low-temperature stability of the electromechanical properties.

Furthermore, the diffuse relaxor ferroelectric phase transitions (from rhombohedral to tetragonal phase between 200 and 320 °C) in BNT is believed to originates from coexistence between long-range ordered ferroelectric domains and polar nanoregions, which are responsible for its frequency-dependent dielectric behaviour and diffuse phase transitions. Therefore, understanding and tailoring this relaxor-ferroelectric crossover is key for achieving large strain, low hysteresis, and improved fatigue resistance. Such engineered phase coexistence enables BNT-BT compositions to perform effectively in energy harvesting and actuator applications, where large strain, low losses, and operational reliability are essential [9].

Electric field-induced strain in binary BNT–BT compositions often observed under relatively high electric fields. Therefore, ternary modifications have been introduced to combine an MPB between a rhombohedral (BNT) and tetragonal (BT) with an orthorhombic (KNN) to ease domain wall movement and polarization rotation. An electric field induced phase transformation and domain reorientation near the MPB composition also enhances strain responses (~0.45%) and improves piezoelectric coefficients [41]. Another ternary system BNT–BT–BKT was also explored

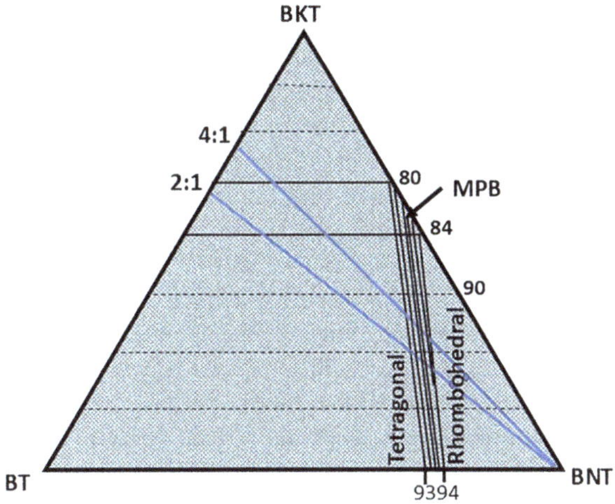

Fig. 8.6 Phase diagram of BNT-BKT-BT ternary system showing MPB between tetragonal and rhombohedral phases [9]

and found to be good enough to replace PZT in specific applications such as ultrasonic wire bonding transducers and accelerometers (Fig. 8.6) [42–45]. The introduction of KNN or BKT into binary BNT-BT system facilitates stabilization of monoclinic or mixed-phase regions, supporting polarization rotation and field-induced phase switching. These systems demonstrate reduced hysteresis, improved fatigue resistance, and enhanced thermal stability compared to binary BNT-based ceramics.

8.4.2.2 Processing and Microstructural Considerations

Bi-based binary and ternary solid solutions are typically synthesized using conventional ceramic route. BNT-based ceramics are susceptible to volatilization of Bi and Na during high-temperature sintering, leading to poor sintered density, non-stoichiometry and secondary phase formation. Hot pressing ensures densification but is rarely practical due to cost. To solve this, chemical processing routes such as sol-gel, hydrothermal, stearic acid gel, and citrate gel methods have been applied. These routes provide molecular-level homogeneity, controlled grain size, high purity, reduced sintering temperatures, and ease of thin-film processing, making them viable for preparation of high-quality BNT and BNT–BT powders [9].

Further improvements in piezoelectric properties can be achieved through the use of textured ceramics techniques, which promote the formation of highly anisotropic microstructures. Such texturing also mitigates the risk of lowered depolarization temperature (T_d) generally associated with compositional modifications. Reactive templated grain growth (RTGG) and templated grain growth with hot pressing (TGG/HP) have been used successfully in BNT-based ceramics, which result in enhanced anisotropic piezoelectric properties [46, 47]. For instance, $Bi_4Ti_3O_{12}$ templates, when combined with Bi_2O_3 or Na_2CO_3 additives, promote preferential orientation in BNT-based ceramics. Grain size refinement and orientation control show decisive role in optimizing strain-electric field behavior and domain wall mobility. Enhanced grain orientation can leads to a substantial improvement in the piezoelectric coefficients and coupling factors compared with randomly oriented ceramics [48, 49].

Compositional modification using doping in BNT-based material are extensively used to tailor their sintering behavior and electromechanical properties. In pure BNT, variety of dopants have been introduced to reduce sintering temperature, achieve optimal poling, and increase T_c. For example, incorporation of Fe^{3+} facilitate densification at temperatures below 900 °C by creating oxygen vacancies that promote diffusion. In contrast, other dopants such as Li, Mn, Ce, Bi_2O_3, $BaCu_{0.5}W_{0.5}O_3$ effectively lower electrical conductivity, thereby enabling efficient poling and improved electromechanical properties, including k_p, d_{33}, and P_r [9]. Introduction of additives such as $BiScO_3$ and $BiFeO_3$ further enhances remnant polarization (P_r) and raises the Curie temperature (T_C) beyond 400 °C. However, these advantages are often accompanied by increased coercive fields (E_c), which significantly make the poling process difficult. Subsequently, BNT has also been modified with other dopants including La, Sr, Ta, $NiNbO_3$, $NaNbO_3$ and KNN, while BNT-6BT has been doped with La, Nb, Mn, Ag, Li, or Ce. Similarly, BNT–BKT with Ce, Sr, or Sr/La and BNT–BKT–BT with Li [9, 12, 36, 50]. While, these compositional modifications give rise to enhanced P_r, k_p, d_{33}, and T_m (commonly taken as T_c) values, they often result in reduced E_c and T_d compared with undoped compositions.

Although pure BNT has certain drawbacks, such as high coercive field and low strain at practical operating voltages, solid-solutions prepared near MPB compositions and within ternary systems demonstrate significantly enhanced performance. By leveraging the relaxor behavior, phase boundary engineering, and advanced processing strategies, BNT-based ceramics can achieve high electromechanical response, good thermal stability, and functional reliability. These advantages of BNT-based materials as an alternative of PZT are well aligned with the global pursuit of environmentally sustainable electronic devices.

8.4.3 Ba(Zr,Ti)O$_3$-(Ba,Ca)TiO$_3$ (BZT-BCT) Systems

Barium titanate ($BaTiO_3$), one of the oldest perovskite structured ferroelectrics developed in 1940s, remains a cornerstone material in functional ceramics. It exhibits a relatively high electromechanical coupling factor ($k_p \sim 0.50$) but only a modest piezoelectric strain coefficient ($d_{33} \sim 190$ pC/N). Owing to its high dielectric constant and low loss characteristics, BT has long been widely investigated for capacitor and multilayered capacitor applications. With the quest for lead-free ferroelectric materials, BT-based systems have regained much attention due to its promising piezoelectric properties. The focus has been compositional modification to obtain a MPB like characteristics, where more than two structural phases can coexist. Recently, Liu and Ren have shown an improved piezoelectric property and enhanced critical temperatures in BT-based solid solution, (1-x(Ba(Zr$_{0.2}$Ti$_{0.8}$)O$_3$-x(Ba$_{0.7}$Ca$_{0.3}$)TiO$_3$ (abbreviated as

BZT-*x*BCT) [13]. Particularly, this MPB composition showed a piezoelectric coefficient d_{33} in the range of 550–620 pC/N which is higher compared to the most popular PZT (d_{33} = 500–600 pC/N).

8.4.3.1 Structure and Phase Diagram of BZT-BCT

Pristine $BaTiO_3$ (BT) exhibits a ferroelectric tetragonal (T-FE) structure at room temperature but undergoes several displacive-type structural phase transitions with changes in temperature. At ~120 °C, the paraelectric cubic phase transforms into a ferroelectric tetragonal (C → T) phase due to the displacement of Ti ions within the TiO_6 octahedra, resulting in an increment in the *c* lattice parameter and reduction in *a* and *b* lattice parameters. Upon further decrease in temperature, the tetragonal phase transforms into an orthorhombic phase at ~5 °C, followed by an orthorhombic to rhombohedral transition at ~ −80 °C. These structural phase transitions give rise to an enhanced dipole moment when changes from tetragonal to orthorhombic and to a rhombohedral phase. Such temperature-dependent phase transitions are referred to as polymorphic phase transition (PPT). As a result, the electromechanical properties become highly sensitive to thermal fluctuations, and cycling across the transition temperature may induce depolarization, thereby compromising performance stability. However, the incorporation of suitable dopants or the formation of solid solutions can shift the T → O phase transition temperature near the room temperature, thereby broadening the usable temperature range.

A chemical substitution at *A*-site (Ba^{2+}) and (or) *B*-site (Ti^{4+}) of the perovskite structure are made to tailor the electromechanical properties in $BaTiO_3$ (BT). Yu and Li modified BT with 5 mol% of Zr^{4+} i.e., $Ba(Zr_{0.05}Ti_{0.95})O_3$ and obtained T → O phase transition temperature around 58 °C and further modified with Ca^{2+} which showed an improved piezoelectric property. Further modification by making solid solution using (Ca, Zr) co-doping at both *A*-site (Ba^{2+}), $(Ba_{0.70}Ca_{0.30})TiO_3$ (BCT) and *B*-site (Ti^{4+}), $Ba(Zr_{0.20}Ti_{0.80})O_3$ (BZT) respectively showed remarkable piezoelectric properties composition near MPB. The phase diagram of $Ba_{1-x}Ca_x)(Zr_yTi_{1-y})O_3$ (BZT-*x*BCT) pseudo-binary system showed symmetry-lowering structural transitions from high-temperature paraelectric cubic phase (C) to the low-temperature ferroelectric rhombohedral (R) and tetragonal (T) phases. Their phase diagram is characterized by the MPB, which corresponds to the coexistence of R-phase (BZT side) and T-phase (BCT side) with spontaneous polarizations along <111> and < 100> respectively. The various phase transition sequences observed with respect to the changing composition and increasing temperature is shown in Fig. 8.7. It is seen that with $x \leq 0.32$ exhibit a single-phase transition from R-phase to C-phase, while composition $0.32 < x < 0.8$ there exists a two-phase transition with R-T-C phase sequence. At $x > 0.8$, a single-phase transition from T-phase to C-phase can be seen. Furthermore, composition with $x = 0.32$ (32BCT), there exists a C-R-T tricritical or triple point, wherein all the first-order phase transitions such as C → T, C → R, and T → R coincide (Fig. 8.7). This type of C-R-T-type triple point characterizes several other Pb-based piezoelectric systems such as PZT and PMN-PT solid solutions,

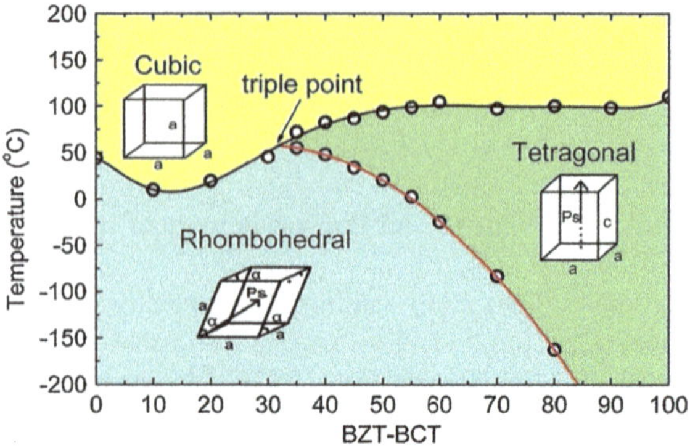

Fig. 8.7 Phase diagram of BZT-BCT system, characterized by a MPB separating a ferroelectric rhombohedral (BZT side) and tetragonal (BCT side) phase. (Reprinted with permission from Ref. [13]. Copyright by (2009) American Physical Society

wherein highest permittivity has been reported. The phase diagram of BZT-xBCT as proposed by Liu and Ren [13] is based on dielectric data, X-ray diffraction (XRD), and transmission electron microscopy (TEM) observations. However, the presence of an orthorhombic ferroelectric (O) phase has not been identified by Liu and Ren. Recently, the temperature-composition structural phase diagram of BZT-xBCT solid solutions has been reinvestigated by Keeble using high-resolution synchrotron X-ray diffraction [51]. They have suggested that phase transition sequence exhibiting R → C type as reported earlier is commonly observed in highly doped BT ceramics ($BaTi_{1-x}B_xO_3$ with B = Zr, Ce, Hf etc.). However, the R-T-C type phase sequence is unusual. Hence, the revised phase diagram has been characterized by the formation of phase convergence region (non-point-like region) that includes R-O-T-C phase sequence (high to low temperature). This particular phase transition sequence observed on cooling the BZT-xBCT compositions with higher x values, for example, $x = 0.5$ and 0.4 is similar to the C-T-O-R transition sequence observed for BT, both in phase sequence and in crystallographic space group symmetry.

8.4.3.2 Processing and Microstructural Optimization

To Optimise the performance of BZT–BCT ceramics, dense and uniform microstructures need to be prepared. Hot pressing, spark plasma sintering (SPS), and conventional solid-state sintering have been employed to prepare fine-grained microstructures with few secondary phases. Synthesis of BZT-BCT using a solid-state route typically requires relatively higher calcination (>1300 °C) and sintering temperatures (>1400 °C). Although extensively used, it produces coarse grains and sometimes secondary phases, thus limiting dielectric and piezoelectric properties. Optimization of the synthesis route begins from the quality of the raw material such as purity, particle size, agglomeration etc. The powder characteristics including optimized calcination and sintering temperature are anticipated to play a critical role in determining the microstructural, electrical, and optical properties of ferroelectric ceramics. The maximum piezoelectric properties (d_{33} ~ 650 pC/N) were obtained at an optimum calcination temperature (1300 °C) and sintering temperature (1540 °C) for 0.5BZT-0.5BCT ceramics [52].

Grain size engineering is of another important strategy for tuning the electromechanical response. Coarse grains generally facilitate domain wall mobility, whereas abnormal grain growth often introduces heterogeneity and deteriorate dielectric properties. To obtain finer particle sizes and improved chemical homogeneity, wet-chemical synthesis routes such as sol–gel, combustion, hydrothermal, and co-precipitation methods have been widely explored. Sol-gel method provides excellent compositional control but extremely sensitive to processing conditions. Combustion method enables synthesis of nano-sized powders at relatively lower temperature, while hydrothermal route yields phase-pure, fine powders under moderate conditions. The co-precipitation method also offers good chemical homogeneity but with strict control of pH and precursor chemistry.

$(1-x)BaTi_{0.8}Zr_{0.2}O_3-xBa_{0.7}Ca_{0.3}TiO_3$ ($x = 0.2$–0.4) powders prepared using sol-gel method and calcined at temperature 900 °C for 2 h, followed by sintering at 1250 °C, exhibit a cubic-rhombohedral-tetragonal (C-R-T) triple point near $x = 0.3$. The composition showed a high piezoelectric coefficient of ~391 pC/N and dielectric constant ~9000 [53]. Similar composition, $(Ba_{0.88}Ca_{0.12})(Zr_{0.12}Ti_{0.88})O_3$ was synthesized using alkoxide and nitrate precursors, obtaining phase-pure powders at 800 °C with nanometric crystallites (~37 nm). The sintered ceramics showed a Curie temperature of ~95 °C, a dielectric constant of ~2588, and d_{33} ~ 215 pC/N [54]. Ceramics produced via citrate or polymeric precursor combustion methods also achieve excellent functional properties at relatively low calcination temperature (~700–1000 °C) and moderate sintering temperatures (~1250–1300 °C). This method could produce nanosized particle ranging from 30 to 60 nm and grain size in the sintered pellet ranging from 0.5 to 20 μm essential for controlling electromechanical properties. These processing routes enable the realization of high dielectric permittivity, strong electromechanical coupling (k_p ~ 50%), and stable piezoelectric response (d_{33} ~ 400 pC/N) [55–58]. Sol–gel and combustion synthesis routes provide an efficient way to produce BZT–BCT ceramics with desired microstructures, small particle sizes, and superior electromechanical performance, making them strong candidates for actuator, transducer, and capacitor applications.

8.5 Applications in New-Generation Devices

The emphasis on environmentally friendly materials and technologies has gain significant momentum in the development of next-generation devices based on lead-free perovskite piezoelectric materials. Despite these efforts, commercial piezoelectric applications still largely rely on lead zirconate titanate (PZT), which accounts approximately 80% of the global piezoelectric ceramics market. While the high lead content in PZT raises severe toxicity concerns, lead-free perovskite systems provide both ecological safety and competitive electromechanical performance. Additionally, their structural flexibility, compositional tunability, and adaptability to diverse processing techniques make them promising candidates for the next generation of smart and sustainable devices.

This growing interest for lead-free piezoelectric ceramics can also be seen in the global market trends, which is projected to grow at a CAGR of 8–12% by 2030. It is expected that the revenues to rise from \$279 million in 2024 to approximately \$550 million by 2030 (Fig. 8.8a) [59]. The increasing demand from consumer electronics, automotive systems, medical devices, and industrial automation, where regulatory compliance and sustainability are becoming decisive factors serves as the key driving force behind this market growth.

The technological appeal of lead-free piezoelectric materials lies in their strength in delivering strong electromechanical responses through the interaction between intrinsic ferroelectric polarization mechanisms and extrinsic contributions such as domain wall motion, phase coexistence, and lattice distortions. In particular, compositions engineered near morphotropic and polymorphic phase boundaries show superior piezoelectric responses, as structural instabilities in these regions facilitate polarization rotation. As a result, the electromechanical properties of such lead-free systems not only match but, sometimes even exceed the performance benchmarks established by lead-based soft PZT ceramics.

Among the various lead-free perovskite systems, barium titanate (BT)-based solid solutions, particularly $Ba(Zr,Ti)O_3$–$(Ba,Ca)TiO_3$ (BZT–BCT)-based materials emerged as promising candidates for wide range of piezoelectric applications. In actuator and transducer applications, textured BZT–BCT ceramics are highly attractive owing to their exceptionally high piezoelectric coefficients and strong electromechanical coupling, while also exhibiting promising displacement characteristics suitable for precision positioning systems. This is why BCT-BZT-based materials have secured a substantial 35–40% share of lead-free piezoelectric ceramics market (Fig. 8.8b). This positions them as a true market competitor to conventional lead-based PZT. Potassium sodium niobate (KNN)-based ceramics represent another rapidly growing segment, with an estimated share of 18–30% (Fig. 8.8b). Their strong

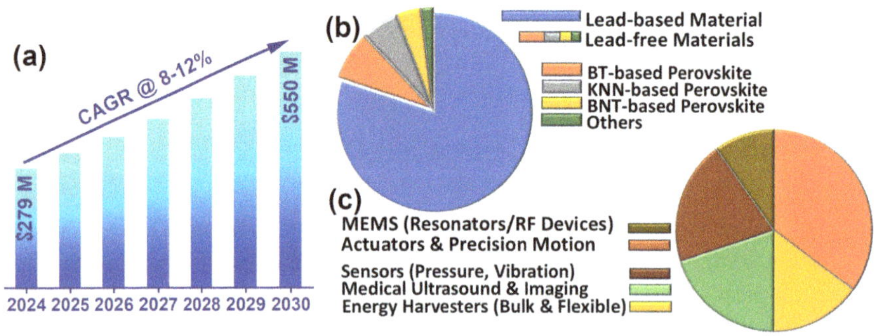

Fig. 8.8 (**a**) Projection of global lead-free piezoelectric ceramic market (**b**) market share of lead-based and various lead-free system (**c**) next-generation application areas of lead-free perovskite materials

growth is due to high performance in high-temperature applications. Their utilization in vibration sensors, accelerometers, and energy-harvesting devices has been increasing steadily, supported by advances in compositional control and processing strategies that improve densification, reliability, and reproducibility. Consequently, KNN-based systems are gaining much attention in industrial and automotive sensing technologies. Bismuth sodium titanate (BNT)-based systems, despite their drawbacks with high coercive fields, processing and reproducibility, remain appealing due to their high Curie temperatures and strong piezoelectric activity, and large remnant polarization. However, using a solid-solution engineering, enhanced strain and reduced hysteresis are possible to achieve, making them suitable for actuators.

From an application point of view, the commercial deployment of lead-free piezoelectrics has so far been concentrated in capacitor dielectrics and simple piezoelectric components such as buzzers and resonators. However, the fastest growth is expected in near future in the areas of microelectromechanical systems (MEMS), ultrasonic motors, thin-film-based sensors, and energy harvesting systems (Fig. 8.8c). It is further anticipated that integrating lead-free perovskite into energy harvesting systems promises sustainable solutions for powering autonomous sensors and wearable electronics.

A rapidly emerging futuristic wearable applications involves the introduction of piezoelectric perovskite nanostructures into flexible polymer matrix, such as poly (vinylidene fluoride) and its co-polymers to form piezoelectric nanogenerators (PENG) [60–62]. These systems with the advantages of mechanical flexibility with efficient energy conversion and are well suited for wearable electronics, electronic skin, soft robotics, and smart textiles [63]. Moreover, the inherent biocompatibility of lead-free perovskites broadens their applicability in biomedical technologies, including implantable sensors, therapeutic actuators, and tissue-engineering platforms. Bone tissue engineering and self-assisted wound healing are some of the emerging areas of active investigations [64, 65].

Despite considerable development in material optimization, successful device integration still demands advances in interface engineering, packaging, and long-term stability testing. Collaborative efforts among materials scientists, device engineers, and system integrators are therefore important to fully comprehend the potential of lead-free perovskite piezoelectrics in sustainable and next-generation devices.

8.6 Summary

Lead-free perovskite piezoelectric materials have become evident as an important class of functional ceramics as an alternative to conventional lead-based systems. Growing awareness of lead toxicity, in conjunction with increasingly stringent restrictions imposed by environment regulatory bodies, has driven considerable research efforts. This research focuses on developing lead-free compositions that can deliver competing electromechanical performance while ensuring long-term reliability and scalability.

This chapter has provided an exhaustive overview of the major lead-free perovskite systems, namely potassium sodium niobate (KNN)-based, bismuth sodium titanate (BNT)-based, and barium titanate (BT)-based solid solutions, with particular emphasis on their structure–property relationships, phase evolution, and performance enhancement strategies. Although each system exhibits distinct merits and demerits, the structural versatility of the perovskite lattice allows effective compositional tuning and phase-boundary engineering.

Advanced processing techniques have played a critical role in enhancing the performance of lead-free perovskite piezoelectrics. Techniques such as spark plasma sintering, sol–gel synthesis, combustion methods, and templated grain growth have enabled improved densification, refined grain size, controlled phase purity, and enhanced domain mobility. In addition to this, domain engineering and texturing strategies have significantly reduced the performance gap between lead-free ceramics and their lead-based counterparts.

In conclusion, lead-free perovskite piezoelectrics offer a technologically versatile and environmental-friendly materials for next-generation devices. Continued progress in compositional design, processing optimization, and domain structure control is expected to further enhance their functional performance and reliability. With continuously rising market demand and expanding applications including sensors, actuators, energy harvesters, biomedical devices, and flexible electronics, these materials are poised to play a defining role in the future of sustainable piezoelectric technologies.

References

1. K. Uchino, *Ferroelectric Devices*, 2nd edn. (CRC Press Taylor & Francis Group, Boca Raton, London, New York, 2010)
2. Off. J. Eur. Union **46**(L37), 24–38 (2003)
3. Off. J. Eur. Union **46**(L37), 19–23 (2003)
4. T.R. Shrout, S.J. Zhang, J. Electroceram. **19**(1), 113–126 (2007)
5. T. Takenaka, H. Nagata, Y. Hiruma, Y. Yoshii, K. Matumoto, J. Electroceram. **19**(4), 259–265 (2007)
6. S. Mitra, A.R. Kulkarni, O. Prakash, J. Appl. Phys. **114**, 6 (2013)
7. S. Mitra, A.R. Kulkarni, W. Jo, J. Am. Ceram. Soc. **99**(3), 888–895 (2015)
8. Y. Guo, K.-i. Kakimoto and H. Ohsato, Appl. Phys. Lett. 85 (18), 4121–4123 (2004)
9. J. Rödel, W. Jo, K.T.P. Seifert, E.-M. Anton, T. Granzow, D. Damjanovic, J. Am. Ceram. Soc. **92**(6), 1153–1177 (2009)
10. B. Jaffe, R.S. Roth, S. Marzullo, J. Appl. Phys. **25**(6), 809–810 (1954)
11. I. Coondoo, N. Panwar, A. Kholkin, J. Adv. Dielectr. **03**(02), 1330002 (2013)
12. D. Xiao, J. Adv. Dielectr. **01**(01), 33–40 (2012)
13. W. Liu, X. Ren, Phys. Rev. Lett. **103**(25), 257602 (2009)
14. J. Rödel, K.G. Webber, R. Dittmer, W. Jo, M. Kimura, D. Damjanovic, J. Eur. Ceram. Soc. **35**(6), 1659–1681 (2015)
15. X. Liu, R. Hong, C. Tian, J. Mater. Sci. Mater. Electron. **20**(4), 323–327 (2008)
16. A. M. Glazae, Simple ways of determining perovskite structures. Acta Crystallogr. A **31**(6), 756–762 (1975). https://doi.org/10.1107/S0567739475001635
17. G. Shirane, R. Newnham, R. Pepinsky, Phys. Rev. **96**(3), 581–588 (1954)

18. Y. Saito, H. Takao, T. Tani, T. Nonoyama, K. Takatori, T. Homma, T. Nagaya, M. Nakamura, Nature **432**(7013), 84–87 (2004)
19. W.R.C. Bernard Jaffe Jr., H. Jaffe, *Piezoelectric Ceramics* (Academic Press Inc. (London) Ltd., London & New York, 1971)
20. K. Wang, J.-F. Li, Appl. Phys. Lett. **91**(26), 262902 (2007)
21. Y. Dai, X. Zhang, G. Zhou, Appl. Phys. Lett. **90**(26), 262903 (2007)
22. D. Damjanovic, J. Am. Ceram. Soc. **88**(10), 2663–2676 (2005)
23. B. Noheda, D.E. Cox, G. Shirane, R. Guo, B. Jones, L.E. Cross, Phys. Rev. B **63**(1), 014103 (2000)
24. B. Noheda, Curr. Opinion Solid State Mater. Sci. **6**(1), 27–34 (2002)
25. M. Ahtee, A.M. Glazer, Acta Crystallogr. A **32**(3), 434–446 (1976)
26. M. Ahtee, A.W. Hewat, Acta Crystallogr. A **34**(2), 309–317 (1978)
27. J. Wu, D. Xiao, J. Zhu, Chem. Rev. **115**(7), 2559–2595 (2015)
28. T. Takenaka, K.-i. Maruyama, K.S. Koichiro Sakata, Jpn. J. Appl. Phys. **30**(9S), 2236 (1991)
29. S. Zhang, R. Xia, H. Hao, H. Liu, T.R. Shrout, Appl. Phys. Lett. **92**(15), 152904 (2008)
30. S. Zhang, R. Xia, T.R. Shrout, Appl. Phys. Lett. **91**(13), 132913 (2007)
31. T. Zheng, J. Wu, D. Xiao, J. Zhu, Prog. Mater. Sci. **98**, 552–624 (2018)
32. V.A.I.G.A. Smolenskii, A.I. Agranovskaya, N.N. Krainik, Soviet Phys.-Solid State **2**, 4 (1961)
33. Y. Li, W. Chen, J. Zhou, Q. Xu, H. Sun, R. Xu, Mater. Sci. Eng. B **112**(1), 5–9 (2004)
34. Y. Hiruma, H. Nagata, T. Takenaka, J. Appl. Phys. **105**(8), 084112 (2009)
35. Y.S. Sung, J.M. Kim, J.H. Cho, T.K. Song, M.H. Kim, T.G. Park, Appl. Phys. Lett. **98**(1), 012902 (2011)
36. D.Q. Xiao, D.M. Lin, J.G. Zhu, P. Yu, J. Electroceram. **21**(1–4), 34–38 (2007)
37. G.O. Jones, P.A. Thomas, Acta Crystallogr. B **58**(2), 168–178 (2002)
38. G.O. Jones, P.A. Thomas, Acta Crystallogr. B **56**(3), 426–430 (2000)
39. V. Dorcet, G. Trolliard, P. Boullay, Chem. Mater. **20**(15), 5061–5073 (2008)
40. G. Trolliard, V. Dorcet, Chem. Mater. **20**(15), 5074–5082 (2008)
41. S.-T. Zhang, A.B. Kounga, E. Aulbach, H. Ehrenberg, J. Rödel, Appl. Phys. Lett. **91**(11), 112906 (2007)
42. Y. Hiruma, Y. Makiuchi, R. Aoyagi, H. Nagata, T. Takenaka, Adv. Dielectr. Mater. Electron. Devices, 139–146 (2006)
43. S. Zhang, T.R. Shrout, H. Nagata, Y. Hiruma, T. Takenaka, IEEE Trans. Ultrason. Ferroelectr. Freq. Control **54**(5), 910–917 (2007)
44. K.W. Kwok, H.L.W. Chan, *In Piezoelectric Ceramics, Edited by E* (Suaste-Gomez (IntechOpen, London, 2010)
45. S.H. Choy, X.X. Wang, H.L.W. Chan, C.L. Choy, Appl. Phys. A Mater. Sci. Process. **82**, 715–718 (2006)
46. T. Kimura, T. Takahashi, T. Tani, Y. Saito, J. Am. Ceram. Soc. **87**(8), 1424–1429 (2004)
47. T. Motohashi, T. Kimura, J. Eur. Ceram. Soc. **27**(13), 3633–3636 (2007)
48. T. Kimura, E. Fukuchi, T. Tani, Jpn. J. Appl. Phys. **44**(11R), 8055 (2005)
49. D.-S. Lee, S. Jeong, E.-C. Park, J.-S. Song, J. Electroceram. **17**, 505–508 (2006)
50. P.K. Panda, J. Mater. Sci. **44**(19), 5049–5062 (2009)
51. D.S. Keeble, F. Benabdallah, P.A. Thomas, M. Maglione, J. Kreisel, Appl. Phys. Lett. **102**(9), 092903 (2013)
52. P. Wang, Y. Li, Y. Lu, J. Eur. Ceram. Soc. **31**(11), 2005–2012 (2011)
53. Z.W. Tan, W.G. Fu, X.Y. Deng, R.B. Yang, X.F. Guan, C. Lu, Y.J. Zhang, L.R. Han, Adv. Mater. Res. **148-149**, 1480–1485 (2011)
54. R.B. Yang, W.G. Fu, X.Y. Deng, Z.W. Tan, Y.J. Zhang, L.R. Han, C. Lu, X.F. Guan, Adv. Mater. Res. **148-149**, 1062–1066 (2011)
55. J.P. Praveen, K. Kumar, A.R. James, T. Karthik, S. Asthana, D. Das, Curr. Appl. Phys. **14**(3), 396–402 (2014)
56. M. Wang, R. Zuo, S. Qi, L. Liu, J. Mater. Sci. Mater. Electron. **23**(3), 753–757 (2012)
57. V.S. Puli, A. Kumar, D.B. Chrisey, M. Tomozawa, J.F. Scott, R.S. Katiyar, J. Phys. D. Appl. Phys. **44**(39), 395403 (2011)

58. A. Reyes-Montero, L. Pardo, R. López-Juárez, A.M. González, M.P. Cruz, M.E. Villafuerte-Castrejón, J. Alloys Compd. **584**, 28–33 (2014)
59. B. Research, (BCC Research, Boston, 2025)
60. C. Mitharwal, S. Geetanjali, A. Malhotra, M.K. Bagla, S.M. Srivastava, C.M.S. Gupta, E. Negi, A.R. Kar, Kulkarni, S. Mitra, J. Alloys Compd. **876**, 160141 (2021)
61. A. Bagla, K. Hembram, F. Rault, F. Salaün, S. Sundarrajan, S. Ramakrishna, S. Mitra, J. Phys. Chem. C **129**(12), 5808–5820 (2025)
62. K. I. Park, M. Lee, Y. Liu, S. Moon, G. T. Hwang, G. Zhu, J. E. Kim, S. O. Kim, D. K. Kim, Z. L. Wang and K. J. Lee, Adv. Mater. 24 (22), 2999–3004, 2937 (2012)
63. A. Bagla, P. Goyel, F. Rault, F. Salaün and S. Mitra, J. Text. Inst., 1–13 (2025)
64. F. C. Kao, H. H. Ho, P. Y. Chiu, M. K. Hsieh, J. C. Liao, P. L. Lai, Y. F. Huang, M. Y. Dong, T. T. Tsai, Z. H. Lin, Self-assisted wound healing using piezoelectric and triboelectric nanogenerators. Sci. Technol. Adv. Mater. **23**(1), 1–16 (2022)https://doi.org/10.1080/14686996.2021.2015249
65. S. Swain, R. D. K. Misra, T. R. Rautray, Nanoscale generators for tissue healing: a perspective. Int. J. Nanomed., 11859–11882 (2024)https://doi.org/10.2147/IJN.S480938

Chapter 9
Stable Lead-Free Perovskite Materials for Hydrogen Production via Water Splitting

Ishika Bhullar, Srimanta Das, Manjunatha Kempasiddaiah, Namrata Pant, and Soumitra Satapathi

Abstract Rapid increase of global population, escalating energy consumption, and severe climate change led to think about future energy security. The greenhouse gas emissions through excessive use of fossil fuels and their limited resources, compelled scientific community to find out replacement of fossil fuels. Molecular hydrogen stands out to be one of the replacements which can revive the threat for future energy security. Among several methods of hydrogen production, water splitting is one of the cleanest ways to produce hydrogen, accomplished through three methodologies: photocatalysis, electrocatalysis, and photoelectrochemical process. This chapter initially elucidates basic mechanisms of these three processes for hydrogen generation through water splitting. It then focuses on the role of lead-free perovskites, exploring both halide and oxide systems across all three methodologies. Thereafter, different strategies including foreign element doping, vacancy engineering, surface engineering, encapsulation strategy are introduced to improve the performance of oxide and halide perovskites separately for water splitting application. Finally, this chapter outlines the potential of lead-free perovskites for hydrogen generation via water splitting. In summary, this chapter offers a thorough analysis of current state of art lead free perovskites and their performances, future design strategies for further improvement in performances for hydrogen generation through water splitting.

Keywords: Hydrogen generation · Oxide perovskites · Halide perovskites · Photocatalysis · Electrocatalysis · Photoelectrocatalysis

---

Ishika Bhullar and Srimanta Das contributed equally to this work and share first authorship.

I. Bhullar · S. Das · M. Kempasiddaiah · N. Pant (✉) · S. Satapathi (✉)
Department of Physics, Indian Institute of Technology Roorkee, Roorkee, India
e-mail: pantnamrata10@gmail.com

S. Satapathi
e-mail: soumitra.satapathi@ph.iitr.ac.in

S. Satapathi
Center for Sustainable Energy, Indian Institute of Technology Roorkee, Roorkee, India

V. Chauhan et al. (eds.), *Sustainable Lead-Free Perovskite Materials*, Topics in Applied Physics 157, https://doi.org/10.1007/978-3-032-24873-2_9

9.1 Introduction

The overexploitation of fossil fuels for energy generation, and the associated emission of greenhouse gases have driven the search for renewable alternatives. Currently, about 90% of global energy demand is met through non-renewable sources e.g. coal, petroleum, and natural gases. However, their constrained availability and severe environmental impact have compelled the scientific community to explore sustainable options. Molecular hydrogen stands out among renewables (solar, wind, hydropower, geothermal, and bioenergy) as a promising energy carrier ascribed to its remarkably high gravimetric energy density (~120 MJ/kg) and its ability to deliver clean energy with negligible greenhouse gas emissions. However, the use of renewable energy resources for hydrogen production is not more than 2%. There are several techniques to produce hydrogen, such as thermo-catalytic process, [1, 2] natural gas reforming, [3] gasification of coal [4] and biomass, [5] electrolysis of water [6] etc. Among them, hydrogen production through water splitting is a widely studied process on a small scale whereas large scale production of hydrogen is limited. Therefore, achieving the net zero by 2050 is a difficult and formidable task. According to the international energy agency (IEA) report on 'Global Hydrogen Review 2025', the global hydrogen demand increased to almost 100 Mt in 2024, which is 2% up from 2023 [7]. Despite of that the uptake of low emission hydrogen is not yet meeting the goal set in recent times. Therefore, there is an urgent need for enhancement in hydrogen demand, infrastructure development as well as low-cost material development. There are three different methods of H_2 generation through water splitting such as photochemical, electrochemical, and photo-electrochemical techniques. In photocatalytic water splitting, solar energy is directly harnessed by semiconductor catalysts to break down H_2O into H_2 and O_2. In contrast, electrochemical water splitting relies solely on external electrical energy, while photoelectrochemical water splitting integrates both solar and electrical energy to drive the reaction. Besides exploring various energy sources for H_2 generation via water splitting, another vital aspect of these processes is development of earth-abundant, and low-cost catalytic materials. Various catalyst materials, such as noble metals noble metal-based catalysts, [8] transition metal compounds, [9, 10] perovskites and other mixed oxides, [11] layered double hydroxides (LDHs), [12] and carbon-based catalysts, [13] have been widely explored over the past decades for H_2 generation via water splitting. Perovskite materials displayed great potential for this application attributable to their exceptional properties for example compositional flexibility, band gap tuneability, exceptional optical properties, good thermal stability, and efficient charge separation. Perovskite materials are depicted by ABX_3 formula, where X corresponds to an anion and A, B are cations having dissimilar radii. They are classified into two categories based on their anion, perovskite oxides (where X is oxygen) and halides (X is halide anions). Each class of materials exhibits different optoelectronic and chemical properties along with varying degree of stability under different operating conditions. Lead (Pb) is a key constituent of perovskite materials; however, it is a known carcinogen, and even minimal exposure has been linked to severe health risks across all age groups. Despite of several advantages in lead-based perovskites, especially stability in ambient environment, the concern of lead toxicity as well as the difficulty of disposal has intensified the role of lead-free perovskites in different application. Hence, the following section systematically highlights recent advancements of lead-free perovskites in photochemical, electrochemical, and photoelectrochemical H_2O splitting for H_2 generation.

9.2 Basic Mechanism of Photocatalytic Water Splitting

Semiconductor photocatalysts act as the core energy converters in photocatalytic water splitting, they facilitate the conversion of solar-to-chemical energy. When photons having energies larger than the bandgap falls on semiconductor, generates electron–hole pair, which can participate in redox reactions for water splitting. However, only a fraction of these carriers can successfully move from the bulk region to the surface without recombination losses. For efficient hydrogen generation, the photocatalyst must possess an appropriate bandgap and suitably aligned band-edge positions [14]. Specifically, conduction band edge must be situated at a potential more negative than 0 V (vs. RHE) to facilitate the hydrogen evolution reaction (HER), whereas valence band edge must lie at a more positive potential than + 1.23 V versus RHE to initiate the oxygen evolution reaction (OER). Figure 9.1 illustrates schematics of (a) a single and (b) a dual absorber photocatalyst with reduction potential required for HER and OER [15]. Various factors are considered to assess the efficiency of hydrogen production of different semiconductor materials. One of them is solar-to-hydrogen (STH) efficiency which is determined by three interrelated factors: $STH = \eta_{\mathrm{la}} \times \eta_{\mathrm{cs}} \times \eta_{\mathrm{sc}}$ where η_{la} are photons that successfully stimulate the semiconductor to create photogenerated carriers, η_{cs} is the number of photogenerated carriers successfully transferred to the surface of catalyst and η_{sc} is the amount of photogenerated carriers contributing to surface catalysis [16]. Since visible region (400–700 nm) constitutes most of the portion of solar spectrum, significant efforts have been directed in designing photocatalysts that are active under visible light [17].

Among various candidates, oxide perovskites and halide perovskites have drawn significant consideration due to their unique optoelectronic and catalytic properties which are discussed in the Sect. 5.1, which distinguish them from conventional photocatalytic materials.

9.3 Basic Mechanism of Electrochemical Water Splitting

Electrochemical water splitting process includes two half-cell reactions: a reduction reaction at the cathode, known as HER, and an oxidation reaction at the anode, called OER. H_2 and O_2 gases are generated at the cathode and anode, respectively, and can

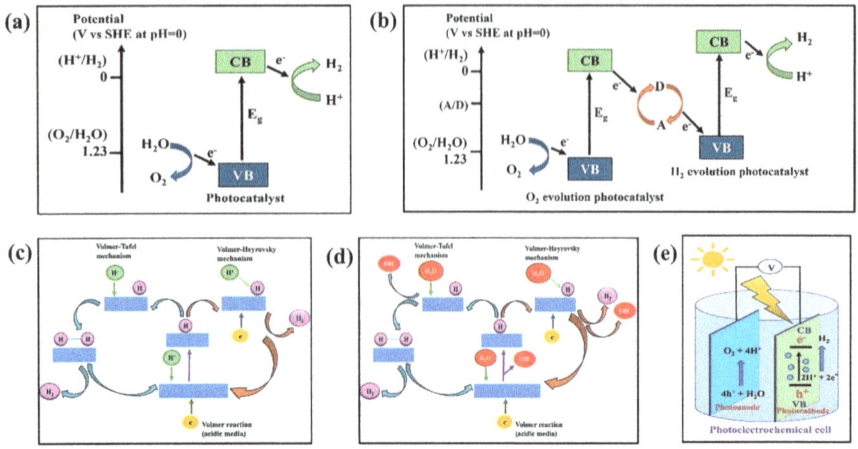

Fig. 9.1 General schematic depiction of photocatalytic water splitting using (**a, b**) photocatalyst having single and dual absorber where A are electron-accepting species, CB is conduction band, D's are electron-donating species, E_g is energy bandgap of semiconductor, VB is valence band. Schematic illustration of mechanism of H_2 formation in (**c**) acidic and (**d**) alkaline medium; and (**e**) PEC water splitting process

be collected and utilized to produce energy on demand through fuel cells. Theoretically, 1.23 eV is the minimum energy required to initiate the overall water splitting process. However, in practical scenarios, a significantly higher energy input is necessary due to kinetic barriers. This additional energy, exceeding the thermodynamic equilibrium value, is known as the overpotential. Numerous efforts have been made to lower the overpotential for both HER and OER by developing various catalysts with different compositions and morphologies. Despite these attempts, achieving low-cost, large-scale hydrogen production remains a significant challenge. Commercial Pt/C is currently the benchmark catalyst used for HER. However, their high cost and scarcity hinder widespread commercial adoption. Consequently, there is a strong push toward identifying low-cost, earth-abundant materials as alternative catalysts for HER. Due to their abundance and low cost, transition metal-based catalysts, including oxides, sulfides, nitrides, phosphides, and selenides, have received a lot of attention in this regard. As shown in Figs. 9.1c and d, the hydrogen evolution reaction follows different pathways based on the electrolyte pH. A multistep process takes place on the electrocatalyst surface, where H^+ in acidic medium or H_2O molecules in alkaline medium are reduced to produce H_2 gas. Typically, the mechanism involves three elementary steps. The first is the Volmer step, in which either a H_3O^+ ion in acidic conditions or a H_2O molecule in alkaline conditions accepts an electron, forming an adsorbed hydrogen intermediate (M-H*) on the catalyst surface.

Volmer mechanism

$$H_3O^+ + M + e^- \rightarrow M - H^* + H_2O \ldots \text{Acidic media}$$

$$H_2O + M + e^- \rightarrow M - H^* + OH^- \ldots \text{ Alkaline media}$$

The Heyrovsky mechanism constitutes the second stage of HER, in which an adsorbed hydrogen intermediate (M-H*) interacts with either a proton in acidic media or a water molecule in alkaline media. With the addition of an electron, this leads to the formation of H_2.

Heyrovsky mechanism

$$H^+ + e^- + M - H^* \rightarrow H_2 + M \ldots \text{Acidic media}$$

$$H_2O + e^- + M - H^* \rightarrow H_2 + OH^- + M \ldots \text{Alkaline media}$$

The third stage of HER is the Tafel mechanism, where two neighbouring adsorbed hydrogen species (M–H*) recombine on the surface of the catalyst to release H_2.

When the Volmer step is fast and the Tafel step acts as rate-determining step, the HER follows Volmer–Tafel pathway. Conversely, if the Volmer step is quick and the Heyrovsky step is rate-limiting, the reaction follows the Volmer–Heyrovsky pathway. As discussed above, the adsorption energy of H^* on the surface of catalyst is a critical factor governing H_2 evolution activity. This adsorption strength is determined by the Gibbs free energy of H^* adsorption (ΔG_{H*}), which serves as a key descriptor which is directly associated with the intrinsic activity of HER catalysts.

9.3.1 Volcano Plots

The Sabatier principle states that the adsorption of H* intermediates and the subsequent desorption of H_2 on a catalyst surface must be optimally balanced. The catalyst should bind H* strongly enough to facilitate the reaction but not so strongly that it hinders product release. For maximum HER efficiency, ΔG_{H*} must be nearly thermoneutral, ideally towards near zero ($\Delta G_{H*} \sim 0$). According to Sabatier principle, the maximum HER exchange current density (j_0) is attained when ΔG_{H*} is zero. Based on this concept, Parsons introduced a distinctive "volcano-type" plot that relates the HER exchange current density (j_0) to the quantum-chemically derived ΔG_{H*}. More recently, this volcano trend has been validated by correlating experimentally measured j_0 values with ΔG_{H*} values computed using density functional theory (DFT). At the apex of the volcano, where $\Delta G_{H*} = 0$, the catalyst exhibits optimal HER activity. If ΔG_{H*} is positive, H* adsorption is too weak, leading to an exponential increase in j_0 as ΔG_{H*} approaches zero. Conversely, if ΔG_{H*} is negative, H* binds too strongly, resulting in a decrease in j_0 with further deviation from the thermoneutral value.

9.4 Basic Mechanism of Photoelectrochemical Water Splitting

Unlike purely photo- or electrochemical processes, PEC water splitting offers a more controlled and efficient platform for solar-driven H_2 production. PEC system integrates semiconductor materials into photoelectrodes, enabling direct transport of charges through an external circuit. The basic configuration of the PEC process is as shown in Fig. 9.1e. PEC process proceeds in three fundamental steps: (1) incident photon absorption by the semiconductor electrode, (2) generation, separation, and transport of photogenerated carriers to surface of the electrode, and (3) injection of these charges into the electrolyte to drive HER and OER. This configuration allows a better control over charge carrier separation and migration along with integration with external bias to improve efficiency and stability. Additionally, PEC systems enable accurate assessment of semiconductor performance using electrochemical methods like chronoamperometry, linear sweep voltammetry, and incident photon-to-current efficiency (IPCE) measurements, providing information about the durability and inherent activity of photocatalysts under working conditions [18].

9.5 Lead Free Perovskite for Water Splitting

9.5.1 Lead Free Perovskite in Photocatalytic Hydrogen Generation

9.5.1.1 Oxide Perovskite

Oxide Perovskite has gained attention as photocatalysts owing to their adjustable electronic structure, compositional flexibility, and chemical stability. With the general formula ABO_3, these compounds comprise B-site cations the center of corner-sharing octahedra and A-site cations occupying the interstitial sites. The first reported application of an oxide perovskite in photocatalytic water splitting was $SrTiO_3$ in 1980, prompting extensive investigation of related materials for photocatalytic H_2 generation via water splitting. $SrTiO_3$ has bandgap of 3.2 eV which absorbs in UV range and by doping the photocatalyst with some foreign element the bandgap of the compound can be reduced to make it absorb in visible range of solar spectrum [19]. Akihiko et al. doped Ru, Rh, Mn, and Ir into $SrTiO_3$ which enhances the solar spectrum to visible region, resulting from transfer of electrons from donor level of dopant to the CB of $SrTiO_3$. Also, Rh doped $SrTiO_3$ shows highest H_2 generation rate among all doped materials due to Rh^{3+} on the B-site, which may operate as both the surface reaction centre and the absorption centre for visible light [19]. To further improve the activity of $SrTiO_3$, codoping with two different elements has also been reported. For example, Yu et al. synthesised Cr and B codoped $SrTiO_3$ with narrow bandgap and enhanced visible-light absorption of visible light ascribed to strong p-d repulsion between B 2p and Cr 3d orbitals. These findings present a novel approach to modify perovskites band structure via co-doping of two elements [20]. Unlike 3D perovskites, layered 2D perovskites exhibit greater resistance to moisture, oxygen, and thermal stress, making them particularly attractive for photocatalytic water-splitting applications. Layered perovskite are 2D perovskites having BO_6 framework with shared edges. These are perovskites that are derivatives of single perovskite oxides made up of two-dimensional stacks. The stacked layer of perovskite is bonded by weak electrostatic attraction which enables easy structure tunability. The interlayer cation can be replaced to increase the activity of perovskites owing to the weaker connection between layers by doping with alternative metal ions that are comparable in size to the interlayer cation. Depending on the stacked layers orientation in the ABO_3 framework layered perovskite oxides are classified into (100), (110), and (111) types. The (100) type includes Aurivillius $[(Bi_2O_2)(A_{n-1}B_nO_{3n+1})]$, Ruddlesden–Popper $[A_{n+1}B_nO_{3n+1}$ or $A'_2A_{n-1}B_nO_{3n+1}]$ and Dion–Jacobsen $[A'(A_{n-1}B_nO_{3n+1})]$ phases. The (111) and (110) types follow general formulas $(A_{n+1}B_nO_{3n+3})$ and $(A_nB_nO_{3n+2})$ respectively. The number of spanning BO_6 octahedra and thickness of a layers are determined by 'n' in the above formulations. Owing to the charge separation driving force produced from their structural laminations, layered compounds have been identified as viable contenders for photocatalytic overall water splitting [21, 22]. Sr_2TiO_4 has drawn a lot of interest since it is a readily available compound made up of inexpensive and non-toxic components. Functional doping offers a simple yet effective approach to boost performance of Sr_2TiO_4 towards photocatalytic H_2O splitting by various methods which will be discussed later in Sect. 6.1. For instance, La/Rh co-doping has been employed in Sr_2TiO_4 to boost HER performance [23]. Recently, nitridation of this perovskite has been found to increase the rate of hydrogen evolution [24]. Sun et al. exfoliated Dion-Jacobson phasic $RbLaTa_2O_{6.77}N_{0.15}$-layered perovskite to create N-loaded $LaTa_2O_{6.77}N_{0.15}$-nanosheets. The produced photocatalyst has demonstrated amazing photocatalytic performance in the effective development of O_2 and H_2, with the maximum activity observed at photon wavelengths of 600 nm. This multilayer perovskite's apparent quantum efficiency (AQE) for O_2 evolution in the visible light area was 3.27%, which was 4–8 times greater than its bulk counterpart [25]. Table 9.1 presents some recently reported perovskite oxide catalysts for photocatalytic hydrogen generation.

9.5.1.2 Halide Perovskite

Halide perovskites are another class of perovskite materials which has garnered attention to the scientific community for photocatalytic hydrogen generation application due to their uniqueness of high value absorption coefficients, variable band gap, long carrier diffusion length, and higher carrier mobility. Their ability to be processed in a solution facilitates the production of photocatalysts in a scalable and economical manner. But there are certain challenges in Halide perovskites in water splitting applications like poor stability in aqueous environments, as most halide perovskites are

Table 9.1 Recently reported perovskite oxide catalysts for photocatalytic hydrogen generation

Sl. no.	Photocatalyst	Reaction conditions	Light source	Yield of H_2	References
1	$LaFeO_3$	Water + Ethanol; 4 h irradiation period	200 W tungsten lamp	3315 $\mu molg^{-1}\ h^{-1}$	[26]
2	$SrTiO_3$	Ultrapure water + methanol; 24 h irradiation period	300 W Xe-lamp	23 $mmolh^{-1}$	[27]
3	$LiTaO_3$	Water	450 W Hg lamp	712.0 $\mu molh^{-1}$	[28]
4	Pt-g-C_3N_4/$KNbO_3$	Water + 10 vol% TEOA; 8 h irradiation period	300 W Xe arc lamp	1019.38 $\mu molg^{-1}\ h^{-1}$	[29]
5	$CsCa_2Nb_3O_{10}$	Water + methanol; irradiation time of 2 h	350 W Hg immersion lamp	450 μmol	[30]
6	$Fe_xLa_{2x}Ti_2O_7$ (0.01–0.05%)	Water + methanol as hole scavenger	UV–visible	45 $\mu molh^{-1}$	[31]
7	Cu-doped $CaTiO_3$	20 vol% methanol	300 W Xe-lamp	140.7 $\mu molg^{-1}\ h^{-1}$	[32]
8	Er-doped $CaTiO_3$	Distilled water and methanol	300 W Xe-lamp	461.25 $\mu molg^{-1}\ h^{-1}$	[33]
9	$Ca_{0.94}Ag_{0.03}La_{0.03}TiO_3$	DI water + methanol	300 W Xe-lamp	1064.2 $\mu molg^{-1}\ h^{-1}$ in UV and 10 $\mu molg^{-1}\ h^{-1}$ in Vis	[34]
10	La-, Rh-*co*-doped $SrTiO_3$	20 vol% methanol; 0.1 g of catalyst;	300 W Xe-lamp	46.2 $\mu molh^{-1}$	[35]
11	$SrTiO_3$:C, N	DI water + methanol; irradiation time of 1 h	300 W Xe-lamp	714.4 μLh^{-1}	[36]
12	Ce, N co-doped $SrTiO_3$	Catalyst powder 50 mg; 10 mL methanol + 50 mL deionized water; catalyst powder 50 mg;	300 W Xe-lamp, ($\lambda \geq 400$ nm)	4.28 $\mu molg^{-1}\ h^{-1}$	[37]
13	$LaFeO_3$/RGO	15 vol% methanol	250 W Xe-lamp	82 $mmolg^{-1}\ h^{-1}$	[38]
14	F/N co-doped Sr_2TiO_4	DI water	300 W Xe-lamp ($\lambda \geq 420$ nm)	12.1 $\mu molh^{-1}$	[24]

highly sensitive to moisture and oxygen, leading to hydrolysis, leaching of the halide, and rapid oxidation of B-site cations such as Sn^{2+} into inactive Sn^{4+} [39]. Another major limitation is their electronic structure of the material while some compositions exhibit conduction band positions suitable for hydrogen evolution, their valence band alignment often remains inadequate for oxygen evolution, making overall water splitting difficult. Overcoming these challenges through targeted strategies are essential to boost the efficacy of halide perovskites in H_2O splitting applications.

Halide perovskites show unique performance in photocatalysis that differs from their behaviour in photovoltaic and light-emitting applications. Their properties and applications significantly depend on the composition of the perovskite materials. For example, Methyl ammonium lead bromide ($MAPbI_3$) perovskite, which displays decent photovoltaic behaviour, might seem like an obvious choice for photocatalytic applications, its inferior stabilities are highly undesirable, especially for photocatalysis. It would quickly decompose in the presence of H_2O and O_2. To overcome this issue, other lead-based perovskites such as $FAPbI_3$ and all inorganic $CsPbI_3$ (both relatively more stable than $MAPbI_3$), were also explored, but concerns on environment and health safety, due to the presence of lead remain an inevitable issue with these compositions [40]. Thus, considering the environmental concerns, lead-free perovskite materials have gained attention. Pb-free halide perovskite nanocrystals offer tunable band gaps (1.4–3.1 eV), [41, 42] high absorption, reduced charge recombination, defect tolerance, long carrier diffusion lengths, efficient charge transport and lower toxicity in comparison to lead-based counterparts. Bi and Sn are common alternatives for Pb in lead-free perovskites. In 2018, Chen et al demonstrated $DMASnI_3$ to be stable in water for 16 h and was tested as a photocatalyst for H_2 production [43]. A very modest HER rate of 0.64 $\mu molh^{-1}$ was reported, but the authors stated that O_2 evolution from water oxidation induced Sn oxidation. In the field of Pb-free HPs, significant potential exists for developing materials via various approaches, as discussed in Sect. 6.2. Table 9.2 lists the recently reported Pb-free HPs for photocatalytic H_2 generation.

9.5.2 *Lead Free Perovskite in Electrochemical Hydrogen Generation*

Lead-free halide perovskites (LFHPs) exhibit poor ambient stability, which has restricted their exploration in hydrogen generation via water splitting. Consequently, most research in this field has shifted toward lead-free oxide perovskites. Beyond simple perovskite oxides (ABO_3), other structural variants such as double perovskites [A-site and B-site ordered, $A_2BB^{'}O_6$], Ruddlesden–Popper phases ($A_{n+1}B_nO_{3n+1}$), and quadruple perovskites ($AA^{'}{}_3B_4O_{12}$), also been investigated for H_2O splitting. These compositional diversity results in an extended range of functional properties. Oxide perovskites were first reported as electrocatalysts for H_2 generation in alkaline medium in 2016. Xu et al. investigated A-site Pr-doped

Table 9.2 Pb-free halide perovskites recently reported in the literature for hydrogen generation

Sl. no.	Photocatalyst	Reaction	Light source	H_2 generation efficiency (μmolg^{-1} h^{-1})	References
1	$DMASnBr_3$	DI water + 10% *v*/*v* TEOA; 2 g/L of photocatalyst	Simulated solar light 500 Wm^{-2} (5 h)	11	[44]
2	$DMASnBr_3$ @g-C_3N_4	DI water + 10% *v*/*v* TEOA	1500 W Xe-lamp	18.6	[45]
3	$DMASnI_3$	100 mL DI water; 200 mg of catalyst	300 W Xe-lamp for 5 h at 10 °C	3	[46]
4	PEA_2SnBr_4/g-C_3N_4	DI water + 10% *v*/*v* TEOA	Solar light 500 Wm^{-2} (6 h)	1613	[47]
5	$Cs_2Pt_{0.05}Sn_{0.95}Cl_6$	Water + 10%(v/v) TEOA; Pt 3wt%	300 W Xe-lamp	16.11	[48]
6	$Cs_2Pt_{0.25}Sn_{0.75}Cl_6$	Water + 10%(v/v) TEOA; Pt 3wt%	300 W Xe-lamp	3.99	[48]
7	$Cs_2Pt_{0.05}Sn_{0.95}Cl_6$	Water + 10%(v/v) TEOA; Pt 0wt%	300 W Xe-lamp	1.91	[48]

$Pr_{0.5}(Ba_{0.5}Sr_{0.5})_{0.5}Co_{0.8}Fe_{0.2}O_{3-\delta}$ (Pr0.5BSCF) as a catalyst material exhibiting significantly enhanced HER activity and stability [49]. The improved HER activity was ascribed to the altered surface electronic structure resulting from Pr doping.

Strontium-based oxide perovskites have gained prominence as potential electrocatalytic materials for HER, owing to their exceptional electrical conductivity, structural stability and flexible oxidation states. Incorporating Sr enhances charge transfer and increases the density of active species, aligning them as promising substitutes to noble-metal-based catalysts [50]. Several synthesis strategies have been reported for strontium-based oxide perovskite electrocatalysts. Recently, Yu et al. developed conductive monoclinic $SrIrO_3$ perovskite as efficient electrocatalytic material for alkaline H_2O splitting [51]. The material which are synthesized under ambient conditions, exhibited notable performance towards both HER and OER. During HER, surface self-reconstruction induced by lattice Sr^{2+} leaching resulted in an ~ 11-fold enhancement in activity compared to the pristine sample. Notably, the study also demonstrated that $SrIrO_3$ activated under HER conditions could be effectively reused as a superior OER catalyst. In another study, Sarmad et al. also investigated praseodymium doping $Sr_2TiFeO_{6-\delta}$ on both A and B site for electrocatalytic H_2 generation via water splitting. The optimized catalyst exhibited HER overpotential of 182.4 mV@5 mV/s and a Tafel slope value of 77 mVdec^{-1}. The catalyst shows an electrochemical stability of 1000 cycles with an overpotential drop from 182.4 mV to 181.9 mV. The boosted catalytic performance was ascribed to the decreased crystallinity of the double perovskite upon Pr doping [52].

Lanthanum-based perovskite oxides represent another promising class of materials for electrocatalytic HER, attributed to their flexible electronic structures, high stability, and strong catalytic activity. The ability of lanthanum and transition metals to exist in various oxidation states boosts their charge transfer capability and catalytic efficiency [53]. The catalytic properties of lanthanide-based materials are heavily dependent on their synthesis conditions. The first reported application of a La-based electrocatalyst in HER dates to the 1970s with $LaNi_5$ [54]. Later, La containing catalysts have been extensively investigated for electro-catalytical applications as they improve the stability and efficiency of electrochemical application by reducing the activation energy barrier, enhancing electron transfer, and modulating the adsorption energy of active cites [53, 55, 56]. The significant differences in atomic radii and electron-withdrawing characteristics amongst La and transition metals allow precise tuning of the electronic structure, thus enhancing electrocatalytic efficiency. Recently, Li et al. introduced a phosphatization strategy for oxide perovskites, demonstrating that P-doped $Pr_{0.5}La_{0.5}BaCo_2O_{5+\delta}$ nanofibers (P-PLBC-F) exhibit electrocatalytic activity equivalent to that of commercial Pt/C electrodes, with an HER overpotential of − 307 mV@500 mA/cm^2 and a Tafel slope of 32.9 mV/dec in alkaline medium [57]. Their study revealed that phosphorus doping effectively modulates the electronic structure of perovskite and optimizes the free energy for HER. Furthermore, they proposed two possible mechanisms for alkaline HER: the Volmer–Heyrovsky and Volmer–Tafel pathways. Based on small Tafel slope (−32.9 mVdec^{-1}), they concluded that P-PLBC-F predominantly follows the Volmer–Tafel mechanism. Besides electronic properties modulation through structural variation, foreign element doping, or surface modification; many other strategies were employed to improve the electrocatalytic performance of H_2 generation. Recently, Khan et al. studied a composite material in which $La_{0.4}Sr_{0.4}Ti_{0.9}Ni_{0.1}O_{3-\delta}$ (LSTN) perovskite nanoparticles are interlaced between 2D nanosheets of $Ti_3C_2T_x$. The combined material exhibits excellent charge transferability because the synthesized oxide perovskite introduces numerous oxygen vacancies and a variable valence state, while the MXene contributes a highly conductive and hydrophilic surface [58].

Barium-based perovskite oxides have also showed promise as HER electrocatalysts because of their tunable electronic structures, and high electrical conductivity. Xu et al. employed thermal reduction assisted phase transformation method to synthesize $BaMoO_3$ from $BaMoO_4$ to investigate the electrocatalytic activity for H_2 generation through water splitting, shown on Fig. 9.2a [59]. The research found that the interconnected MoO_6 octahedra in $BaMoO_3$ are better at catalysing the HER than isolated MoO_4 tetrahedra. The enhancement of activity in $BaMoO_3$ is ascribed to stabilisation of tetravalent state of Mo and enhanced concentration of oxygen vacancy, thereby increasing the electrical conductivity of $BaMoO_3$ shown in Fig. 9.2b. Table 9.3 lists the lead-free oxide perovskites and their catalytical performances in electrochemical H_2 production.

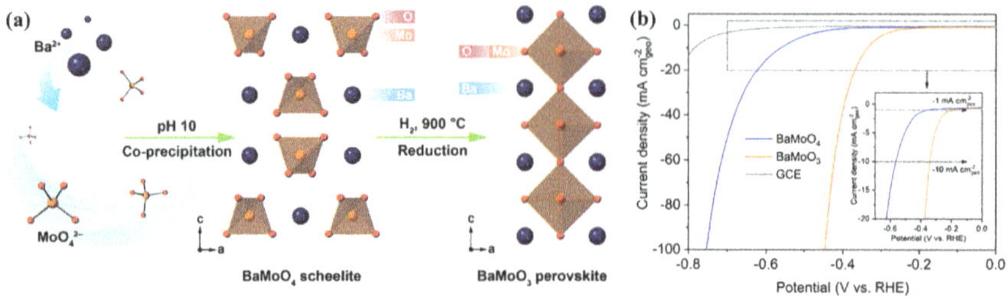

Fig. 9.2 (**a**) Formation of $BaMoO_3$ via transformation of $BaMoO_4$ (**b**) Electrochemical LSV measurements in alkaline medium [59]

9.5.3 Lead Free Perovskite in PEC Hydrogen Generation

Photocatalytic and PEC approaches both rely on optimal band alignment, effective charge separation, and strong catalytic activity. Yet, the electrode-based architecture of PEC systems offers greater design flexibility and diagnostic control, making them a valuable complement to photocatalytic studies for advancing solar water splitting technologies.

9.5.3.1 Halide Perovskite

Halide perovskites have shown great promise for PEC water splitting by reason of their outstanding optoelectronic properties. Their structural flexibility allows band edge positions to be engineered for favourable alignment with the hydrogen and oxygen evolution reaction potentials, presenting them as good candidates for driving overall water splitting. Researchers have explored lead-free halide perovskites for PEC H_2O splitting applications as lead-based perovskites have the challenge of toxicity and instability. Tin based perovskite have been extensively reported by many researchers for PEC water splitting applications. Romani et al. developed 2D tin-based perovskite PEA_2SnBr_4 with exceptional water stability showing exciting prospects for generation of hydrogen by water splitting [47]. Furthermore, by incorporating triethanolamine (TEOA) as a sacrificial agent, PEA_2SnI_4 was combined with g-C_3N_4 for H_2O splitting, achieving a remarkable H_2 evolution rate of 1600 $\mu molg^{-1}\,h^{-1}$. Bismuth based perovskite also have been extensively employed in PEC water splitting applications. For example, Nandigana et al. studied bismuth-based double perovskites in which $Cs_2AgBiCl_6$ showed photocurrent density of 10 μA/cm^2 and $Cs_3Bi_2Cl_9$ showed 6 μAcm^{-2} at 0.9 V (vs Ag/AgCl), respectively with excellent photostability [72]. In addition, indium, antimony-based perovskites have been reported for PEC water splitting. Sikarwar et al. developed Cs_2AgMCl_6 (M = Bi, In, Sb) exhibited remarkable stability and have been examined as photoanodes for PEC water splitting [73]. Out of which, $Cs_2AgInCl_6$ achieves a photocurrent density

Table 9.3 Oxide perovskites recently reported in the literature for electrochemical H_2 generation

Si. no.	Electrocatalysts	Electrolyte	Current density mA/cm^2	Overpotential (mV)	Tafel slope (mV/dec)	Stability	References
1	Ce-doped $LaCoO_3$	1 M KOH	10	305	144	1000 CV cycles	[60]
2	$LaCoO_3$	0.1 M KOH	0.36	238	–	1000 CV cycles	[61]
3	3 wt% g-C_3N_4 @LSTN ($La_{0.4}Sr_{0.4}Ti_{0.9}Ni_{0.1}O_{3-\&}$)	1 M KOH	10	148	132	1000 CV cycles	[62]
4	$La_{0.6}Sr_{0.4}CoO_{3-\delta}$@NiMn$(OH)_2$	1 M KOH	10	33	81	~ 20 h	[63]
5	$La_{0.9}Co_{0.4}Fe_{0.6}O_{3-\delta}$ nickel–iron alloy foam	1 M KOH	10	160.5	160.5	~ 20 h	[64]
6	$SrTiFe_{0.9}Pr_{0.1}O_{6-\delta}$	1 M KOH	10	277.6	77	1000 CV cycles	[52]
7	$RuO_2/BaRuO_3$	1 M KOH	10	16	33.37	~ 200 h stability test	[65]
8	$La_{1-x}Sr_xCoO_3/Ti_3C_2T_x$ MXene/Ni	1 M KOH	10	279	74.3	~ 200 h stability test	[66]
9	$SrFe_{1-x}Ru_xO_{3-\delta}$	1 M KOH	10	41	58.7	~ 96 h stability test	[67]
10	$LaCo_{0.67}Cu_{0.33}O_3$	1 M KOH	10	298	94.7	–	[68]
11	$BaNiO_3$ nanoparticles	1 M KOH	10	427	64	~ 24 h	[69]
12	$SrFeO_{3-\delta}$/Ti-MXene	1 M KOH	10	163	101.55	~ 45 h	[70]
13	Sr_2FeNbO_6	1 M KOH	10	176	128.7	~ 15 h	[71]

Table 9.4 Lead free halide perovskites recently reported in the literature for PEC hydrogen generation

Sl. no.	Photoelectrode	Electrolyte	Photocurrent ($\mu A/cm^2$)	Stability	References
1	Cs_2PtI_6 photoanode	pH 1–13	800 at 1.23 V_{RHE}	Stable for 12 h	[74]
2	Cs_2OsX_6 (X = Br, Cl, I)	0.05 M KCl	400 at 0.384 $V_{Ag/AgCl}$	Stable for 12 h	[75]
3	Cs_2PdX_6 (X = Cl, Br)	KOH (pH = 11) + 0.05 M KCl	450, 100 (X = Cl, Br)	–	[76]
4	$Cs_2AgBiCl_6$ and $Cs_3Bi_2Cl_9$	1 M KOH (pH∼13.7)	10 and 6 at 0.9 $V_{Ag/AgCl}$	Stability over 10 h	[72]
5	$Cs_2Ru_mPt_{1-m}X_6$ (X = Cl^-, Br^-) (x = 0.5)	0.05 M KCl (pH = 11)	560, 320 (X = Cl^-, Br^-)	Stable for 30 min	[77]
6	$Cs_2AgInCl_6$	Acetonitrile water mixture (9:1 volume ratio)	290	100 CV cycles within the potential range of 0 to 1	[73]
7	$Cs_2AgBiCl_6$		340		
8	$Cs_2AgSbCl_6$		160		
9	$Cs_2AgBi_{0.5}In_{0.5}Cl_6$		260		
10	Cs_2ReCl_6	Water, 0.38 $V_{Ag/AgCl}$	140	2 h under light without any degradation	[78]
11	Cs_2ReBr_6		200		
12	Cs_2PdBr_6	0.1 $molL^{-1}$ PB solution (pH = 8.0)	2.9	Stable for 10 h	[79]
13	Cs_2PtBr_6		335		

of 0.5 mA/cm^2 at 1.23 V_{RHE}, having conversion efficiency around 60%. Table 9.4 lists the Pb-free halide perovskites for PEC hydrogen generation.

9.5.3.2 Oxide Perovskite

In addition to its role in photocatalysis, $SrTiO_3$ is also considered as potential photoelectrode for PEC H_2O splitting. However, most $SrTiO_3$ photoelectrodes are typically prepared in powder or bulk form, which suffers from severe charge carrier recombination. This drawback mainly arises from numerous grain boundaries and poor particle-to-substrate contact, both of which hinder efficient charge transport. Such limitations can be mitigated by fabricating $SrTiO_3$ as thin-film photoelectrodes. Moreover, its light absorption can be extended through strategies such as localized surface plasmon resonance (LSPR)-induced absorption [80, 81]. For instance, Shi et al. reported ordered and periodic nanoporous $SrTiO_3$ photoelectrodes as well as reduced $SrTiO_3$ (r-$SrTiO_3$) films with crystalline-core@amorphous-shell structure containing oxygen vacancies, enabling visible-light absorption via LSPR effect [82].

The r-$SrTiO_3$ films exhibited significantly improved PEC water-splitting activity than the pristine $SrTiO_3$. Furthermore, several visible-light-responsive perovskite oxides, including $LaFeO_3$, [83, 84] doped $LaFeO_3$, [85] and Fe-doped $SrTiO_3$ [86] have demonstrated notable photo-electrocatalytic activity toward both HER and OER, highlighting their dual potential as photoelectrodes and cocatalysts. Notably, Ibrahim et al. fabricated an n-type $LaFeO_3$ perovskite photoanode that delivered a photocurrent density of 8.2 $mAcm^{-2}$ at 1.7 $V_{Ag/AgCl}$ [84]. The impact of factors such as calcination temperature, electrolyte species, concentration, and pH on their photocatalytic performance is assessed under AM 1.5G light irradiation. Table 9.5 lists the lead-free halide perovskites for PEC hydrogen generation.

Table 9.5 Oxide perovskites recently reported in the literature for PEC hydrogen generation

Sl. no.	Photoelectrode	Electrolyte	Photocurrent density	STH conversion efficiency/ stability	References
1	$MgFeO_3$	0.5 M Na_2SO_4	9.1 mA/cm^2 at 1.2 V_{RHE}	0.97% / stable upto 10 h	[13]
2	$MgMnO_3$	0.5 M H_2SO_4 (pH = 1)	13.21 mA/cm^2 at 1.2 V_{RHE}	2.45%/ stable for 9 h	[87]
3	$BiFeO_3$	1 M NaOH aqueous solution (pH of ~ 13.6)	0.16 mA/cm^2 at 1.23 V_{RHE}	–	[88]
4	$CaTiO_3$-TiO_2	1 M KOH solution	1.81 mA/cm^2 at 1.23 V_{RHE}	0.55% at 0.74 V_{RHE}	[89]
5	$NiFeO_x$@Ge-PH	1 M NaOH (pH = 13.6)	4.6 mA/cm^2 at 1.23 V_{RHE}	4.8%/stable for 5 h	[90]
7	$LaFeO_3$	1 M Na_2SO_4	8.2 mA/cm^2 at 1.7 $V_{Ag/AgCl}$	50% retention rate at 30 min	[84]
8	$CeFeO_3$	1 M NaOH	6.9 mA/cm^2 at 1.5 $V_{Ag/AgCl}$	–	[91]
9	Mesoporous $SrTiO_3$	0.5 M NaOH	0.3 mA/cm^2 at 1.0 $V_{Ag/AgCl}$	–	[80]
10	$LaFeO_3$	0.1 M KOH	0.30 mA/cm^2 at 1 $V_{Ag/AgCl}$	–	[92]
11	$LaFe_{0.9}Cu_{0.1}O_3$		0.99 mA/cm^2 at 1 $V_{Ag/AgCl}$	–	
12	$LaFe_{0.9}Co_{0.1}O_3$		0.85 mA/cm^2 at 1 $V_{Ag/AgCl}$	–	
13	$LaFe_{0.9}Mn_{0.1}O_3$		0.52 mA/cm^2 at 1 $V_{Ag/AgCl}$	–	

9.6 Strategies to Enhance the Performance of Hydrogen Generation

9.6.1 Oxide Perovskite

In general, oxide perovskites are not ideal catalysts for hydrogen generation owing to their high energy barrier for the adsorption of hydrogen species, resulting in the Volmer step slow, a rate limiting step for H_2 generation. Therefore, to enhance the activity of oxide perovskite for hydrogen generation, many strategies are employed such as elemental doping, surface modification, oxygen vacancy creation etc. These strategies help to control the electronic structure of oxide perovskite materials which is known to have a linear combination of atomic orbitals, helps to make direct correlation among electronic properties of the materials and crystal chemistry. For transition metal oxide perovskite, the electronic structure is strongly affected by the hybridization between d-orbitals of transition metal and oxygen 2p orbitals within the framework of BO_6 octahedra, leading to the formation of σ (sigma) and π (pi) orbitals that dictates the catalytic properties. The following section describes different strategies to modulate the structural, electronic, and optical properties of oxide perovskite, thereby enhancing the water splitting performance, hence, hydrogen generation efficiency. Subsequent sections detail approaches for tailoring the structural, electronic and optical properties of oxide perovskites to optimize water-splitting efficiency and hydrogen evolution activity.

9.6.1.1 Doping

Elemental doping has emerged as a promising approach to modulate the electronic structure of oxide perovskites, thereby improving intermediate adsorption and enhancing hydrogen evolution efficiency.

There are many strategies of doping in oxide perovskite such as A-site substitution, B-site substitution, codoping strategy etc. Among them, A-site substitution, guided by the Goldschmidt tolerance factor, has been widely explored to improve the electrical conductivity. Recently, Hussen et al. demonstrated that partial substitution of Sm with Bi in $SmFeO_3$ lowered the bandgap from 2.06 eV to 2.02 eV (Fig. 9.3a, b), decreased dislocation density due to ionic compatibility, and significantly improved carrier mobility [93]. PEC measurements revealed that Bi-doped $SmFeO_3$ achieved a photocurrent density of 2.5 $mAcm^{-2}$, surpassing that of pristine $SmFeO_3$ (2.17 $mAcm^{-2}$) at – 0.387 V_{RHE} under 1 sun illumination (Fig. 9.3c, d).

Numerous studies have examined the influence of B-site substitution in perovskites by transition metal ions on HER activity. B-site cations in perovskites possess a greater tendency for valence state variation, enabling direct modulation of the transition metal's oxidation state upon doping at B site. Recently, Kartigueyane et al. reported Mn-doped $SrFeO_{3-\delta}$ perovskite supported on Ti-MXene as a bifunctional electrocatalytic material for water splitting [70]. Here, Fe at the B-site was

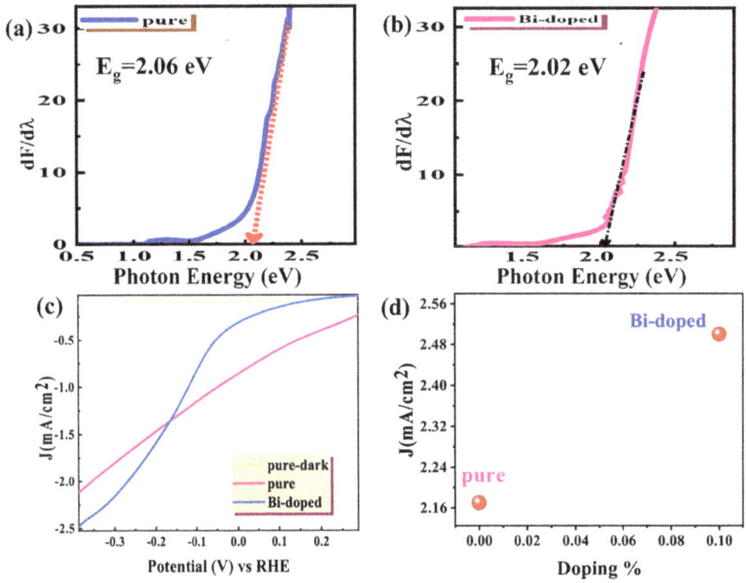

Fig. 9.3 **(a, b)** Photonic band gaps of pure and Bi-doped $SmFeO_3$, **(c)** photocurrent density versus voltage relative to RHE under dark and illumination condition, and **d** correlation between Bi doping % and generated photocurrent density [93]

partially substituted with Mn, disrupting the crystal lattice, altering the electronic structure, and facilitating fast electron transport while increasing the number of active sites. Mn incorporation also generated oxygen vacancies and surface defects, contributing to higher active site density. In a related work, Zhang et al. investigated Ru doping in $SrFeO_{3-\delta}$ to replace Fe for enhanced alkaline water splitting [67]. Ru substitution elevated the *d*-band energy levels of Fe active sites, improving the reaction intermediates adsorption, and thereby accelerating electrocatalytic reaction kinetics.

Chandrappa et al. proposed an Ir-doping strategy at the Ti site of $BaTiO_3$ (BTO) to induce visible-light absorption and p-type conductivity in the material [94]. As shown in Fig. 9.4a, the pristine BTO exhibits an absorption edge around 390 nm, corresponding to the electronic transitions from VB (dominated by O 2p orbitals) to the CB (composed of Ti 3d orbitals). In contrast, Ir-doped BTO shows a prominent red shift in the absorption edge to ~ 600 nm, indicating enhanced utilization of the solar spectrum. Computational analysis (Fig. 9.4c, d) revealed that Fermi level shifted in a downward direction significantly from 4.36 eV to 3.18 eV towards the VB upon Ir incorporation which is consistent with increased p-type character. These findings highlight that targeted substitution at the A or B sites of oxide perovskites can effectively modulate their electronic structures, thereby providing a rational pathway to tailor catalytic performance for improved hydrogen evolution efficiency via water splitting.

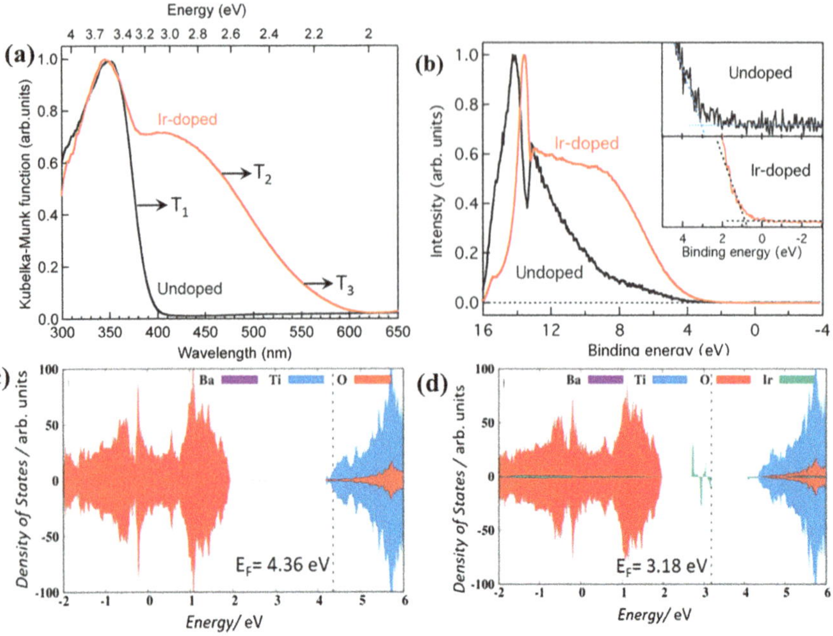

Fig. 9.4 **(a)** Diffuse reflection spectra covering the UV–visible range **(b)** VB-XPS spectra, DOS of **(c)** pristine BTO, and **(d)** Ir-doped BTO with Ir substitution at the Ti site [94]

9.6.1.2 Oxygen Vacancy

Vacancy engineering has emerged as another effective strategy for tuning the electronic structure of oxide perovskites, thereby improving their catalytic performance,. Extensive investigations over recent decades have revealed the impact of oxygen vacancies on modulating the electronic states of such materials. In a recent study, Alom et al. investigated the effect of oxygen vacancies in La_2FeNiO_6 for both the HER and OER [95]. DFT investigations reveale that incorporation of oxygen vacancies alters key electronic parameters, such as: shift in the d-band center relative to the Fermi level, reduced separation between the O 2p and transition-metal d states, and enhanced orbital hybridization (Fig. 9.5a, b). In particular, while the Fe d-band center remained largely unchanged, the Ni d-band center shifted nearer to the Fermi level, from − 4.216 eV in $LaFe_{0.5}Ni_{0.5}O_3$ to − 4.000 eV in the oxygen-deficient $LaFe_{0.5}Ni_{0.5}O_3$-R. Similarly, the O 2p–Ni 3d band separation decreased from 1.428 eV to 1.156 eV, whereas the O 2p-Fe 3d separation was nearly unaffected. These shifts suggest an enhanced electrocatalytic activity for the oxygen-deficient phase. Electrochemical measurements (Fig. 9.5c, d) were in line with the computational predictions. HER polarization curves, recorded in 1 M KOH electrolyte, showed overpotentials (η_{10}) of 520 mV for $LaFe_{0.5}Ni_{0.5}O_3$ and 500 mV for $LaFe_{0.5}Ni_{0.5}O_3$–R. Furthermore, the oxygen-deficient sample exhibited a lower Tafel slope (165 $mVdec^{-1}$) compared to the pristine material which is and indication of

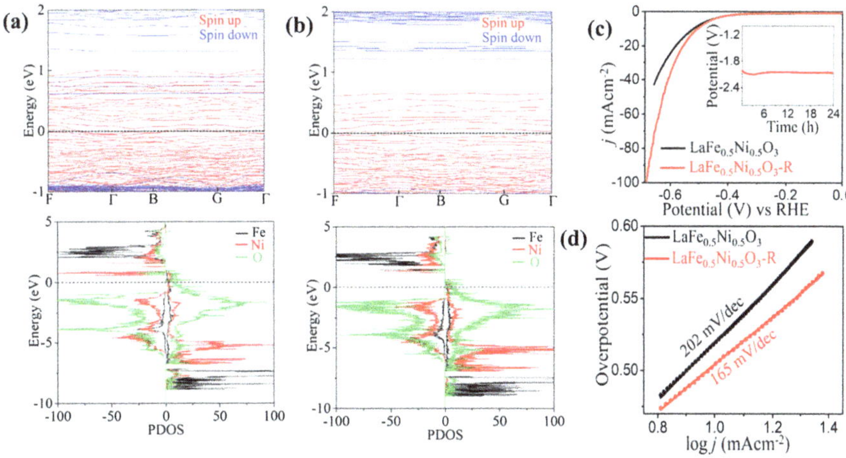

Fig. 9.5 DFT calculated band structure and PDOS of (**a**) $LaFe_{0.5}Ni_{0.5}O_3$ and (**b**) $LaFe_{0.5}Ni_{0.5}O_3$-R with fermi level set at 0 (**c**) HER polarization curves (**d**) Tafel plots and slopes indicating HER kinetics [95]

accelerated reaction kinetics. These findings confirm that oxygen-vacancy modulation is an effective route to enhance the catalytic efficacy of oxide perovskites for H_2O splitting reactions.

The relationship between concentration of oxygen vacancy and catalytic activity in oxide perovskites is not linear, rather an optimal vacancy level often yields the highest performance in water-splitting applications. Ran et al. investigated this phenomenon in doped $SrTiO_3$ photoanodes for PEC water splitting, synthesized via a nonstoichiometric route [96]. LSV measurements which are conducted over 1.0 − 1.5 V versus RHE revealed a distinct performance maximum. In the range 1.23–1.4 V versus RHE, $Sr_{0.95}TiO_{3-\delta}$ exhibited a substantially higher photocurrent density than other samples, reaching rapid saturation, indicative of an optimal oxygen vacancy concentration for PEC activity. At current densities more than 10 $mAcm^{-2}$, both $Sr_{0.95}TiO_{3-\delta}$ and $Sr_{0.8}TiO_{3-\delta}$ displayed markedly reduced overpotentials relative to pristine $SrTiO_3$, signifying superior OER kinetics. Tafel slope analysis further supported this trend, with values of 434 and 428.8 mV dec^{-1} for $Sr_{0.95}TiO_{3-\delta}$ and $Sr_{0.8}TiO_{3-\delta}$, respectively-substantially lower than the 708.4 mV dec^{-1} measured for undoped $SrTiO_3$. These enhancements mainly arise from optimally distributed oxygen vacancies, which act as electron donors, increasing carrier density and facilitating charge transport. In contrast, excessive or poorly distributed vacancies introduce defect states that serve as charge recombination centers, thereby impairing carrier mobility and reducing catalytic efficiency.

9.6.1.3 Surface Modification

Surface modification is also an effective approach to enhance the catalytic efficiency of oxide perovskites for water-splitting applications. Zong et al. employed a surface defect engineering strategy to optimize the electrocatalytic performance of $La_{0.6}Sr_{0.4}Co_{0.2}Fe_{0.8}O_{3-\delta}$ (LSCF) for water electrolysis [97]. This was achieved through selective acid etching, wherein the A-site Sr atoms in the surface were preferentially removed, yielding the LSCF-30 composition. The etching process increased the exposure and reduced the coordination of B-site transition metals, resulting in a modulated electronic structure. The removal of Sr created surface cation vacancies, which are known to promote catalytic activity. The selective leaching behavior was attributed to the relative bond strengths within the ABO_3 perovskite lattice, where the Sr–O bond is theoretically the weakest and thus more susceptible to cleavage compared to the stronger B–O bonds.

Electrochemical measurements revealed that the surface-treated LSCF exhibited a pronounced enhancement in current density for both the OER and HER, as evidenced by the LSV curves in Fig. 9.6a, b. The optimized catalyst which is LSCF-30, demonstrated the lowest overpotentials of 360 mV for OER and 261 mV for HER. Electrochemical impedance spectroscopy (EIS) further revealed that LSCF-30 possessed the smallest charge-transfer resistance for both reactions (Fig. 9.6c, d) which is an indication of a significantly accelerated electron-transfer rate. These results demonstrate the significant role of surface modification in altering the catalytic properties of oxide perovskites. The LSCF-30 system provides a model framework for designing next-generation catalysts with enhanced water-splitting performance.

9.6.2 Halide Perovskites

For addressing the intrinsic limitations of lead-free halide perovskites, several strategies have been explored to improve their efficiency in water-splitting applications. Compositional engineering such as cation/anion alloying and heterovalent doping, are mostly used for tuning the band structures, suppressing defect formation, and improving phase stability. Dimensionality control, such as designing of 2D/3D heterostructures or reduced-dimensional perovskite, improves moisture tolerances. Modification at the surface/interface level by passivation with organic ligands, polymers, or inorganic oxides, effectively reduces trap density and mitigates non-radiative recombination losses. Also, the construction of heterojunctions with suitable semiconductors promotes charge separation and accelerates carrier transport. Defect engineering has been shown to be very successful in improving stability and catalytic performance through efficient passivation and regulated vacancy regulation. Another method for altering the catalyst's performance is morphology tuning. While their intrinsic instability in aqueous conditions has been a fundamental restriction, recent developments in surface passivation, encapsulation methods, and the creation of more water-resistant compositions (such as all-inorganic or 2D perovskites)

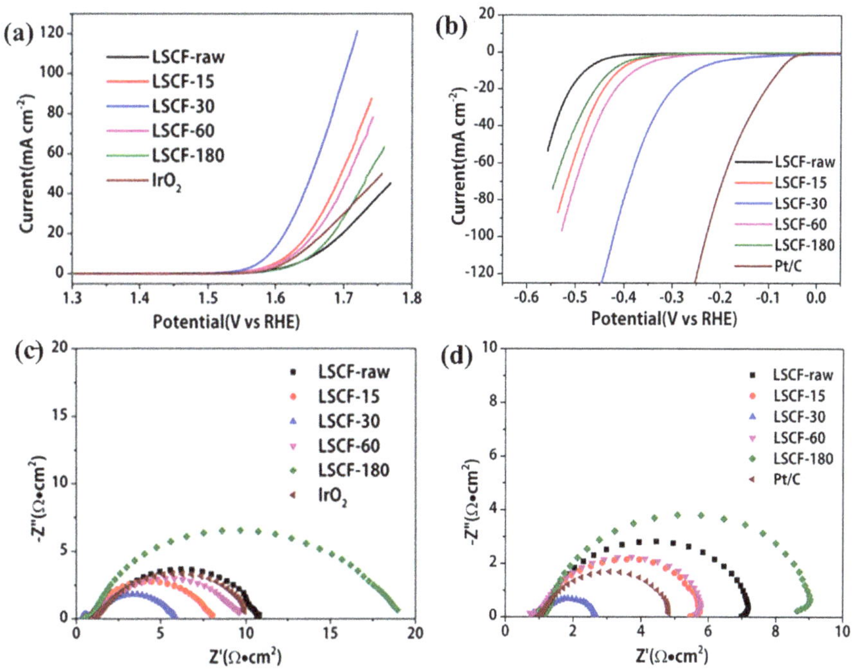

Fig. 9.6 Polarization curves (IR-corrected) of (**a**) OER and (**b**) HER of LSCF samples, nyquist plot for (**c**) OER and HER of LSCF samples [97]

have substantially improved their endurance. These attributes collectively make halide perovskites attractive candidates for next generation photoelectrochemical and photocatalytic water splitting systems. Collectively, these strategies demonstrate how important structural and chemical modifications are to enhancing the effectiveness and long-term robustness of lead-free halide perovskites.

9.6.2.1 Compositional Engineering

Compositional engineering is mostly used for tuning the band structures, suppressing defect formation, and improving phase stability. Functional doping provides a simple yet effective approach to optimize the bandgap, adjust electronic structures, and regulate defect concentrations (e.g., oxygen vacancies), thereby boosting charge separation and overall photocatalytic performance. The crystal and electronic structure of the catalytic materials can be altered which modulates the properties of the catalyst, which helps in boosting the performance of the catalysts for water splitting applications. A-site cation alloying has been shown to significantly improve air stability and moisture stability of lead-free halide perovskites. Pisanu et al. incorporated larger organic cations such as dimethylammonium (DMA) into $MASnBr_3$ which stabilizes the structure and suppresses the rapid oxidation of Sn(II), tuning the bandgap from ~ 2.0 eV to ~ 2.9 eV [97]. This stabilization not only improves the operational lifetime of perovskites under ambient conditions but also enables efficient hydrogen photogeneration in aqueous environments. This highlights the importance of balancing structural robustness with favourable electronic and optical properties. Yin et al. reported a strategy of tailoring vacancy-ordered lead-free halide perovskites through ion substitution [48]. They synthesized $Cs_2Pt_xSn_{1-x}Cl_6$ solid solutions with tunable optical and catalytic properties by substituting Sn^{4+} with Pt^{4+}. Incorporation of Pt introduced a new absorption band around 460 nm and sub-band-gap states that governed the competition between radiative recombination and charge transfer. At low Pt loading ($x = 0.05$), $Cs_2Pt_xSn_{1-x}$ exhibited excellent water stability and activated photocatalytic H_2 evolution in deionized water, with sub-bandgap states serving as active sites for efficient electron transfer. The results showed a H_2 production rate of 16.11 μmolg^{-1} h^{-1}, 3.99 μmolg^{-1} h^{-1} and close to 0 μmolg^{-1} h^{-1}, for $Cs_2Pt_{0.05}Sn_{0.95}Cl_6$, $Cs_2Pt_{0.50}Sn_{0.50}Cl_6$ and $Cs_2Pt_{0.75}Sn_{0.25}Cl_6$ respectively. The hydrogen evolution rate exhibits an inverse relationship with the PL intensity and PLQY. The presence of sub-band-gap states suppresses the self-trapped exciton (STE) through thermal de-trapping. At higher Pt loading ($x \geq 0.5$), the materials showed strong broadband STE emission. This study indicates a route to modulate radiative recombination pathways by Pt substitution in $Cs_2Pt_xSn_{1-x}Cl_6$. This highlights a composition-dependent switch between photocatalysis and photoluminescence. This work demonstrates that controlling the density and energy levels of sub-band-gap states via cation substitution offers a powerful approach to develop multifunctional, water-stable lead-free perovskites for photocatalytic hydrogen evolution.

9.6.2.2 Structural Engineering

Structural engineering is also a powerful strategy to improve water-splitting performance, as it promotes efficient charge transfer, more stability in aqueous environments, and significantly boosts their photocatalytic and photoelectrochemical activity. The development of water stable, lead-free halide perovskites has gained interest, with $DMASnBr_3$ emerging as promising photocatalyst with structural modification for solar-driven hydrogen generation. Romani et al. reported a highly crystalline, air- and water-stable perovskite synthesized by partially replacing methylammonium (MA^+) with dimethylammonium (DMA^+), yielding a distorted 1D chain structure [45]. This structural engineering enhanced the material's stability under aqueous conditions without any compromise in its optoelectronic properties. They synthesised a composite of $DMASnBr_3$ and $g\text{-}C_3N_4$ which showed high hydrogen evolution activity > 1700 μmolg^{-1} h^{-1} under illumination of visible light for the optimised catalyst 33% $DMASnBr_3$@$g\text{-}C_3N_4$ and using Pt as the co-catalyst. The rate of hydrogen evolution with 33 wt % $DMASnBr_3$@$g\text{-}C_3N_4$ composites with 3wt% Pt as the cocatalyst displays high apparent quantum efficiency and excellent long-term operational durability. The VB edge and CB edge of $DMASnBr_3$ ($g\text{-}C_3N_4$) are placed 6.5 eV and 3.65 eV (6 eV and 3.3 eV)) under the vacuum level. Since both materials absorb in the visible region, the resulting band alignment allows efficient transfer of photoexcited electrons from the CB of g-C_3N_4 to the perovskite, while holes in the VB of $DMASnBr_3$ can similarly migrate to g-C_3N_4. The findings emphasize how cation substitution is crucial for tuning stability but also reinforce the viability of tin-based halide perovskites, which confirms their potential as a sustainable, lead-free option for water splitting.

In a similar study, Romani et al. fabricated a catalyst by coupling PEA_2SnBr_4 with g-C_3N_4 for H_2 photogeneration by H_2O splitting to address the poor stability and the suboptimal charge dynamics [47]. One approach is structural engineering of perovskite into two-dimensional frameworks, where bulky hydrophobic cations (phenylethyl ammonium, PEA) protect the inorganic layers from oxidation and hydrolysis, which ensures long-term water stability. Beyond the intrinsic stability, heterojunction design offers another pathway to boost the activity. Coupling lead-free perovskites with semiconductors such as g-C_3N_4 forms a type-II band alignment which promotes directional charge transfer, suppresses charge carrier recombination, and broadens visible-light absorption, leading to hydrogen evolution rates exceeding 1600 $\mu molg^{-1}$ h^{-1}.

9.6.2.3 Defect Engineering

By defect engineering, surface defects can enhance the charge separation and modulate the coordinated active species which are favourable for photocatalytic water splitting [98, 99]. He et al. studied the role of defect in $Cs_2AgBiBr_6$ by inducing surface defects which facilitated surface charge separation by modulating the local atomic configuration and electronic structure, lead to an optimized surface environment that significantly enhanced the photocatalytic efficiency of defect-rich $Cs_2AgBiBr_6$ [100]. Consequently, defect-rich $Cs_2AgBiBr_6/Mo_3S_{13}{}^{2-}$ photocatalyst achieved an H_2 evolution rate of 24.7 μmol/g after 10 h of visible-light illumination, representing a 5.5-fold increase over the pristine material shown in (Fig. 7a, b). Thus, introducing the surface defects can enhance photocatalytic stability of $Cs_2AgBiBr_6$ with the mechanism shown in Fig. 9.7c.

9.6.2.4 Encapsulation of the Photoelectrodes

To protect the photoelectrode from degradation induced by the electrode permeation during electrochemical reactions in water-splitting devices, epoxy resin has been widely adopted as a blocking layer in perovskite-based photoelectrode research [101, 102]. It has several advantages like strong mechanical stability, chemical inertness across a wide pH condition, and cost-effectiveness. However, epoxy resin inevitably encounters the charge transport layer or top electrode of the photoactive area. So, careful consideration must be given to the fabrication process to avoid undesirable interfacial reactions or delamination. A key drawback of epoxy resin is its

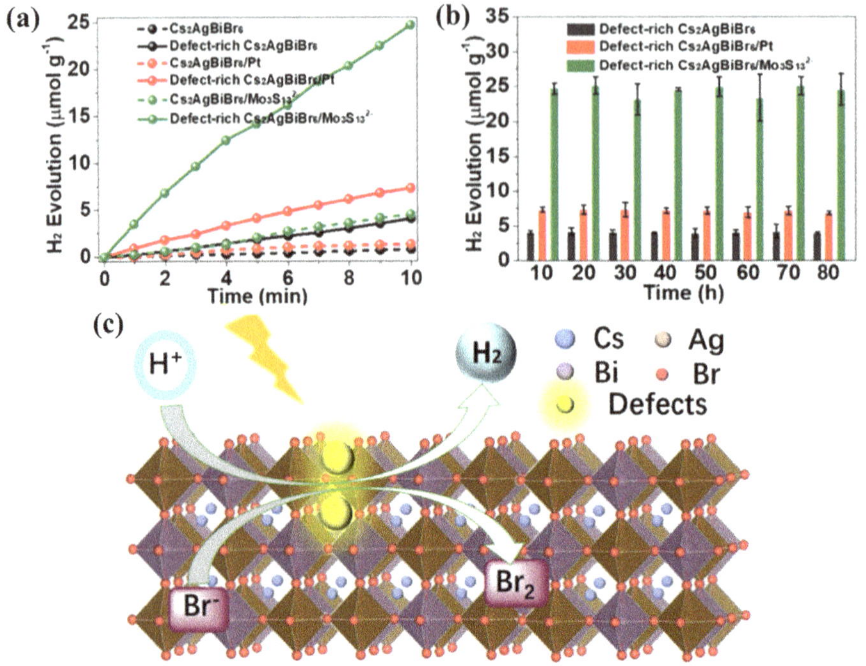

Fig. 9.7 (**a**) H_2 generation activity under visible light irradiation, (**b**) photocatalytic activity of defect-rich $Cs_2AgBiBr_6$, $Cs_2AgBiBr_6$/Pt, and $Cs_2AgBiBr_6/Mo_3S_{13}^{2-}$, and (**c**) schematic diagram of H_2 evolution on defect-rich $Cs_2AgBiBr_6$ [100]

intrinsic insulating nature, which restricts its direct application on the active photoelectrode surface. To address this drawback, epoxy is typically used to encapsulate the entire PEC device while leaving the active region exposed, this keeps water out without compromising photoelectrochemical performance. For the exposed photoactive area, encapsulation materials must facilitate effective charge transport in addition to preventing electrolyte penetration. Consequently, several conductive encapsulation strategies have been developed to protect the active region of perovskite photoelectrodes while maintaining their functional performance [103]. Recently, Lee et al. utilized glycidyltrimethylammonium chloride (GTMCl) treatment to modify the SnO_2 interface in immersed perovskite-based anodes, thereby enhancing operational stability through the suppression of nonradiative recombination [104].

9.7 Conclusions

Overall, lead free perovskites has arisen as a potential candidate for H_2-generation through electrocatalytic, photocatalytic and photo-electrocatalytic H_2O splitting. Their structural flexibility, tunable bandgap, high light absorption coefficient, and cost effectiveness makes them suitable for one of the alternatives of noble metal-based catalyst materials for H_2O splitting applications. Various strategies like doping, compositional engineering, defect engineering, morphology regulations have paved the way closer to the practical implementation. Despite substantial advancements, there are still various challenges. Oxide perovskites possess larger value of bandgap and lower electron–hole mobility, necessitating further improvement in the engineering of optoelectronic properties. Lattice doping and defect engineering have proven effective in enhancing photoinduced carrier density and electrical conductivity, whereas improved band bending further facilitates electron–hole separation in oxide perovskite. Halide perovskites possess better optoelectronic properties, but their stability in ambient environment is a big challenge for hydrogen generation through water splitting. Although encouraging progress has been made to combat the issues of stability for halide perovskite through quasi-2D structure, surface passivation techniques, encapsulation strategy to improve the stability under operating conditions, still there is a need for extensive fundamental studies for making them water stable. Long term stability of the catalyst materials is another significant aspect which need to be addressed for practical implementation of the technology. Consequently, deactivation mechanism study of the materials is of great importance to further improve the catalyst durability. Therefore, looking forward, the integration of experimental techniques with computational and advanced theoretical modelling through machine learning can accelerate progress in a much faster way by tailored design strategy of the catalyst and can play a vital role in the practical implementation of the technology. In summary, while significant challenges are still there, perovskite materials are the forefront of the hydrogen generation technology through water splitting.

References

1. P. Vedele, E. Sartoretti, G. Torretti, C. Novara, F. Salomone, F. Giorgis, M. Antonini, S. Bensaid, Thermocatalytic methane pyrolysis over iron-based catalysts for turquoise hydrogen production: activity and kinetic studies. Chem. Eng. J. 163392 (2025)
2. K. Harun, S. Adhikari, H. Jahromi, Hydrogen production via thermocatalytic decomposition of methane using carbon-based catalysts. RSC Adv. **10**(67), 40882–40893 (2020)
3. I. Martínez, M.C. Romano, P. Chiesa, G. Grasa, R. Murillo, Hydrogen production through sorption enhanced steam reforming of natural gas: Thermodynamic plant assessment. Int. J. Hydrogen Energy **38**(35), 15180–15199 (2013)
4. R. Lan, H. Jin, L. Guo, Z. Ge, S. Guo, X. Zhang, Hydrogen production by catalytic gasification of coal in supercritical water. Energy Fuels **28**(11), 6911–6917 (2014)
5. S.N. Reddy, S. Nanda, A.K. Dalai, J.A. Kozinski, Supercritical water gasification of biomass for hydrogen production. Int. J. Hydrogen Energy **39**(13), 6912–6926 (2014)
6. J.M.V. Nsanzimana, L. Cai, A. Djire, G. Pagot, P. Mattana, K. Vezzù, E. Negro, B.Y. Xia, V. Di Noto, Tailoring chemical microenvironment of iron-triad electrocatalysts for hydrogen production by water electrolysis. Adv. Energy Mater. **15**(25), 2501686 (2025)
7. IEA, G. H. R., IEA, Paris (2025), https://www.iea.org/reports/global-hydrogen-review-2025
8. F. Yang, Y. Pan, J. Zhu, Enhanced catalytic activity of noble metal@ borophene/WS2 heterojunction for hydrogen evolution reaction. Appl. Surf. Sci. **680**, 161321 (2025)

9. Y. Huang, Y. Chen, Y. Zhang, H. Zhang, K. Xiong, X. Ye, Q. Liu, J. Zhu, Transition-metal atoms embedded MoTe2 single-atom catalyst for efficient electrocatalytic hydrogen evolution reaction. Appl. Surf. Sci. **680**, 161335 (2025)
10. J. Cheng, Y. Yu, Z. Sui, S. Lei, Q. Duan, Y. Zhao, K. Liu, L. Zhang, F. Guo, Transition metal co-induced phase transition of Co x Mo1−x Se2 to promote a hydrogen evolution reaction. ACS Sustain. Chem. Eng. **13**(5), 2176–2187 (2025)
11. M. Aleen, N. Ahmed, U.M. Khan, S. Bashir, R. Subramaniam, G. Ali, Bifunctional Mg doped SrCo0. 7Fe0. 3O3-δ perovskites for enhanced oxygen and hydrogen evolution reactions for Green hydrogen production applications. Int. J. Hydrog. Energy **177**, 151432 (2025)
12. X. Li, S. Xu, J. Li, S. Zhang, B. Zhang, R. Zhao, D. Zhao, F. Wu, NiFe-LDH as a bifunctional electrocatalyst for efficient water and seawater electrolysis: enhanced oxygen evolution and hydrogen evolution reactions. Nanoscale Adv. (2025)
13. K. Ahmed, S. Hameed, K. Patchigolla, N. Dawood, Z.K. Ghouri, Carbon-based electrocatalysts for hydrogen evolution reaction. Energy Convers. Manag.: X **26**, 100892 (2025)
14. C. Acar, I. Dincer, G.F. Naterer, Review of photocatalytic water-splitting methods for sustainable hydrogen production. Int. J. Energy Res. **40**(11), 1449–1473 (2016)
15. J. Cui, C. Li, F. Zhang, Development of mixed-anion photocatalysts with wide visible-light absorption bands for solar water splitting. Chem. Sus. Chem. **12**(9), 1872–1888 (2019)
16. K. Takanabe, Solar water splitting using semiconductor photocatalyst powders. Sol. Energy Fuels, 73–103 (2015)
17. F. Li, Y. Liu, R. Liu, Z. Sun, D. Zhao, C. Kou, Preparation of Ca-doped $LaFeO_3$ nanopowders in a reverse microemulsion and their visible light photocatalytic activity. Mater. Lett. **64**(2), 223–225 (2010)
18. S. Nishioka, F.E. Osterloh, X. Wang, T.E. Mallouk, K. Maeda, Photocatalytic water splitting. Nat. Rev. Methods Prim. **3**(1), 42 (2023)
19. K. Iwashina, A. Kudo, Rh-doped $SrTiO_3$ photocatalyst electrode showing cathodic photocurrent for water splitting under visible-light irradiation. J. Am. Chem. Soc. **133**(34), 13272–13275 (2011)
20. H. Yu, J. Wang, S. Yan, T. Yu, Z. Zou, Elements doping to expand the light response of $SrTiO_3$. J. Photochem. Photobiol. A **275**, 65–71 (2014)
21. M.R. Allen, A. Thibert, E.M. Sabio, N.D. Browning, D.S. Larsen, F.E. Osterloh, Evolution of physical and photocatalytic properties in the layered titanates $A_2Ti_4O_9$ (A= K, H) and in nanosheets derived by chemical exfoliation. Chem. Mater. **22**(3), 1220–1228 (2010)
22. I.A. Rodionov, I.A. Zvereva, Photocatalytic activity of layered perovskite-like oxides in practically valuable chemical reactions. Russ. Chem. Rev. **85**(3), 248 (2016)
23. X. Sun, X. Xu, Efficient photocatalytic hydrogen production over La/Rh co-doped Ruddlesden-Popper compound Sr_2TiO_4. Appl. Catal. B **210**, 149–159 (2017)
24. J. Yu, J. Huang, R. Li, Y. Li, G. Liu, X. Xu, Fluorine-expedited nitridation of layered perovskite Sr_2TiO_4 for visible-light-driven photocatalytic overall water splitting. Nat. Commun. **16**(1), 361 (2025)
25. M. Lv, X. Sun, S. Wei, C. Shen, Y. Mi, X. Xu, Ultrathin lanthanum tantalate perovskite nanosheets modified by nitrogen doping for efficient photocatalytic water splitting. ACS Nano **11**(11), 11441–11448 (2017)
26. S.N. Tijare, M.V. Joshi, P.S. Padole, P.A. Mangrulkar, S.S. Rayalu, N.K. Labhsetwar, Photocatalytic hydrogen generation through water splitting on nano-crystalline $LaFeO_3$ perovskite. Int. J. Hydrogen Energy **37**(13), 10451–10456 (2012)
27. M. Qureshi, A.T. Garcia-Esparza, G. Jeantelot, S. Ould-Chikh, A. Aguilar-Tapia, J.-L. Hazemann, J.-M. Basset, D. Loffreda, T. Le Bahers, K. Takanabe, Catalytic consequences of ultrafine Pt clusters supported on $SrTiO_3$ for photocatalytic overall water splitting. J. Catal. **376**, 180–190 (2019)
28. S. Takasugi, K. Tomita, M. Iwaoka, H. Kato, M. Kakihana, The hydrothermal and solvothermal synthesis of $LiTaO_3$ photocatalyst: suppressing the deterioration of the water splitting activity without using a cocatalyst. Int. J. Hydrogen Energy **40**(16), 5638–5643 (2015)

29. D. Xu, L. Li, T. Xia, W. Fan, F. Wang, H. Bai, W. Shi, Heterojunction composites of g-C_3N_4/$KNbO_3$ enhanced photocatalytic properties for water splitting. Int. J. Hydrogen Energy **43**(34), 16566–16572 (2018)
30. N. Kulischow, C. Ladasiu, R. Marschall, Layered Dion-Jacobson type niobium oxides for photocatalytic hydrogen production prepared via molten salt synthesis. Catal. Today **287**, 65–69 (2017)
31. D.W. Hwang, H.G. Kim, J.S. Lee, J. Kim, W. Li, S.H. Oh, Photocatalytic hydrogen production from water over M-doped $La_2Ti_2O_7$ (M= Cr, Fe) under visible light irradiation (λ> 420 nm). J. Phys. Chem. B **109**(6), 2093–2102 (2005)
32. S.N. Lim, S.A. Song, Y.-C. Jeong, H.W. Kang, S.B. Park, K.Y. Kim, H2 production under visible light irradiation from aqueous methanol solution on $CaTiO_3$: Cu prepared by spray pyrolysis. J. Electron. Mater. **46**(10), 6096–6103 (2017)
33. L. Meng, K. Zhang, K. Pan, Y. Qu, G. Wang, Controlled synthesis of $CaTiO_3$: Ln_{3+} nanocrystals for luminescence and photocatalytic hydrogen production. RSC Adv. **6**(7), 5761–5766 (2016)
34. H. Zhang, G. Chen, X. He, J. Xu, Electronic structure and photocatalytic properties of Ag–La codoped $CaTiO_3$. J. Alloy. Compd. **516**, 91–95 (2012)
35. Y. Wei, J. Wan, J. Wang, X. Zhang, R. Yu, N. Yang, D. Wang, Hollow multishelled structured $SrTiO_3$ with La/Rh Co-doping for enhanced photocatalytic water splitting under visible light. Small **17**(22), 2005345 (2021)
36. I. Tamiolakis, D. Liu, F.-X. Xiao, J. Xie, I.T. Papadas, T. Salim, B. Liu, Q. Zhang, S.A. Choulis, G.S. Armatas, Mesoporous implantable Pt/$SrTiO_3$: C, N nanocuboids delivering enhanced photocatalytic H2-production activity via plasmon-induced interfacial electron transfer. Appl. Catal. B **236**, 338–347 (2018)
37. L. Wang, L. Wang, K. Zhao, D. Cheng, W. Yu, J. Li, J. Wang, F. Shi, Hydrogen production performance of active Ce/N co-doped $SrTiO_3$ for photocatalytic water splitting. Int. J. Hydrogen Energy **47**(92), 39047–39057 (2022)
38. S. Samajdar, S. Bera, P.S. Das, H. Finch, V.R. Dhanak, S. Chakraborty, T. Maiyalagan, K. Annapurna, S. Ghosh, Exploration of 1D–2D $LaFeO_3$/RGO S-scheme heterojunction for photocatalytic water splitting. Int. J. Hydrogen Energy **48**(47), 17838–17851 (2023)
39. E.W.-G. Diau, E. Jokar, M. Rameez, Strategies to improve performance and stability for tin-based perovskite solar cells. ACS Energy Lett. **4**(8), 1930–1937 (2019)
40. A. Dubey, S.L. Sanchez, J. Yang, M. Ahmadi, Lead-free halide perovskites for photocatalysis via high-throughput exploration. Chem. Mater. **36**(5), 2165–2176 (2024)
41. X. Li, X. Gao, X. Zhang, X. Shen, M. Lu, J. Wu, Z. Shi, V.L. Colvin, J. Hu, X. Bai, Lead-free halide perovskites for light emission: recent advances and perspectives. Adv. Sci. **8**(4), 2003334 (2021)
42. Y. Gao, Y. Pan, F. Zhou, G. Niu, C. Yan, Lead-free halide perovskites: a review of the structure–property relationship and applications in light emitting devices and radiation detectors. J. Mater. Chem. A **9**(20), 11931–11943 (2021)
43. G. Chen, P. Wang, Y. Wu, Q. Zhang, Q. Wu, Z. Wang, Z. Zheng, Y. Liu, Y. Dai, B. Huang, Lead-free halide perovskite Cs3Bi2xSb2–2xI9 (x$\approx$ 0.3) possessing the photocatalytic activity for hydrogen evolution comparable to that of (CH3NH3) PbI3. Adv. Mat. **32**(39), 2001344 (2020)
44. A. Pisanu, A. Speltini, P. Quadrelli, G. Drera, L. Sangaletti, L. Malavasi, Enhanced air-stability of Sn-based hybrid perovskites induced by dimethylammonium (DMA): synthesis, characterization, aging and hydrogen photogeneration of the MA 1–x DMA x SnBr 3 system. J. Mat. Chem. C **7**(23), 7020–7026 (2019)
45. L. Romani, A. Speltini, F. Ambrosio, E. Mosconi, A. Profumo, M. Marelli, S. Margadonna, A. Milella, F. Fracassi, A. Listorti, Water-stable DMASnBr 3 lead-free perovskite for effective solar-driven photocatalysis. Angew. Chem. **133**(7), 3655–3662 (2021)
46. D. Ju, X. Zheng, J. Liu, Y. Chen, J. Zhang, B. Cao, H. Xiao, O.F. Mohammed, O.M. Bakr, X. Tao, Reversible band gap narrowing of Sn-based hybrid perovskite single crystal with excellent phase stability. Angew. Chem. Int. Ed. **57**(45), 14868–14872 (2018)

47. L. Romani, A. Bala, V. Kumar, A. Speltini, A. Milella, F. Fracassi, A. Listorti, A. Profumo, L. Malavasi, PEA 2 SnBr 4: a water-stable lead-free two-dimensional perovskite and demonstration of its use as a co-catalyst in hydrogen photogeneration and organic-dye degradation. J. Mat. Chem. C **8**(27), 9189–9194 (2020)
48. H. Yin, J. Chen, P. Guan, D. Zheng, Q. Kong, S. Yang, P. Zhou, B. Yang, T. Pullerits, K. Han, Controlling photoluminescence and photocatalysis activities in lead-free Cs2PtxSn1−xCl6 perovskites via ion substitution. Angew. Chem. Int. Ed. **60**(42), 22693–22699 (2021)
49. X. Xu, Y. Chen, W. Zhou, Z. Zhu, C. Su, M. Liu, Z. Shao, A perovskite electrocatalyst for efficient hydrogen evolution reaction. Adv. Mater. **28**(30), 6442–6448 (2016)
50. N.F. Atta, A. Galal, S.M. Ali, The catalytic activity of ruthenates $ARuO_3$ (A= Ca, Sr or Ba) for the hydrogen evolution reaction in acidic medium. Int. J. Electrochem. Sci. **7**(1), 725–746 (2012)
51. J. Yu, X. Wu, D. Guan, Z. Hu, S.-C. Weng, H. Sun, Y. Song, R. Ran, W. Zhou, M. Ni, Monoclinic $SrIrO_3$: an easily synthesized conductive perovskite oxide with outstanding performance for overall water splitting in alkaline solution. Chem. Mater. **32**(11), 4509–4517 (2020)
52. Q. Sarmad, U.M. Khan, M.M. Baig, M. Hassan, F.A. Butt, A.H. Khoja, R. Liaquat, Z.S. Khan, M. Anwar, M.A. SA, Praseodymium-doped Sr_2TiFeO_6-δ double perovskite as a bi-functional electrocatalyst for hydrogen production through water splitting. J. Environ. Chem. Eng. **10**(3), 107609 (2022)
53. H. Zhu, P. Zhang, S. Dai, Recent advances of lanthanum-based perovskite oxides for catalysis. ACS Catal. **5**(11), 6370–6385 (2015)
54. Singh, B, Lanthanide-based catalysts for electrochemical water splitting: unraveling the interplay of structure, properties, and performance. Chem. Commun. (2025)
55. S.G. Peera, T.G. Lee, A.K. Sahu, Pt-rare earth metal alloy/metal oxide catalysts for oxygen reduction and alcohol oxidation reactions: an overview. Sustain. Energy Fuels **3**(8), 1866–1891 (2019)
56. M. Tayebi, Z. Masoumi, B. Seo, C.-S. Lim, C.H. Hong, H.J. Kim, D. Kyung, H.-G. Kim, Production of H2 and glucaric acid using electrocatalyst glucose oxidation by the Ta NiFe LDH electrode. ACS Appl. Mater. Interfaces **16**(20), 26107–26120 (2024)
57. S. Li, T. Xia, Y. Dou, Y. Xie, J. Wang, H. Zhao, L. Huo, Phosphatizing engineering of perovskite oxide nanofibers for hydrogen evolution reaction to achieve extraordinary electrocatalytic performance. Adv. Func. Mater. **32**(26), 2112164 (2022)
58. R. Khan, M.T. Mehran, S.R. Naqvi, A.H. Khoja, M.M. Baig, M.A. Akram, F. Shahzad, S. Hussain, A highly efficient A-site deficient perovskite interlaced within two dimensional MXene nanosheets as an active electrocatalyst for hydrogen production. Int. J. Hydrogen Energy **47**(88), 37476–37489 (2022)
59. X. Xu, Y. Pan, Y. Zhong, L. Ge, S.P. Jiang, Z. Shao, From scheelite BaMoO4 to perovskite $BaMoO_3$: Enhanced electrocatalysis toward the hydrogen evolution in alkaline media. Compos. B Eng. **198**, 108214 (2020)
60. D. Ji, C. Liu, Y. Yao, L. Luo, W. Wang, Z. Chen, Cerium substitution in LaCoO 3 perovskite oxide as bifunctional electrocatalysts for hydrogen and oxygen evolution reactions. Nanoscale **13**(22), 9952–9959 (2021)
61. C. Wu, Y. Sun, X. Wen, J.-Y. Zhang, L. Qiao, J. Cheng, K.H. Zhang, Adjusting oxygen vacancies in perovskite $LaCoO_3$ by electrochemical activation to enhance the hydrogen evolution reaction activity in alkaline condition. J. Energy Chem. **76**, 226–232 (2023)
62. U.A. Asif, T. Noor, E. Pervaiz, N. Iqbal, N. Zaman, LSTN (La0. 4Sr0. 4Ti0. 9Ni0. 1O3-&) perovskite and graphitic carbon nitride (g-C3N4) hybrids as a bifunctional electrocatalyst for water-splitting applications. J. Alloy. Compd. **939**, 168668 (2023)
63. M. Christy, H. Rajan, S.S. Subramanian, S. Choi, J. Kwon, S.A. Patil, K. Lee, H.B. Park, T. Song, U. Paik, Tuning the surface chemistry of La0. 6Sr0. 4CoO3-δ perovskite via in-situ anchored chemical bonds for enhanced overall water splitting. Int. J. Hydrogen Energy **51**, 685–699 (2024)
64. X. Xu, K. Guo, X. Yu, Controllable La deficiency engineering within perovskite oxides for enhanced overall water splitting. Molecules **29**(6), 1342 (2024)

65. Y. Yan, T. Meng, Y. Chen, Y. Yang, D. Wang, Z. Xing, X. Yang, Atomic-level coupled RuO_2/$BaRuO_3$ heterostructure for efficient alkaline hydrogen evolution reaction. J. Energy Chem. **88**, 356–362 (2024)
66. Y. Lu, H. Zhang, Y. Wang, X. Zhu, W. Xiao, H. Xu, G. Li, Y. Li, D. Fan, H. Zeng, Solar-driven interfacial evaporation accelerated electrocatalytic water splitting on 2D perovskite oxide/MXene heterostructure. Adv. Func. Mater. **33**(21), 2215061 (2023)
67. W. Zhang, M. Xue, X. Zhang, C. Si, C. Tai, Q. Lu, M. Wei, X. Han, J. Ma, S. Chen, Boosting oxygen/hydrogen evolution catalysis via ruthenium doping in perovskite oxide for efficient alkaline water splitting. Appl. Surf. Sci. **664**, 160278 (2024)
68. J.-H. Liu, D. Yan, S.-F. Li, Surface engineering of perovskite oxide LaCo0. 67Cu0. 33O3 for improved overall water splitting activity. Inorg. Chem. **64**(21), 10533–10541 (2025)
69. J. Junita, D. Jayalakshmi, J.D. Rodney, Combustion-derived $BaNiO_3$ nanoparticles as a potential bifunctional electrocatalyst for overall water splitting. Int. J. Hydrogen Energy **48**(38), 14287–14298 (2023)
70. S. Kartigueyane, A.K. SP, M.S. AlSalhi, S.P. Rajendra, I. In, S.J. Lee, C. Yan, S. Angaiah, Electrocatalytic activity of multifunctional Mn-Doped $SrFeO_3$-δ/Ti-MXene nanohybrid for efficient overall water splitting. ACS Appl. Energy Mater. **8**(8), 5004–5016 (2025)
71. C. Avcıoğlu, M.F. Bekheet, S. Avcıoğlu, F. Kaya, B.C. Kim, C. Kaya, A. Gurlo, Unveiling the electrocatalytic activity of bifunctional iron-niobium double perovskites for overall water splitting: a-site cation influence. Adv. Mater. Interfaces **12**(1), 2400559 (2025)
72. P. Nandigana, S. Pari, D. Sujatha, M. Shidhin, C. Jeyabharathi, S.K. Panda, Lead-free bismuth-based halide perovskites with excellent stability for visible-light-driven photoelectrochemical water splitting. Chem. Select **8**(4), e202204731 (2023)
73. P. Sikarwar, I.T. Koneri, T. Appadurai, A.K. Chandiran, Highly efficient photoelectrochemical water oxidation using Cs 2 Ag M Cl 6 (M= In, Bi, Sb) halide double perovskites. Phys. Rev. Appl. **19**(4), 044083 (2023)
74. H. Dong, C. Ran, W. Gao, M. Li, Y. Xia, W. Huang, Metal halide perovskite for next-generation optoelectronics: progresses and prospects. ELight **3**(1), 3 (2023)
75. M. Hamdan, A.K. Chandiran, Cs2PtI6 halide perovskite is stable to air, moisture, and extreme pH: application to photoelectrochemical solar water oxidation. Angew. Chem. **132**(37), 16167–16172 (2020)
76. M. Manoj, J.S. Halpati, A.K. Chandiran, Tuneable vacancy-ordered Halide perovskites Cs 2 PdX 6 (X= Cl, Br) for photoelectrochemical water splitting. Chem. Commun. (2025)
77. J.S. Halpati, N. Shanmugam, M. Manoj, M. Hamdan, A.K. Chandiran, Mixed-Tetravalent Cs2RumPt1− mX6 (X= Cl−, Br−)-based vacancy-ordered halide double perovskites for enhanced solar water oxidation. Solar RRL **8**(5), 2300924 (2024)
78. N.V.P. Chandra, M. Hamdan, A.K. Chandiran, Stable Cs 2 ReX 6 (X–Cl, Br) vacancy-ordered perovskites for solar water splitting. Sustain. Energy Fuels **7**(4), 949–955 (2023)
79. H. Peng, L. Xu, Y. Sheng, W. Sun, Y. Yang, H. Deng, W. Chen, J. Liu, Highly conductive ligand-free Cs2PtBr 6 perovskite nanocrystals with a narrow bandgap and efficient photoelectrochemical performance. Small **17**(38), 2102149 (2021)
80. A.L. Sangle, S. Singh, J. Jian, S.R. Bajpe, H. Wang, N. Khare, J.L. MacManus-DriscollJ, Very high surface area mesoporous thin films of $SrTiO_3$ grown by pulsed laser deposition and application to efficient photoelectrochemical water splitting. Nano Lett. **16**(12), 7338–7345 (2016)
81. C. He, X. Bu, S. Yang, P. He, G. Ding, X. Xie, Core-shell $SrTiO_3$/graphene structure by chemical vapor deposition for enhanced photocatalytic performance. Appl. Surf. Sci. **436**, 373–381 (2018)
82. L. Shi, W. Zhou, Z. Li, S. Koul, A. Kushima, Y. Yang, Periodically ordered nanoporous perovskite photoelectrode for efficient photoelectrochemical water splitting. ACS Nano **12**(6), 6335–6342 (2018)
83. J. Fan, X. Ran, H. Fu, J. Sun, X. An, X. Yang, The roles of oxygen vacancies in $LaFeO_3$ photocathodes in photoelectrochemical water splitting. Appl. Surf. Sci. **649**, 159011 (2024)

84. I.M. Nassar, S. Wu, L. Li, X. Li, Facile preparation of n-type $LaFeO_3$ perovskite film for efficient photoelectro chemical water splitting. Chem. Select **3**(3), 968–972 (2018)
85. Y. Zhu, W. Zhou, J. Yu, Y. Chen, M. Liu, Z. Shao, Enhancing electrocatalytic activity of perovskite oxides by tuning cation deficiency for oxygen reduction and evolution reactions. Chem. Mater. **28**(6), 1691–1697 (2016)
86. S. Piskunov, O. Lisovski, J. Begens, D. Bocharov, Y.F. Zhukovskii, M. Wessel, E.J.T. Spohr, C-, N-, S-, and Fe-doped TiO_2 and $SrTiO_3$ nanotubes for visible-light-driven photocatalytic water splitting: prediction from first principles. J. Phys. Chem. C **119**(32), 18686–18696 (2015)
87. H.S. Anusha, S.P. Jijoe, T. Tenzin, V. Divya, K.M. Anilkumar, S.R. Yashas, K. Sumana, K. Wantala, H.P. Shivaraju, Enhanced photoelectrochemical water splitting and photocatalytic decomposition performance of visible light active $MgMnO_3$ perovskite nanostructure. Int. J. Hydrogen Energy **77**, 1074–1089 (2024)
88. G. Liu, S.K. Karuturi, H. Chen, D. Wang, J.W. Ager, A.N. Simonov, A. Tricoli, Enhancement of the photoelectrochemical water splitting by perovskite BiFeO3 via interfacial engineering. Sol. Energy **202**, 198–203 (2020)
89. N. Wei, Y. Chen, X. Wang, M. Kan, T. Zhang, Y. Zhao, Solution chemistry quasi-epitaxial growth of atomic $CaTiO_3$ perovskite layers to stabilize and passivate TiO_2 photoelectrodes for efficient water splitting. Fundam. Res. **3**(6), 918–925 (2023)
90. K.-Y. Yoon, J. Park, M. Jung, S.-G. Ji, H. Lee, J.H. Seo, M.-J. Kwak, S. Il Seok, J.H. Lee, J.-H. Jang, NiFeOx decorated Ge-hematite/perovskite for an efficient water splitting system. Nat. Commun. **12**(1), 4309 (2021)
91. N.R. Manwar, R.G. Borkar, R. Khobragade, S.S. Rayalu, S.L. Jain, A.K. Bansiwal, N.K. Labhsetwar, Efficient solar photo-electrochemical hydrogen generation using nanocrystalline $CeFeO_3$ synthesized by a modified microwave assisted method. Int. J. Hydrogen Energy **42**(16), 10931–10942 (2017)
92. Q. Peng, B. Shan, Y. Wen, R. Chen, Enhanced charge transport of $LaFeO_3$ via transition metal (Mn Co, Cu) doping for visible light photoelectrochemical water oxidation. Int. J. Hydrogen Energy **40**(45), 15423–15431 (2015)
93. M. Hussein, S.R. Qutb, S. El-Dek, W.Z. Tawfik, M.S. Abdel-wahab, Exploring perovskite $SmFeO_3$ and Bi-Doped $SmFeO_3$ for hydrogen generation in green technologies. Electrochim. Acta, 146379 (2025)
94. S. Chandrappa, S.J. Galbao, P. Sankara Rama Krishnan, N.A. Koshi, S. Das, S.N. Myakala, S.-C. Lee, A. Dutta, A. Cherevan, S.J.T. Bhattacharjee, Iridium-doping as a strategy to realize visible-light absorption and p-type behavior in BaTiO3. J. Phys. Chem. C. **127**(25), 12383–12393 (2023)
95. M.S. Alom, N. Acharya, A.N. Andriotis, M. Menon, F. Ramezanipour, Highly active water-splitting electrocatalyst developed by the creation of oxygen vacancies in a perovskite oxide. ACS Appl. Energy Mater. **8**(6), 3369–3378 (2025)
96. X. Ran, Y. Bai, H. Zeng, J. Zhang, H. Fu, X. An, X. Yang, Manipulating oxygen vacancy in $SrTiO_3$ nanoparticles to achieve enhanced photoelectrochemical performance in water splitting. ACS Appl. Nano Mater. **7**(23), 27543–27554 (2024)
97. R. Zong, Y. Fang, C. Zhu, X. Zhang, L. Wu, X. Hou, Y. Tao, J. Shao, Surface defect engineering on perovskite oxides as efficient bifunctional electrocatalysts for water splitting. ACS Appl. Mater. Interfaces **13**(36), 42852–42860 (2021)
98. H. Ali, Z. Ajmal, A.Y.A. Alzahrani, M.H. Al Mughram, A.M. Abu-Dief, R. Al-Faze, H.M. Hassan, S.R. Al-Mhyawi, Y. Al-Hadeethi, Y.J.I. Orooji, Defect-driven innovations in photocatalysts: pathways to enhanced photocatalytic applications. InfoMat, e70040
99. M. Wang, J. Chen, C. Zhang, H. Ding, H. Wu, X. Li, S. Huai, Z. Tang, X. Zhao, H. Liu, Engineering surface defect active sites in SnS_2 nanosheets with electron-donating groups for efficient photoelectrochemical water splitting. J. Catal. **446**, 116087 (2025)
100. Z. He, Q. Tang, X. Liu, X. Yan, K. Li, D. Yue, Lead-free Cs2AgBiBr 6 perovskite with enriched surface defects for efficient photocatalytic hydrogen evolution. Energy Fuels **35**(18), 15005–15009 (2021)

101. J. Zhang, T. Wu, J. Duan, M. Ahmadi, F. Jiang, Y. Zhou, B. Hu, Exploring spin-orbital coupling effects on photovoltaic actions in Sn and Pb based perovskite solar cells. Nano Energy **38**, 297–303 (2017)
102. J. Park, J. Lee, H. Lee, H. Im, S. Moon, C.S. Jeong, W. Yang, J. Moon, Hybrid perovskite-based wireless integrated device exceeding a solar to hydrogen conversion efficiency of 11%. Small **19**(27), 2300174 (2023)
103. W. Jeong, J. Yun, G. Jang, C.S. Jeong, J.J.A.M. Moon, Immersed halide perovskite-based electrochemical cells for stable solar water splitting: achievements, opportunities, and prospects. Adv. Mat., e09817 (2025)
104. H. Choi, Y.Y. Kim, S. Seo, Y. Jung, S.M. Yoo, C.S. Moon, N.J. Jeon, S. Lee, K. Lee, F.M. Toma, Suppression of undesired losses in organometal halide perovskite-based photoanodes for efficient photoelectrochemical water splitting. Adv. Energy Mater. **13**(31), 2300951 (2023)

Chapter 10
Opportunities of Lead Free Perovskites Materials for Memristive Applications

Yogesh Khrolia, Anamika, and Abhimanyu Singh Rana

Abstract The growing demand for low-power electronics has generated strong interest in memristors and neuromorphic architectures. Perovskite-based memristors are promising due to their flexibility and compatibility with existing semiconductor technologies. Organic–inorganic lead-free perovskites are environmentally friendly while offering low operating voltages, high retention and endurance. $Cs_2AgBiBr_6$ is the first lead-free halide perovskite memristor showing excellent performance. Similarly, 2D-layered ferroelectric $CuInP_2S_6$ is scalable and suitable for flexible neuromorphic devices. Antimony-based perovskites ($Cs_3Sb_2Br_9$) are easy to synthesize exhibiting good ON/OFF ratios. Tin-based halide perovskites are shown to give high mobility. The metal–insulator-metal (MIM) structure is typically fabricated by depositing electrodes in a crossbar geometry using chemical and physical thin-films deposition methods combined with lithography. This chapter discusses the operating principles, materials selection, devices structures, and fabrication methods of perovskite-based memristive devices.

Keywords Memristors · Perovskites · Resistive switching · Neuromorphic computing · Artificial synapses · Brain-like computing

10.1 Introduction

From the last decade it is clear that only transistor scaling is not enough for modern computation requirements. Moore's law [1] which states that by downscaling the transistor size the number of transistors per unit area doubles (roughly) every twenty four months upheld very well for decades but started becoming irrelevant after 2008. The amount of data generated per year is increasing exponentially but not the computation efficacity due to transistors/area saturation limits [2]. Statistics show that nowadays computing performance is doubling every two months comparative to doubling

Y. Khrolia · Anamika · A. S. Rana (✉)
Centre for Advanced Materials and Devices, School of Engineering and Technology, BML Munjal University (Hero Group), Gurugram, India
e-mail: abhimanyu.rana@bmu.edu.in

V. Chauhan et al. (eds.), *Sustainable Lead-Free Perovskite Materials*, Topics in Applied Physics 157, https://doi.org/10.1007/978-3-032-24873-2_10

every twenty-four months in Moorse's era. A year wise compression of Moore's law (transistors/area), data generated (demands for compute), and computing efficiency is shown in Fig. 10.1a.

Unfortunately, this increasing demand requires a lot of energy. According to International Energy Agency's 2024 report in 2022 data centers consumed 460TWh of electricity worldwide which is projected upto 1000 TWh by 2026 [3]. This is equivalent Japan's total annual energy consumption. The AI models require energy tens of GWh to train.

This much energy consumption is majorly due to separation of memory and computation units i.e., von Neumann architecture. Here more energy is consumed in transferring the data between the memory and processor. The heat produced in this process slows down the computation. Modern computer processors are capable of performing high computation, but they slow down when they wait for data transfer between memory and processor creating a memory wall, this is called as von Neumann bottleneck [4] which was named after mathematician and physicist John von Neumann. In 1945 he drafted the idea for a stored program in computer 'First draft of a report on the EDVAC' where EDVAC stands for Electronic Discrete Variable Automatic Computer [5]. Figure 10.1b demonstrates the latency and loss in

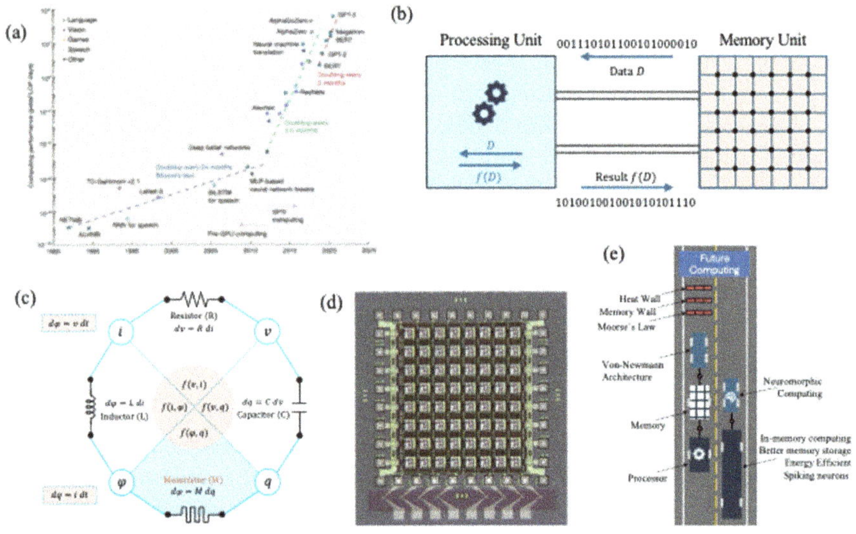

Fig. 10.1 **a** Increasing power demands over time in petaflop days. Till 2012 the data demands were doubling every two years; recent trends show that it's doubling every two months **b** Conventional computing system i.e., Von Neumann architecture shows the separation of processing and memory units where data D is transferred from memory to processing unit to perform operation f, and then back to memory unit. This causes a latency and lot of power consumption **c** The four fundamental two terminal circuit elements—resistor R, capacitor C, inductor L, and memristor M **d** An optical micrograph of an integrated memristive neural network, consisting of an 8 × 8 memristive synapse crossbar interfaced with eight memristive artificial neurons **e** Representation of comparison between conventional and in memory computation

energy due to data transfer between memory and processing unit via bus. To run a program f the processor has to fetch data D from the memory then after computation the result f(D) is sant back to the memory unit via bus. This process becomes very inefficient for large programs.

10.1.1 Introduction to Memristors

Memristor (memory + resistor) a two terminal device which can memories its resistance based on history of current and voltage passing through it; has potential to overcome these issues with conventional computing architecture. It shows many unique properties which could not be realized with RLC networks alone. In 1971 Leon O. Chua presented a paper "Memristor—The Missing Circuit Element" was the first theoretical model of memristor [6]. The four fundamental two terminal circuit elements are resistor R, capacitor C, inductor L and memristor M are defined by variables voltage v, current i, charge q and flux φ. Chua noticed that there four variables form six combinations out of which five $\{(v, i), (v, q), (v, \varphi), (i, q), (i, \varphi)\}$ are known by mathematical relations, so using axiomatic point of view there should be existence of (φ, q) relationship. Figure 10.1c represents these four two terminal elements and their relations. The memristance—M can be defined by the relationship charge and flux as

$$d\varphi = Mdq$$

where memristance $M(q)$ is a function of charge, φ is flux and q is charge.

2008 was the year when researchers at Hewlett Packard Labs (HP labs) reported a physical model of titanium dioxide based 8 × 8 crossbar array memristor. Figure 10.1d shows a cross-section view of titanium dioxide sandwiched between top and bottom platinum electrode [7]. TiO_2 layer acts as storage layer and can reconfigure its resistance based on history of current and voltage which represents the memory effect. This memorized state can't be destroyed by removing the power supply, thus makes it a non-volatile memory. This behavior of memristance cannot be reproduced by any combination of fundamental circuit elements—resistor, capacitor and inductor. This was the first major milestone after the theoretical model of memristor, and HP named it "The missing memristor found" [8].

These memristors can mimic the biological synapses in our brain and thus represent artificial synapses [9]. One of the major applications of this synapse behavior is in neuromorphic computing for spike based neural networks (SNNs) and deep neural networks (DNNs) [10, 11], electronic skin sensors [12], sensory receptors [13] and artificial vision system [14].

Our brain has nearabout 10^{15} synapses and 10^{11} which consumes ~20 W of power. Similar digital artificial neural network consumes 7.9 MW (approximately 10^6 times) of power. This is because our brain does not separate memory from processing. Mimicking this brain neural network, the neuromorphic computer can overcome the issue with traditional computers [4].

In 2008 memory devices were prepared that show reproducable resistive switching effect using perovskite oxide based materials [15]. Further silicon oxides SiO_2 [16], metal oxides (like ZeO_2/TiO_2 [17], TiO_x [18, 19], HfO_2 [20], $SrTiO_{3-x}$ [21]), metal nitrides and sulphides (like TiN [22], AiN [23], MoS_2 [24]), graphene [25], organic bioinspired electronic materials [26–28] and MXenes [29, 30] have been studied for memory devices. The oxide based memristors are naturally stable but there inherent oxgen vacency defects limits there memristance behivour. The fabrication of these materials often requires large equipments like supttering etc. The metal sulfides, organic electronic materials and MXenes are flexible but show low retention and poor durability with complex fabrication processes.

Although the early reaserch on halide perovskites (HP) are focused on solar cells but the resistive switching of HP [31] makes them strong contender for memristors. Long career diffusion length makes them ideal for low voltage low energy switching that supports energy efficient in memory computing [31, 32]. HP have high defect tollerence so they have very low mobility loss makes them purposeful for defect tuning and syneptic weights [33, 34]. HP can show reconfigurable behivour between volatile and non volatile (diffusive and drift) modes by electrochemical reactions which [35] is not possible by existing memristive devices. In addition HP show good flexibility, adjustable band gap, strong light absorption. Although HP have potential to provide in memory computing solutions, a lot research in this noval devis has to be done.

10.2 Need for Perovskite Memristors

Perovskites share a crystal structure of ABX_3 where A is a cation like Methylammonium (MA), Formamidinium (FA), Cesium etc., B is a metal cation like Pb which is toxic, Sn or Ge etc., X is anion like I, Cl, Br (halides) or oxygen.

Recent advances in Lead halide perovskites based resistive switching device, such as ferroelectric tunnel junctions, artificial synapse based memristors or neuromorphic devices over the fast few decades, and new advanced integrated circuit technologies have given us computation with high speed/high density memories and processors. Which overcome the challenges facing Von Neumann architecture based technologies. The Pb based halide perovskites facing issues as it is hazards to human health as well as enviromant due to its high toxicity, and unstabilty in ambient environment. Lead leakage during fabrication, disposal of devices, and for large commercialization applications posses long term risks environmently. Recently, Lead-free halide perovskites gain significant attention for several memristive applications due to their unique properties such as tunable band gap, non-toxicity, ease of fabrication, optoelectronic behaviour, mechanical flexibility, reproducibility, good reliability and stablity. Toxic behaviour of lead based perovskites has driven the development of several kind lead- free halides perovskites, suitable candidate for memristors, phovolatics, solar cells, LEDs, etc. These can be categorized based on substitution of B-site cation, structure, and its switching behaviour. Zhao et al. reported that tin-based ($CH_3NH_3SnI_3$) lead-free perovskites have same conduction properties as Pb-based, filamentry type switching, but Sn^{2+} is easily oxidised to form Sn^{4+} [36]. Bismuth-based ($Cs_2AgBiBr_6$) is first lead- free double-perovskites having excellent performance such as high endurance, and low power operation [37]. Wang Y. et al. studied that Cu/Sb based air stable resistive switching device having low operating voltage < 0.6Volt, stable endurance > 400 cycles, good flexibilty, and > 10^4 s retention time making it suitable for high performance RRAM applications [38]. These materials include binary, ternary (Sn, Bi/Sb, Cu/Ag) based metal halides, quarternary (double perovskites) each having stablity and switching uniformity that offers a sustainable, eco-friendly, and safer alternatives for future non- volatile memory applications.

The lead free perovskites can be catagrised as

- Bi/Sb-based 0D or 2D perovskite memristors ($A_3B_2X_9$)
- Double perovskite memristors ($A_2B^+B^{3+}X_6$)
- Cu/Ag-based perovskite memristors
- Layerd inorganic perovskite memristors
- Mixed cation memristors

10.2.1 Bi/Sb-Based Perovskite Memristors

Its commonly known structure is $A_3B_2X_9$ derived from ABX_3 by replacing divalent cation B^{2+} like Pb^{2+} or Sn by trivalent cation having similar ionoc radius like Bi^{2+} or Sb. A is big cation like MA^+ or FA^+ or Cs^+ and X is halide anion (Cl^-, Br^-, I^-). It can form 0D (dimer) for example $Cs_3Bi_2I_9$ having wide bandgap around ~2.1–2.2 eV [39] or 2D (layered) structure for example $Cs_3Sb_2I_9$ [40] having bandgap around ~2.04 eV. Although they have poor optoelectronic properties as compared to Pb/Sn based perovskites but have better environmental impact.

10.2.2 Double Perovskite Memristors

Also known as elpasolite, the double perovskite memristors have structure of $A_2BB'X_6$ derived from ABX_3 by replacing B site by monovalent and trivalent cation (for example Ag^+/Bi^{3+} [41]) arranged in rock salt pattern. A is big cation like MA^+ or FA^+ or Cs^+ and X is halide anion (Cl^-, Br^-, I^-). These are the most explored lead free perovskite memristors due to there thermal stability, chemical tunability and favourable optoelectronic properties. The most studied double perovskite is $Cs_2AgBiBr_6$ which has a indirect bandgap of ~2.1 eV and strong electron–phonon coupling [42, 43] which is highly stable in environmental conditions.

10.2.3 Copper and Silver Based Halide Perovskite Memristors

The Ag/PMMA/$Cs_3Cu_2I_5$/ITO memristor can be prepared using one step solution process where ultrathin polymethylmethacrylate (PMMA) is employed to stop direct contact between Ag electrode and iodine ions so it doesn't react. It was observed that this memristor shows low operating voltage of $< \pm 1$ V, show large ON/OFF ratio of 20^2, have high retention of 10^4 s and stable endurance upto 100 cycles [44]. Earlier in 2007 $RbAg_4I_5$ sandwiched between Ag and Pt electrodes showed switching behaviour. The $RbAg_4I_5$ film of 300 nm was prepared using pulsed laser deposition and using ion beam lithography and shows ON/OFF ratio of 10^4 [45]. In a recent study Ag/$Cs_2Cu_2I_5$/ITO memristor was prepared using dual-source vapor deposition, which show low power consumption of 0.013 W, having operating voltage of 1.2 V, retention time of $> 10^4$ s and endurance of > 300 cycles [46].

10.2.4 Mixed Cation Based Memristors

Isoelectronic properties of Sn with Pb make it the most promising candidate for the B site cation in Pb based perovskite memristors. But in presence of oxygen and moisture the Sn^{2+} is a unstable and it oxidises into Sn^{4+} which further results in nonradiative recombinatin causes poor power conversion efficiency [47]. The band gap of (TMPDA)$SnCl_4$ is found to be of 3.5 eV and for (TMPDA)$SnBr_4$ it is found to be if 3.19 eV [48], where TMPDA refers to tetramethyl-1,3-propanediamine.

10.3 Different Types of Memristive Devices

On the basis of switching mechanisms the memristors can be classified majorly in four catagries [49].

- Resistive Switching (RS)
- Phase change memory (PCM)
- Spintronic memristor
- Ferroelectric Tunnel Junction Memristors

An insulating layer which is responsible for memory storage is sandwiched between two conducting electrodes i.e., metal–insulator-metal (MIM) stack. Generally, the electrode which is grounded is called bottom electrode (BE) and the electrode which is connected to current/voltage terminal is called top electrode (TE). The resistance of the sandwiched layer can be switched by applying bias to MIM stack.

Based on sudden or gradual switching between low resistive state LRS and high resistive stage HRS, RS phenomenon can be classified in digital or analog switching respectively.

The applied voltage at which HRS state changes to LRS state is called SET voltage V_{SET} (Fig. 10.2a) and the applied voltage at which the LRS changes to HRS is called RESET voltage V_{RESET}.

Digital switching is used for memory devices like RRAMs—Resistive Random-Access Memories. The LRS state can be considered as ON state or "digital 1" state and the HRS state can be considered as OFF state or "digital 0" state. Further it can be classified into unipolar, bipolar, and threshold RS. Unipolar switching is due to Joule heating induced rapture in conducting filaments. In this type of switching the switching is independent of polarity of applied potential i.e., V_{SET} and V_{RESET} can be from same polarity of voltage. Usually, V_{RESET} is smaller than V_{SET} except for SiO_x [50]. It is generally observed in structures having same material for top and bottom electrode like Pt/TiO_2/Pt [51], Al/ZrO_2/Al [52], Pt/HfO_2/Pt [53], and Pt/NiO/Pt [54]. But in case of bipolar switching the switching depends upon the polarity of applied potential so there V_{SET} and V_{RESET} are always from opposite polarity. It is usually due to asymmetry of device because of different electrode materials which causes different interfacial energy barriers. Some examples of bipolar switching are Pt/TiO_2/TiN [55], TiN/HfO_2/ITO [56], TiN/HfO_2/Pt [57], TiN/ZnO/Pt [58]. Some

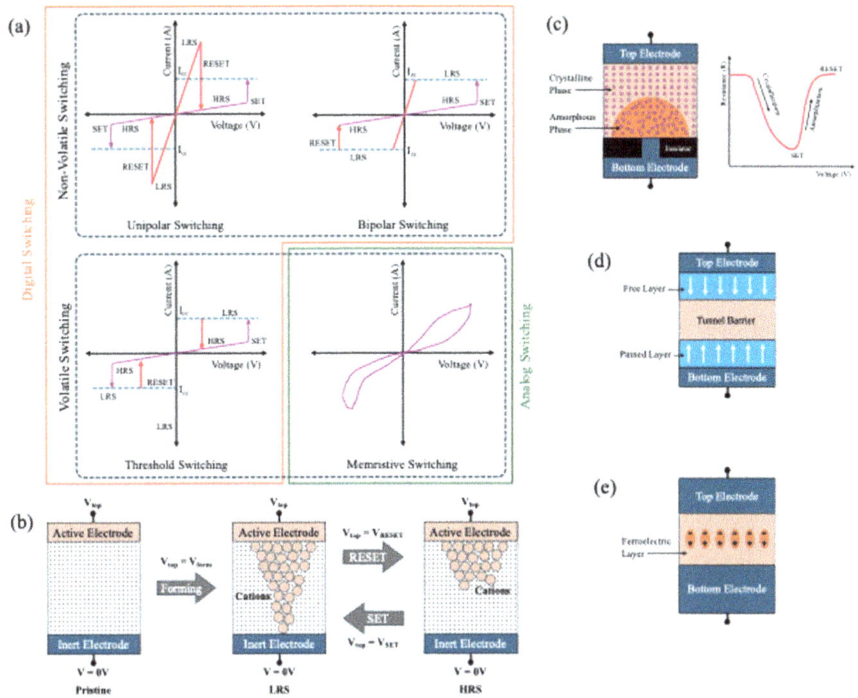

Fig. 10.2 **a** IV curves for unipolar, bipolar, threshold and memristive switching **b** Bipolar resistive switching representing the forming process and transition between SET and RESRT **c** phase change memory and its resistance voltage characteristics **d** spintronic memristor **e** ferroelectric tunnel junction memristor

materials for example Al/NiO/ITO [59]. TiO_2 [60, 61], TaO_x [62], $BGeO_3$ [63] show coexistence of unipolar/bipolar RS simultaneously. Larger compliance current I_{CC} creates unipolar switching while minor compliance current creates bipolar switching [60, 61]. Both unipolar and bipolar type of switching retains their resistive state after power removal thus inherits the memory effect and are non-volatile memory devices.

In threshold switching the device rests to HRS state whenever the applied voltage drops below the certain threshold voltage. Thus, it is unable to retain the LRS state after removing the applied potential, which puts it into volatile resistive switching category.

In case of analog switching the resistive state changes continuously not instantly, which mimics our biological system thus suits for neuromorphic computing.

10.3.1 Resistive Switching

In this type of RS formation and rapture of conductive filament makes the memristor to switch between LRS and HRS states. As shown in Fig. 10.2b, the metallic or non-metallic conductive filament is formed by redox rection due to anion or metallic cation migration. The ions can be oxygen vacancies (e.g., $Pt/TiO_2/Pt$, $Pt/SiO_2/Ta$) [8, 64], active metal anions or cations such as sulfurs (e.g., $Au/MoS_2/Au$) [65] or halide perovskites (e.g., $ITO/Cs_2AgBiBr_6/Al$, $ITO/Cu_2AgBiI_6/Ag$) [42] which belongs to dielectric layer or electrodes. Depending upon the type of ion migration the resistive switching is classified into electrochemical metallization mechanism (ECM) also known as conductive bridging random access memory (CBRAM), valance-change mechanism (VCM) and thermochemical mechanism (TCM).

Electrochemical metallization mechanism (cation based) a low power consuming, show fast but unstable switching is used for non-volatile memory applications. It has the active electrode/solid electrolyte/inert electrode configuration where Ag, Cu (say M) are generally used for active and Pt, Ir or TiN are used for inert electrode. Upon the application of forwarding potential on active electrode (anode) the metal M oxidizes as $M \rightarrow M^{n+} + ne^-$. The oxidized metal cation M^{n+} moves towers the inert electrode (cathode) via functional layer under the influence of electric field and these metal cations are reduced ($M^{n+} + ne^- \rightarrow M$) and deposited at cathode. Gradually a conductive wire filament is formed and the memristor achieves LRS state. This is called forming process. Application of reverse potential causes oxidation and raptures the conductive path creating HRS state [66]. ECM is observed in various oxides e.g., ZrO_x [67], MoO_x/GdO_x [68], Al_2O_3 [69] and chalcogenides [70] e.g., GeS_2.

Valance-change mechanism [71] (anion based) is somewhat like ECM where generally both electrodes are inert such as Al, Pt, Au, indium tin oxide (ITO), and fluorine doped tin oxide (FTO). Positively charged oxygen vacancies or oxygen ions can be used for nonmetallic conductive filament formation. There are some chances that oxygen can react with surrounding atmosphere, moisture [72]. Thermal stability is required at higher temperatures to reduce the influence of Joule heating effect.

Binary oxides like HfO_2, Ta_2O_5 and AlO_x are some of the most promising materials for this RS where HfO_x based materials show high-K dielectric [73].

Thermochemical Mechanism, the forming process is carried out by thermal decomposition of conduction filament (CF) introduced by Joule heating [74, 75]. Since thermal changes are unaffected by the potential polarity, TCM RS usually show unipolar switching. It's usually an inert electrode/metal oxide semiconductor/inert electrode structure. Some examples for TCM RS materials are ZnO [76], CoO [77] NiO [77, 78].

10.3.2 Phase Change Memory (PCM)

Crystalline and amorphus phase are responsible for hystrasis formation in PCM memristors, Fig. 10.2 (c). The phase change is encouraged by Joule heating [79, 80]. Initially on a amorphus state, application of lower amplitude voltage for sufficientrises the temprature to crystalization temperature time casuses it to melt and then cool slowly and finally creating crystaline phase (SET) resulting in LRS. To change the phase back to amorphus (RESET) a higher voltage is applied for shorter time period which melts the material and then cools repidly resulting in HRS [80]. These materials can retain there amorphous/crystalline state for many years making them ideal for stand-alone storage class memory (SCM) [81] industry. The reaserch on PCM based devises is motivated by chalcogenide like $Ge_2Sb_2Te_5$ [82] or Ag- and In- doped Sb_2Te [83]. Some other exampleas are AgInSbTe [84], GeTe [85]

10.3.3 Spintronic Memristor

It consists of two ferromagnetic metal layes in MIM structure works on the phenominon on tunneling magnetoresistance which is the buildig block for MRAM devices. One of the ferromagnetic layer has fixed magnetic polarization for refrence and it is reffered as pinned layer, Fig. 10.2d. Another layer known as free layer can change polarization via programing. When both of the ferromagnetic layers have same direction of polarization (parellel) then it's the LRS, and opposite direction (antiparellel) is HRS.

10.3.4 Ferroelectric Tunnel Junction Memristors

It consistis of a ferroelectric layer in meral MIM structure having different thickness electrodes works by switching the polarization of ferroelectric material such as perovskites. Appling voltage to MIM structure the charge is induced on electrodes that forces to aline the rendomly oriented ferroelectic dipoles present in the middle ferroelectric layer Fig. 10.2e.

10.4 Performance Parameters of Memristive Devices

In memristors SET/RESET voltage play key role in the device performance, referes as the threshold voltage at which the device operation from high resistance state to low resistance state and from low resitance state to high resistance state is called as V_{set}, V_{reset} respectivly. These measurements with low set/reset values, high ON/OFF ratios prefferable for low power, stable and reliable devices. These ratios can be calculated from I-V graphs at specific voltages sweep cycles, and spiking voltages.

10.4.1 ON/OFF Ratio

From the I-V graph I_{ON}/I_{OFF} or R_{OFF}/R_{ON} is referes as the ratio of current flow or resistance changes from HRS to LRS and vice-versa. In memristors, high conductance state (ON) as logic '1' and low conductance state (OFF) as logic '0' with seperable resistance states work as a switching window. Depending upon materials, fabrication processes, device structures it ranges from 10^2 to 10^6. ON/OFF ratio of lead-free perovskites Tsung et al. reported Ag/MoOx/CsI/Ag electrode tuning with ITO, Pt, Au, Cu bottom electrodes in 1T1R crossbar network having ~ 10^{10}, and C. Cuhadar et al. studied that Pt/$Rb_3Bi_2I_9$/Au, Pt/$Rb_3Bi_{1.8}Na_{0.2}I_9$/Au based devices having and ~ 10^7 [86]. It can be enhance by optimizing thickness of switching layer, electrode tuning, doping, deposition conditions, fabrication techniques and interface engineering.

10.4.2 Retention

In the I-V graph retention refers to the time span for a device to hold its resistance state, it is measured in seconds. It measures its volatile or non-volatile memory properties. A small voltage is applied as read voltage and current value is calculated at that particular point to find out its retention value, it is typically ranges from 10^3 to 10^4 s. In case of perovskites retention plays critical role for device performance testing because sometimes there is diffusion of filaments take place, so low resistance state returns to high resistance state it effects stabilty of device. In 2020 Shuaipeng Ge et al. reported that 0-D Ag(AgO_x)/$Cs_3Bi_2I_9$/FTO device shows high reset voltage and larger retention time than 2-Ag(AgO_x)/$Cs_2Bi_3I_{10}$/FTO device [87]. For 0-D device it shows retention of ~ 10^3 s than 2-D shows ~ 4×10^2 s, dimenssionality also play important role to enhance device performance.

10.4.3 Endurance

In the I-V curve endurance is measured by the number of cycles in which the device can switch between LRS and HRS successfully, also having a distinguishable ON/OFF ratio. The endurance can be measured by applying multiple voltage sweeps cycles or continuous voltage spikes cycles untill device failure. For lead free perovskites it is one of the important parameter to testing the device stability, technological requirements in devices are of > 10^9 cycles. In 2021 Y. Zhang et al. repoerted three-diamensional halide perovskites nanowires sandwitches in between Ag and Al shows endurance of ~ 3×10^6 cycles, and long retention time, [88]. Lee et al. reported that Pt/Ta_2O_{5-x}/TaO_{2-x}/Pt asymmetric bilayer device shows endurance ~ 10^{12} cycles [89]. The enhanced stablity of a device dependent on cystal structure as Y. Zhang reported in article, and also on number of layers of switching material.

10.4.4 Switching Speed

Switching speed refers to the time taken for a memory device to switch from HRS to LRS on applied different voltage sweeps or spike pulses. For calculating switching speed it required, different voltage sweeps or spike pulses with certain width at which change in resistance states recorded from I-V graph. As the technological requirements for switching are < 10 ns. Lee explored that Ta_2O_{5-x}, and TaO_{2-x} bilayers-based device shows 10 ns switching time [89]. Zhang et al. reported that in three dimensional perovskite nanowires show ultra high switching speed ~ 200 ps for monoclinic $MAPbI_3$ based device [88]. Switching speed is a important parameter for testing device performance, and less switching speed is prefferable. It can be dependent on many parameters like bilayers, deposition parameters, and also on dimension of switching material.

10.4.5 Power Consumption

Power consumption is also a key parameter in memristor, it refers to the operation power is used for a device to switch from HRS to LRS [90]. In 2020 Haibo Ye et al. reported that ITO/$AgBiI_4$/PMMA/Ag device in which Ag work as active metal and through ECM mechanism shows non-volatile switching at ultra low switching voltage ~ 0.16 V [91]. Wang K. et al. developed Al/Ti_3C_2: Ag/Pt device which operates it in one spike unit in as less as 0.35 pJ energy consumption [92]. Wang et al. fabricate bilayer Ta_2O_5/Al_2O_3 gate dielectric synaptic transistor-based device. It consumes an ultra-low power as19.9 aJ per synaptic event 0.001 Vdrain voltage and a short 1 ms pulse width [93]. In perovskites, power consumption is also a key challenge to enhance the performance of devices, we can tune this using different deposition technologies, fabrication architectures, and different lead-free materials etc.

All these parameters required to analyse device performance has unique advantages compared with traditional memory technologies.

10.5 Comparision of Performance Matrics of Different Perovskite Materials

10.5.1 Bi/Sb-Based Perovskite Memristor

Using density functional theory (DFT) on 696 compositions in four different crystal structures were investigated by choosing different combinations of A, B and X site elements for predicting high switching speed memory based on migration barrier and defect formation energy. it was found that dimer $Cs_3Sb_2I_9$ had much lower formation energy and comparable migration energy as of layered $Cs_3Sb_2I_9$. On exprimental verification it was found that the dimer $Cs_3Sb_2I_9$ has a faster switching speed of 20 ns while the layered $Cs_3Sb_2I_9$ has slower switching speed of 100 ns [94]. An exprimental study between dimer $Cs_3Sb_2I_9$ and layered $Cs_3Sb_2I_9$ it was found that layered $Cs_3Sb_2I_9$ has better ON/OFF ratio, retention and endurence and exhibites energy efficient artificial synaptic functions for in memory computing [95]. In this expriment it was found that the dimmer phase is forming free phase as pristine state was initially in LRS state before applying external bias. When using Ag/Au electrodes ECM is reported in dual phase $AgBi_2I_7$–$Cs_3Sb_2I_9$ [97] but VCM is more pronounced while using inert electodes like Au, Pt etc. [96, 97].

10.5.2 Double Perovskite Memristors

$Cs_2AgBiBr_6$ which is non toxic and super stable was found having ON/OFF ratio of 479, retention time 10^4 s, retention of 1000 cucles at low oprating voltage (+4 V, −0.3 V) and high HRS to LRS ratio ∽441 even after 30 days of storage in environmental conditions [37]. Few studys show unability to RESET after HRS to LRS switching making write once read many behivour (WORM) [98]. Doping or partial substitution can increase the switching properties of double perovskite memristors. A study show substitution of Cl on place of Be in ITO/$Cs_2AgBiBr_{6-x}Cl_x$/Al (x = 0, 0.2, 0.4, 0.8, 1.0) shifts the XRD peak (004) plane, the ON/OFF ratio increases from 10^3 to 10^4, SET voltage decreases from −0.40 V to −0.21 V and the RESET voltage decreases from 1.55 V to 1.34 V with incresing Cl concentration. This is SET/RESET voltage drop is due to the increased defect density and the increase in SET/RESRT ratio is due to larger HRS because of lower number of free carrers. Another study shows doping of rare earth element Nd in $Cs_2AgBiBr_6$ increse ON/OFF ratio by 1000 times [98]. A and B site substitution have also been studied.

10.5.3 Copper and Silver Based Halide Perovskite Memristors

The Ag/PMMA/$Cs_3Cu_2I_5$/ITO memristor can be prepared using one step solution process where ultrathin polymethylmethacrylate (PMMA) is employed to stop direct contact between Ag electrode and iodine ions so it doesn't react. It was observed that this memristor shows low operating voltage of $< \pm 1$ V, show large ON/OFF ratio of 20^2, have high retention of 10^4 s and stable endurance upto 100 cycles [44]. Earlier in 2007 $RbAg_4I_5$ sandwiched between Ag and Pt electrodes showed switching behivour. The $RbAg_4I_5$ film of 300 nm was prepared using pulsed laser deposition and usning ion beam lithography and shows ON/OFF ratio of 10^4 [45]. In a recent study Ag/$Cs_2Cu_2I_5$/ITO memristor was prepared using dual-source vapor deposition, which show low power consumption of 0.013 W, having operating voltage of 1.2 V, retention time of $> 10^4$ s and endurance of > 300 cycles [46].

10.5.4 Layerd/2D Perovskite Memristors

Layered, 2D perovskite or inorganic perovskite memristors have wider bandgap then there 3D counterparts. The comparison between a quasi-two-dimensional (2D) halide perovskite $(PEA)_2Cs_3Pb_4I_{13}$, and a 3D $CsPbI_3$ shows that $(PEA)_2Cs_3Pb_4I_{13}$ has a ON/OFF ratio of 10^9 which is three order of magnitude of that for $CsPbI_3$. The $(PEA)_2Cs_3Pb_4I_{13}$ retained its ON/OFF current ratio for 2 weeks while $CsPbI_3$ could not sustain even for five days. This proofs the long term stability of quasi-2D halide perovskites [99]. Another study showing two-dimensional (2D)-layered single-crystal (SC) $CsPb_2Br_5$ microsheet-based RS device shows ON/OFF ratio of upto 10^8 having retention time of $> 2.5 \times 10^4$ s, ow-operating voltage of < 0.32 V [100].

10.5.5 Mixed Cation Memristors

The intrensic oxidation tendensy of Sn makes it challenging to make a stable device. SnF_2 addtitive can be added to $CsSnI_3$ based resistive switching memory devices to decrease the oxidation [101] (Table 10.1 and Fig. 10.3).

Table 10.1 Comparison of different device structures, using different mechanisms, ON/OFF ratios, V_{SET}, V_{RESET}, retention time, endurance that required to check the overall performance and stability of device

Device structure	Mechanism	V_{SET} (V)	V_{RESET} (V)	ON/OFF ratio	Endurance (no. of cycles)	Retention time (s)	References
Ag/MoOx/CsI/Ag	Ag (ECM)	− 0.171	+ 0.057	10^{10}	10^5 (pulse)	2×10^6	[102]
Ag/PMMA/BiI_3/PMMA/Au	Bi (VCM)	+ 0.26	− 0.23	10^8	1.6	5×10^4	[112]
Ag/MoOx/CsBr/Ag	Ag (ECM)	− 0.161	+ 0.046	–	–	–	[102]
Cu foil/hBN/BiI_3/Au	I (VCM)	+ 0.26	− 0.18	10^7	2.5×10^3	10^4	[104]
Pt/$CsSnBr_3$/Ag	Ag (ECM)	+ 0.25	− 0.48	10^2	–	–	[105]
Pt/$CsSnBr_3$/Pt	Pt (ECM) & Br (VCM)	+ 0.69	− 0.56	–	–	–	[105]
PI/Pt/$CsSnBr_3$/Pt	Br (VCM)	+ 0.20	− 0.15	10^5	50	10^4	[106]
ITO/$MA_3Bi_2I_9$/Cu	Cu (ECM)	+ 1.44	− 6.9	10^4	1730 (Pulse)	3×10^5	[107]
Ag(AgO_x)/$Cs_3Bi_2I_9$/FTO	I (VCM)	− 0.12	+ 1.00	10^6	250	10^3	[87]
Ag(AgO_x)/$CsBi_3I_{10}$/FTO	I (VCM)	− 0.14	+ 0.55	10^6	250	4×10^2	[87]
ITO/$AgBiI_4$/PMMA/Ag	Ag (ECM)	+ 0.16	− 0.16	10^4	700 (Pulse)	10^4	[91]
Pt/$Rb_3Bi_2I_9$/Au	I (VCM)	+ 0.09	–	2.9×10^7	200 (Pulse)	1000	[86]
Pt/$Rb_3Bi_{1.8}Na_{0.2}I_9$/Au	I (VCM)	+ 0.3	–	–	100 (Pulse)	1000	[86]
ITO/Layered-$Cs_3Sb_2I_9$/Al	I (VCM)	+ 0.5	− 2.3	10^4	2000 (Pulse)	10^4	[95]
ITO/$Cs_3Sb_2I_9$/Al	I (VCM)	+ 0.4	− 2.9	10^4	100 (Pulse)	10^4	[95]
ITO/$Cs_3Bi_2Br_9$/Ag	Br (Interface)	− 0.5	+ 0.4 to + 1	10	3200 (Pulse)	10^3	[108]
ITO/PEDOT:PSS/$Cs_3Bi_2Cl_9$/PMMA/Au	Cl (VCM)	+ 0.45	–	< 10	10^4	10^5	[109]
ITO/PEDOT:PSS/$Cs_3Bi_2I_9$/PMMA/Au	I (VCM)	+ 0.73	–	8×10^3	10^4	10^5	[109]

(continued)

Table 10.1 (continued)

Device structure	Mechanism	V_{SET} (V)	V_{RESET} (V)	ON/OFF ratio	Endurance (no. of cycles)	Retention time (s)	References
ITO/PEDOT:PSS/$Cs_3Bi_2Br_9$/PMMA/Au	Br (VCM)	+ 0.64	–	150	10^4	10^5	[109]
ITO/$Cs_3Cu_2I_5$/PMMA/Ag	Ag (ECM)	+ 0.6	– 0.44	100	100	10^5	[44]
ITO/$Cs_2AgBiBr_6$/Au	Br & Ag (VCM)	+ 1.53	– 3.4	10	10^3	10^5	[41]
ITO/$Cs_2AgBiBr_6$:Nd/Au	Br & Ag (VCM)	+ 0.9	–	10^5	–	1.2×10^4	[110]
ITO/$Cs_2AgBiBr_{6-x}Cl$/Al	Br & Cl (VCM)	– 0.21	+ 1.34	10^4	100	10^4	[111]
ITO/$Cs_2AgEuBr_6$/Au	Br & Ag (VCM)	+ 3.5	– 1.5	10^4	100	1.2×10^4	[112]
ITO/Cu_2AgSbI_6/PMMA/Ag	Ag (ECM) & I (VCM)	+ 0.72	– 0.74	~ 10^2	1600	10^4	[113]

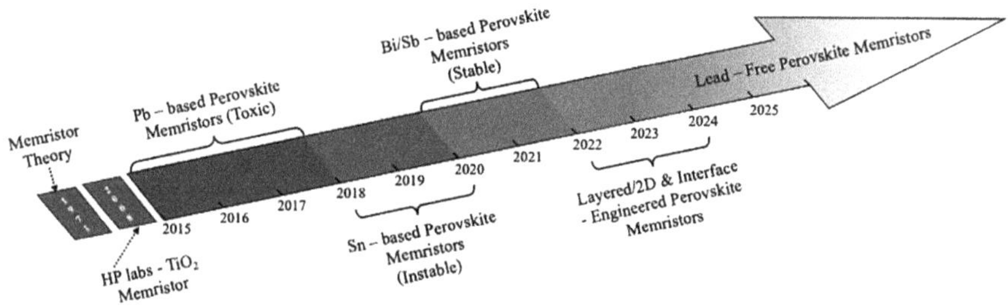

Fig. 10.3 Evolution of lead free memristors

10.6 Device Fabrication Process

The fabrication of the devices based on the growth of active layers as a lead-free perovskite using different deposition methods to achieve suitable phase and its interface quality, choice of substrates for different applications, top and bottom electrode tunning with active layer to form metal-insular-metal (MIM) device for charge transport, and electrodes materials impacts on mechanisms. So, for the fabrication we need to carefully optimize all these things as discussed above. For fabrication of memristors we need different steps i.e. Substrate choice and preparation, Deposition methods, Bottom electrode tuning, Active layer deposition, Post deposition annealing, Top electrode tuning, Device architecture.

10.6.1 Substrate Choice

The choice of substrates plays crucial role in thermal stability, interface quality, fabrication and performance of lead-free perovskites-based devices. It not only provides the mechanical support it is also crucial for growth of films and orientation of crystal planes. Different substrates can be choose based on different properties and applications. For different memories and optical devices, we need FTO fluorine tin oxide, indium tin oxide (ITO) coated substrates are commonly used. For uniform growth of films Silicon (Si) or SiO_2, flexible, and glass substrates are used. Some lead-free perovskites for suitable phase transition required post deposition annealing above 300o C, s required suitable substrate with high thermal resistance. Quartz, Alumina and sapphire (Al_2O_3) are mostly used. For transparent electronic and optical properties ITO and FTO on glass are used for transparency and conductivity as bottom electrodes, memristors need metal coated substrates for bottom/top electrodes such as Pt/Si, and Au/glass. Flexible substrate like polyethylene terephthalate (PET) are being explored for wearable and portable electronic devices.

The carefully choosing suitable substrates based on these feature helps to enhance the long- term performance, and efficiency of devices.

10.6.2 Top and Bottom Electrode Materials

Optimizing the device properties, electrode tuning is the critical step, because suitable electrode selection and its tuning impacts the charge transport, interface quality, and stability of devices. Suitable bottom electrode like Pt, Au, ITO, and FTO in perovskites helps to provide stable growth, lattice matching with switching layer, structural quality that enhance overall performance of device. The top electrode materials control the switching mechanisms, barrier asymmetry, and tunnelling behaviour. In symmetric device configuration like TiN/TiN exhibit stable, and reproducible switching, but in case of asymmetric device configuration TiN/Pt, ITO/TiN, Al/TiN, and Al/ITO exhibit multifunctional optoelectronic, asymmetric tunneling response. There are so many flexible substrates have been explored with switching materials polyimide (PI) [114], cellulose paper [115], polydimethylsiloxane (PDMS) [116], polyester sulfone (PES) [117], polyethylene terephthalate (PET) [118], and polyvinyl pyrrolidone-blended chitosan (CHP) [119] etc. These devices face some challenges of stability, reliability and bending failure etc. that can be improved by different electrode tuning.

10.6.3 Deposition Methods

Lead based perovskites are lead-free perovskites are gaining significant attention due its environmentally friendly, non-toxicity, low power consumption, and easily scalable properties suitable for different optical and electric applications as solar cells, memristors, photovoltaics etc. To scalable fabrication of these devices, we need different deposition methods based on different industrial applications. There are several deposition techniques for lead free perovskites are listed below in Table 10.2.

10.6.4 Device Structure

In the recent advancements of memristive devices, there are several device structure has been introduced to optimize switching performance, stability, scalability, flexibility, and integration with existing technologies. Memristors are designed in metal–insulator-metal architecture in which switching or active layer is sandwiched in between the two metal electrodes. For comercially designed memristors, need a large number of smaller memristors cells are arranged in integrated structure of MIM. Several structures have been devoloped to form MIM architectures. These structures are categorize as vertically stacked or single cell, planer or lateral, cross-point, and cross-bar array have been illustrated in Fig. 10.4.

10.7 Conclusion and Outlook

The rapid progress of memory devices based on perovskites, metal halides, and metal oxides switching materials has opened multiple opportunities for fundamental research as well as practical applications. In the era of massive data creation, high power consumption, delay in big data transfer and complex tasks to handle this there is a urgent need of better computing systems that are not only fast, reliable as well as highly energy efficient. Conventional CMOS SiO_2 based technology are reaching their physical limits in continous miniaturization of devices, high power consumption, massive heat generation and data transfer bottoleneck. For this emerging lead-free perovskite materials are highly promising candidate for next generation energy efficient neuromorphic and sustainable devices.

Among all perovskite materials $Cs_2AgBiBr_6$ based devices gained significant attention as first lead-free perovskites memristors that exhibit excellent performance. Other advantage of this material is ferroelectric switching behaviour. Ferroelctric perovskites with reversible polarizing switching enables data storage with minimum energy usage address the challenge of high power usege in conventional DRAMs and flash memories. This also highlights the potential of double perovskites materials to

Table 10.2 Deposition techniques, there advantages and disadvantages

Deposition techniques	Materials used	Advantages/ disadvantages	Top/bottom electrodes	References
Spin coating	$CH_3NH_3SnI_3$, $FASnI_3$	Advantages simple, fast, lab-scale device fabrication, Compatible with various substrates Disadvantages not ideal for large-scale production	Top electrode Ag, Au Bottom electrode ITO/PEDOT: PSS	[120]
Drop casting	$Cs_3Sb_2I_9$, $CsPbBr_3$, $MASnI_3$	Advantages simple, minimum precursor using, does not require high-temperature annealing, can produce high phase purity films, used for photodetectors Disadvantages limited scalability, not uniform coating over substrate	Top electrode Ag, Au Bottom electrode ITO, FTO	[121, 122]
Thermal evaporation	$CsSnI_3$, $MASnI_3$	Advantages better film uniformity Disadvantages reactivity issues with Sn-based thin films	Top electrode Ag, Au Bottom electrode ITO/PEDOT: PSS	[123, 124]
Electron-beam evaporation	Sn-based perovskites, Metal oxides	Advantages controlled thickness, and composition Disadvantages Costly, high-energy consumption	Top electrode Au, WO_{3-x} Bottom electrode FTO, Glass	[125, 126]
Sol–gel method	$Na_{0.5}K_{0.5}NbO_3$, $BiFeO_3$, $CaTiO_3$,	Advantages environmentally friendly process, better compositional uniformity Disadvantages requires high-temperature annealing, cracking and porosity issues	Top electrode Au Bottom electrode Pt/Ti/Si, Glass	[127]

(continued)

Table 10.2 (continued)

Deposition techniques	Materials used	Advantages/ disadvantages	Top/bottom electrodes	References
DC/RF sputtering	Metal electrodes, metal oxides, inorganic perovskites	Advantages highly uniform, better adhesion Disadvantages high cost, lower throughput than solution-based methods	Top electrode Au, Ag Bottom electrode Glass, ITO	[128]
Lithography	Bi-based films, $(C_6H_5CH_2NH_3)_2PbI_4$	Advantages enabling micropattern, device integration Disadvantages complex, more expensive, time-consuming	Top electrode Patterned Au/Ag Bottom electrode glass	[129]

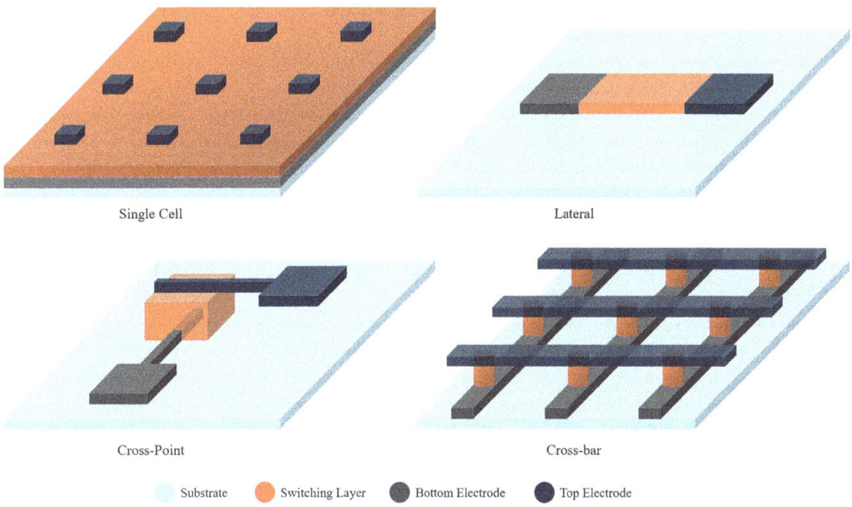

Fig. 10.4 Schametic arrangement of commonly studied memristors

address the challenge of toxicity issue in lead based perovskites devices. In neuromorphic computing devices, the ability of perovskites based hetrostructures to mimic synaptic behaviour openes opportunities for developing artificial intellegence based hardwares that can process large data in real time with lesser enrgy consumption compared to conventional systems.

The continuous downscaling of these materials in large scale industrial level production, achieving long term stability, reliability, and ensuring environment stability are the key challenges. Advancements in deposition techniques such as sputtering deposition, thermal evaporation, atomic layer deposition, pulse laser deposition along with lithogrpahy, will be crucial in achieving uniformity, stability and reliability. Electrode engeeniring, substrate selection, bilayer/multiyer stacking, interface quality, and suitable deposition methods provides key opportunities for scaling down of devices without compromising performance.

In nutshell, lead free perovskites metal halides materials have significant potential applications in resistive switching memories, and neuromorphic computing systems. However, studies are required to unlock the potential of emerging lead free perovskites halide materials for information storage proccesing and efficient neuromorphic computing applications.

References

1. G.E. Moore, Cramming more components onto integrated circuits, Reprinted from Electronics, volume 38, number 8, pp. 114, (1965) ff. IEEE Solid-State Circuits Soc. Newsl. **11**(3), 33–35 (2009). https://doi.org/10.1109/N-SSC.2006.4785860
2. J.D. Kendall, S. Kumar, The building blocks of a brain-inspired computer. Appl. Phys. Rev. **7**(1). https://doi.org/10.1063/1.5129306
3. https://iea.blob.core.windows.net/assets/6b2fd954-2017-408e-bf08-952fdd62118a/Electricity2024-Analysisandforecastto2026.pdf
4. G. Indiveri, S.C. Liu, Memory and information processing in neuromorphic systems. Proc. IEEE **103**(8), 1379–1397 (2015). https://doi.org/10.1109/JPROC.2015.2444094
5. J. Von Neumann, First draft of a report on the EDVAC. IEEE Ann. Hist. Comput. **15**(4), 27–75 (1993). https://doi.org/10.1109/85.238389
6. L. Chua, Memristor-the missing circuit element. IEEE Trans. Circuit Theory **18**(5), 507–519 (1971). https://doi.org/10.1109/TCT.1971.1083337
7. The memristor revisited. Nat. Electron **1**, 261 (2018). https://doi.org/10.1038/s41928-018-0083-3
8. D.B. Strukov, G.S. Snider, D.R. Stewart, R.S. Williams, The missing memristor found. Nature. **453**(7191), 80–83 (2008). https://doi.org/10.1038/nature06932
9. S.H. Jo, T. Chang, I. Ebong, B.B. Bhadviya, P. Mazumder, W. Lu, Nanoscale memristor device as synapse in neuromorphic systems. Nano letters. **10**(4), 1297–1301 (2010). https://doi.org/10.1021/nl904092h
10. I. Boybat, M. Le Gallo, S.R. Nandakumar, T. Moraitis, T. Parnell, T. Tuma, E. Eleftheriou, Neuromorphic computing with multi-memristive synapses. Nat. Commun. **9**(1), 2514 (2018). https://doi.org/10.1038/s41467-018-04933-y
11. S. Kumar, X. Wang, J.P. Strachan, Y. Yang, W.D. Lu, Dynamical memristors for higher-complexity neuromorphic computing. Nat. Rev. Mater. **7**(7), 575–591 (2022). https://doi.org/10.1038/s41578-022-00434-z
12. Y. Zhang, M. Qiu, X. Zhang, G. Zheng, K. Dai, C. Liu, C. Shen, Skin-inspired high-performance E-skin with interlocked microridges for intelligent perception. Adv. Funct. Mater. **35**(1), 2412065 (2025). https://doi.org/10.1002/adfm.202412065
13. J.H. Yoon, Z. Wang, K.M. Kim, H. Wu, V. Ravichandran, Q. Xia, J.J. Yang, An artificial nociceptor based on a diffusive memristor. Nat. Commun. **9**(1), 417 (2018). https://doi.org/10.1038/s41467-017-02572-3
14. L. Xu, W. Wang, Y. Li, Y. Lin, W. Yue, K. Qian, G. Shen, N: ZnO/MoS_2-heterostructured flexible synaptic devices enabling optoelectronic co-modulation for robust artificial visual systems. Nano Res. **17**(3), 1902–1912 (2024). https://doi.org/10.1007/s12274-023-6004-4

15. N. Tulina, I.Y. Borisenko, V.V. Sirotkin, Reproducible resistive switching effect for memory applications in heterocontacts based on strongly correlated electron systems. Phys. Lett. A **372**(44), 6681–6686 (2008). https://doi.org/10.1016/j.physleta.2008.09.015
16. D. Liu, H. Cheng, R. Peng, Y. Yin, Two resistive switching behaviors in Ag/SiO_2/Pt memristors. In: 2016 IEEE 16th International Conference on Nanotechnology (IEEE-NANO) (pp. 651–654) (2016). IEEE. https://doi.org/10.1109/NANO.2016.7751368
17. Y. Li, P. Yuan, L. Fu, R. Li, X. Gao, C. Tao, Coexistence of diode-like volatile and multilevel nonvolatile resistive switching in a ZrO_2/TiO_2 stack structure. Nanotechnology. **26**(39), 391001 (2015). https://doi.org/10.1088/0957-4484/26/39/391001
18. A. Prakash, D. Deleruyelle, J. Song, M. Bocquet, H. Hwang, Resistance controllability and variability improvement in a TaOx-based resistive memory for multilevel storage application. Appl. Phys. Lett. **106**(23) (2015). https://doi.org/10.1063/1.4922446
19. X. Lian, M. Wang, M. Rao, P. Yan, J.J. Yang, F. Miao, Characteristics and transport mechanisms of triple switching regimes of TaOx memristor. Appl. Phys. Lett. **110**(17) (2017). https://doi.org/10.1063/1.4982593
20. K.H. Xue, B. Traore, P. Blaise, L.R. Fonseca, E. Vianello, G. Molas, Y. Nishi, A combined Ab initio and experimental study on the nature of conductive filaments in Pt/HfO_2/Pt resistive random access memory. IEEE Trans. Electron Devices **61**(5), 1394–1402 (2014). https://doi.org/10.1109/TED.2014.2312943
21. H. Nili, T. Ahmed, S. Walia, R. Ramanathan, A.E. Kandjani, S. Rubanov, S. Sriram, Microstructure and dynamics of vacancy-induced nanofilamentary switching network in donor doped $SrTiO_{3-x}$ memristors. Nanotechnology **27**(50), 505210 (2016). https://doi.org/10.1088/0957-4484/27/50/505210
22. B.J. Choi, J.J. Yang, M.X. Zhang, K.J. Norris, D.A. Ohlberg, N.P. Kobayashi, R.S. Williams, Nitride memristors. Appl. Phys. A **109**(1), 1–4 (2012). https://doi.org/10.1007/s00339-012-7052-x
23. B.J. Choi, A.C. Torrezan, J.P. Strachan, P.G. Kotula, A.J. Lohn, M.J. Marinella, J.J. Yang, High-speed and low-energy nitride memristors. Adv. Func. Mater. **26**(29), 5290–5296 (2016). https://doi.org/10.1002/adfm.201600680
24. B. Tang, H. Veluri, Y. Li, Z.G. Yu, M. Waqar, J.F. Leong, A.V. Thean, Wafer-scale solution-processed 2D material analog resistive memory array for memory-based computing. Nat. Commun. **13**(1), 3037 (2022). https://doi.org/10.1038/s41467-022-30519-w
25. B. Walters, M.V. Jacob, A. Amirsoleimani, M. Rahimi Azghadi, A review of graphene-based memristive neuromorphic devices and circuits. Adv. Intell. Syst. **5**(10), 2300136 (2023). https://doi.org/10.1002/aisy.202300136
26. Y. Qi, B. Sun, G. Fu, T. Li, S. Zhu, L. Zheng, Y. Chen, A nonvolatile organic resistive switching memory based on lotus leaves. Chem. Phys. **516**, 168–174 (2019). https://doi.org/10.1016/j.chemphys.2018.09.008
27. T. Fu, X. Liu, H. Gao, J.E. Ward, X. Liu, B. Yin, J. Yao, Bioinspired bio-voltage memristors. Nat. Commun. **11**(1), 1861 (2020). https://doi.org/10.1038/s41467-020-15759-y
28. M.H. Kim, H.L. Park, M.H. Kim, J. Jang, J.H. Bae, I.M. Kang, S.H. Lee, Fluoropolymer-based organic memristor with multifunctionality for flexible neural network system. NPJ Flex. Electron. **5**(1), 34 (2021). https://doi.org/10.1038/s41528-021-00132-w
29. H. Teixeira, C. Dias, A.V. Silva, J. Ventura, Advances on MXene-based memristors for neuromorphic computing: a review on synthesis, mechanisms, and future directions. ACS Nano **18**(33), 21685–21713 (2024). https://doi.org/10.1021/acsnano.4c03264
30. L. Lv, P. Zhang, X. Yang, H. Wan, G. Ma, H. Wang, Novel 2D MXene-based materials in memristors: fundamentals, resistive switching properties and applications. Surf. Interfaces **44**, 103678 (2024). https://doi.org/10.1016/j.surfin.2023.103678
31. K. Kang, W. Hu, X. Tang, Halide perovskites for resistive switching memory. J. Phys. Chem. Lett. **12**(48), 11673–11682 (2021). https://doi.org/10.1021/acs.jpclett.1c03408
32. S. Satapathi, K. Raj, Yukta, M.A. Afroz, Halide-perovskite-based memristor devices and their application in neuromorphic computing. Phys. Rev. Appl. **18**(1), 017001. https://doi.org/10.1103/PhysRevApplied.18.017001

33. R.A. John, N. Shah, S.K. Vishwanath, S.E. Ng, B. Febriansyah, M. Jagadeeswararao, N. Mathews, Halide perovskite memristors as flexible and reconfigurable physical unclonable functions. Nat. Commun. **12**(1), 3681 (2021). https://doi.org/10.1038/s41467-021-24057-0
34. D. Das, D.S. Assi, S. Kazim, V.A. Roy, S. Ahmad, Decoding halide perovskite for neuromorphic and memristive devices. Mater. Horiz. (2025). https://doi.org/10.1039/D5MH00534E
35. R.A. John, Y. Demirağ, Y. Shynkarenko, Y. Berezovska, N. Ohannessian, M. Payvand, M.V. Kovalenko, Reconfigurable halide perovskite nanocrystal memristors for neuromorphic computing. Nat. Commun. **13**(1), 2074 (2022). https://doi.org/10.1038/s41467-022-29727-1
36. D. Zhao, Y. Zhu, W. Cheng, W. Chen, Y. Wu, H. Yu, Cellulose-based flexible functional materials for emerging intelligent electronics. Adv. Mater. **33**(28), 2000619 (2021). https://doi.org/10.1002/adma.202000619
37. F. Zeng, Y. Tan, B. Sun, W. Hu, H. Yin, X. Tang, M. Tang, Stable bipolar resistive switching in lead-free $Cs_2AgBiBr_6$ memristors for neuromorphic computing. ACS Omega (2025). https://doi.org/10.1021/acsomega.5c04702
38. Y. Wang, Y. Yuan, W. Zhang, F. Qi, N. Zhang, T. Liu, X. Tang, Flexible resistive switching device based on air-stable lead-free Cu_3SbI_6 perovskite film for nonvolatile memory application. Appl. Phys. Lett. **120**(26) (2022). https://doi.org/10.1063/5.0093943
39. B. Saparov, F. Hong, J.P. Sun, H.S. Duan, W. Meng, S. Cameron, D.B. Mitzi, Thin-film preparation and characterization of $Cs_3Sb_2I_9$: a lead-free layered perovskite semiconductor. Chem. Mater. **27**(16), 5622–5632 (2015). https://doi.org/10.1021/acs.chemmater.5b01989
40. S. Paramanik, A. Maiti, S. Chatterjee, A.J. Pal, Large resistive switching and artificial synaptic behaviors in layered Cs3Sb2I9 lead-free perovskite memory devices. Adv. Electron. Mater. **8**(1), 2100237 (2022). https://doi.org/10.1002/aelm.202100237
41. S. Zhai, J. Gong, Y. Feng, Z. Que, W. Mao, X. He, L. Chu, Multilevel resistive switching in stable all-inorganic nip double perovskite memristor. iscience. **26**(4) (2023). https://doi.org/10.1016/j.isci.2023.106461
42. B. Wu, W. Ning, Q. Xu, M. Manjappa, M. Feng, S. Ye, T.C. Sum, Strong self-trapping by deformation potential limits photovoltaic performance in bismuth double perovskite. Sci. Adv. **7**(8), eabd3160 (2021). https://doi.org/10.1126/sciadv.abd3160
43. H. Lei, D. Hardy, F. Gao, Lead-free double perovskite $Cs_2AgBiBr_6$: fundamentals, applications, and perspectives. Adv. Func. Mater. **31**(49), 2105898 (2021). https://doi.org/10.1002/adfm.202105898
44. F. Zeng, Y. Guo, W. Hu, Y. Tan, X. Zhang, J. Feng, X. Tang, Opportunity of the lead-free all-inorganic $Cs_3Cu_2I_5$ perovskite film for memristor and neuromorphic computing applications. ACS Appl. Mater. Interfaces **12**(20), 23094–23101 (2020). https://doi.org/10.1021/acsami.0c03106
45. X.F. Liang, Y. Chen, L. Chen, J. Yin, Z.G. Liu, Electric switching and memory devices made from RbAg4I5 films. Appl. Phys. Lett. **90**(2) (2007). https://doi.org/10.1063/1.2431438
46. Y. Zhang, X. Mao, X. Chang, J. Xie, Y. Niu, S. Gong, M. Qian, Resistive switching and artificial synapses performance of co-evaporated $Cs_3Cu_2I_5$ films. App. Phys. Lett. **125**(22) (2024). https://doi.org/10.1063/5.0235918
47. L. Lanzetta, T. Webb, N. Zibouche, X. Liang, D. Ding, G. Min, S.A. Haque, Degradation mechanism of hybrid tin-based perovskite solar cells and the critical role of tin (IV) iodide. Nat. Commun. **12**(1), 2853 (2021). https://doi.org/10.1038/s41467-021-22864-z
48. R. Yao, J. Lin, K. Liu, Y. Xu, B. Xiao, J. Zhao, W. Yuan, Structure and optical properties of SN-based halide perovskites ($C_{10}H_{18}N_2$) SnX_4 (X= Cl, Br, I). ACS Omega **9**(20), 22352–22359 (2024). https://doi.org/10.1021/acsomega.4c01835
49. I.H. Im, S.J. Kim, H.W. Jang, Memristive devices for new computing paradigms. Adv. Intell. Syst. **2**(11), 2000105 (2020). https://doi.org/10.1002/aisy.202000105
50. J. Yao, L. Zhong, D. Natelson, J.M. Tour, In situ imaging of the conducting filament in a silicon oxide resistive switch. Sci. Rep. **2**(1), 242 (2012). https://doi.org/10.1038/srep00242

51. C. Rohde, B.J. Choi, D.S. Jeong, S. Choi, J.S. Zhao, C.S. Hwang, Identification of a determining parameter for resistive switching of TiO_2 thin films. App. Phys. Lett. **86**(26) (2005). https://doi.org/10.1063/1.1968416
52. C.Y. Lin, C.Y. Wu, C.Y. Wu, T.C. Lee, F.L. Yang, C. Hu, T.Y. Tseng, Effect of top electrode material on resistive switching properties of Zro_2 film memory devices. IEEE Electron Device Lett. **28**(5), 366–368 (2007). https://doi.org/10.1109/LED.2007.894652
53. Y.M. Kim, J.S. Lee, Reproducible resistance switching characteristics of hafnium oxide-based nonvolatile memory devices. J. Appl. Phys. **104**(11) (2008). https://doi.org/10.1063/1.3041475
54. S. Seo, M.J. Lee, D.H. Seo, E.J. Jeoung, D.S. Suh, Y.S. Joung, B.H. Park, Reproducible resistance switching in polycrystalline NiO films. Appl. Phys. Lett. **85**(23), 5655–5657 (2004). https://doi.org/10.1063/1.1831560
55. C. Yoshida, K. Tsunoda, H. Noshiro, Y. Sugiyama, High speed resistive switching in Pt/TiO_2/TiN film for nonvolatile memory application. App. Phys. Lett. **91**(22) (2007). https://doi.org/10.1063/1.2818691
56. C. Ye, T.F. Deng, J. Wu, C. Zhan, H. Wang, J. Zhang, Role of ITO electrode in the resistive switching behavior of TiN/HfO_2/ITO memory devices at different annealing temperatures. Jpn. J. Appl. Phys. **54**(5), 054201 (2015). https://doi.org/10.7567/JJAP.54.054201
57. L. Goux, Y.Y. Chen, L. Pantisano, X.P. Wang, G. Groeseneken, M. Jurczak, D.J. Wouters, On the gradual unipolar and bipolar resistive switching of TiN\HfO_2\Pt memory systems. Electrochem. Solid-State Lett. **13**(6), G54 (2010). https://doi.org/10.1149/1.3373529
58. N. Xu, L.F. Liu, X. Sun, C. Chen, Y. Wang, D.D. Han, B. Yu, Bipolar switching behavior in TiN/ZnO/Pt resistive nonvolatile memory with fast switching and long retention. Semicond. Sci. Technol. **23**(7), 075019 (2008). https://doi.org/10.1088/0268-1242/23/7/075019
59. X.C. Yuan, J.L. Tang, H.Z. Zeng, X.H. Wei, Abnormal coexistence of unipolar, bipolar, and threshold resistive switching in an Al/NiO/ITO structure. Nanoscale Res. Lett. **9**(1), 268 (2014). https://doi.org/10.1186/1556-276X-9-268
60. D.S. Jeong, H. Schroeder, R. Waser, Coexistence of bipolar and unipolar resistive switching behaviors in a Pt/TiO_2/Pt stack. Electrochem. Solid-State Lett. **10**(8), G51 (2007). https://doi.org/10.1149/1.2742989
61. H. Schroeder, D.S. Jeong, Resistive switching in a Pt/TiO_2/Pt thin film stack–a candidate for a non-volatile ReRAM. Microelectron. Eng. **84**(9–10), 1982–1985 (2007). https://doi.org/10.1016/j.mee.2007.04.042
62. T. Liu, M. Verma, Y. Kang, M.K. Orlowski, Coexistence of bipolar and unipolar switching in Cu/TaOx/Pt resistive devices for Cu and oxygen vacancy nanofilaments. ECS Trans. **45**(3), 279 (2012). https://doi.org/10.1149/1.3700893
63. L. Liu, S. Zhang, Y. Luo, G. Yuan, J. Liu, J. Yin, Z. Liu, Coexistence of unipolar and bipolar resistive switching in $BiFeO_3$ and Bi0. 8Ca0. $2FeO_3$ films. J. Appl. Phys. **111**(10) (2012). https://doi.org/10.1063/1.4716867
64. B.J. Choi, A.C. Torrezan, K.J. Norris, F. Miao, J.P. Strachan, M.X. Zhang, R.S. Williams, Electrical performance and scalability of Pt dispersed SiO_2 nanometallic resistance switch. Nano Lett. **13**(7), 3213–3217 (2013). https://doi.org/10.1021/nl401283q
65. W. Wang, G.N. Panin, X. Fu, L. Zhang, P. Ilanchezhiyan, V.O. Pelenovich, T.W. Kang, MoS_2 memristor with photoresistive switching. Sci. Rep. **6**(1), 31224 (2016). https://doi.org/10.1038/srep31224
66. I. Valov, R. Waser, J.R. Jameson, M.N. Kozicki, Electrochemical metallization memories—Fundamentals, applications, prospects. Nanotechnology **22**(25), 254003 (2011). https://doi.org/10.1088/0957-4484/22/25/254003
67. Q. Liu, J. Sun, H. Lv, S. Long, K. Yin, N. Wan, M. Liu, Real-time observation on dynamic growth/dissolution of conductive filaments in oxide-electrolyte-based ReRAM. Adv. Mater. **24**(14), 1844 (2012). https://doi.org/10.1002/adma.201104104
68. J. Yoon, H. Choi, D. Lee, J.B. Park, J. Lee, D.J. Seong, H. Hwang, Excellent switching uniformity of Cu-doped MoO_x GdO_x bilayer for nonvolatile memory applications. IEEE Electron Device Lett. **30**(5), 457–459 (2009). https://doi.org/10.1109/LED.2009.2015687

69. A. Belmonte, W. Kim, B.T. Chan, N. Heylen, A. Fantini, M. Houssa, L. Goux, A thermally stable and high-performance 90-nm $Al_2 O_3$\Cu-based 1T1R CBRAM Cell. IEEE Trans. Electron Devices **60**(11), 3690–3695 (2013). https://doi.org/10.1109/TED.2013.2282000
70. C. Gopalan, Y. Ma, T. Gallo, J. Wang, E. Runnion, J. Saenz, S. Hollmer, Demonstration of conductive bridging random access memory (CBRAM) in logic CMOS process. Solid-State Electron. **58**(1), 54–61 (2011).https://doi.org/10.1016/j.sse.2010.11.024
71. S. Munjal, N. Khare, Valence change bipolar resistive switching accompanied with magnetization switching in $CoFe_2O_4$ thin film. Sci. Rep. **7**(1), 12427 (2017). https://doi.org/10.1038/s41598-017-12579-x
72. T. Heisig, C. Baeumer, U.N. Gries, M.P. Mueller, C. La Torre, M. Luebben, R. Dittmann, Oxygen exchange processes between oxide memristive devices and water molecules. Adv. Mater. **30**(29), 1800957 (2018). https://doi.org/10.1002/adma.201800957
73. G.D. Wilk, R.M. Wallace, J. Anthony, High-κ gate dielectrics: current status and materials properties considerations. J. Appl. Phys. **89**(10), 5243–5275 (2001). https://doi.org/10.1063/1.1361065
74. X. Zhang, L. Xu, H. Zhang, J. Liu, D. Tan, L. Chen, W. Li, Effect of joule heating on resistive switching characteristic in AlOx cells made by thermal oxidation formation. Nanoscale Res. Lett. **15**(1), 11 (2020). https://doi.org/10.1186/s11671-019-3229-y
75. P. Sun, L. Li, N. Lu, Y. Li, M. Wang, H. Xie, M. Liu, Physical model of dynamic Joule heating effect for reset process in conductive-bridge random access memory. J. Comput. Electron. **13**(2), 432–438 (2014). https://doi.org/10.1007/s10825-013-0552-x
76. J.Y. Chen, C.L. Hsin, C.W. Huang, C.H. Chiu, Y.T. Huang, S.J. Lin, L.J. Chen, Dynamic evolution of conducting nanofilament in resistive switching memories. Nano Lett. **13**(8), 3671–3677 (2013). https://doi.org/10.1021/nl4015638
77. K. Nagashima, T. Yanagida, M. Kanai, U. Celano, S. Rahong, G. Meng, T. Kawai, Carrier type dependence on spatial asymmetry of unipolar resistive switching of metal oxides. Appl. Phys. Lett. **103**(17) (2013). https://doi.org/10.1063/1.4826558
78. K. Jung, H. Seo, Y. Kim, H. Im, J. Hong, J.W. Park, J.K. Lee, Temperature dependence of high-and low-resistance bistable states in polycrystalline NiO films. App. Phys. Lett. **90**(5) (2007). https://doi.org/10.1063/1.2437668
79. Z. Yang, D. Zhang, J. Cai, C. Gong, Q. He, M. Xu, X. Miao, Joule heating induced non-melting phase transition and multi-level conductance in $MoTe_2$ based phase change memory. App. Phys. Lett. **121**(20) (2022). https://doi.org/10.1063/5.0127160
80. F. Marrone, J. Secco, B. Kersting, M. Le Gallo, F. Corinto, A. Sebastian, L.O. Chua, Experimental validation of state equations and dynamic route maps for phase change memristive devices. Sci. Rep. **12**(1), 6488 (2022). https://doi.org/10.1038/s41598-022-09948-6
81. D. Kau, S. Tang, I.V. Karpov, R. Dodge, B. Klehn, J.A. Kalb, G. Spadini, A stackable cross point phase change memory. In: 2009 IEEE International Electron Devices Meeting (IEDM), (pp. 1–4) (2009). IEEE. https://doi.org/10.1109/IEDM.2009.5424263
82. N. Yamada, E. Ohno, K. Nishiuchi, N. Akahira, M. Takao, Rapid-phase transitions of GeTe-Sb_2Te_3 pseudobinary amorphous thin films for an optical disk memory. J. Appl. Phys. **69**(5), 2849–2856 (1991). https://doi.org/10.1063/1.348620
83. J. Tominaga, T. Kikukawa, M. Takahashi, R.T. Phillips, Structure of the optical phase change memory alloy, Ag–V–In–Sb–Te, determined by optical spectroscopy and electron diffraction. J. Appl. Phys. **82**(7), 3214–3218 (1997). https://doi.org/10.1063/1.365627
84. J.J. hang, H.J. Sun, Y. Li, Q. Wang, X.H. Xu, X.S. Miao, AgInSbTe memristor with gradual resistance tuning. App. Phys. Lett. **102**(18) (2013). https://doi.org/10.1063/1.4804983
85. M. Suri, O. Bichler, D. Querlioz, B. Traoré, O. Cueto, L. Perniola, B. DeSalvo, Physical aspects of low power synapses based on phase change memory devices. J. Appl. Phys. **112**(5) (2012). https://doi.org/10.1063/1.4749411
86. C. Cuhadar, S.G. Kim, J.M. Yang, J.Y. Seo, D. Lee, N.G. Park, All-inorganic bismuth halide perovskite-like materials $A_3Bi_2I_9$ and A_3Bi1. 8Na0. 2I8. 6 (A= Rb and Cs) for low-voltage switching resistive memory. ACS Appl. Mater. & Interfaces **10**(35), 29741–29749 (2018). https://doi.org/10.1021/acsami.8b07103

87. S. Ge, X. Guan, Y. Wang, C.H. Lin, Y. Cui, Y. Huang, T. Wu, Low-dimensional lead-free inorganic perovskites for resistive switching with ultralow bias. Adv. Func. Mater. **30**(25), 2002110 (2020). https://doi.org/10.1002/adfm.202002110
88. Y. Zhang, S. Poddar, H. Huang, L. Gu, Q. Zhang, Y. Zhou, Z. Fan, Three-dimensional perovskite nanowire array–based ultrafast resistive RAM with ultralong data retention. Sci. Adv. **7**(36), eabg3788 (2021). https://doi.org/10.1126/sciadv.abg3788
89. M.J. Lee, C.B. Lee, D. Lee, S.R. Lee, M. Chang, J.H. Hur, K. Kim, A fast, high-endurance and scalable non-volatile memory device made from asymmetric Ta_2O_5-x/TaO_2-x bilayer structures. Nat. Mater. **10**(8), 625–630 (2011). https://doi.org/10.1038/nmat3070
90. H. Guan, J. Sha, Z. Zhang, Y. Xiong, X. Dong, H. Bao, Y. Wang, Optical and oxide modification of CsFamapbibr memristor achieving low power consumption. J. Alloy. Compd. **891**, 162096 (2022). https://doi.org/10.1016/j.jallcom.2021.162096
91. H. Ye, B. Sun, Z. Wang, Z. Liu, X. Zhang, X. Tan, G. Liao, High performance flexible memristors based on a lead free AgBiI 4 perovskite with an ultralow operating voltage. J. Mater. Chem. C **8**(40), 14155–14163 (2020). https://doi.org/10.1039/D0TC03287E
92. K. Wang, J. Chen, X. Yan, MXene Ti_3C_2 memristor for neuromorphic behavior and decimal arithmetic operation applications. Nano Energy **79**, 105453 (2021). https://doi.org/10.1016/j.nanoen.2020.105453
93. J. Wang, J. Wang, J. Zhang, W. Huang, X. Wang, M. Zhang, Ultralow-power synaptic transistors based on Ta_2O_5/Al_2O_3 bilayer dielectric for algebraic arithmetic. Adv. Elect. Mat. **8**(5), 2100922 (2022). https://doi.org/10.1002/aelm.202100922
94. Y. Park, S.H. Kim, D. Lee, J.S. Lee, Designing zero-dimensional dimer-type all-inorganic perovskites for ultra-fast switching memory. Nat. Commun. **12**(1), 3527 (2021). https://doi.org/10.1038/s41467-021-23871-w
95. S. Paramanik, A.J. Pal, Dimensionality-dependent resistive switching in 0D and 2D Cs3Sb2I9: energy-efficient synaptic functions with the layered-phase. Adv. Elect. Mat. **8**(10), 2200211 (2022). https://doi.org/10.1002/aelm.202200211
96. S.Y. Kim, H. Zhang, J. Rubio-Magnieto, Operating mechanism principles and advancements for halide perovskite-based memristors and neuromorphic devices. J. Phys. Chem. Lett. **15**(40), 10087–10103 (2024). https://doi.org/10.1021/acs.jpclett.4c02170
97. G.S.H. Thien, M. Ab Rahman, B.K. Yap, N.M.L. Tan, Z. He, P.L. Low, K.Y. Chan, Recent advances in halide perovskite resistive switching memory devices: a transformation from lead-based to lead-free perovskites. ACS Omega **7**(44), 39472–39481 (2022). https://doi.org/10.1021/acsomega.2c03206
98. C. Lyu, C. Liu, H. Min, X. Shi, R. Jiang, Z. Ao, L. Wang, Rare earth Nd-doping lead-free double perovskite Cs_2AgBiBr 6 films with improved resistive memory performance. J. Alloy. Compd. **913**, 165300 (2022). https://doi.org/10.1016/j.jallcom.2022.165300
99. H. Kim, M.J. Choi, J.M. Suh, J.S. Han, S.G. Kim, Q.V. Le, H.W. Jang, Quasi-2D halide perovskites for resistive switching devices with ON/OFF ratios above 109. NPG Asia Mater. **12**(1), 21 (2020). https://doi.org/10.1038/s41427-020-0202-2
100. U. Jung, D.S. Woo, S. Kim, Z. Tan, J. Park, Anisotropic resistive switching of 2D-layered single crystal halide perovskite CsPb 2 Br 5-based memristor. Nano Res. **18**(1) (2025). https://doi.org/10.26599/NR.2025.94907023
101. J.S. Han, Q.V. Le, J. Choi, H. Kim, S.G. Kim, K. Hong, H.W. Jang, Lead-free all-inorganic cesium tin iodide perovskite for filamentary and interface-type resistive switching toward environment-friendly and temperature-tolerant nonvolatile memories. ACS Appl. Mater. Interfaces **11**(8), 8155–8163 (2019). https://doi.org/10.1021/acsami.8b15769
102. T.K. Su, W.K. Cheng, C.Y. Chen, W.C. Wang, Y.T. Chuang, G.H. Tan, H.W. Lin, Room-temperature fabricated multilevel nonvolatile lead-free cesium halide memristors for reconfigurable in-memory computing. ACS Nano **16**(8), 12979–12990 (2022). https://doi.org/10.1021/acsnano.2c05436
103. M.H. Chen, P.F. Lin, B.Y. Chen, J.F. Cheng, Reliable BiI_3-based resistive random-access memory devices with a high on/off ratio. ACS Appl. Electron. Mater. **5**(1), 255–264 (2022). https://doi.org/10.1021/acsaelm.2c01316

104. C.S. Li, Y.T. Wu, P.F. Lin, B.Y. Chen, F.Y. Fu, J.F. Cheng, M.H. Chen, Enhance the properties of BiI3-based resistive switching devices via mixing Ag and Au electrodes. Adv. Mater. Interfaces **10**(8), 2202188 (2023). https://doi.org/10.1002/admi.202202188
105. H. Wang, H. Wei, X. Zeng, J. Zhou, Y. Zhu, Conductive path mechanism for bipolar resistive switching characteristics in lead-free perovskite CsSnBr3-based nonvolatile memories. Phys. Status Solidi (A). **218**(21), 2100501 (2021). https://doi.org/10.1002/pssa.202100501
106. Q. Cui, X. Liu, N. Li, H. Zeng, D. Chu, H. Li, S. Liu, A new metal-free molecular perovskite-related single crystal with quantum wire structure for high-performance x-ray detection. Small **20**(13), 2308945 (2024). https://doi.org/10.1002/smll.202308945
107. S. Poddar, Y. Zhang, Y. Zhu, Q. Zhang, Z. Fan, Optically tunable ultra-fast resistive switching in lead-free methyl-ammonium bismuth iodide perovskite films. Nanoscale **13**(12), 6184–6191 (2021). https://doi.org/10.1039/D0NR09234G
108. J. Li, Y. Zhang, C. Yao, N. Qin, R. Chen, D. Bao, Optoelectronic modulation of interfacial defects in lead-free perovskite films for resistive switching. Adv. Elect. Mat. **8**(4), 2101094 (2022). https://doi.org/10.1002/aelm.202101094
109. Z. Liu, P. Cheng, R. Kang, J. Zhou, X. Wang, X. Zhao, Z. Zuo, Inorganic lead-free and bismuth-based perovskite nanoscale-thick films for memristors and artificial synapse applications. ACS App. Nano Mat. **6**(22), 21000–21015 (2023).https://doi.org/10.1021/acsanm.3c04006
110. B.W. Zhang, C.H. Lin, S. Nirantar, E.Q. Han, Y. Zhang, Z. Wang, L. Wang, Lead-free perovskites and metal halides for resistive switching memory and artificial synapse. Small Struct. **5**(6), 2300524 (2024). https://doi.org/10.1002/sstr.202300524
111. C. Sun, F. Luo, L. Ruan, J. Tong, L. Yan, Y. Zheng, X. Zhang, Enhanced memristive performance of double perovskite $Cs_2AgBiBr$ 6-xClx devices by chloride doping. Chem. Plus Chem. **86**(11), 1530–1536 (2021). https://doi.org/10.1002/cplu.202100404
112. T. Tong, C. Liu, J. Xu, H. Min, S. Chen, Y. Lyu, C. Lyu, Rare earth halide double perovskites for high-performance resistive random access memory. J. Mat. Chem. C **11**(14), 4946–4952 (2023). https://doi.org/10.1039/D3TC00350G
113. Y. Yuan, Y. Wang, X. Tang, N. Zhang, W. Zhang, Enhanced resistive switching performance through air-stable Cu_2AgSbI_6 thin films for flexible and multilevel storage application. ACS Appl. Mater. Interfaces **14**(48), 53990–53998 (2022). https://doi.org/10.1021/acsami.2c15332
114. H.R. Fischer, K. Tempelaars, A. Kerpershoek, T. Dingemans, M. Iqbal, H.V. Lonkhuyzen, C. Semprimoschnig, Development of flexible LEO-resistant PI films for space applications using a self-healing mechanism by surface-directed phase separation of block copolymers. ACS Appl. Mater. Interfaces **2**(8), 2218–2225 (2010). https://doi.org/10.1021/am100223v
115. L. Lei, Y. Tan, X. Yuan, W. Dou, J. Zhang, Y. Wang, D. Tang, Flexible electric-double-layer thin film transistors based on a vertical InGaZnO 4 channel. RSC Adv. **11**(29), 17910–17913 (2021). https://doi.org/10.1039/D1RA02155A
116. J. Chen, J. Zheng, Q. Gao, J. Zhang, J. Zhang, O.M. Omisore, H. Li, Polydimethylsiloxane (PDMS)-based flexible resistive strain sensors for wearable applications. Appl. Sci. **8**(3), 345 (2018). https://doi.org/10.3390/app8030345
117. A. Dindar, J.B. Kim, C. Fuentes-Hernandez, B. Kippelen, Metal-oxide complementary inverters with a vertical geometry fabricated on flexible substrates. App. Phys. Lett. **99**(17). https://doi.org/10.1063/1.3656974
118. M. Asad, M.H. Sheikhi, M. Pourfath, M. Moradi, High sensitive and selective flexible H2S gas sensors based on Cu nanoparticle decorated SWCNTs. Sens. Actuators, B Chem. **210**, 1–8 (2015). https://doi.org/10.1016/j.snb.2014.12.086
119. R. Kumar, S. Ranwa, G. Kumar, Biodegradable flexible substrate based on chitosan/PVP blend polymer for disposable electronics device applications. J. Phys. Chem. B **124**(1), 149–155 (2019). https://doi.org/10.1021/acs.jpcb.9b08897
120. S. Imani, S.M. Seyed-Talebi, J. Beheshtian, E.W.G. Diau, Simulation and characterization of $CH_3NH_3SnI_3$-based perovskite solar cells with different Cu-based hole transporting layers. Appl. Phys. A **129**(2), 143 (2023). https://doi.org/10.1007/s00339-023-06428-0

121. Z. Lai, Y. Meng, F. Wang, X. Bu, W. Wang, P. Xie, J.C. Ho, Direct drop-casting synthesis of all-inorganic lead and lead-free halide perovskite microcrystals for high-performance photodetectors. Nano Res. **15**(4), 3621–3627 (2022). https://doi.org/10.1007/s12274-021-3907-9
122. C. Zuo, A.D. Scully, W.L. Tan, F. Zheng, K.P. Ghiggino, D. Vak, M. Gao, Crystallisation control of drop-cast quasi-2D/3D perovskite layers for efficient solar cells. Commun. Mater. **1**(1), 33 (2020). https://doi.org/10.1038/s43246-020-0036-z
123. M. Wang, W. Wang, B. Ma, W. Shen, L. Liu, K. Cao, W. Huang, Lead-free perovskite materials for solar cells. Nano-Micro Lett. **13**(1), 62 (2021). https://doi.org/10.1007/s40820-020-005 78-z
124. Y. Yu, D. Zhao, C.R. Grice, W. Meng, C. Wang, W. Liao, Y. Yan, Thermally evaporated methylammonium tin triiodide thin films for lead-free perovskite solar cell fabrication. RSC Adv. **6**(93), 90248–90254 (2016). https://doi.org/10.1039/C6RA19476A
125. F. Ali, N.D. Pham, L. Fan, V. Tiong, K. Ostrikov, J.M. Bell, T. Tesfamichael, Low hysteresis perovskite solar cells using an electron-beam evaporated WO3–x thin film as the electron transport layer. ACS Appl. Energy Mater. **2**(8), 5456–5464 (2019). https://doi.org/10.1021/acsaem.9b00547
126. B. Jin, D. Zhao, F. Liang, L. Liu, D. Liu, P. Wang, M. Qiu, Electron-beam irradiation induced regulation of surface defects in lead halide perovskite thin films. Research (2021). https://doi.org/10.34133/2021/9797058
127. S. Hosseini, A. Gholizadeh, Optimizing physical properties of BiFe1-xMnxO_3 perovskite thin films via a low-temperature sol-gel method. Ceram. Int. **49**(24), 40258–40267 (2023). https://doi.org/10.1016/j.ceramint.2023.09.361
128. J.M.C.D. Silva Filho, V.A. Ermakov, F.C. Marques, Perovskite thin film synthesised from sputtered lead sulphide. Sci. Rep. **8**(1), 1563 (2018). https://doi.org/10.1038/s41598-018-197 46-8
129. M.E. Kamminga, H.H. Fang, M.A. Loi, G.H. Ten Brink, G.R. Blake, T.T. Palstra, J.E. Ten Elshof, Micro patterned 2D hybrid perovskite thin films with enhanced photoluminescence lifetimes. ACS Appl. Mater. Interfaces **10**(15), 12878–12885 (2018). https://doi.org/10.1021/acsami.8b02236

Chapter 11
Advancement in Thermoelectric Performance in Lead Free Perovskites

Arvind Kumar, Komal Verma, Brijmohan Prajapati, and Manish Kumar

Abstract There is an increasing demand of renewable energy resources due to various factors that affect our environment significantly. Thermoelectric energy can be one such potential resource for us to circumvent the need of energy demand and to improve our lifestyles while preserving the environment. Perovskites materials are interesting in sense to show good thermoelectric performance. Most of the lead-based perovskites are the key elements to show reasonable thermoelectric performance. But due to their toxic nature and less reliability, researchers are now tending to look for lead-free materials. Lead free perovskites such as $CsSnI_3$, double perovskite (e.g. A_2GeSnF_6, Cs_2MGaBr_6 etc.), anti-perovskite (e.g. X_3BN) and its nanostructures are showing interest due to its potential application for thermoelectric materials [1–4]. The observed zT parameters for such perovskites are found to be in the range of 0.8–1.0 and some lead-free perovskites found to shows zT > 1 [1, 2]. The key point in having large zT, i.e., high conversion efficiency of waste heat into electrical energy is the basic requirement of the thermoelectric material. The high conversion efficiency of thermoelectric materials depends on several factors such as Seebeck coefficient, electrical conductivity and thermal conductivity etc. In this chapter, we will discuss in details about the different aspects, advancement, its basics and the future scope of lead-free perovskites materials for its potential applications in thermoelectric.

Keywords Perovskites materials · Structural properties · Electrical conductivity · Thermal conductivity · Thermoelectricity

---

A. Kumar (✉) · K. Verma
School of Physical Sciences, Jawaharlal Nehru University, New Delhi, India
e-mail: arvind@jnu.ac.in

K. Verma
e-mail: verma.komal.9044@gmail.com

B. Prajapati (✉)
Department of Physics, Deshbandhu College, University of Delhi, New Delhi, India
e-mail: brijmohan@db.du.ac.in

M. Kumar
Department of Physics, Atma Ram Sanatan Dharma College, University of Delhi, New Delhi, India
e-mail: mkumar2@arsd.du.ac.in

V. Chauhan et al. (eds.), *Sustainable Lead-Free Perovskite Materials*, Topics in Applied Physics 157, https://doi.org/10.1007/978-3-032-24873-2_11

11.1 Introduction

The global demand for sustainable and clean energy technologies have intensified researches into materials capable of harvesting energy from waste heat. Among these, thermoelectric (TE) materials have secured great attention due to their unique ability to directly convert thermal energy into electrical energy through the Seebeck effect and vice versa via the Peltier effect [1]. This bidirectional energy conversion potential has enabled thermoelectric materials to play a crucial role in niche applications such as spacecraft power sources, waste heat recovery systems, and microelectronic cooling. However, the widespread adoption of thermoelectric technologies remains hindered by the limited efficiency and environmental concerns associated with traditional TE materials.

Historically, high-performance thermoelectric materials have predominantly included lead (Pb)- and tellurium (Te)-based compounds which possess favorable electrical conductivity and low thermal conductivity [2]. Nonetheless, their high toxicity, scarcity, and environmental hazards have driven the scope for alternative materials that are both efficient and environmentally benign. In this context, lead-free perovskites have emerged as a promising class of materials for next-generation thermoelectric applications. Perovskite materials, with the general formula ABX_3, offer a versatile platform for material design due to their tunable electronic structure, structural flexibility, and compositional diversity [3]. While they have been extensively studied for their photovoltaic performance, recent research has highlighted their potential in thermoelectric applications, especially when modified to be lead-free. Lead-free perovskites not only mitigate toxicity concerns but also exhibit intrinsic properties favorable for thermoelectric performance, such as moderate bandgaps, low thermal conductivity, and high Seebeck coefficients [4]. One of the major challenges in developing efficient lead-free thermoelectric perovskites lies in optimizing the dimensionless figure of merit (zT), defined as;

$$zT = \frac{S^2 \sigma T}{\kappa} \tag{11.1}$$

where S is the Seebeck coefficient, σ is the electrical conductivity, T is the absolute temperature, and κ is the total thermal conductivity. Achieving a high zT requires a delicate balance of these interdependent parameters, often referred to as the "phonon-glass electron-crystal" paradigm [5]. While traditional strategies to enhance zT, such as doping, band structure engineering, and nano-structuring remain relevant, novel methods, to tailor the unique structure and chemistry of lead-free perovskites are increasingly being explored.

Recent advancements have shown that incorporating elements such as tin (Sn), bismuth (Bi), cesium (Cs) and antimony (Sb) in the perovskite lattice can significantly improve thermoelectric performance. For instance, Sn-based perovskites like

$CH_3NH_3SnI_3$ have demonstrated promising electrical properties and are considered as potential replacements for Pb-based analogues [6]. Furthermore, low-dimensional perovskite structures and hybrid organic-inorganic perovskites have exhibited reduced lattice thermal conductivity due to enhanced phonon scattering, making them suitable candidates for high zT applications [7]. Another important direction involves defect and interface engineering to suppress thermal conductivity while maintaining good carrier transport. The application of external fields, strain engineering, and the integration of nanostructures have also been shown to enhance carrier mobility and phonon scattering in lead-free perovskite systems [8]. Additionally, computational approaches, including density functional theory (DFT) and machine learning-assisted materials discovery, are being increasingly employed to screen and predict high-performance thermoelectric perovskite candidates. Despite these promising developments, the field of lead-free thermoelectric perovskites is still in its nascent stage. Many challenges remain, including long-term material stability, scalability of synthesis methods, and integration into practical devices. Continued interdisciplinary research combining materials science, chemistry, physics, and engineering is essential to overcome these barriers.

This chapter provides a comprehensive overview of the recent advancements in the thermoelectric performance of lead-free perovskites. We have explored the fundamental principles governing thermoelectric behavior, review notable materials and their performance metrics, discuss state-of-the-art strategies for performance enhancement, and highlight current challenges and prospects in the field.

11.2 Perovskite Materials and their Classification

Perovskites are a broad class of materials that share the same crystal structure as the mineral calcium titanate ($CaTiO_3$), which was first discovered in the Ural Mountains in 1839 and named after Russian mineralogist Lev Perovski. Due to their isotropic nature in three dimensions, they are also called as 3D perovskite. Remarkably one can transform these higher dimensions' structure to its lower counter part by adding the bulk cations. These materials have attracted substantial scientific interest due to their unique structural flexibility and remarkable range of functional properties, including ferroelectricity, piezoelectricity, superconductivity, magnetoresistance, and photovoltaic efficiency. They may exist in orthorhombic/tetragonal/cubic crystal structures. Perovskites are generally represented by the formula ABX_3, where:

- **A** is a large cation (usually a monovalent or divalent ion)
- **B** is a smaller metal cation (usually a transition metal),
- **X** is typically an anion such as oxygen or a halide (O^{2-}, Cl^-, Br^-, I^-).

In the ideal cubic perovskite structure, the A cation is located at the corners of the unit cell, the B cation at the body center, and the X anions at the face centers. The stability of the perovskite structure is often described using the Goldschmidt tolerance factor (t):

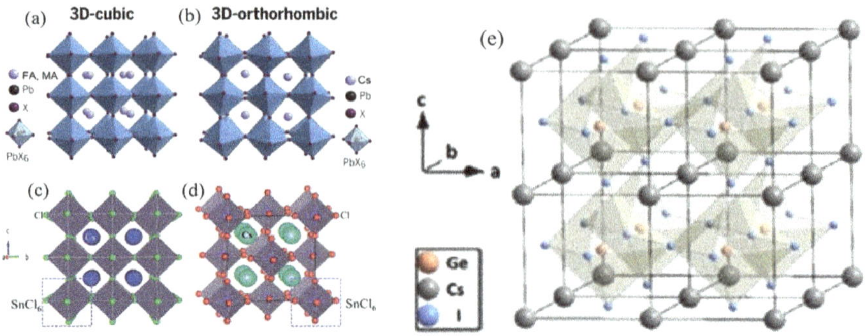

Fig. 11.1 **a** The cubic LHP structure (A = MA, FA). **b** The orthorhombic LHP structure (A = Cs). **a** and **b** both reproduced with permission of ref. [11]. **c** Cubic structure of $CsSnCl_3$ NCs. Reproduced with permission of ref. [12]. **d** Crystal structure of Cs_2SnCl_6 NCs. Reproduced with permission of ref. [13]. **e** Rhombohedral structure of $CsGeI_3$ NCs. Reproduced with permission of ref. [14]

$$t = \frac{r_A + r_X}{\sqrt{2}(r_B + r_X)} \tag{11.2}$$

where r_A, r_B and r_X are the ionic radii of the A, B, and X ions, respectively. A tolerance factor close to 1 suggests a stable cubic perovskite. Deviations often lead to distortions such as orthorhombic, tetragonal, or rhombohedral symmetries [9].

Perovskites can be classified based on various criteria, primarily the nature of their A, B, and X components. Figure 11.1a–e shows the crystal structure of halide-based perovskites [10].

Basically, lead based halide perovskite (LHPs) having structural formula, $APbX_3$ crystallizes in cubic and/or orthorhombic structure have been as shown in Fig. 11.1a, b, respectively. This structure consists of three-dimensional (3D) network of corner-sharing $[PbX_6]^{4-}$ octahedral. Pb^{2+} ions in the structure are located at the center of the octahedral and A^+ ions occupied the space between the adjacent octahedra. In the case of lead-free halide perovskite (LFHP) nano crystals (NCs) (Fig. 11.1c–e), the crystal structure depends primarily on substitutions of B sites (e.g., Pb sites as in LHP). Figure 11.1c–e shows the crystal structure of cubic $CsSnCl_3$ nanocrystals (NCs), crystal structure of Cs_2SnCl_6 NCs and rhombohedral structure of $CsGeI_3$ NCs, respectively. Various substitutions at B site in the perovskite structures offer a variety of crystal structures for LFHP NCs. The main classifications of the perovskite structures are shown in Fig. 11.2.

Oxide perovskites

These are the most studied compounds, where X = O in the formula ABX_3. They have known to show diverse physical properties and applications. For example, $BaTiO_3$ is a kind of ferroelectric material that have been used in capacitors and while mostly studied compound viz. $LaMnO_3$ exhibits a colossal magnetoresistance (CMR) which are extensively used in electronics, sensors, and actuators [15].

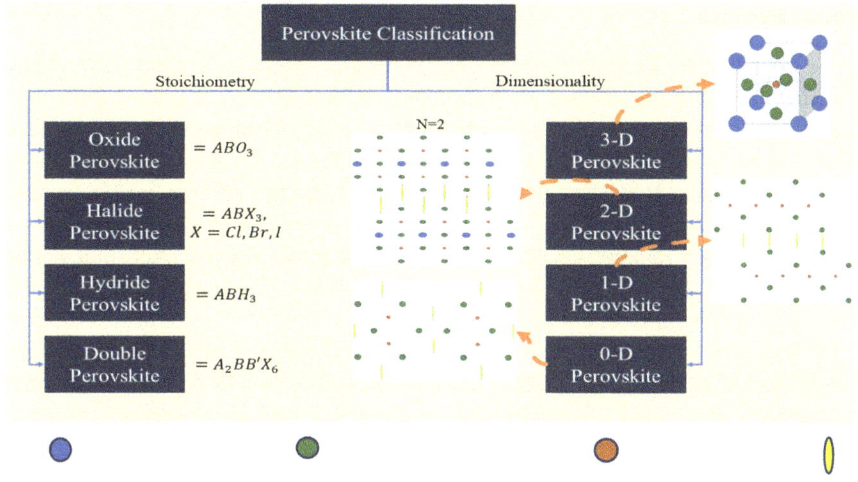

Fig. 11.2 Classification of perovskites in terms of stoichiometry and dimensionality. Here the A cation represented by blue circle, X anion represented by green circle and B cation represented by red circle. Yellow symbol in the figure is the bulky cation present in 2D, 1D and 0D perovskites

Halide perovskites

Another extensively studied pervoskite are halogens (e.g. Cl, Br, I etc.) incorporated elements at X sites in the formula ABX_3. They have gained prominence utility in photovoltaic applications, especially hybrid organic-inorganic pervoskites (HOIPs) such as; $CH_3NH_3PbI_3$/$MAPbI_3$; A benchmark material in perovskite solar cells. These materials exhibit high absorption coefficients, long carrier diffusion lengths, and tunable band gaps [3, 16]. These structures involve two different cations at the B-site, providing increased chemical flexibility. The ultralow thermal conductivity is observed in the organic-inorganic/all inorganic halide perovskite (e.g. $MAPbI_3$, $CsPbI_3$, $CsSnI_3$). Pisoni et al. observed that $MAPbI_3$ has a lattice thermal conductivity 0.5 $Wm^{-1}K^{-1}$ at room temperature [17]. Guo and Wang also predicted that lattice thermal conductivity of $CsPbI_3$ and $CsSnI_3$ is 0.54 $Wm^{-1}K^{-1}$ and 0.25 $Wm^{-1}K^{-1}$, respectively at room temperature [18]. In all inorganic halide perovskite, the organic component does not contribute and thus they possess greater stability for a wider temperature range.

Hydride Perovskite

Perovskites containing hydrogen, where X = H are belongs to this category. MAH Shah et al. have shown that $LiAeH_3$ possesses a zT value of about 0.5 [19]. Hydride perovskite is mainly studied for hydrogen storage application. They are also gaining recognition in the field of thermoelectricity.

Double Perovskites

Pervoskites having a composition of $A_2BB'X_6$ are known as double perovskites. Varieties of double perovskites were explored and studied for the potential thermoelectric applications in the last few years [20, 21]. Sofi et al. studied the lead-free halide Cs_2MGaBr_6 (M = Li, Na) double perovskites for sustainable energy. They have found significant figure of merit (zT) values of 1.08 and 1.04 for $Cs_2LiGaBr_6$ and $Cs_2NaGaBr_6$, respectively [20]. SA Khandy et al. have observed a zT of ~ 1.0 foroxide based Sr_2HoNbO_6 double perovskite [21]. Haque et al. also reported that lattice thermal conductivity value of double halide perovskite, $Cs_2InAgCl_6$ is ~ 0.1–0.2 $Wm^{-1}K^{-1}$ with zT ~ 0.94 at 700K [22].

In the last few decades, perovskite compounds explored much for majorly in optoelectronic devices. Due to their complex structure and its low lattice thermal conductivity, they could be the potential candidates for the thermoelectric applications. Therefore, researchers are now looking towards them with the lens of thermoelectricity. Figure 11.2 also show the categorization of perovskites based on stoichiometry and dimensionality. In terms of stoichiometry, we have tried to incorporate the major classification of perovskites. If we dive deep into this; we find organic, inorganic, organic-inorganic, intrinsic, doped, organometallic perovskites also. Perovskite also exists in nano-structures. The 0D, 1D and 2D perovskites are made from 3D perovskites by introducing bulky cations in between the layers/wires/clusters to isolated them from certain directions. The perovskites are formed by A site cation interacting with the BX_6 octahedra. In 3D Perovskites, they have fully connected 3D network (e.g., $MAPbI_3$). Whereas in 2D Perovskites, layered structures have alternating organic and inorganic layers (e.g., Ruddlesden-Popper phases). It is described by the formula $A'_mA_{n-1}B_nX_{3n+1}$ where A′ is the cation connecting the 2D sheets and $A_{n-1}B_nX_{3n+1}$ is the octahedron forming 2D sheet, 'n' denotes the layer thickness [23]. If n = 1 then it forms a 2D perovskite whereas for $1 < n < 6$ it forms quasi-2D perovskite [24]. In Fig. 11.2, a quasi-2D perovskite for n = 2 has been shown. In 1D Perovskites, they have structured chain/wire like connections. The growth is restricted to one direction only by the presence of bulky organic cation as shown in the Fig. 11.2 For example [DMEDA]$PbCl_4$ is a one-dimensional organic lead halide perovskite [25]. The chemical formula is A_3BX_5. The 1D perovskites are more theoretically stable compared to 2D or 3D perovskites. In 0D Perovskites, isolated clusters (quantum dots) with a chemical formula A_4BX_6 or $A_3B_2X_9$ are identified the 0D perovskites are known to shown good optical properties. For examples the optical properties of Cs_4PbBr_4 is comparatively better than its 3D $CsPbBr_3$ [26]. Dimensional engineering in the perovskite materials allows tuning of material properties for specific applications, especially in light-emitting diodes (LEDs) and solar cells etc. The versatility of perovskite materials is rooted in their ability to accommodate a wide range of ionic substitutions and structural distortions, leading to diverse functionalities [27]. Additionally, Fig. 11.3 also shows different perovskite applications [28]. The halide-based perovskites offer impressive optoelectronic properties but their practical application faces challenges such as: Moisture sensitivity, Lead toxicity and Long-term stability. Strategies to overcome these issues include

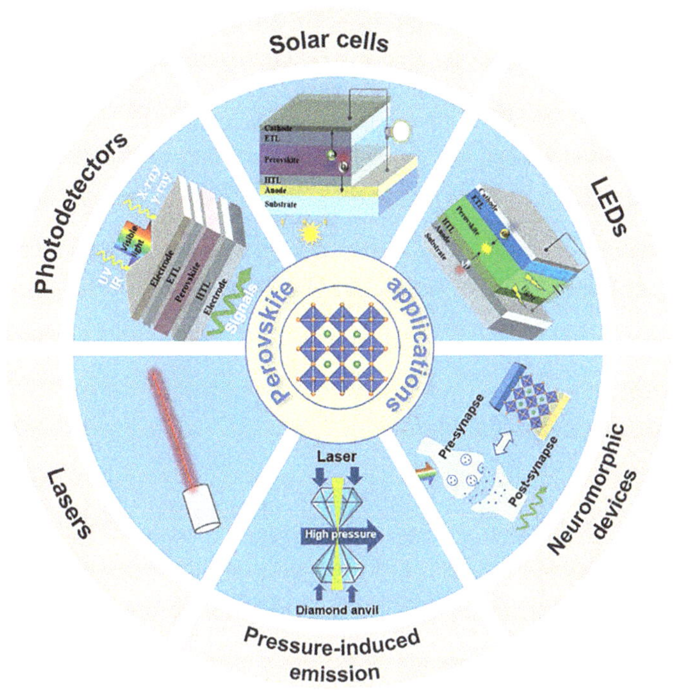

Fig. 11.3 Diagram showing different perovskite applications. Reprinted with permission of ref. [28]

compositional engineering (e.g., partial substitution of Pb^{2+}), encapsulation, and development of lead-free alternatives like tin-based or bismuth-based perovskites. Perovskite materials, owing to their structural diversity and tunable properties, have emerged as a central class in modern materials science. Their classification based on composition, dimensionality, and functional behavior provides a systematic understanding critical for developing next-generation technologies in energy, electronics, and photonics.

11.3 Synthesis Techniques for Lead Free Perovskites

The increasing concern over lead toxicity and environmental impact has driven the development of lead-free perovskite materials. Substituting lead (Pb^{2+}) with non-toxic cations such as tin (Sn^{2+}), germanium (Ge^{2+}), bismuth (Bi^{3+}), antimony (Sb^{3+}), and silver (Ag^{+}) has yielded promising alternatives. However, the synthesis of lead-free perovskites presents new challenges, including oxidation instability (e.g., $Sn^{2+} \rightarrow Sn^{4+}$), low crystallinity, and reduced carrier lifetimes. Therefore, it becomes crucial to search methods that enable the development of lead-free halide perovskites without compromising device performance. In broader aspect people opt generally the two most common process to developed the lead-free perovskites. The two categories are; (i) solution approach and the next one is (ii) vapor deposition process. The former one is employed to those processes where materials have been prepared by incorporating the organic and inorganic solvent however in the later one people use the anti-solvent in the synthesis process to accelerate the precipitation. Moreover, various others synthesis techniques also have been developed and optimized to address these issues and achieve high-purity, phase-stable, and high-performance lead-free perovskites. In the section we will discuss alle the preparation techniques one by one.

11.3.1 Solution-Based Methods

It is the most used method to prepare the high-quality perovskites at different scale for the applications in optoelectronics devices. The precursor solution has been prepared by adding certain organic solvents like hydrohalic acid and the inorganics solvent like N, N-dimethylformamide (DMF), dimethyl sulfoxide (DMSO), and γ-butyrolactone (GBL). Then after the prepared solution undergoes a spin coating treatment. Finally, to get the crystallites into the solid films form one may go through the thermal annealing. The two very common approach namely one step spin coating and two step deposition are in trends now a days for the preparation of such type materials.

11.3.1.1 One-Step Spin Coating

Single step spin coating method is one of the most widely used technique to get the uniform films on the substrate having the film thickness below 10 nm. People use this technique predominantly due its low cost, less hazardous and the controlled thickness by steering the speed or viscosity. A basic pictorial representation of one step spin coating has been shown in Fig. 11.4. The perovskite precursor solution has been prepared by dissolving inorganic halides into a prepared solution of organic halides. The small fluid of the above prepared solution has been dropped on to the center of substrate followed by the spinning with the rpm around 3000 units. To get the pure phase perovskite annealing treatment has-been given to the deposited sample on the substrate. Despites of outstanding merits, spin coating have certain limitations also.

The use of large substrate seeks the challenge as it ceases the speed to high enough rates. Only few percent of the solution were found to adheres to the substrate while the majority is rushed to the coating bowl and subsequently discarded. This simply affects the cost and increase the disposal expenses. Hong et al. has synthesized the $CsSnI_3$ through the one step spin coating approach. They have use CsI and SnI_2 as inorganic halides and dimethyl sulfoxide (DMSO) as organic precursor. The precursor solution prepared by mixing above to being coated on an indium tin oxide (ITO) glass at a

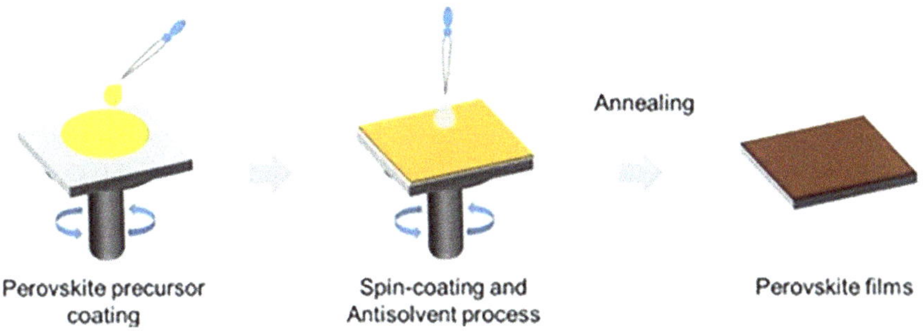

Fig. 11.4 Basic steps involve in one step spin coating method. Reprinted with permission of ref. [29]

speed around 4000 rpm. To improve film uniformity and crystallinity, anti-solvent dripping (e.g., toluene, chlorobenzene etc.) has been dropped on substrate during spin-coating followed by subsequent annealing at 90 °C for 10 min [30].

11.3.1.2 Two-Step Deposition

This approach improves stoichiometric control and as well as the better results in the fabrication of perovskite film even high humid conditions which generally leads to better phase purity. In contrast to one step spin coating, this process work well for Sn-based perovskites because of the solubility of SnI_2 or Sn halides in most solvents (such as 2-propanol etc.). This method is commonly used for the low oxidation tin-based perovskites. One of the potential candidates studied from various researchers for lead free perovskite is Phenethylammonium tin iodide (PEA_2SnI_4). This perovskite has been obtained by two step spin coating approaches. First the reprecipitation then after the crystallization process of PEA_2SnI_4 will occur. Normally in first step a high concentration precursor mixture has been formed by compounding the Phenylethylammonium iodide (PEAI), tin (II) iodide (SnI_2), and tin powder in a solvent namely γ-butyrolactone (GBL). After this precipitation will occur by mixing above prepared solution to the dichloromethane (DCM). Second step precursor has been produced by redissolving the above obtained microcrystals of PEA_2SnI_4 to N, N-dimethylformamide (DMF). Now the subsequent spin coating of this solution enables the recrystallization of PEA_2SnI_4 and hence formation of the good quality thin film [31]. In a similar consequence, another lead-free perovskite viz. $(CH_3NH_3)_3Bi_2I_9$ thin films has also been prepared by Ahmad et al. by two-step spin coating method. They observed that by switching from one step spin coating to modified two step coating one can obtain the remarkable stability as well as the device performance [32].

11.3.2 Vapor Deposition Methods

We generally see basically two methods by vapor deposition technique in the preparation of lead-free perovskites thin films. First one is Thermal evaporation and the second one is Chemical vapor deposition. In following section, we will discuss each one briefly.

11.3.2.1 Thermal Evaporation

In contrast to spin coating method, the thermal evaporation technique provides the smoother surface as well as high quality thin films of lead-free perovskites. Due to its independency on solvent this method covers the wider range of advantages over other methods. Possibility of traps/defects can be avoided due to the use of solvent which is challenging to the spin coating method. However, despite various superiority, there are some challenges were seen to this method. It seeks the potential breakdown of the organic precursors. It also suffers during the calibration of deposition rate especially for dual source evaporation. Sometimes, due to high partial pressure of organic precursors, one may get the risk of significant contamination in the evaporation chamber. In this method one may use two ways to get the high-quality films. (i): perovskite precursor materials are simultaneously or sequentially evaporated under vacuum and condensed on a heated substrate or (ii): one can evaporate the perovskite powder separately by putting it into the tungsten boat at ambient temperature. It is very important to calibrate the internal temperature of evaporator, source pressure, and evaporation rate to get the high-quality perovskite thin films. In this method the perovskite precursor may be used as single or multiple sources. Nominally the thermal evaporation has been under the high vacuum. Zeng et al. successfully synthesized the cesium-based lead-free perovskite e.g. $CsGeI_3$. In their study the dense, uniform, good crystallinity, and minimal pinholes nature of synthesized $CsGeI_3$ have been employed to the photodetection successfully [33].

11.3.2.2 Chemical Vapor Deposition (CVD)

Chemical vapor deposition (CVD) is one of another technique for the preparation of lead-free perovskite films under the category of vapor deposition. This method can be used to produce the perovskites with different dimensionality and few defects. The perovskite precursors (at high temperature) and the substrate (at low temperature) have been placed at calibrated separation within the chamber. A subsequent inert air flow has been made from high temperature region to the low temperature region for the proper deposition of the desired perovskite precursor to that on substrate. One lead free perovskite film namely $CsSnX_3$ (where X=Cl, Br, or I.) have been synthesized by Han et al. by CVD technique [34]. A schematic representation of CVD technique has been shown in Fig. 11.5. From the figure it has observed that the

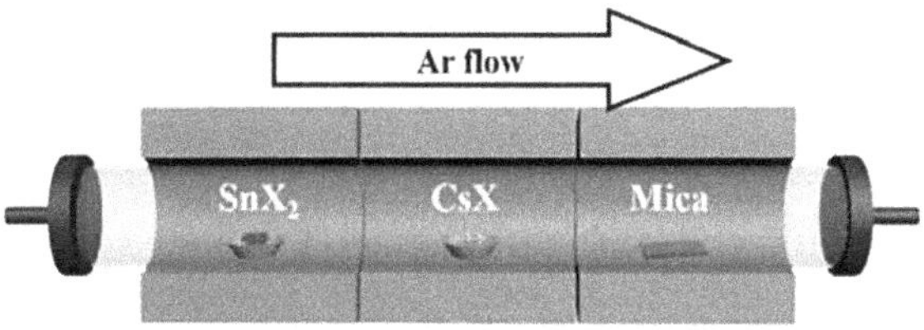

Fig. 11.5 Schematic of $CsSnX_3$ nanowires prepared by CVD. Reprinted with permission of ref [35]

CVD is divided into three regions. The SnX_2 and CsX_2 powders have been placed at the upstream in the air flow and midstream respectively. On the downstream a mica substrate has been placed for the film deposition. The three regions were heated to 220–350, 610–670 and 200–260 °C respectively in an argon flow maintaining the tube pressure at 0, 16 Torr for 20 min. The use of substrate is arbitrary. For example, one may use porous alumina template as substrate as well.

11.3.3 Solid State Reaction

Solid state reaction is one of the most traditional and widely used method to prepare the perovskites. Material migration and the interface reaction are the two basic mechanism involves to this method. In a good quality agate mortar pestle, all the required reactants' particles were grinded evenly. Grinding should be done as per demand so that the by putting the grinded powder to high temperature, chemical reaction may occur on the surface of the reactants to produce a new phase containing structural defects. To obtain the desired phase, grinding and the sintering process may repeat several times. In this way one can get the structural adjustment and crystal growth of the required phase sample. Using a solid-state reaction, a lead-free perovskite namely $CsSnI_3$ has been synthesized by Chung et al. successfully. The mixed the precursors of CsI and SnI_2 properly in an evacuated Pyrex tube and calcined the mixed precursor at 550 °C for an hour. Then after a subsequent cooling of the heated tube has been made for six hours to grow the perovskite [36].

11.3.4 Hydrothermal Method

This is one of the very common methods to synthesize the perovskite of different shape and size by controlled use of the capping agent. The stoichiometric proportions of the metal oxide precursor have been prepared carefully. To get a sol one may use the desirable solvent and mixed the metal oxide precursor to this solvent. In this way the precursors used for the synthesis of perovskite are exposed to a calibrated pressure and temperature (typically between the boiling point of water and critical temperature of material) in an autoclave. After this the sealed autoclave has been put into the oven/or furnace for the required time. Different chemical reaction may take place within the autoclave to get the desired phase perovskite. After cooling the autoclave up to room temperature, we can separate out the wet prepared powder from the solvent. Moreover, subsequent ultrasonication can also be used to separate the unwanted weakly bounded reactants from the perovskites. The wet perovskite powder has been dried in an oven after being centrifuged. By adjusting the temperature, pH, time, capping agent, and reactant proportionality one can control the particle growth, phase, shape, and crystallinity of the final product. Though various reports are available for the perovskite preparation via hydrothermal method, one of which be chosen here as curtesy. This is Zhou et al. who has synthesized the $Cs_2AgInCl_6$ by hydrothermal method. They have made the precursor solution by taking HCL, CsCl, AgCl and $InCl_3$ in stoichiometric proportion in a high-pressure reactor (i.e. the autoclave). Autoclave has been exposed to a temperature of 150 °C for a desired time and then cooled to room temperature to nucleate into the product. Subsequent filtration and washing have been achieved to get the white $Cs_2AgInCl_6$ crystals [37].

11.3.5 Hot Injection Method

Hot injection method is one of the best used methods of material scientist to synthesize the perovskite nanocrystals. Initially, a cold precursor solution required for the perovskite has been prepared under the nitrogen environment. After this the cold precursor solution is injected into a hot precursor solution containing a surfactant and solvent for the rapid nucleation and growth of nanocrystals. This has been ensured that there is no time lag during the injection and the precipitation process so that the reaction occurs promptly and efficiently upon mixing the precursor solutions. Following the cooling steps in ice bath the prepared nano perovskite has been recovered by means of centrifugation. It is worth noting that the temperature during the injection and the reaction time after injection time would play a crucial role for the dimensionality growth of the nanocrystals. The first lead free perovskite nanocrystal by this methos have been produced by Tom C. Jellicoe et al. as the $CsSnX_3$ (X = Cl, $Cl_{0.5}$ $Br_{0.5}$, Br, $Br_{0.5}$ $I_{0.5}$, I) [37]. Simialrly using this method one more study have been done by Jellicoe et al. regarding $CsSnX_3$(X = halogen except lead) [12]. Additionally, it was Liu et al. who have prepared $Cs_2AgInCl_6$ quantum dots (QDs) successfully by this method.

In summary, the synthesis of lead-free perovskites involves a diverse range of techniques tailored to balance stability, performance, and scalability. While solution-based methods remain dominant for thin-film devices, solid-state and vapor-phase methods are gaining attention for improving structural and optoelectronic qualities. Continued optimization of these techniques, especially for Cs-based, **Sn-based** and **Bi-based** perovskites, will be critical to achieving sustainable, high-efficiency, and environmentally benign perovskite materials.

11.4 Thermoelectricity and its Various Aspects

Thermoelectricity refers to the direct conversion between thermal and electrical energy through thermoelectric effects. These phenomena offer promising approaches for waste heat recovery, solid-state refrigeration, and power generation, especially in remote or portable applications. The three primary thermoelectric effects are:

(i) **Seebeck effect**—Generation of voltage due to a temperature gradient.
(ii) **Peltier effect**—Absorption or emission of heat when an electric current passes through the junction of two materials.
(iii) **Thomson effect**—Heat absorption or release in a conductor carrying current with a temperature gradient.

11.4.1 Seebeck Effect

It was first discovered by Alessandro Volta, an Italian scientist in 1774. Later the independent work was carried out by Thomas Johann Seebeck (German physicist) in 1821, and thus the name Seebeck effect. This effect is observed when the material (metallic or semiconducting) is kept between the hot and the cold source. A voltage difference is observed due to the existence of temperature gradient [38, 39]. The induced voltage (V) is proportional to the temperature gradient (ΔT):

$$V = S.\Delta T$$

where **S** is the **Seebeck coefficient** (V/K). Materials with high Seebeck coefficients are essential for thermoelectric applications.

11.4.2 Peltier and Thomson Effects

Peltier Effect (Jean Peltier, 1834) describes heat absorption or release at a junction due to current flow. This effect can also be described as an effect whose working is opposite to the Seebeck effect. Here, the heat is released or absorbed at the junction by passing the current through the material.

Thomson Effect (William Thomson, 1851) refers to heating or cooling of a homogeneous conductor with a temperature gradient under current flow. In this effect the Seebeck coefficient is not constant.

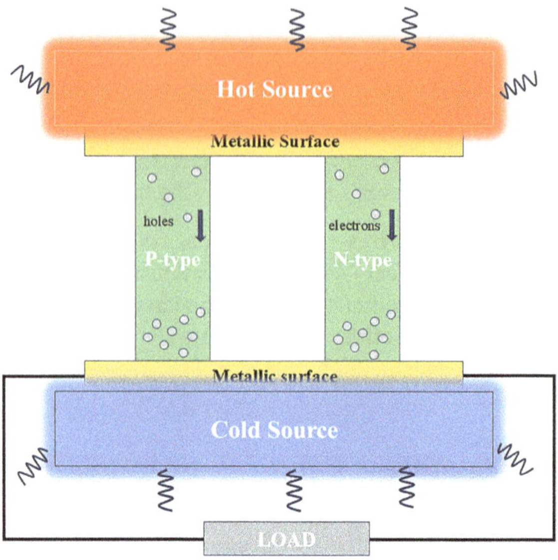

Fig. 11.6 A figure representing the fundamentals of Seebeck effect

11.4.3 Working and Fabrication of Seebeck/Peltier Effect

In order to fabricate a thermoelectric device, numerous numbers of p-n diodes are connected in series through metallic surfaces and ceramic insulating plates, but they are thermally in parallel. The p-diode is filled with holes whereas the n-diode is filled with electrons, together they form a thermocouple as shown in Fig. 11.6. In Seebeck effect, both the sides of the junctions are at different temperature, i.e., at different thermal energy. The hotter side creates a deficiency of carriers in both p-and n-type semiconductor resulting in a net positive charge near the hotter region and a negative charge near the cold surface. When a wire is attached to the load and to the metallic surface, a net voltage difference is observed across the load, resulting in a flow of current through the wire. Whereas in Peltier effect when the similar thermoelectric module is connected to the DC source, the electron in the n-type and hole in the p-type starts moving. The electron moves from n-type to p-type resulting it to jump from higher energy level to lower, releasing thermal energy [40]. These effects are **reversible** and form the basis of **thermoelectric cooling** technologies.

11.4.4 Thermoelectric Materials and Properties

Thermoelectric materials basically based on the above discussed three effects, i.e., Seebeck, Peltier and Thomson effects and are classified into three main categories based on their working temperature range such as (i) Low-temperature thermoelectric materials, (ii) Mid-temperature thermoelectric materials, and (iii) High-temperature thermoelectric materials [41].

(i) Low-temperature thermoelectric materials also called near room temperature thermoelectric (TE) materials and work under the temperature 500 K. Tetradymites, alloys like Bismuth Chalcogenides (Bi_2Se_3, Bi_2Te_3, Bi_2S_3, and Antimony Chalcogenides (Sb_2Se_3, Sb_2Te_3, Sb_2S_3) belong to the family of low-temperature thermoelectric materials [42–47].
(ii) Mid-temperature thermoelectric materials work efficiently between 500 and 900 K. Lead Chalcogenides (PbTe, PbSe, PbS, Tin Chalcogenides (SnSe, SnTe, SnS), Indium Chalcogenides (InSe, InTe, InS), and Germanium Chalcogenides (GeTe, GeSe, GeS) and many more belong to the group of mid-temperature range thermoelectric materials [48–58].
(iii) High-temperature thermoelectric materials are those materials which are known to work at higher temperature, i.e., at more than 900 K. SiGe, La_3Te_4, $Yb_{14}MnSb_{11}$, ZrNiSn, and FeNbSb-based half-Heusler alloys belong to high-temperature thermoelectric materials [59–63].

As an alternative, thermoelectric materials can also be classified as; (i) Inorganic semiconductors; (ii) Oxide thermoelectrics and (iii) organic and hybrid materials. In inorganic semiconductors, Bismuth telluride (Bi_2Te_3), Lead telluride (PbTe): and Silicon-germanium alloys (SiGe) are the best examples. They worked at near room temperature, in mid temperature range and at high temperature range, respectively. Oxides like $Ca_3Co_4O_9$ and $NaCo_2O_4$ are the examples for the oxide based thermoelectric materials. They offer high thermal stability and oxidation resistance, making them suitable for harsh environments [64]. Organic semiconductors (e.g., PEDOT: PSS) are flexible and lightweight but typically offer lower ZT values whereas Hybrid perovskites and metal-organic frameworks (MOFs) are emerging materials for thermoelectric research due to their tunability and low thermal conductivity.

Oxide Thermoelectric

Oxides like $\mathbf{Ca_3Co_4O_9}$ and $\mathbf{NaCo_2O_4}$ offer high thermal stability and oxidation resistance, making them suitable for harsh environments [64].

Organic and Hybrid Materials

For a good thermoelectric material, thermoelectric materials must exhibit a unique balance between electrical conductivity (σ), Seebeck coefficient (S), and low thermal conductivity (κ). The key parameters in case of efficient thermoelectric or to characterize the thermoelectric materials are electrical conductivity (σ), Seebeck coefficient (S), thermal conductivity (κ), power factor (PF) and figure of merit (zT). Seebeck Coefficient (S) and Electrical Conductivity (σ) should be high to reduce Joule heating for good thermoelectricity whereas thermal Conductivity (κ) should be low to maintain temperature gradients. This includes the thermal conductivity caused by electrons and lattice. Very important parameters, i.e., Power factor (PF) and figure of merit (zT) for measurement of thermoelectric efficiency can be defined as;

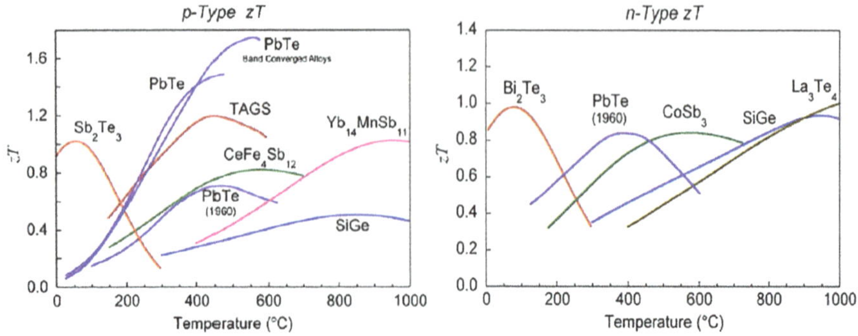

Fig. 11.7 zT for common thermoelectric materials. Reprinted with permission of ref. [66]

$$\mathrm{PF} = \mathrm{S}^2\sigma \text{ and } zT = \frac{S^2\sigma T}{\kappa_l + \kappa_e} \tag{11.3}$$

where, the symbols have their usual meaning (refer Sect. 11.1). $\kappa = \kappa_l + \kappa_e$. κ_l stands for lattice thermal conductivity and κ_e stands for electronic thermal conductivity. zT a dimensionless quantity of a thermoelectric materials, basically describe the conversion efficiency of heat into a useful form of energy. Generally, zT > 1 is generally desired for practical applications. Various materials with higher zT have been studied and reported in this field of research which is rapidly advancing as shown in fig. 11.7. This rapidly research field in thermoelectric needs some concern needs to be addresses regarding the stability and measurement issues [65].

Procedure to Calculate Device ZT from materials properties

Conversion efficiency of a thermoelectric device, describes the amount of waste heat energy converted to useful electrical energy or vice versa, is typically reported as a function of zT (device figure of merit). It is given as:

$$\eta = \frac{T_h - T_c}{T_h} \frac{\left(\sqrt{1 + zT} - 1\right)}{\sqrt{1 + zT} + \left(\frac{T_c}{T_h}\right)} \tag{11.4}$$

where T_h =temperature of hot source and T_c =temperature of cold source.

zT plays a very important role in the conversion efficiency and is much of a talk amongst researchers, since it is a materials property. This property can be enhanced only by changing the material's characteristics. One should keep in mind the difficulty faced by researches when it comes to enhancing the *zT* parameter. Since this property depends on S, σ, κ which are inter-dependent onto each other. Changing one parameter brings the change to other parameters which makes it difficult to increase *zT* above certain extent. Although there have been some advances in this field to go beyond this level.

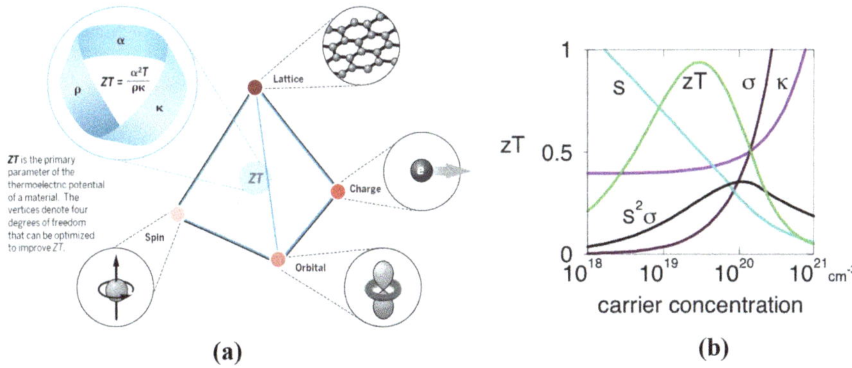

Fig. 11.8 **a** Various features that affect zT effect zT. Reprinted with permission of ref. [8]; **b** Variation of different parameters (S, σ, **κ** and zT value) with concentration. Reprinted with permission of ref. [1]

These parameters (S, σ, **κ** and zT) are inter-dependent on each other and oppose the change in zT. For example, a material with large bandgap, which results in high temperature gradient across the material and hence high *S*. The high *S* is a consequence of large bandgap. The large bandgap results in low conductivity since higher energy is required for the carrier to jump from valence to conduction band. This results in low σ and hence low κ_e which is a consequence of Wiedemann-Franz Law. So, even if one tries to enhance zT, these inter-dependent quantity resists the overall change. The second, contribution in the thermal conductivity is due to κ_l which is a lattice property and its behaviour is associated with the lattice dynamics. However, from the rest three parameter, it was observed or it is a rough estimate on how altering one parameter brings a significant change in the others. Figure 11.8a shows the variation of very fundamental parameters (lattice, charge, spin, orbital) that play a decent role in affecting the overall behavior of material. They significantly alter mechanical, magnetic, optical, thermoelectric, etc. properties. Figure 11.8b represents the variation of primary parameters and its interdependency on each other for a thermoelectric material. It also shows that how one can maximized the value of zT by balancing the parameters for certain concentration. In the next sections of the chapter, more details are included in details related to these key parameters and strategies to enhance the figure of merit (zT).

11.4.5 Applications of Thermoelectric Materials

11.4.5.1 Power Generation

- **Automotive exhaust systems**: Recover waste heat to power vehicle electronics. A large amount of heat is released from automobiles on daily basis and if we

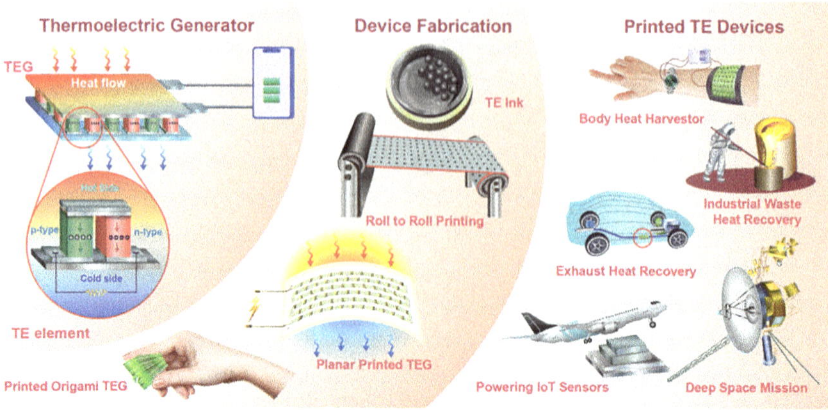

Fig. 11.9 Schematic illustration of the printed TE materials processing, device fabrication, and the application of the thermoelectric devices (TED). Reprinted with permission of ref. [67]

use thermoelectric devices in such automobiles then that waste energy could be utilized by the automobile reducing the heat released in the environment and also reducing the fuel needed to run the automobile.

- **Industrial waste heat recovery**: Improve overall energy efficiency. The heat released by steel industries or other metal industries is huge. There are some thermoelectric materials that prove to provide good thermoelectricity at higher temperature. Utilizing them in big industries will prove to be a good initiative towards healthier environment and lifestyle.
- **Remote space missions**: Radioisotope thermoelectric generators (RTGs) power spacecraft like Voyager and Mars rovers. For remote operations, where there is a limited supply of energy, thermoelectric devices can be an excellent choice to draw energy and utilize them appropriately.
- **Health Sector/Bio-medical application:** Non-toxic and stable thermoelectric devices can be used in health sector. One such example is a pace-maker, where up until now it works on a battery which needs to be replaced in almost 10–15 years. The battery can be replaced by a thermoelectric device, which can draw heat energy from the body itself and work for a longer time without replacing.

There are many other applications of thermoelectric devices that can be used in our daily life which in a way improves our daily lifestyle and causes much less damage to the environment. Figure 11.9 shows the schematic diagram of TE material processing.

11.4.5.2 Thermoelectric Cooling

- **Portable coolers**: Compact and silent solid-state refrigerators. In automobiles thermoelectric cooling (TEC) is needed for ventilation, air-conditioning. These coolers help in maintaining the temperature without producing noise. However, there is still much to develop in this field to improve its efficiency till mechanical refrigerators.

- **Laser and detector cooling**: It is a method where a low energy laser is used to excite material which later releases higher energy by de-exciting utilizing the heat present in the material. It requires precise thermal management. Materials like $MAPbI_3$ and $PhEPbI_4$ can be utilized as a laser cooling device [68].
- **Bio-medical application:** Cold wrapper consisting of TEC can be applied to treat swelling and inflammation [69]. They can also be utilized in tissue cooling. It was reported that a cooling application based on integrating polydimethylsiloxane and TECs can treat the reversible brain deactivation in animals ranging from rodents to primates [70]. This also make them very efficient in health sector field.

11.4.5.3 Challenges and Future Directions

Despite advances, several challenges remain:

- Low ZT at room temperature in many materials.
- Non-constant zT over a wide range of temperature.
- Material scarcity and toxicity (e.g., Te, Pb).
- Mechanical stability under thermal cycling.
- Cost-effectiveness and scalability for real-world deployment.

Future research is focusing on:

- Earth-abundant materials (e.g., Mg_3Sb_2, SnSe).
- Flexible and wearable thermoelectrics.
- Synergistic integration with photovoltaics and batteries.
- Easily scalable, durable, sustainable, non-toxic, less costly and environment friendly materials.

Thermoelectric materials represent a compelling solution for sustainable energy conversion and thermal management. A deeper understanding of transport phenomena, coupled with advances in material design, is key to realizing efficient, scalable, and environmentally friendly thermoelectric devices.

11.5 Thermoelectric Characterizations

To evaluate the performance and applicability of thermoelectric materials in the device applications, precise characterizations of their physical and transport properties are essential. Thermoelectric characterization involves measuring the parameters that contribute to the material's figure of merit (zT), which determines its efficiency in converting heat into electricity or vice versa. Each component that contributes to the zT parameter, i.e., with Sebeck coefficient, electrical conductivity, thermal conductivity and temperature (mentioned in Sect. 11.4 above) must be accurately measured, typically over a range of temperatures, to understand materials behavior in real-world conditions.

However, it is of critical importance to study the other parameters on which these parameters basically dependent. In this section, detailed discussion on those parameters are done to understand the impact of these parameters on the entire processing, behavior and physics of a thermoelectric material. This section focused on the experimental and theoretical methods taken to predict the value of these quantities.

11.5.1 Seebeck Coefficient (S)

The **Seebeck coefficient** quantifies the voltage generated per unit temperature gradient in a material and given as;

$$S = \frac{\Delta V}{\Delta T} \tag{11.5}$$

The Seebeck coefficient also depend on other parameter such as carrier concentration (n), effective mass (m^*) and carrier mobility, etc. as given by Mott's Formula;

$$S = \frac{8\pi^2 K_B^2}{3eh^2} m^* T \left(\frac{\pi}{3n}\right)^{\frac{2}{3}} \tag{11.6}$$

where the constants have their usual meaning. From this formula it is clear that a material with high effective mass and low carrier concentration will possess larger Seebeck coefficient (S). It is also well known that mobility indirectly affects the Seebeck coefficient through carrier concentration and scattering mechanism. Mobility is influenced by the effective mass (m*) and scattering time (τ) as $\mu = \frac{e\tau}{m^*}$ [2], i.e., in case of high mobility typically corresponds to a lower effective mass, which lead to lower S and vice versa. In case of high carrier concentration, the Fermi level lies deeper in the conduction or valance band, which can reduce the Seebeck coefficient. For the low carrier concentration, Fermi level is closer to the band edge, leading to higher Seebeck coefficient. The negative/positive value of the Seebeck coefficient indicates the type of majority charge carriers, i.e., they are behaving like are n-type/p-type, respectively. Therefore, in short, during the thermoelectric measurements, one has to keep in mind the following points;

- Two heat arrangement should be used to create a temperature gradient (ΔT) across the sample.
- Usually, a high impedance voltmeter has been used to measure the thermoelectric voltage (ΔV).
- One should calculate the Seebeck coefficient using voltage versus temperature plot.
- For the calibration constantan, Cu etc. has been used as reference material.

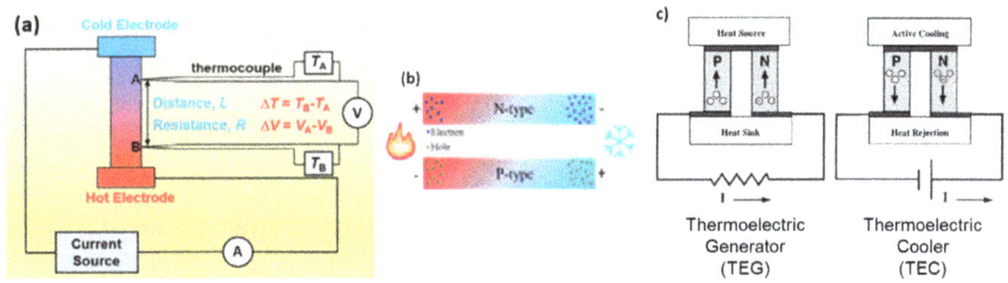

Fig. 11.10 **a** Schematic depiction of S and σ measurements. Reproduced with permission of ref. [71]. **b** Schematic illustration of charge carrier diffusion process under temperature gradient for n- and p-type semiconductors, **c** Structure of thermoelectric coolers (TECs) and thermoelectric generators (TEGs). Reproduced with permission of ref. [72]

- During the measurements contact resistance, temperature stability and measurement directions (longitudinal vs transverse) also played a crucial role and can affect the accuracy.

Figure 11.10a shows the systematic representation of the commercial instruments used to measure Seebeck coefficient and electrical conductivity of the compounds [71]. The Seebeck effect can be seen in both conductors and semiconductors. This happens when charge carriers move from the hot side to the cold side as shown in Fig. 11.10b [72].

Thermoelectric devices are operational systems capable of converting heat into electrical energy and vice versa. These devices can be categorized into two groups based on their distinct ways of functioning: thermoelectric generators (TEGs), which can produce electricity from a difference in temperature, and thermoelectric coolers (TECs), which can turn electrical current into a temperature difference. The structure of TECs and TEGs is shown in Fig. 11.10c. TECs can function as electronic coolers to maintain the surface temperature of an electronic device below a dangerous threshold. Additionally, they can function as refrigerators or air conditioners for cooling. TEGs can produce power from heat and be utilized as an energy source for aerospace, deep ocean, and military biomedical applications [73]. TEGs offer the benefits of lower volume, lighter weight, reduced noise and pollution, and simpler maintenance over conventional electricity generators [74, 75].

11.5.2 Electrical Conductivity (σ)

Second very important parameters to characterize any thermoelectric material is electrical conductivity. This is a measure of a material's ability to conduct electric current. Conductivity of any materials has the inverse relation with the resistivity of the material, and directly proportional to the carrier concentration and carrier mobility (as given below). It was well known that metals possess highest conductivity followed by semiconductors and insulators.

$$\sigma = \frac{1}{\rho} \tag{11.7}$$

$$\sigma = ne\mu = \frac{n\tau e^2}{m^*} \tag{11.8}$$

where ρ is the resistivity of a material, n is the carrier concentration, τ is the relaxation time and μ stands for carrier mobility. The above relation is derived from Drude's model. An essential step for controlling electrical conductivity is to achieve an appropriate n (concentration of electrons in an n-type semiconductor) or p (concentration of holes in a p-type semiconductor). σ typically decreases with temperature in metals and increases in semiconductors. Thus, temperature-controlled measurements are essential for thermoelectric applications. ρ follows inverse temperature coefficient for semiconductor and insulator. Thus, there is a rise in σ with increasing temperature. It is mainly due to the reason that the carriers at higher temperature have enough energy to cross the valence band and move towards conduction band. Therefore, with increasing temperature electrical conductivity increases. Whereas in metals due to thermal agitation at higher temperature the flow of carriers is restricted and resisted resulting in decrease of electrical conductivity.

There are several methods to determine the resistivity and thus the conductivity of a material but the most used methods are Four-point probe method which is widely used method to determine the resistivity of a material. It consists of four probes system as shown in the Fig. 11.11a. The outer probes are used to pass current and the inner probe is used to measure voltage. In the four probes arrangement, all the probes are equally spaced. The four-probe arrangement eliminates contact resistance and suitable to determine the resistivity more accurately. Figure 11.11 draws a clear image of the four-probe method and resistivity of the materials can be estimated using the following relations;

$$\rho = \frac{\pi}{\ln(2)}\frac{Vtk}{I} \tag{11.9}$$

Where, ρ is the resistivity of the sample, V is the voltage, I is the current, t is sample thickness, k is a correction factor based on the ratio of the probe to wafer diameter and on the ratio of wafer thickness to probe separation.

Van der Pauw method is another very important method used to measure the resistivity and Hall mobility of irregularly shaped samples. In this method to measure these properties four small contacts were placed on the periphery of a sample of irregular shape as shown in Fig. 11.11b. Figure 11.11c represents the diagram to measure the electrical conductivity of a sample which is equipped with four electrical contacts E1 to E4 by means of the van der Pauw method. In this arrangement, there are four contacts, driving a current from E2 to E1 and measuring the potential difference between E4 and E3. This provides the 4-point resistance, i.e., $R_{21,43} = U_{43}/I_{21}$; similarly, $R_{31,42}$ 1/4 $= U_{42}/I_{31}$. Using these two values into the van der Pauw equation (given below), electrical conductivity of the sample can be estimated having

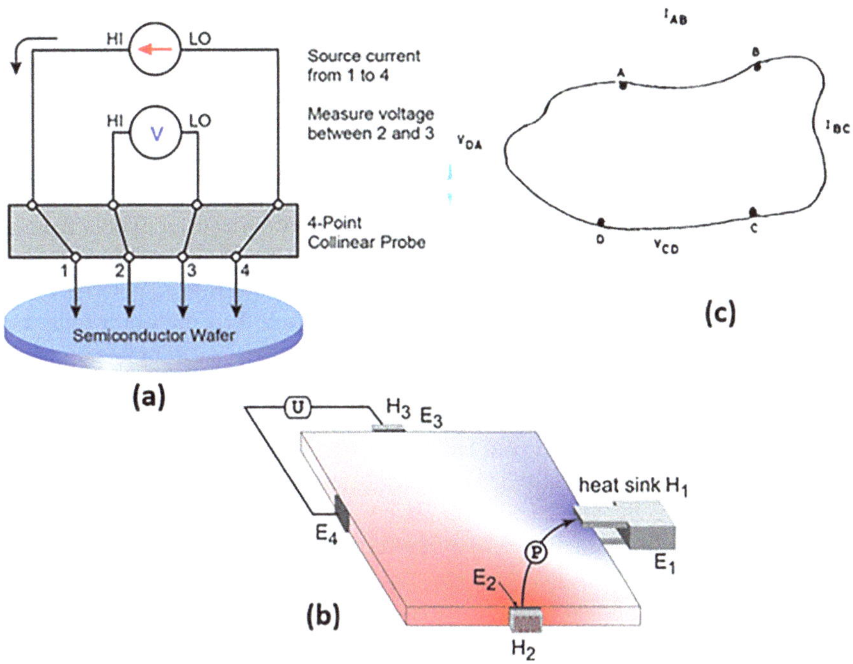

Fig. 11.11 **a** Shows schematic diagram of four-probe method. Reproduced with permission of ref. [76]. **b** shows the schematic diagram of Van Der Pauw method. Reproduced with permission of ref. [77]. **c** Sample of irregular shape with four contacts at arbitrary places along the circumference. Reproduced with permission of ref. [78]

the known sample thickness value ‘d’.

$$\exp\left(-\pi\sigma R_{21,43}\right) + \exp\left(-\pi\sigma R_{31,42}\right) = 1 \tag{11.10}$$

The suspended sample (Fig. 11.11c is also equipped with two heaters H_2 and H_3 and mechanically and thermally connected to the heat sink H_1. If heat is created in H_2 or H_3, a temperature gradient and, hence, a measurable Seebeck voltage can be detected between E_4 and E_3 or E_4 and E_2, respectively. This voltage is used to determine the ratio of the thermal conductivity and Seebeck coefficient of the sample. In combination with the measured σ and ZT, k and S can be calculated individually and the sample is completely characterized with respect to its thermoelectric properties.

Hall coefficient ($R_H = \frac{1}{nq}$) measurement technique is also useful technique which provides the information about the carrier concentration (n) and carrier mobility $\left(\mu = \mu = \frac{\sigma}{nq}\right)$ of the compound. With the help of these parameters, electrical conductivity ($\sigma = nq\mu$) and the origin of the Seebeck coefficient can be interpreted.

11.5.3 Thermal Conductivity (κ)

Thermal conductivity (κ) is a very significant parameter to characterize thermoelectric materials It can be describe in terms of two parts, first one is the lattice (phonon) contribution (κ_L) and second one is the electronic contribution (κ_e) in total thermal conductivity of the sample. Basically, $\kappa = \kappa_L + \kappa_e$ is the measure of heat transferred per unit time per unit cross section. This process involves the transfer of heat through free carriers (e.g. electrons) and through lattice phonons. The electronic contribution (κ_e) of thermal conductivity could be understood and determined in terms of the well-known Wiedemann-Franz Law as; $\kappa_e = L\sigma T$, Where **L** is the Lorenz number (~ 2.45×10^{-8} W·Ω·K^{-2} for metals) and σ, T are in their usual meanings at discussed above.

According to the Wiedemann-Franz Law, κ_e is directly proportional to electrical conductivity (σ) value. Estimation of lattice thermal conductivity is a little bit tricky although κ_l can be theoretically achieved by using the phonon gas model, $\kappa_l = \frac{1}{3} C_v v_g a$, where, C_v is the heat capacity of phonon, v_g is the group velocity of phonon and a is the average mean free path of phonon [79]. For a good thermoelectric material, it should have intrinsically low thermal conductivity. In the recent past halide perovskites are found to show ultralow thermal conductivity, with different ion compositions or dimensions.

Experimentally to measure the thermal transport properties of the compounds, different approaches are developed. Figure 11.12 shows the schematic diagram for the thermal conductivity measurements systems and related results. Figure 11.12 (a) represent the **two-probe configuration** for measurement of low-temperature thermal conductivity (k) using TTO/PPMS (top panel), and variation of heat power and signal versus time (bottom panel) (a). The variation of low-temperature k for skutterudites compounds with and without thermometer calibrations shown in Fig. 11.12b.

Figure 11.12c shows the principle of thermal diffusivity (l) measurement by the **laser flash method (LFA)**. In this method an optical fiber laser beam acts as a light

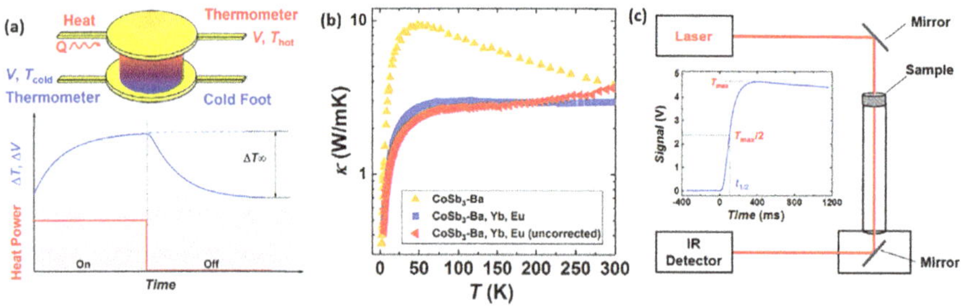

Fig. 11.12 **a** Two-probe configuration for measurement of low-temperature thermal conductivity (k) using TTO/PPMS (top panel), and variation of heat power and signal versus time (bottom panel) (a). **b** and low-temperature k for skutterudites with and without thermometer calibrations. **c** Principle of thermal diffusivity (l) measurement by the laser flash method (LFA). Reproduced with permission of ref. [71]

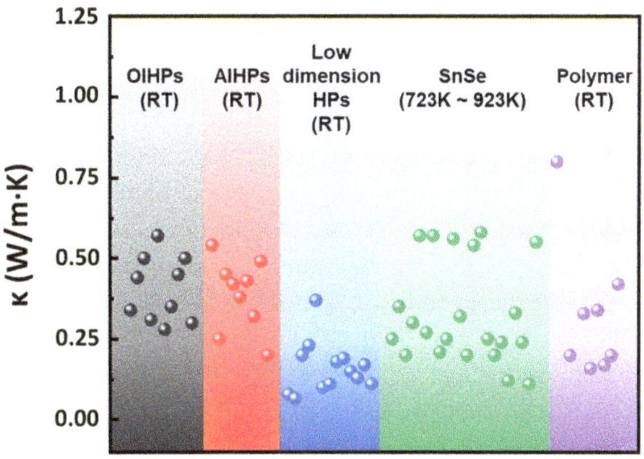

Fig. 11.13 Variation of thermal conductivity (k) for different materials. Reproduced with permission of ref. [72]

pulse to supply an energy source and temperature rise is measured by a thermocouple. When the high-intensity laser beam heats the front surface of a sample, it generates a transient temperature rise which is measured with the help of thermocouple. This temperature rise profile, recorded with high precision, reveals the crucial information about the material's thermal properties. This method basically measures the thermal diffusivity (α) and specific heat capacity (C_p) and density (ρ) are used to calculate thermal conductivity using the following relation;

$\alpha = \frac{\kappa}{\rho.C_p}$, where is the thermal diffusivity, κ is known for thermal conductivity, is the density of the material and is the specific heat capacity of the material. Inset of Fig. 11.12c shows the correct signal shape to measure the thermal diffusivity, κ for the accurate estimation of the thermal conductivity of the samples.

Before the Laser Flash Method, thermal conductivity measurement relied on techniques like the **steady-state method**, which involved establishing a constant temperature difference across a sample and measuring the heat flow. While accurate, this method was time-consuming and required specific sample configurations. Figure 11.13 shows the variation of thermal conductivity (k) for different materials. It clearly demonstrates the reported values for the thermal conductivities of organic-inorganic halide perovskites (OIHPs), all-inorganic halide perovskites (AIHPs), low dimension halide perovskites, SnSe and polymers systems.

11.5.4 Power Factor

The **Power Factor (PF)** is a key metric for assessing thermoelectric performance, independent of thermal conductivity. It is the product of electrical conductivity and Seebeck coefficient, i.e., $PF = S^2\sigma$. A high-power factor indicates effective electrical performance and is critical in optimizing ZT.

11.6 Halide (and Hybrid) Perovskites as Thermoelectric Materials

Halide and hybrid perovskites have gained significant attention not only for their success in photovoltaics but also for their emerging potential in thermoelectric (TE) applications. Their soft lattice structures, low thermal conductivity, and tunable electronic properties make them appealing candidates for energy conversion at low to medium temperatures. While traditional TE materials such as Bi_2Te_3 and PbTe have dominated the field, the advent of halide-based perovskites, particularly hybrid organic–inorganic halide perovskites (HOIPs) and inorganic cesium-based halide perovskites, presents new opportunities for sustainable and tunable thermoelectric. Figure 11.14 shows the possible applications and thematic evolution of lead-free perovskites in computational studies.

As already discussed in the previous section, halide perovskites are known to experience ultralow thermal conductivity and when this feature is combined with high Seebeck coefficient, a large zT is expected. It makes them a promising candidate for the study of thermoelectric properties. However, the electrical conductivity remains a major problem for halide perovskites.

11.6.1 *Experimental Studies*

Saini et al. studied the hybrid-halide perovskite thin films for thermoelectric applications. They studied the thermoelectric parameters in the temperature range of 30 °C to 50 °C at different annealing time (5 to 15 min) [81]. Figure 11.15a–c shows the variation of electrical conductivity, Seebeck coefficient and power factor for thin

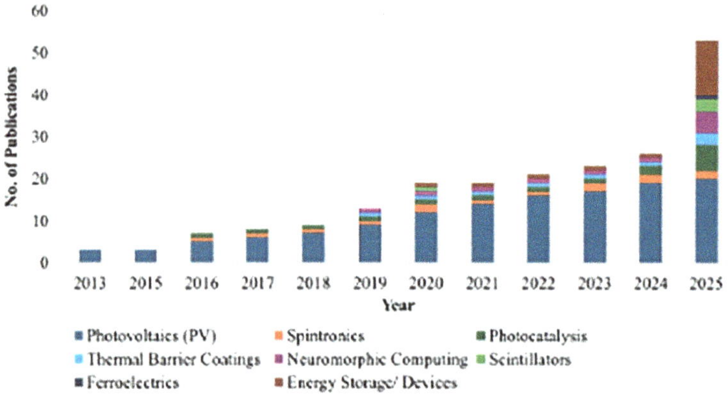

Fig. 11.14 Possible applications and thematic evolution of lead-free perovskites in computational studies. Reproduced with permission of ref. [80]

films of halide-based perovskite, $MASnI_3$. They were found that $MASnI_3$ thin films annealed for 5 min at 100 °C shows the best thermoelectric performance.

In the recent past several experimental studies were performed on halide-based lead-free perovskite materials for their applications in thermoelectric devices. Asker et al. performed the study the doping effect on zero-dimensional inorganic halide

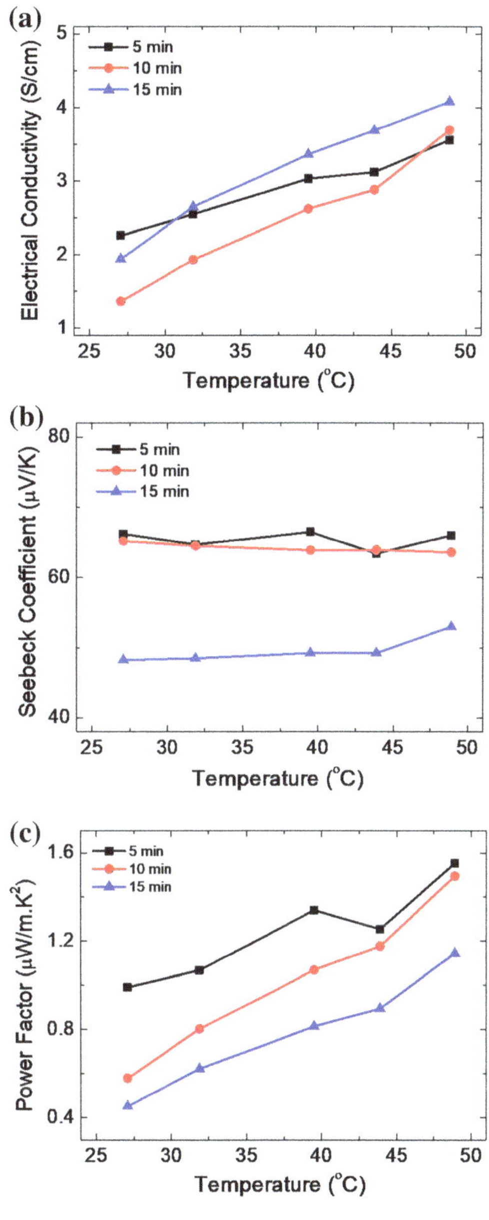

Fig. 11.15 Temperature dependence of thermoelectric parameters of $MASnI_3$ thin films annealed at 100 °C for 5 min, 10 min, and 15 min: **a** electrical conductivity; **b** Seebeck coefficient and **c** power factor. Reproduced with permission of ref. [81]

perovskite derivative, $Cs_3Cu_2I_5$ to enhance the zT parameter. They found that thermal conductivity is in the ultralow regime and decreases with doping [82].

11.6.2 Theoretical Studies

As mentioned, there are techniques to measure these properties experimentally. We can also approximate the value of these parameters using theoretical predictions. Density functional theory provides us a way to approximate and predict the behavior of the materials and determine some key parameters. Thermoelectric properties can be determined using BoltzTraP code package. It provides us the information of the electronic part using the semi-classical Boltzmann transport equation. The lattice information is given by Gibbs package, where with the help of Slack's formula, we can determine the lattice thermal conductivity to approximate the zT parameter. To accurately measure zT, we can also use phono3py package which calculates the κ_l by introducing the lattice vibrations. Figures 11.16, 11.17 and 11.18 are the theoretical plots generated for a double halide perovskite Cs_2IngCl_6 by Haque et al. These figures show us how Seebeck coefficient, electrical conductivity, thermal conductivity (electronic and lattice), power factor, and figure of merit varies with temperature for a specified carrier concentration. This material has an experimental bandgap of ~3.2 eV whereas theoretical calculations suggest it to be ~3.67eV by using TB-mBJ [22]. The experimental and theoretical measurements are in close approximations.

It can be clearly observed in Fig. 11.16, that the κ_l is decreasing with increase in temperature, which usually happens, since at high temperature the contribution of lattice diminishes, whereas the κ_e dominates at higher temperature due to its semiconducting nature. In Fig. 11.16, the S trends opposite to the σ, κ_e as already discussed they show opposite behaviour. Although due to the presence of large bandgap the seebeck coefficient is significantly large at lower temperatures. σ, κ_e seems to dominate at higher temperatures. In Fig. 11.17, the PF is the product of S and σ and it increases rapidly at higher temperature. The zT turns out to be 0.94 for TB-mBJ potential at 700 K.

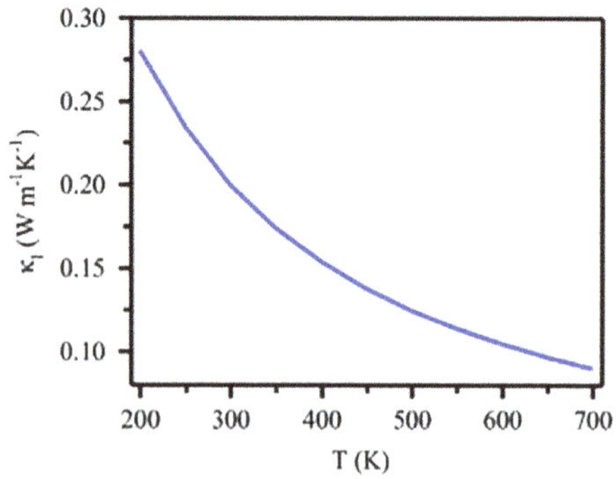

Fig. 11.16 Lattice thermal conductivity calculated using Phono3py for $Cs_2InAgCl_6$. Reproduced with permission of ref. [22]

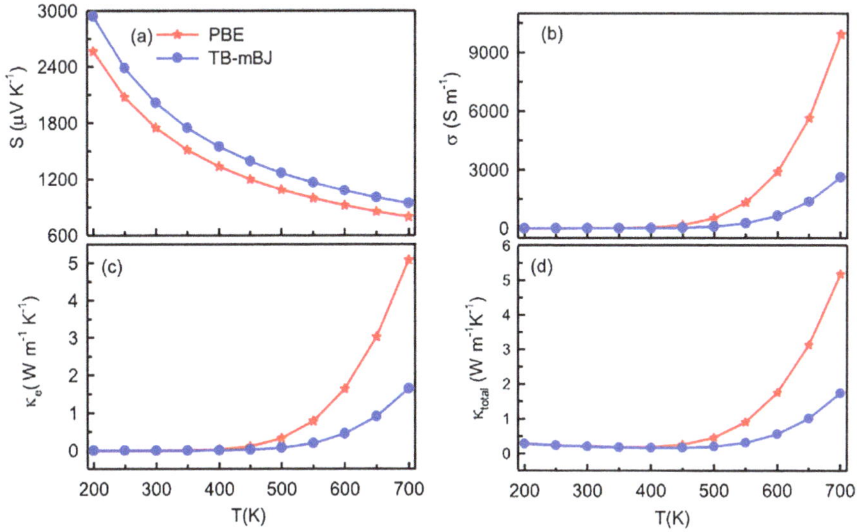

Fig. 11.17 Temperature dependence of thermoelectric transport properties of Cs2InAgCl6 at intrinsic carrier concentration ~3×10^{16} cm^{-3}: **a** Seebeck coefficient (S), **b** electrical conductivity (s/t), **c** electronic part of the thermal conductivity (ke/t), and **d** total thermal conductivity (k). Reproduced with permission of ref. [22]. These plots are generated using WIEN2k DFT simulation package along with BoltzTraP code

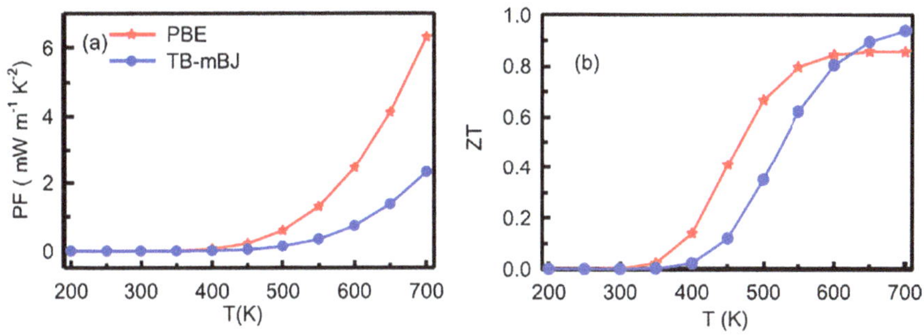

Fig. 11.18 Temperature dependence of **a** Power factor (S2s) and **b** thermoelectric figure of merit of Cs2InAgCl6 at intrinsic carrier concentration ~3×10^{16} cm^{-3}. Reproduced with permission of ref. [22]. These plots are generated using WIEN2k DFT simulation package along with BoltzTraP code

11.6.3 Comparison Between Oxide Versus Halide Perovskites

While oxide perovskites dominate in high-temperature, durable applications, halide and hybrid perovskites are emerging for low-temperature, lightweight, and flexible devices, albeit with significant challenges in stability, conductivity, and toxicity (Table 11.1).

Table 11.1 Tabulates the comparison between oxide and halide based perovskies in view of different parameters to characterize the good thermoelectric materials [80–85]

Property	Oxide Perovskites (e.g., $Ca_3Co_4O_9$)	Halide Perovskites (e.g., $MAPbI_3$, $CsSnI_3$)
Thermal conductivity (κ)	Moderate to high (~ 1–10 W/m·K)	Very low (~ 0.2–0.5 W/m·K)
Seebeck coefficient (S)	Moderate to high (~ 100–200 μV/K)	High (~ 200–400 μV/K)
Electrical conductivity (σ)	High (metallic behavior in some)	Low–moderate (requires doping)
Stability	Excellent (high T, oxidative stability)	Poor–moderate (sensitive to heat/ moisture)
Toxicity	Typically low (depending on elements)	Lead toxicity concern (Pb-based variants)
Flexibility/Processability	Brittle ceramics, high-temperature sintering	Solution-processed, flexible films possible
Band structure tunability	Limited	Highly tunable (via halide/cation substitution)
Applications	High-temp TEGs, industrial recovery	Low-temp TEGs, wearable thermoelectrics

Halide and hybrid perovskites exhibit a unique set of properties desirable for thermoelectric applications: ultralow thermal conductivity, tunable electronic structure, and solution-process ability. Experimental and theoretical studies demonstrate their potential, especially in the low-to-mid temperature range. However, improvements in material stability, carrier mobility, and non-toxic alternatives are critical for real-world applications. Comparative assessments with oxide perovskites highlight distinct advantages and limitations, making each class suitable for different thermoelectric regimes.

11.7 Strategies to Enhance the Thermoelectric Performance

The thermoelectric performance of a material is governed by its dimensionless figure of merit (zT). Maximizing zT is challenging due to the interdependent nature of these parameters. Hence, tailored strategies are essential to decouple or optimize them simultaneously. Several attempts have been made to enhance the zT, which ensures the amount of energy conversion from heat to electrical energy. Researchers still face certain limitations while dealing with bulk materials. Though there are certain ways which acts as a game changer in the field of thermoelectricity. This section outlines the most effective strategies for enhancing zT in perovskite-based and other

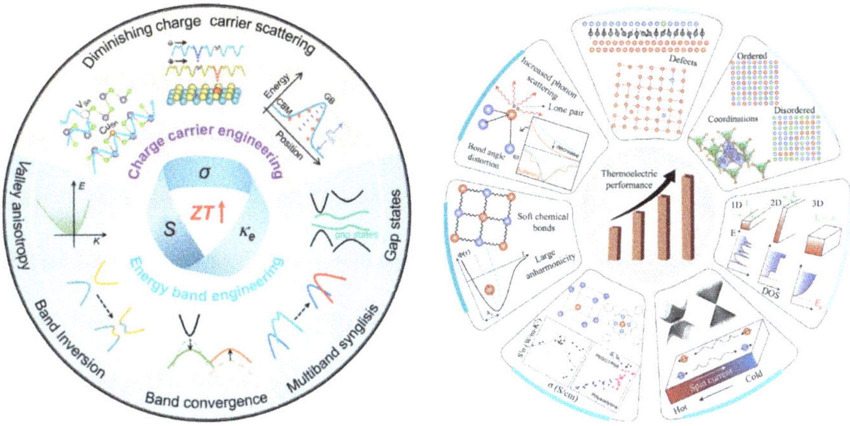

Fig. 11.19 Schematic showing different strategies to enhance the thermoelectric performance. Reproduced with permission of ref. [85, 86]

thermoelectric materials. Figure 11.19 shows the schematic for different strategies to enhance the thermoelectric performance.

11.7.1 Band Engineering (Doping and Alloying)

Band engineering involves modifying the electronic band structure to optimize carrier concentration, mobility, and density of states (DOS), thereby improving both Seebeck coefficient and electrical conductivity. To enhance the zT parameter, doping, i.e., introducing the extrinsic elements to control carrier concentration and mobility in the host material is the efficient and significant technique. This is responsible for the modification of the DOS profile, phonon scattering, carrier concentration process in the host materials. By introducing such type of elements (*n-type or p-type dopant) basically* adjusts the Fermi level position in the host material by enhancing the hole/electron concentration. Similarly, by substituting the atoms on the A, B, or X-sites in perovskites (e.g., halide or metal site substitution) thermoelectric properties can be tuned. It causes reduction in lattice thermal conductivity by introducing mass fluctuation and strain scattering. It also responsible to converge the multiple conduction bands, which enhances density-of-states effective mass without compromising mobility.

11.7.2 Nano-structuring and Nanocomposites

Nano structuring and nanocomposites formation is another important technique to enhance the thermoelectric performance. Such type of modulation, i.e., reducing grain size or introducing nanoscale features increases phonon scattering at grain boundaries or interfaces, which significantly lowers the value of lattice thermal conductivity (κ_l). Overall, nano-structuring is one such method which breaks the inter-dependency of the S, σ *and* κ. Now, it becomes much easier to alter a particular parameter without affecting the other.

11.7.3 Dimensional and Structural Control

Dimensional modification via reducing the dimensionality (from 3D to 2D, 1D, or 0D) of the prepared compounds can increase carrier confinement and density of states which enhances the Seebeck coefficient. In low dimensional materials, sharp density of states near the fermi level was found which attributed to improved S. In these structures due to classical and quantum size effects, reduced κl can be observed which brings significant improvement to the zT parameter [85].

11.7.4 Defect Engineering and Vacancy Control

Point defects, such as vacancies, antisites, or interstitials, can be deliberately introduced or controlled to modify both electronic and thermal transport. Defects can be intrinsic or extrinsic. Though intrinsic defect will not have any effect on the carrier concentration but it will certainly break the lattice symmetry and thus reduce lattice thermal conductivity. Adding extrinsic defect to the materials results in increased phonon scattering essential for low thermal conductivity, high carrier concentration which is required for high conductivity. However, introducing defect to a material may increase or decrease the Seebeck coefficient. Thus, we have to be very specific with our choices in introducing defects to a crystal. Vacancy generation also affects the band-structure and DOS profile by introducing degenerate bands near the fermi level, which significantly improves the carrier mobility and bandgap [85, 86].

Overall, to enhance the zT parameter of thermoelectric materials, requires a multi-pronged strategy by combining both electronic and phononic optimizations as tabulated in the table below:

Strategy	zT Component Improved	Typical Methods
Band engineering	↑ S, ↑ σ	Doping, alloying
Nano-structuring	↓ κ	Nano-inclusions, superlattices

(continued)

(continued)

Strategy	zT Component Improved	Typical Methods
Dimensional control	↑ S, ↓ κ	2D/1D perovskites, layer stacking
Defect engineering	↑ S, ↓ κ	Controlled vacancy creation
High-pressure processing	↑ σ, ↓ κ	SPS, hot pressing, phase control

These approaches are often complementary. For halide perovskites, integrating dimensional control, doping, and defect tuning under mild synthesis conditions may offer the most promising route toward high-efficiency, low-cost thermoelectric materials.

11.8 Future Scope and Challenges

The growing emphasis on environmentally friendly and sustainable technologies has driven research into lead-free perovskite materials for thermoelectric (TE) applications. Lead-free alternatives such as tin (Sn)-based, bismuth (Bi)-based, and double perovskite structures present lower toxicity and competitive thermoelectric properties, opening new avenues in energy conversion. The focus is driven towards such materials which remain unharmed by the presence of heat, light, moisture, oxygen. Since these parameters have the potential to degrade a material and reduce its life expectancy and also effective conversion efficiency.

These lead-free perovskites are useful and offer promising solutions for harvesting waste heat from low-to-medium temperature sources such as industrial exhausts, vehicle engines, and electronic devices. Their relatively low processing temperatures and solution-process ability enable integration into flexible and unconventional substrates deployment in wearable thermoelectric harvesters/generators (TEGs). Their compact, lightweight, and efficient thermoelectric modules could power sensors and communication devices in remote or off-grid locations, eliminating dependence on batteries. Although still underexplored, lead-free perovskites could potentially be engineered for thermoelectric cooling applications, providing localized temperature control in electronics or medical devices without harmful refrigerants.

Despite promising experimental and theoretical advancements, several challenges must be addressed to realize the full potential of lead-free perovskite thermoelectric.

1. **Stability Issues**:

- Environmental sensitivity (moisture, oxygen, thermal stress) remains a major hurdle, especially for Sn-based perovskites, which are prone to oxidation ($Sn^{2+} \rightarrow Sn^{4+}$).
- The degradation of such materials can also cause the release of harmful gases which is the cause of concern for environment, children, and every other living being.

- Developing encapsulation techniques, more robust crystal chemistries (e.g., double perovskites like $Cs_2AgBiBr_6$), and all-inorganic compositions can improve long-term stability.

2. **Optimizing Carrier Transport**:

- Strategies to enhance electrical conductivity without compromising the Seebeck coefficient are essential.
- Controlled doping, alloying, and defect engineering remain key areas of research to achieve high carrier mobility and optimize carrier concentration.

3. **Reducing Thermal Conductivity**:

- Despite intrinsically low lattice thermal conductivity, further reductions are possible through nanostructuring and compositional disorder.

4. **Eco-Friendly and Scalable Synthesis**:

- Developing green synthesis routes that avoid toxic solvents and high-energy processes will aid commercial viability.
- Large-area, solution-processable films or bulk materials compatible with roll-to-roll manufacturing are desirable for cost-effective mass production.

5. **Materials Discovery and Design**:

- Machine learning and high-throughput computational screening are expected to accelerate discovery of novel lead-free perovskites with tailored thermoelectric properties.
- Exploring new A, B-site cations and halide anions combinations, including double and layered perovskites, expands the compositional space.

The quest for non-toxic, efficient, and versatile thermoelectric materials positions lead-free perovskites as a frontier of materials science. There is still a lot to develop and outlook in the field of perovskite based thermoelectric materials. With rapid advances in understanding their structure-property relationships and processing methods, they hold great promise for future sustainable energy technologies. With the pace researchers are looking for new ways of enhancing zT by nano-structuring, alloying, defect production will definitely provide us materials with high conversion efficiency. Synergistic collaboration between experimentalists, theorists, and engineers will accelerate the translation of lead-free perovskite thermoelectric from the lab to practical devices, contributing to energy-efficient electronics, wearable technologies, and smart environmental sensors.

11.9 Conclusions

Recent advancements in lead-free perovskite materials have significantly advanced the field of thermoelectric energy conversion by addressing the critical challenges of toxicity, environmental stability, and performance optimization. Tin-based, bismuth-based, and double perovskite compositions have emerged as promising candidates demonstrating improved Seebeck coefficients, electrical conductivity, and inherently low thermal conductivity—key attributes for achieving high thermoelectric figure of merit (zT). Key strategies such as band engineering (doping and alloying), nano-structuring, dimensional control, and defect engineering have been effectively employed to optimize the intricate balance among electrical and thermal transport properties. These approaches have enabled enhanced carrier mobility, reduced lattice thermal conductivity, and favorable electronic band structures tailored for thermoelectric performance. Despite notable progress, challenges related to material stability, reproducibility, and scalable eco-friendly synthesis persist and must be overcome to translate laboratory success into practical applications. Continued efforts in understanding fundamental charge-phonon interactions, improving environmental robustness, and integrating computational materials design will accelerate the discovery of novel lead-free perovskites with superior thermoelectric efficiency. Looking forward, the multifunctional nature of lead-free perovskites—combining optoelectronic, thermoelectric, and mechanical properties positions them uniquely for next-generation wearable, flexible, and sustainable energy-harvesting devices. With coordinated interdisciplinary research, lead-free perovskites hold substantial promise to contribute to efficient, low-cost, and environmentally benign thermoelectric technologies that support global energy sustainability goals.

References

1. G.J. Snyder, E.S. Toberer, Complex thermoelectric materials. Nat. Mater. **7**(2), 105–114 (2008)
2. J.P. Heremans, V. Jovovic, E.S. Toberer, A. Saramat, K. Kurosaki, A. Charoenphakdee, S. Yamanaka, G.J. Snyder, Enhancement of thermoelectric efficiency in PbTe by distortion of the electronic density of states. Science **321**(5888), 554–557 (2008)
3. A. Kojima, K. Teshima, Y. Shirai, T. Miyasaka, Organometal halide perovskites as visible-light sensitizers for photovoltaic cells. J. Am. Chem. Soc. **131**(17), 6050–6051 (2009)
4. Z. Xiao, W. Meng, J. Wang, D.B. Mitzi, Y. Yan, Searching for promising new perovskite-based thermoelectric materials. Mater. Horiz. **5**(3), 500–509 (2018)
5. G.A. Slack, New materials and performance limits for thermoelectric cooling. In *CRC Handbook of Thermoelectrics* (CRC Press, (1995)) (pp. 407–440)
6. C.C. Stoumpos, C.D. Malliakas, M.G. Kanatzidis, Semiconducting tin and lead iodide perovskites with organic cations: phase transitions, high mobilities, and near-infrared photoluminescent properties. Inorg. Chem. **52**(15), 9019–9038 (2013)
7. L.D. Zhao, G. Tan, S. Hao, J. He, Y. Pei, H. Chi, H. Wang, S. Gong, H. Xu, V.P. Dravid, C. Uher, J. Snyder, C. Wolverton, M.G. Kanatzidis, Ultrahigh power factor and thermoelectric performance in hole-doped single-crystal SnSe. Science **351**(6269), 141–144 (2016)
8. J. He, T.M. Tritt, Advances in thermoelectric materials research: looking back and moving forward. Science **357**(6358), eaak9997 (2017)

9. O. Muller, R. Roy, *The Major Ternary Structural Families* (Springer-Verlag, 1974)
10. Q. Fan et al., Lead-free halide perovskite nanocrystals: crystal structures, synthesis, stabilities, and optical properties. Angew. Chem. Int. Ed. **59**(3), 1030–1046 (2020)
11. M.V. Kovalenko, L. Protesescu, M.I. Bodnarchuk, Properties and potential optoelectronic applications of lead halide perovskite nanocrystals. Science **358**(6364), 745–750 (2017)
12. T.C. Jellicoe et al., Synthesis and optical properties of lead-free cesium tin halide perovskite nanocrystals. J. Am. Chem. Soc. **138**(9), 2941–2944 (2016)
13. Z. Tan et al., Highly efficient blue-emitting bi-doped Cs_2SnCl_6 perovskite variant: photoluminescence induced by impurity doping. Adv. Funct. Mater. **28**(29), 1801131 (2018)
14. X. Wu et al., Synthesis of lead-free CsGeI3 perovskite colloidal nanocrystals and electron beam-induced transformations. Chem. Asian J. **13**(13), 1654–1659 (2018)
15. M.E. Lines, A.M. Glass, *Principles and Applications of Ferroelectrics and Related Materials* (Oxford University Press, 2001)
16. M.A. Green, A. Ho-Baillie, H.J. Snaith, The emergence of perovskite solar cells. Nat. Photonics **8**(7), 506–514 (2014)
17. A. Pisoni et al., Ultra-low thermal conductivity in organic–inorganic hybrid perovskite CH3NH3PbI3. J. Phys. Chem. Lett. **5**(14), 2488–2492 (2014)
18. S.-D. Guo, J.-L. Wang, Potential thermoelectric materials $CsMI_3$ (M=Sn and Pb) in perovskite structures from first-principles calculations. RSC Adv. **6**(103), 101552–101559 (2016)
19. M.A.H. Shah et al., Exploring pressure induced thermoelectric properties of $LiAeH_3$ (Ae=Ca, Sr, Ba) perovskite hydrides along with optoelectronic features. Mater. Sci. Semicond. Process. **137**, 106238 (2022)
20. M.Y. Sofi et al., Exploring the lead-free halide Cs_2MGaBr_6 (M= Li, Na) double perovskites for sustainable energy applications. Sci. Rep. **14**(1), 5520 (2024)
21. S.A. Khandy, D.C. Gupta, Electronic structure, magnetism and thermoelectric properties of double perovskite Sr2HoNbO6. J. Magn. Magn. Mater. **458**, 176–182 (2018)
22. E. Haque, M. Anwar Hossain, Electronic, phonon transport and thermoelectric properties of Cs2InAgCl6 from first-principles study. Comput. Condens. Matter **19**, e00374 (2019)
23. F. Zhang et al., Advances in two-dimensional organic–inorganic hybrid perovskites. Energy Environ. Sci. **13**(4), 1154–1186 (2020)
24. F. Cao, P. Zhang, L. Li, Multidimensional perovskite solar cells. Fundam. Res. **2**(2), 237–253 (2022)
25. C. Sun et al., [DMEDA] PbCl 4: a one-dimensional organic lead halide perovskite with efficient yellow emission. Cryst. Eng. Comm. **22**(8), 1480–1486 (2020)
26. Z. Gan et al., The optical properties of Cs_4PbBr_6–$CsPbBr_3$ perovskite composites. Nanoscale **11**(31), 14676–14683 (2019)
27. B. Saparov, D.B. Mitzi, Organic–inorganic perovskites: structural versatility for functional materials design. Chem. Rev. **116**(7), 4558–4596 (2016)
28. L. Zhang et al., Advances in the application of perovskite materials. Nano Micro Lett. **15**(1), 177 (2023)
29. H. Wang et al., Recent progress in large-area perovskite photovoltaic modules. Trans. Tianjin Univ. **28**(5), 323–340 (2022)
30. W.-L. Hong et al., Efficient Low-temperature solution-processed lead-free perovskite infrared light-emitting diodes. Adv. Mater. (Deerfield Beach, Fla.) **28**(36), 8029–8036 (2016)
31. V. Chauhan et al., Prospects for lead free perovskite for photovoltaic applications and biological impacts: challenges and opportunities. Inorg. Chem. Commun. **157**, 111421 (2023)
32. K. Ahmad et al., A (CH3NH3) $3Bi_2I_9$ Perovskite based on a two-step deposition method: lead-free, highly stable, and with enhanced photovoltaic performance. Chem. Electro. Chem. **6**(4), 1192–1198 (2019)
33. H. Zeng et al., Thermal evaporation of lead-free inorganic perovskite CsGeI3 for photodetection. Appl. Phys. Lett. **121**, 10 (2022)
34. M. Han et al., Controllable growth of lead-free all-inorganic perovskite nanowire array with fast and stable near-infrared photodetection. J. Phys. Chem. C **123**(28), 17566–17573 (2019)

35. F. Yang et al., Lead-free perovskites: growth, properties, and applications. Sci. China Mater. **64**(12), 2889–2914 (2021)
36. I. Chung et al., CsSnI3: semiconductor or metal? High electrical conductivity and strong near-infrared photoluminescence from a single material. High hole mobility and phase-transitions. J. Am. Chem. Soc. **134**(20), 8579–8587 (2012)
37. J. Zhou et al., Composition design, optical gap and stability investigations of lead-free halide double perovskite $Cs_2AgInCl_6$. J. Mater. Chem. A **5**(29), 15031–15037 (2017)
38. X. Zhu, Y. Yue, F. Li, A review on thermoelectric energy harvesting from asphalt pavement: Configuration, performance and future. Constr. Build. Mater. **228**, 116818 (2019)
39. T.J. Seebeck, *Magnetische polarisation der metalle und erze durch temperatur-differenz*, vol 70 (Engelmann, 1895)
40. S. Baru, S. Bhatia, A review on thermoelectric cooling technology and its applications. IOP Conf. Ser. Mater. Sci. Eng. **912**(4) (2020)
41. EDM Rowe, D. Ph, D. Sc, F. Group, *Handbook* (2006)
42. Y.H. Choi et al., Simple tuning of carrier type in topological insulator Bi_2Se_3 by Mn doping. Appl. Phys. Lett. **101**, 15 (2012)
43. E. Ashalley et al., Bismuth telluride nanostructures: preparation, thermoelectric properties and topological insulating effect. Front. Mater. Sci. **9**(2), 103–125 (2015)
44. Y. Bai et al., Boosting the thermoelectric performance of n-type Bi_2S_3 by compositing rGO. J. Alloys Compd. **933**, 167814 (2023)
45. M. Kim, D. Park, J. Kim, Self-healing boron-doped Sb_2Se_3 thermoelectric materials prepared using liquid metallic Ga–Sn alloys. J. Mater. Sci. **58**(22), 9251–9263 (2023)
46. H. Qin et al., Achieving a high average zT value in Sb_2Te_3-based segmented thermoelectric materials. ACS Appl. Mater. Interfaces **12**(1), 945–952 (2019)
47. T. Nasr, H.M.-M. Ben, N. Kamoun-Turki, First-principles study of electronic, thermoelectric and thermal properties of Sb_2S_3. J. Alloys Compd. **663**, 123–127 (2016)
48. O. Falkenbach et al., Thermoelectric properties of nanostructured bismuth-doped lead telluride $Bi_x(PbTe)_{1-x}$ prepared by co-ball-milling. J. Electron. Mater. **43**(6), 1674–1680 (2014)
49. J. Androulakis et al., High-temperature charge and thermal transport properties of the n-type thermoelectric material PbSe. Phys. Rev. B Condens. Matter Mater. Phys. **84**(15), 155207 (2011)
50. Y. Li et al., Enhanced thermoelectric performance of hot-press Bi-doped n-type polycrystalline PbS. Mater. Sci. Semiconductor Process. **121**, 105393 (2021)
51. R. Moshwan et al., Eco-friendly SnTe thermoelectric materials: progress and future challenges. Adv. Funct. Mater. **27**(43), 1703278 (2017)
52. Y.-M. Han et al., Thermoelectric performance of SnS and SnS–SnSe solid solution. J. Mater. Chem. A **3**(8), 4555–4559 (2015)
53. Y. Zhai et al., Thermoelectric properties of $In_{1.3-x}Sn_xSe$ prepared by spark plasma sintering method. J. Alloys Compd. **553**, 270–272 (2013)
54. J.Y. Cho et al., Thermoelectric properties of spark plasma-sintered In_4Se_3–In_4Te_3. J. Electron. Mater. **40**(5), 1024–1028 (2011)
55. J. Tabernor, P. Christian, P. O'Brien, A general route to nanodimensional powders of indium chalcogenides. J. Mater. Chem. **16**(21), 2082–2087 (2006)
56. D. Guo et al., The n-and p-type thermoelectricity property of GeTe by first-principles study. J. Alloys Compd. **810**, 151838 (2019)
57. S. Hao et al., Computational prediction of high thermoelectric performance in hole doped layered GeSe. Chem. Mater. **28**(9), 3218–3226 (2016)
58. S. Roychowdhury et al., Germanium chalcogenide thermoelectrics: electronic structure modulation and low lattice thermal conductivity. Chem. Mater. **30**(17), 5799–5813 (2018)
59. T. Noguchi, Powder processing of thermoelectric materials-focusing on SiGe with new sintering technique, in *XVI ICT'97. Proceedings ICT'97. 16th International Conference on Thermoelectrics (Cat No. 97TH8291)*, (IEEE, 1997)
60. D. Cheikh et al., Praseodymium telluride: a high-temperature, high-ZT thermoelectric material. Joule **2**(4), 698–709 (2018)

61. U. Aydemir et al., Thermoelectric properties of the Zintl phases $Yb_5M_2Sb_6$ (M=Al, Ga, In). Dalton Trans. **44**(15), 6767–6774 (2015)
62. N.S. Chauhan et al., Compositional tuning of ZrNiSn half-Heusler alloys: thermoelectric characteristics and performance analysis. J. Phys. Chem. Solids **123**, 105–112 (2018)
63. T. Zhu et al., Compromise and synergy in high-efficiency thermoelectric materials. Adv. Mater. **29**(14), 1605884 (2017)
64. I. Terasaki, Y. Sasago, K. Uchinokura, Large thermoelectric power in $NaCo_2O_4$ single crystals. Phys. Rev. B **56**(20), R12685 (1997)
65. K.A. Borup et al., Measuring thermoelectric transport properties of materials. Energy Environ. Sci. **8**(2), 423–435 (2015)
66. G.J. Snyder, E.S. Toberer, Comply thermoelectric materials. Nat. Mater. *7*, 105–114 (2008)
67. A. Sarbajna et al., Inorganic-based printed thermoelectric materials and devices. Adv. Eng. Mater. **25**(2), 2200980 (2023)
68. S.-T. Ha et al., Laser cooling of organic–inorganic lead halide perovskites. Nat. Photonics **10**(2), 115–121 (2016)
69. Z. Slanina, M. Uhlik, V. Sladecek, Cooling device with peltier element for medical applications. IFAC-PapersOnLine **51**(6), 54–59 (2018)
70. A.T. Recanzone et al., Fabrication of an inexpensive, implantable cooling device. J. Neurophysiol. **107**, 3543–3558 (2012)
71. T.R. Wei, M. Guan, J. Yu, T. Zhu, L. Chen, X. Shi, How to measure thermoelectric properties reliably. Joule *2*(11), 2183–2188 (2018)
72. Y. Zhou et al., Recent progress of halide perovskites for thermoelectric application. Nano Energy **94**, 106949 (2022)
73. R.A.Ş.İ.T. Ahiska, S. Dislitas, G. Omer, A new method and computer-controlled system for measuring the time constant of real thermoelectric modules. Energy Convers. Manag. **53**(1), 314–321 (2012)
74. S.B. Riffat, X. Ma, Thermoelectrics: a review of present and potential applications. Appl. Therm. Eng. **23**(8), 913–935 (2003)
75. X. Niu, Y. Jianlin, S. Wang, Experimental study on low-temperature waste heat thermoelectric generator. J. Power Sour. **188**(2), 621–626 (2009)
76. M.A. Tupta, Resistivity measurements using a four-point collinear probe. EE Eval. Eng. **55**, 11 (2016)
77. A.A. Ramadan, R.D. Gould, A. Ashour, On the Van der Pauw method of resistivity measurements. Thin Solid Films **239**(2), 272–275 (1994)
78. J. de Boor, V. Schmidt, Efficient thermoelectric van der Pauw measurements. Appl. Phys. Lett. **99**, 2 (2011)
79. W.-L. Ong et al., Orientational order controls crystalline and amorphous thermal transport in superatomic crystals. Nat. Mater. **16**(1), 83–88 (2017)
80. S.K. Fatima et al., Lead-free perovskites for next-generation applications: a comprehensive computational and data-driven review. Mater. Adv. **6**(21), 7634–7661 (2025)
81. S. Saini et al., Hybrid-halide perovskite thin film growth for thermoelectric applications. J. Electron. Mater. **49**(5), 2890–2894 (2020)
82. C. Asker et al., Doping and thermoelectric properties of the zero-dimensional inorganic halide perovskite derivative, $Cs_3Cu_2I_5$. J. Mater. Chem. A **13**(31), 26009–26018 (2025)
83. M.W. Gaultois et al., Data-driven review of thermoelectric materials: performance and resource considerations. Chem. Mater. **25**(15), 2911–2920 (2013)
84. Y. Shao, Y. Yuan, J. Huang, Correlation of energy disorder and open-circuit voltage in hybrid perovskite solar cells. Nat. Energy **1**(1), 1–6 (2016)
85. H. Ming et al., Strategies and prospects for high-performance Te-free thermoelectric materials. Chem. Rev. **125**(7), 3932–3975 (2025)
86. M. Mukherjee, A. Srivastava, A.K. Singh, Recent advances in designing thermoelectric materials. J. Mater. Chem. C **10**(35), 12524–12555 (2022)

Chapter 12
Progress in Lead Free Perovskite Materials for Magnetic Applications

Rekha Gupta and Anurag Tyagi

Abstract Environment friendly lead-free multifunctional perovskites have gained significant research interest recently due to potential magnetic applications. Substitution of toxic Pb^{2+} with nontoxic magnetically active transition metal ions can give rise to a wide range of magnetic properties including ferromagnetism, ferrimagnetism, and spin glass behavior. Inorganic oxide perovskites having general formula ABO_3 provides liberty to accept different cations at A and B site to give rise robust magnetic ordering and magnetoelectric coupling for applications in magnetic sensors and spintronics. In parallel, double perovskites having formula $A_2BB'O_6$ represent half metallic magnetic properties essential for spin filters and high sensitivity magnetic sensors. Lanthanum based manganite perovskites have been explored for their robust magnetoresistive properties ideal for magnetic field sensor applications. Recent advances in synthesis, characterization techniques and theoretical modelling have enhanced the performance and stability of lead-free perovskite materials. Even though the current research is going on to address the challenges like low Curie temperature and stability, lead free magnetic perovskites emerge as promising materials for sustainable and innovative applications. In this chapter, a comprehensive study of various classes of lead-free magnetic perovskites have been presented. Lead free transition metal oxides, multiferroic perovskites, manganites and double perovskites have been studied in detail with their potential applications.

Keywords Perovskites · Exchange interaction · Multiferroics · Magnetic refrigeration · Magnetoresistance

R. Gupta (✉) · A. Tyagi
Department of Physics, Noida Institute of Engineering and Technology, Greater Noida, Affiliated to Dr. A. P. J. Abdul Kalam Technical University, Lucknow, Uttar Pradesh, India
e-mail: rkh.gpt01@gmail.com

V. Chauhan et al. (eds.), *Sustainable Lead-Free Perovskite Materials*, Topics in Applied Physics 157, https://doi.org/10.1007/978-3-032-24873-2_12

12.1 Introduction

Lead free perovskites have gained lot of research attention recently due to their versatile magnetic properties and applications. Traditionally, lead-based perovskites have dominated the research landscape, particularly in the field of optoelectronics and photovoltaics. However, growing environmental and health concerns associated with lead toxicity have shifted the research focus toward lead-free alternatives for technological applications. Lead-free perovskites, which maintain the characteristic ABO_3 or $A_2BB'O_6$ structure, offer an eco-friendly route to explore the magnetic behaviour of multifunctional materials. These compounds represent a broad range of phase diagrams and magnetic ground states due to a competing exchange interaction between localized spins of transition metal ions in which orbital and lattice degree of freedoms are coupled [1]. On the other hand, ferromagnetic interactions are mediated by itinerant charge carriers via double-exchange mechanism [2]. These interactions results in various magnetic phenomena, such as ferromagnetism, antiferromagnetism, ferrimagnetism, and spin-glass behaviour in lead free perovskites. Additionally, magnetic properties of these perovskites can be tuned using chemical substitution, strain engineering, or dimensional control. Lead-free magnetic perovskites exist in various classes including multiferroics, half metallic perovskites, double perovskites with ordered B cations and hybrid organic-inorganic perovskites. The potential applications of these materials exist in data storage, magnetoresistive sensors, magnetic refrigeration, and magnetic tunnelling devices. Their study not only addresses environmental safety but also opens new frontiers in designing next-generation multifunctional materials. In this chapter, potential lead-free magnetic perovskites have been focused to understand their structural and magnetic aspects that underlie their diverse functionalities. Different classes of magnetic perovskites, their magnetic interactions and applications have been comprehended to understand the key factors influencing their technological applications. A brief discussion on the challenges and future directions in the field of lead-free magnetic perovskites has been presented aiming to inspire further research into the design and application of these materials.

12.2 Basic Structure of Perovskite

The basic structure of all perovskite compounds is represented as ABX_3, where A, B are the cations of the different ionic radius such that the ionic radius of A is larger than B and X represent an anion. Here, A, B and X can be the mixture of various organic and inorganic ions which are generally oxide, sulfide, halide, and nitride [3]. Primarily perovskites can be divided into three categories: organic metal halide perovskites, alkaline metal halide perovskites, and inorganic oxide perovskites [3]. Inorganic perovskites, also referred as transition metal perovskite having general formula ABO_3 is the most significant class of ideal perovskite structure. Here, cation

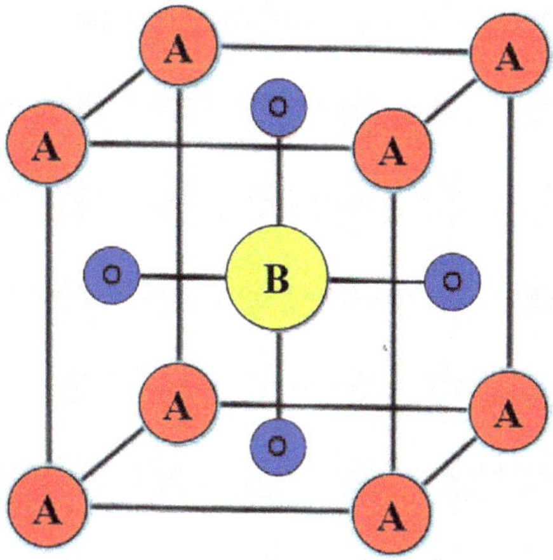

Fig. 12.1 Ideal structure of perovskite

A is an alkali earth metal ion (Ca, Ba, Sr) or Lanthanides (La, Ce, Nd) with slightly larger ionic radius than B cation, while cation B is a transition metal ion (Mn, Ti, Fe) and O is an oxygen anion in the composition of 1:1:3 [3]. Due to the intrinsic capacity of ABO_3 structure to accept alternative replacements at the cation sites, different possible ionic candidates permit the development of various perovskite structures with partial substitutions at both the A and B sites resulting in a broad variety of compounds with distinct cations [4]. The ABO_3 structure can be summarized as a cubic unit cell with B cation occupy at the body centre position (1/2, 1/2, 1/2) (yellow), A ions at the corner positions (0, 0, 0) (red), and oxygen atoms at the face centre of the edges (1/2, 1/2, 0) (blue), shown in Fig. 12.1. In a perovskite cubic unit cell, A cations have 12-fold coordination while B have six-fold coordination surrounded with oxygen anions forming a network of regular corner linked BO_6 octahedra in a stable cubic perovskite structure. The interrelated BO_6 octahedral pattern is recurring throughout the structure and plays a significant role in determining the ferromagnetic properties of perovskite structure [5].

The ideal cubic structure in perovskites as discussed above is rarely observed in perovskites due to structural distortion caused by various factors including, cation displacement, tolerance factor, Jahn-Teller distortion, and octahedral tilting [6]. The structural distortion occurs to accommodate the size mismatch between A and B site cations so that the BO_6 octahedra tilt to reduce symmetry from cubic to orthorhombic, tetragonal, or rhombohedral structure [6]. In the distorted octahedral structure, A-O distances are not same, that decreases the effective coordination of A cations. The degree of distortion in the perovskite structure is measured by tolerance factor or t-factor as given in eq. (12.1):

$$t = \frac{r_A + r_O}{\sqrt{2} r_B + r_O} \tag{12.1}$$

where r_A, r_B and r_O are the ionic radii of cations A, B, and anions O respectively. It can be evidenced from eq. (12.1) that t-factor depends on the ionic radius of A and B cations. Tolerance factor enhances with the increment in ionic radius of A cation (r_A) and reduce with the decreases in ionic radius of B cation (r_B). The symmetry of the crystal is predicted by the value of t-factor. The crystal adopts cubic symmetry only if the t-factor is close to unity ($0.9 < t < 1$) [7]. The symmetry reduces to tetragonal or rhombohedral to orthorhombic or monoclinic as the value of t-factor lowers from unity ($0.75 < t < 0.9$). The stable perovskite structure is found in the range $0.75 < t < 1$ and the crystal adopt hexagonal symmetry if the tolerance factor is greater than unity [7].

Apart from the cation-anion bond strength and relative ionic size, the Jahn-Teller effect has been identified as a major electronic contributor to the structural deformation in oxide perovskites. Jahn-Teller distortion occurs primarily because spontaneous tetragonal elongation can drastically reduce the energy of transition-metal ion d-orbital electrons in a way that eliminates the degeneracy of the d-orbital levels [6]. For example, the B site cations such as Mn^{3+} and Cu^{2+} are commonly known as Jhan-Teller cations. These cations have one unpaired electron in e_g orbit, Mn^{3+} has high spin ($^3d_4$) while Cu^{2+} has low spin ($^3d_9$) and represents splitting of $d_{x^2-y^2}$ and d_{z^2} levels of the e_g subshells followed by the energy stabilization through tetragonal distortion of oxygen octahedra [6]. This distortion affects the physical, chemical, electrical, and magnetic characteristics of perovskite materials and lead to a reduction in the symmetry of their crystal structures. Similarly, a strong Jahn–Teller distortion due to the Mn^{3+} ion (d^4 configuration) in $LaMnO_3$, results in orbital ordering and cooperative distortions of the MnO_6 octahedra which is essential to its antiferromagnetic and insulating behavior [8].

12.3 Double Perovskite

Double perovskites are getting attention in the field of magnetic research due to structural flexibility and interchange between spin and charge of material leading to different magnetic properties. Double perovskites emerge from the ordered arrangement of two different B-site cations. The chemical formula for double perovskite is $A_2B_{1-x}B'_xO_6$ ($0 < x < 1$), where A is divalent or trivalent cation occupied by the rare earth metals/alkaline ion while B or B′ cations are occupied by d-block transition metal ions or lanthanides having the small ionic radius [9]. The unit cell is doubled along all three axes in $A_2BB'O_6$ double perovskite structure shown in Fig. 12.2, which results from the arrangement of two distinct octahedral cations in a three-dimensional pattern, known as rock salt ordering [9]. In $A_2BB'O_6$ double perovskite structure, B site ion is partially filled while B′ ion has a vacant orbital and vice versa. The structural distortion occurs due to the different ionic radius of A and B or B′ sites, which control the bond angles as well as the magnetic transition temperature. The structural distortion further affects the bond lengths of B–O–B′–O–B that enhances the functionality of double perovskites due to active interaction between charge and

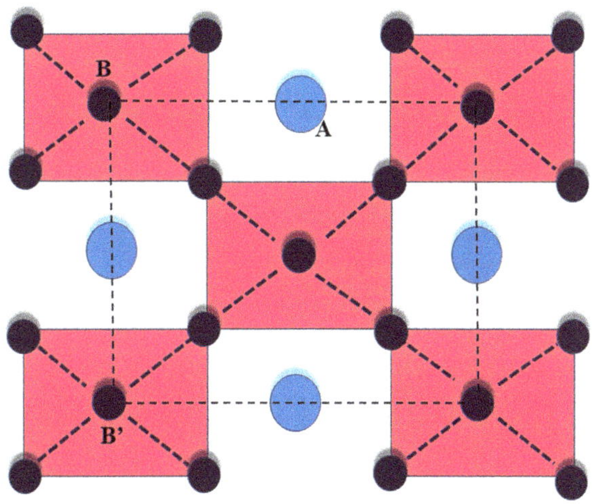

Fig. 12.2 Double perovskite structure

spin [9]. The electronic arrangements at B site cation, the octahedral distortion and selection of non-magnetic cations at A, B and B′ site regulates the magnetic ordering and applications of double perovskites. A crucial feature of double perovskites is 180° B–O–B′ interaction between different d-metal cations which can be exploited to study various magnetic properties.

Another important magnetic property of $A_2BB'O_6$ double perovskites is half metallicity wherein a material behaves as conductor for electrons having one kind of spin orientation e.g. spin up (+ 1/2) and insulator for the opposite orientation spin down (− 1/2) [10]. This property is crucial for applications in spintronic devices where a control over spin currents is desired. A primary half metallic double perovskite is Sr_2FeMoO_6 (SFMO), where the half metallicity arises from the interplay between the Fe and Mo ions and the surrounding oxygen atoms [11]. The Fe ions in SFMO are typically in a high-spin Fe^{3+} ($3d^5$, S = 5/2) state, while the Mo ions are in a strongly covalent Mo^{5+} ($4d^1$, S = 1/2) state [11]. Electron hopping occurs between the Fe and Mo sites, mediated by the intermediate oxygen ions, primarily involving the t_{2g} orbitals of both Fe and Mo. This hopping occurs predominantly in minority spin band, resulting in negative spin polarization of conduction electrons. Hence the carriers at Fermi level are primarily from the spin-down channel and give rise to overall metallic nature of SFMO [11].

12.4 Origin of Magnetism in Perovskites

Magnetism in perovskite materials originates from the magnetic interactions between transition metal ions at B-site and their interactions with surrounding oxygen ions primarily determine the magnetic properties of these materials. Magnetism in these compounds arises from the interactions between unpaired d-electrons of transition

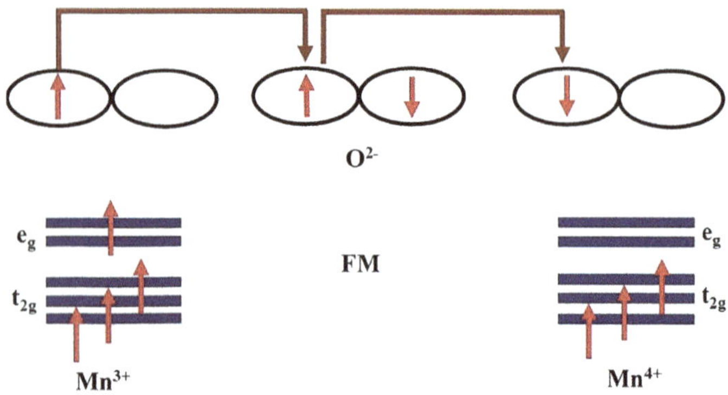

Fig. 12.3 Double exchange interaction

metals mediated through oxygen ions or via direct overlap. Understanding magnetic interactions in perovskites is crucial for tailoring their magnetic and multiferroic properties for applications in spintronics, memory devices, and sensors.

The magnetic interactions in perovskites can be classified as the following types.

12.4.1 Double Exchange Interaction

This type of interaction arises in materials having ions with different oxidation states. For example, in $La_{1-x}Sr_xMnO_3$ (LSMO) the Mn ions can exist in mixed valence states, typically Mn^{3+} and Mn^{4+}. Magnetic interaction involves the transfer of electrons between mixed-valent cations (e.g., Mn^{3+} and Mn^{4+}) through oxygen, aligning spins for efficient hopping of electrons [12]. This spin alignment is energetically favorable and leads to the ferromagnetic alignment of neighboring Mn ions shown in Fig. 12.3. The ferromagnetic alignment of Mn spins due to the double-exchange interaction contributes to the overall ferromagnetic behavior of LSMO. The ability of charge carriers to hop between Mn ions leads to high electrical conductivity, especially in the ferromagnetic metallic state.

12.4.2 Superexchange Interaction

In the case of perovskite manganites, e.g. $LaMnO_3$, the FM interaction is governed by superexchange interaction. If two adjacent atoms have one orbital fully occupied, and other orbital is vacant and they point towards each other. The electron will spend part of the time at the empty orbital of the other ion shown in Fig. 12.4. These interactions are termed as superexchange interactions [13]. $LaMnO_3$ has Mn^{3+} ions with one electron in the e_g orbitals, the e_g orbitals can overlap with oxygen 2p

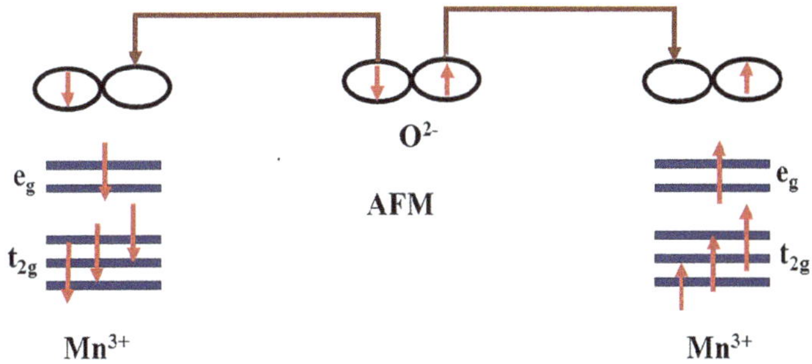

Fig. 12.4 Super exchange interaction

orbitals, leading to a transfer of electrons. This interaction is strongest when the Mn–O–Mn (B–O–B′) bond angle is close to 180 degrees. This is because the electron transfer through the oxygen 2p orbital prefers an antiparallel spin configuration to minimize energy and leads to A-type AFM ordering [13]. Superexchange interaction can be ferromagnetic or antiferromagnetic depending upon the type of interacting orbitals of magnetic ion. According to Goodenough-Kanamori-Anderson (GKA) rule, if the interacting orbitals are half filled and (B–O–B′) is 180°, the exchange is antiferromagnetic and strong shown in Fig. 12.4. On the other hand, if interacting orbitals overlap with different orthogonal p-orbitals and B–O–B′ bond is 90°, the exchange interaction is weak ferromagnetic [13].

12.4.3 Synthesis and Characterization Techniques

The synthesis and characterization of lead-free magnetic perovskites play a critical role in determining their structural integrity, magnetic performance, and functional applicability. To optimize crystallinity, phase purity, and magnetic ordering, various synthesis methods have been adopted while their magnetic and microscopic properties have been characterized by advanced characterization tools.

12.5 Synthesis Techniques

The magnetic properties of the lead-free perovskite materials can be significantly controlled by their microstructure, crystallographic orientation, and cation distribution. The selection of suitable synthesis method is crucial for tailoring magnetic properties of perovskites. Some of the commonly used synthesis methods are discussed below:

12.5.1 Solid-State Reaction Method

This method is the most popular route for synthesis of magnetic perovskite oxides due to its ease of use, scalability, and affordability. The facile synthesis steps involve first, mixing of high purity precursor powders usually metal oxides, carbonates, or nitrates in stochiometric ratio. The homogeneous mixing of the precursors is ensured by long hours manual grinding or ball milling. After through mixing, the obtained powder is calcined at high temperature, often between 800 °C and 1300 °C, to promote solid-state reactions and diffusion. After initial calcination step, the obtained powder is again reground and recalcined to obtain desired phase and crystallinity. Solid state synthesis route is mostly used for synthesis of bulk magnetic perovskites such as ABO_3 and $A_2BB'O_6$ structures [14]. The drawbacks of this method are, course particle size and inhomogeneities. This method also requires hours of grinding and longer reaction time and high temperature for desired phase formation compared to other methods.

12.5.2 Sol-Gel and Pechini Methods

The sol gel reaction method is widely used chemical route for synthesis of perovskite oxide structures due to low temperature facile synthesis steps, and excellent composition control over other chemical methods. This method involves, first the formation of a homogeneous mixture known as "Sol" by dissolving metal nitrates and oxides in a suitable solvent. To promote complexation and control the hydrolysis and condensation, suitable chelating agents like citric acid or ethylene glycol are used during sol formation. Under controlled heating, the gelation of sol results into the formation of the formation of three-dimensional polymeric network formation known as "Gel." This gel is further dried and calcined at moderate temperature to obtain fine particles. This method provides homogeneous distribution of ions at atomic level so that to synthesize doped or mixed-metal perovskites with tailored structural and magnetic properties [15].

12.5.3 Hydrothermal and Solvothermal Synthesis

The other widely used chemical methods for synthesis of magnetic perovskites particularly hybrid compounds in a solvent are Hydrothermal and solvothermal synthesis methods [16]. In this method aqueous solvents (water for hydrothermal) and (organic solvent for solvothermal) are used to dissolve and recrystallize materials. The reactions take place in a sealed autoclave at high temperature and pressure condition. With increased temperature of water, the dielectric constant of water decreases. In a critical region the decreased solubility of solute promotes high supersaturation in solution that results in nucleation and growth of crystals. From this method thermodynamically stable nanoparticles, complex oxides can be synthesized. However, requirement of expensive autoclave and safety during the reaction limits the application of these methods.

12.5.4 Combustion and Auto-Ignition Methods

The auto-ignition method, also known as combustion method is a facile and low-cost technique for synthesizing perovskite materials. In this method the precursor materials such as oxalates, nitrates, acetates are used to get easily dissolved in water. Organic fuels such as citric acid, glycine, urea are used to make a fuel-air mixture. In combustion methods, fuel-air mixture is deliberately ignited using an external energy source, such as a spark or flame, and the temperature, emission products, and heat release are measured to assess combustion efficiency [17]. These methods are critical in studying the thermal stability, reactivity, and ignition delay characteristics of fuels and catalytic materials.

12.6 Thin Film Deposition Techniques

12.6.1 Pulsed Laser Deposition

PLD is a frequently used synthesis technique for growing high-quality perovskite thin films. In this method, a high-power laser pulse is focused on a desired solid perovskite target fixed on a target holder placed in a low-pressure chamber. The most employed laser used in PLD are excimer or Nd:YAG due to high photonic energy and non-thermal ablation. Each laser pulse ablates material from the target, creating a high-energy plasma plume that travels toward a heated substrate, where the atoms deposit to form a crystalline film. PLD offers superiority over other thin film deposition techniques for creating high quality, multicomponent oxides thin films. This technique also offers advantage of stoichiometric transfer of target components, uniform film thickness, and crystallinity [18]. However, depositing uniform film over large area, high energy consumption and complex equipment setup remains major challenges of this technique.

12.6.2 Magnetron Sputtering

Another widely used method for high quality magnetic perovskite thin film synthesis is magnetron sputtering. In this method the ceramic target material of desired stoichiometry is placed in a low-pressure chamber where energetic ions (usually Ar^+) in a plasma environment collide with the target [18]. The high energy ions transfer their energy to the target and atoms are ejected from the target. These atoms get later deposited on a heated substrate placed in the chamber at suitable temperate and pressure conditions. The magnetron arrangement near the target, confines the electrons and enhances the ionization and deposition rates. The film thickens can be precisely controlled by controlling deposition parameters alike temperature and pressure. This technique provides uniform deposition over a larger area, excellent composition control, and high crystallinity. The electronic, magnetic, and structural properties can also be tailored by tuning deposition parameters such as power, gas pressure, and substrate temperature, making this technique suitable for synthesis of lead-free perovskite thin films.

12.6.3 Molecular Beam Epitaxy (MBE)

MBE is thermal vaporization deposition technique for synthesis of high-quality epitaxial perovskite materials thin films. This method is particularly used when atomic-level control over composition and thickness is required [18]. In this method molecules of desired composition are heated to high temperatures so that they are present in gaseous state. The gaseous precursor is then focused in the form of beams on to heated substate in an ultra-high vacuum chamber. The atoms from the beam condensate on the heated substrate to form high quality, epitaxial thin films of uniform thickness. This method is used for growing epitaxial thin films of complex perovskites with sharp interfaces and controlled stoichiometry. The real time feedback of in situ deposition process and film growth is provided by Reflection High-Energy Electron Diffraction (RHEED) technique. This technique has advantage of growing high quality heterostructures and thin films and superlattices to study interface driven magnetism and magnetoresistance.

12.6.4 Chemical Vapor Deposition (CVD)

Magnetic perovskite thin films can be synthesized using Chemical Vapor Deposition (CVD). It is a chemical route process and magnetic perovskites thin film is synthesized by chemical reaction of precursors on a heated substrate placed in a chamber at suitable temperature and pressure. The chemical vapor deposition process is a non-equilibrium thermally driven process where a precise control over the precursor vapor and solution determines crystallinity and phase composition of deposited films. This method is useful for scalable, large area thin film deposition with excellent control over morphology.

12.7 Structural and Magnetic Characterization

To understand the functionalities of magnetic perovskites the following characterization techniques are commonly used.

12.7.1 X-Ray Diffraction (XRD)

X-ray Diffraction (XRD) is the most fundamental characterization techniques for structural and phase purity analysis bulk as well as thin film samples. In this method the incident X-rays interact with the periodic lattice at the sample surface and get diffracted. The diffracted rays satisfy the Braggs equation given by: $n\lambda = 2d \sin\theta$, where λ represents the wavelength of the X-ray used, θ is the angle of diffraction and d represents the interplanar distance. The diffracted rays consist of a set of diffracted intensities at different angle of diffraction, that gives the fingerprint of the lattice structure and crystallinity [19]. This method is also widely employed to identify, lattice strain, lattice impurities, octahedral tilting in perovskites and degree of cation ordering. Furthermore, precise determination of atomic positions and bonding environment can also be analyzed by Rietveld refinement of XRD patterns. These advantages of XRD makes it an indispensable tool in the development and optimization of perovskite-based materials for magnetic, electronic, and optoelectronic applications. However, requirement of some tenth gram of sample, lower detection limit, and requirement of standard reference are some limitations of XRD technique.

12.7.2 Scanning and Transmission Electron Microscopy (SEM/TEM)

Scanning Electron Microscopy (SEM) and Transmission Electron Microscopy (TEM) are powerful techniques widely used to investigate the surface/interface features such as morphology, phase distribution, and presence of defects at atomic level resolution. SEM includes irradiation of sample with electron beam of energy ranging from 0.5 KeV to 40 KeV. The scattered electrons from the sample change their trajectory and provides high-resolution surface imaging due to difference in surface topology. SEM is particularly useful for assessing the surface and interface quality of thin films and heterostructures. On the other hand, TEM can offer both an image and diffraction from the sample which makes it an essential technique in the study of nanomaterials. TEM also offers much higher spatial resolution (better than 0.2 nm) and allows direct imaging of the internal crystal structure, lattice fringes, and defects such as dislocations, stacking faults, and antiphase boundaries. Additionally, High-resolution TEM (HRTEM) and selected area electron diffraction (SAED) can be employed to determine crystallinity, phase purity, and epitaxial relationships in perovskites structures. Energy Dispersive X-ray Spectroscopy (EDS) coupled with TEM, enables elemental mapping and compositional analysis at the nanoscale. In this way, utilizing SEM and TEM together one can get critical insights for understanding the structure–property relationships in perovskite nanostructures to optimize their performance [19].

12.7.3 X-Ray Photoelectron Spectroscopy (XPS)

X-ray Photoelectron Spectroscopy (XPS) is used for the study of elemental composition and oxidation states of the constituent atoms within the top few nanometers of the material. This is surface sensitive technique and works on the principle of photoelectric effect. In this method, the sample surface is irradiated by soft monochromatic X-ray beam (Mg K_α or Al K_α), resulting in excitation of the electrons of the sample and ejecting them from atoms. The kinetic energy of the ejected electrons in definite time (number/sec) is detected by a detector to give rise peaks in spectrum at definite kinetic energies. The position of peak is used to identify the elemental composition. In perovskites structures XPS is particularly useful in identifying the stoichiometry of A-site and B-site cations. The change in valence of cations such as $Fe^{2+}/Fe^{3+}$and Mn^{3+}/Mn^{4+} in perovskites can also be observed that is crucial to determine the electron configuration and magnetic properties of material. The degree of cation ordering in perovskites can also be determined by XPS by analyzing the core-level binding energy shifts and peak intensities [19].

12.8 Raman and FTIR Spectroscopy

Raman and Fourier Transform Infrared (FTIR) spectroscopy are two complementary nondestructive techniques that are used for probing the vibrational properties of perovskites. These techniques are crucial for analyzing the chemical structure and bonds and phase of perovskite structures. Raman spectroscopy is based on the inelastic interaction of light with the chemical bonds of the materials. The inelastically scattered photons containing stocks and ant stocks Raman lines gives fingerprints of the chemical structure of the material. Raman spectroscopy is particularly sensitive to symmetry changes during distortion and crucial to identify octahedral distortions and cation ordering in perovskites. On the other hand, FTIR spectroscopy is used to provide information about vibrational modes related to functional groups and chemical composition and bond (metal–oxygen) stretching. FTIR spectroscopy is based upon the absorption and transmission of infrared light by a sample. The fingerprint absorption spectra generated in FTIR is particularly useful for identifying bond strengths, impurities, and the presence of organic or volatile components in hybrid or doped perovskites [19].

12.9 Magnetic Measurements

The magnetic measurements play a crucial role in determining magnetic properties of lead-free magnetic perovskite. The fundamental magnetic properties like spin ordering, magnetic nature and magnetic transition are characterized by magnetic measurements. The key parameters like magnetization, coercivity, Curie or Neels transition temperature are determined by magnetometry techniques such as vibrating sample magnetometry (VSM) and superconducting quantum interference device (SQUID) magnetometry.

Magnetic measurements play a crucial role in characterizing the magnetic properties of perovskite materials, providing insights into their magnetic ordering, spin dynamics, and electronic interactions. Techniques such as vibrating sample magnetometry (VSM) and superconducting quantum interference device (SQUID) magnetometry are widely used to determine key parameters like magnetization (M), coercivity (H_c), remanence (M_r), and Curie or Néel temperatures (T_c/T_N) [19]. VSM allows the measurement of induced magnetization as a function of applied magnetic field when the sample vibrates in uniform applied magnetic field. By analyzing M-H loop one can determine the magnetic behavior (ferro, ferri or antiferromagnetic) of the given sample. This technique also allows measurement of magnetization as a function of temperature over broad temperature range (low as well as high) to determine magnetic transition temperature and spin glass behavior of the perovskite samples. VSM offers advantages of high accuracy, wide temperature range of measurement and fast data collection. However, VSM cannot detect extremely weak magnetic signals which can be measured by SQUID magnetometry. A squid magnetometer consists of superconducting transducer based on Josephson's junction which can detect very small changes in magnetic field providing ultra-high sensitivity. Magnetic characterization techniques are essential to understand the fundamental magnetic properties and specific applications of magnetic perovskites in spintronics, magnetoresistance, magnetic refrigeration etc.

12.10 Electron Spin Resonance (ESR) and Nuclear Magnetic Resonance (NMR)

The local magnetic properties of perovskites particularly magnetic ordering, defects and vacancies, spin dynamics are probed by electron spin resonance (ESR) spectroscopy. ESR spectroscopy works on the principle of absorption of electromagnetic radiation by one or more unpaired electrons in a magnetic sample when placed in external magnetic field [19]. This technique is also known as Electron Paramagnetic Resonance (EPR) and it is highly selective towards unpaired electrons or paramagnetic materials. Hence ESR is widely employed to study the dynamics of unpaired electrons or defect states within the perovskite lattice. It also provides insights for detection of paramagnetic centers such as Mn^{2+}, Fe^{3+}, or oxygen vacancies in magnetic perovskites. On the other hand, Nuclear magnetic resonance spectroscopy (NMR) is based on the absorption of radio frequency waves by certain atomic nuclei (e.g., ^{1}H, ^{93}Nb, ^{207}Pb), placed strong magnetic field. This technique is useful in determination of A site and B site cation distribution in perovskite structures. Additionally, It can be used to identify, phase segregation, halide mixing and spin dynamics to determine fundamental magnetic properties of materials.

12.11 Neutron Diffraction

Neutron diffraction technique is used for determination of magnetic ordering of cations, which is crucial for determination of magnetic structure of perovskites. This technique offers advantage over X ray diffraction that neutrons interact with atomic nuclei and possess magnetic moment that enables them to scatter off the magnetic moment of electrons, so one can accurately determine atomic and magnetic structure [19]. This technique is used for precise determination of element position (Oxygen), bond length, and octahedral distortions. In double perovskites neutron diffraction is instrumental in resolving B/B' site ordering and detecting subtle magnetic transitions or spin canting phenomena that are invisible to conventional X-ray methods. Moreover, neutrons carry a magnetic moment, making neutron diffraction especially valuable for probing magnetic structures and ordering in perovskites, including antiferromagnetic, ferromagnetic, and canted spin arrangements. The technique thus plays a critical role in correlating structure-property relationships in functional perovskite oxides used in magnetism, catalysis, and multiferroics.

12.12 Magnetoresistance Measurements

The change in the resistance of the material due to the applied magnetic field is known as Magnetoresistance. Lead free magnetic perovskites offer potential applications for magnetoresistive sensors. The experimental measurement of magnetoresistance contains a four-probe sample holder, measures the voltage across the inner probe when constant current is passed through outer probe. The sample is placed on the sample holder and the sample holder is placed in the applied magnetic field either transverse or in longitudinal positions. The magnetic field is varied and the resistance is recorded at different strengths of applied magnetic field. The calculated magnetoresistance (MR) is given by the equation: $MR = \frac{R(H)-R(0)}{R(0)} \times 100\%$, where R(H) represents the material resistance at applied magnetic field H while R(0) is the material resistance in the absence of field. The magnetoresistance measurements are crucial for determining applications of lead-free perovskites in magnetoresistive sensors, and data storage.

12.13 Lead Free Transition Metal Oxide Perovskite

Transition metal oxide perovskites are the most promising class of perovskite materials that offer unique magnetic properties due to effective charge associated with displacement of ions. In case of transition metal perovskites, the p-orbital is fully filled, while the transition metal d-orbital is partially filled or vacant. As the d-orbital is partially filled, material shows magnetic ordering [20]. The origin of magnetism in transition metal oxide perovskites arises primarily from the interactions between the unpaired d-electrons of transition metal cations and the surrounding oxygen anions in the crystal lattice. In these ABO_3-structured materials, the B-site is typically occupied by a transition metal ion (such as Mn, Fe, Co, or Ni), which can exhibit various oxidation states and electron configurations. Magnetic behavior of these materials is largely governed by super exchange or double exchange interactions between transition metal ions mediated by the oxygen ions, depending upon valence states and orbital overlaps. Superexchange interaction occurs when the transition metal ions are in different spin states and separated by a nonmagnetic anion like oxygen resulting in an antiferromagnetic coupling mechanism. In contrast, double exchange interaction is observed when mixed-valence states (e.g., Mn^{3+}/Mn^{4+}) allow electrons to hop between sites, aligning spins to minimize energy often leading to ferromagnetism [13]. The strength and nature of these magnetic interactions are highly sensitive to factors like bond angles, lattice distortions and doping, making perovskite oxides a rich platform for exploring tunable magnetic phenomena e.g., colossal magnetoresistance, electronic phase separation, multiferroics and spin-glass behavior. In this context, we have summarized some of the extensively investigated transition metal oxides for magnetic applications.

$LaMnO_3$ is the parent compound of colossal magnetoresistance (CMR) manganites. $LaMnO_3$ serves as a prototype material for studying strong correlations among charge, lattice degrees of freedom, spin and orbital. $LaMnO_3$ crystallizes in orthorhombically distorted perovskite structure (space group P_{nma}) [21]. Each Mn ion is octahedrally coordinated by oxygen in the form of corner sharing MnO_6 octahedra. The Mn ions are in +3 oxidation state (Mn^{3+}, $3d^4$ configuration) with one electron in the doubly degenerate e_g orbital. The strong electron-lattice coupling caused by the Jahn-Teller distortion lifts orbital degeneracy and leads to orbital ordering [21]. The electronic structure is strongly influenced by electron correlation effects, making undoped $LaMnO_3$ a Mott insulator. The combination of orbital ordering and superexchange interactions dictates the magnetic ordering of the system. In its undoped form, $LaMnO_3$ is an antiferromagnetic insulator with an orthorhombic structure at room temperature. The Mn^{3+} ions possess a high-spin state, and the magnetic ordering arises due to superexchange interactions between neighboring Mn ions via oxygen atoms shown in Fig. 12.3. The magnetic ordering temperature (Néel temperature) of $LaMnO_3$ is around 140 K. When $LaMnO_3$ is doped with divalent cations like Sr^{2+} or Ca^{2+} (e.g., $La_{1-x}Sr_xMnO_3$), some Mn^{3+} ions are oxidized to Mn^{4+}, and double-exchange interactions between Mn^{3+} and Mn^{4+} lead to ferromagnetism and metallic conductivity [12]. These mixed-valence manganites exhibit colossal magnetoresistance, making them highly useful in magnetic sensors and memory technologies. $La_{1-x}Sr_xMnO_3$ (LSMO) is a widely studied mixed-valence manganite perovskite known for exhibiting colossal magnetoresistance (CMR), a drastic reduction in electrical resistance upon the application of a magnetic field. The origin of this magnetoresistance lies in the double exchange mechanism between Mn^{3+} and Mn^{4+} ions, mediated by oxygen atoms. When LSMO is doped with Sr at $x = 0.3$ concentration it shows, strong spin polarization and half metallic ferromagnetic character below Curie temperature (~ 370 K). $La_{0.7}\ Sr_{0.3}MnO_3$ shows strong CMR effect near Curie temperature for promising applications in magnetic sensors, data storage. Other magnetoresistive effects such as low field magnetoresistance due to grain boundary tunnelling and anisotropic magnetoresistance can be induced in thin films and nano particles of LSMO. However, challenges remain in maintaining high CMR at room temperature under low magnetic fields and integrating LSMO with other materials in device architectures.

Other than magnetoresistive manganites, a novel perovskite ferrite, Bismuth ferrite ($BiFeO_3$) is a prominent multiferroic material exhibiting both ferroelectric and antiferromagnetic orderings at room temperature. Due to unique coexistence of electric and magnetic properties in $BiFeO_3$, it has attracted considerable interest in fundamental research and potential applications in spintronics, sensors and memory devices. $BiFeO_3$ crystallizes in a distorted perovskite structure with rhombohedral symmetry (space group R3c) [22]. The magnetic ordering arises from the Fe^{3+} ions, which exhibit G-type antiferromagnetic (AFM) structure. In this configuration each Fe^{3+} spin is antiferromagnetically coupled with its six nearest neighbors. It results in a long-wavelength cycloidal spiral spin modulation with ~ 62 nm period, suppressing the macroscopic magnetization of $BiFeO_3$ [23]. Efforts such as doping, synthesizing nanoparticles or composites, strain engineering can suppress the antiferromagnetic spiral spin structure of $BiFeO_3$, enhancing ferromagnetism. Weak ferromagnetism originates from the Dzyaloshinskii–Moriya (DM) interaction due to the broken inversion symmetry of the lattice [23]. Due to high antiferromagnetic Neel's transition temperature (643 K), $BiFeO_3$ exhibits unique existence of room temperature ferromagnetism and ferroelectric polarization suitable for applications in magnetoelectric sensors and devices. The antiferromagnetic ordering in $BiFeO_3$ is coupled with the ferroelectric polarization giving rise to magnetoelectric coupling i.e., electric control of magnetization and vice versa. The novel existence of room temperature magnetoelectric (ME) coupling in $BiFeO_3$ has stimulated extensive research to obtain high room temperature ME coupling in $BiFeO_3$ and its derivatives. Magnetoelectric coupling in $BiFeO_3$ has been potentially applied in magnetic sensors, memory devices, magnetic energy harvesters, high frequency FMR devices and multiferroic photovoltaics [22]. Perovskite oxides with the formula $LaRO_3$ (where R is a transition metal like Mn, Cr, or Fe) have garnered significant interest due to their diverse magnetic properties and potential for various applications. These materials exhibit orthorhombic distortion due to size mismatch between La and R ion which causes magnetic interactions. Depending upon the choice of R ion, $LaRO_3$ can represent ferromagnetic, antiferromagnetic or spin glass magnetic behavior [24, 25]. These compounds are of profound interest for magnetocaloric applications.

12.14 Lead Free Double Perovskites

Another class of potential lead-free magnetic perovskites has emerged in the form of double perovskite structure having formula $A_2BB'O_6$, where A is a large cation (such as Sr^{2+}, Ba^{2+}, or La^{3+}), and B and B′ are transition metal ions. The ordered arrangement of two different B site cations in these materials offer tunable magnetic properties due to exchange interaction between B and B′ ions [26]. These eco-friendly materials offer structure flexibility and tunable magnetism suitable for next generation magnetic sensors and components.

Recently, R_2NiCrO_6 (R is rare-earth elements) double perovskites have gained much attention due to complex interplay between the magnetic moments of the transition metals (Ni^{2+} and Cr^{3+}) and the rare-earth ions [27]. These compounds typically show ferrimagnetic ordering arising from antiferromagnetic super exchange interactions between Ni^{2+} (d^8, S = 1) and Cr^{3+} (d^3, S = 3/2) via oxygen anions. The magnetic behavior is further influenced by the 4f electrons of the R^{3+} ions, which can contribute to low-temperature magnetic transitions, spin-glass-like behavior, or enhanced magnetic anisotropy depending on the specific rare-earth element. The degree of B-site ordering also significantly affects magnetic interactions, often resulting in variations in Curie temperature and saturation magnetization. Synthesis of R_2NiCrO_6 (R = Ce, Pr and Nd) double perovskites was carried by Irfan et al. by hydrothermal method and ferromagnetic ordering was observed in R_2NiCrO_6 (RE = Ce, Pr and Nd) double perovskites [27]. Ferrimagnetic spin glass property of La_2CrNiO_6 has been evidenced by many research groups [28]. The half metallic nature and spin polarized electronic band structure of Gd_2MnFeO_6 has also been observed by Ramesh et al. [29]. Similarly, ferromagnetic ordering has also been observed in La_2NiCrO_6, Nd_2CoMnO_6, Pr_2FeCrO_6, Gd_2FeCrO_6, Pr_2FeMnO_6 double perovskites [30–34]. The tunable magnetic properties of R_2NiCrO_6 double perovskites are applicable to spintronic and magnetic tunneling devices.

Materials such as R_2FeCrO_6 (R is rare-earth ions), R_2NiMnO_6 have demonstrated significant magnetocaloric properties near room temperature, due to their tunable magnetic ordering, high magnetic entropy change, and structural flexibility. Recently giant magnetocaloric effect with $-\Delta S_M$ of ~ 38.6 J/kg K and RCP of ~ 418 J/kg at 70 kOe magnetic field was observed in Gd_2FeCrO_6 in cryogenic temperature range [35]. Magnetocaloric properties of R_2NiMnO_6 (R = Pr, Nd, Tb, Ho and Y) synthesized via solid state sintering have been studied by Chakraborty et al [36]. They reported the highest value in Ho_2NiMnO_6 for $\Delta S_M \sim 8.4$ J/kg K at 70 kOe and $RCP \sim 59.18$ J Kg^{-1} at 20 kOe respectively [36]. Room temperature magnetic refrigeration has been reported by Ndiaye et al. in La_2MnRuO_6 and $LaAMnFeO_6$ (A = Ba, Sr, Ca) [37].

12.15 Applications

12.15.1 Magnetoresistance

Magnetoresistance is the property of a material and associated with change in electrical resistance when exposed to the external magnetic field [38]. Magnetoresistance in perovskite thin films and heterostructures is crucial for applications in sensors and magnetic memory devices. Depending upon the structure of nanoparticle and thin films, different types of magnetoresistances can be observed e.g. colossal magnetoresistance (CMR), giant magnetoresistance (GMR) and ordinary magnetoresistance (OMR). GMR is typically observed in multilayer thin films of alternating ferromagnetic and non-magnetic metallic layers. The key functioning of GMR devices is due to spin dependent resistive switching, i.e. electrons can travel through the alternating layers when the magnetic layers are aligned in same direction while a high resistance is offered when the magnetism in the alternating layers is not aligned in same direction. GMR is particularly utilized in hard disc drives and magnetic sensors and it laid the foundation of spintronics. The other prominent effect is colossal magnetoresistance (CMR) observed primarily in manganese-based perovskites. These compounds show a sudden drop in electrical resistance upon application of magnetic field due to alignment of electron spins by double exchange interaction between neighboring spins. The specific oxides that have high magnetoresistive behaviors are $La_{1-x}Sr_xMnO_3$ (LSMO), $La_{1-x}Sr_xMnO_3$ (LCMO), Sr_2FeMoO_6 (SFMO) and $PrBaCo_2O_{5+\delta}$ (PBCO) [39, 40].

12.15.2 Magnetic Refrigeration

Magnetic refrigeration phenomenon is a cooling effect based upon magnetocaloric effect in certain perovskites. Magnetocaloric effect (MEC) is when a certain materials heat when upon application of magnetic field and cool down when magnetic field is removed [41]. Unlike conventional gas-compression refrigerators, magnetic refrigeration is environmentally friendly as it eliminates the use of harmful refrigerants like hydrofluorocarbons (HFCs). The process involves cyclic magnetization and demagnetization of magnetocaloric material, typically rare-earth compounds, or specially engineered perovskites near its magnetic phase transition temperature (often the Curie temperature). This results in a temperature change that can be harnessed for cooling. Typical perovskite materials that show high magnetocaloric effects are $GdMnO_3$, $La_{1-x}Sr_xMnO_3$ (A=Ca, Sr, Ba), Y_2NiMnO_6, Sr_2FeMoO_6 and Gd_2CoMnO_6 [42].

12.15.3 Multiferroic Memories and Spin Valves

Multiferroics materials exhibit two or more primary ferroic properties simultaneously namely ferroelectricity, ferromagnetism or ferroelasticity in a single phase [22]. These materials are of great interest due to the possibility of coupling between electric, magnetic, and structural order parameters, leading to novel functionalities such as magnetoelectric coupling. The magnetoelectric coupling (MEC) in multiferroic materials is generation of dielectric polarization in the presence of external magnetic field or conversely induction of magnetic moment under external electric field. This interactive control of magnetization and electric polarization give rise to applications of multiferroic materials in low power magnetoelectric memories, sensors, electrically tunable microwave device [22]. Numerous perovskite multiferroics and composite multiferroic structures have been investigated since the last decade due to their potential applications in next-generation memory devices (such as magnetoelectric RAM), spintronics, sensors, and actuators. Prominent multiferroic materials include $BiFeO_3$, $TbMnO_3$, and $DyMnO_3$.La_2NiMnO_6, Lu_2CoMnO_6 and Pr_2FeCrO_6 [22]. Some other multiferroic compounds involves K_2SeO_4, Cs_2CdI_4, $LuFe_2O_4$, $CoCrO_4$, etc., but the observed ME coupling is either very small or appear only at low temperatures [43, 44]. High values of magnetoelectric coupling for potential device applications is only observed in multiferroic composites, nano/heterostructures and thin films.

12.15.4 Magnetic Sensing

Magnetic sensing refers to the detection and measurement of magnetic fields using specialized materials or devices known as magnetic sensors [45]. Magnetic sensors can detect the magnitude, direction, and variation of magnetic fields, and they are essential in a wide range of applications, from navigation and automotive systems to biomedical devices and industrial automation. These sensors operate based on various physical principles, such as the Hall effect, magnetoresistance, and magnetic induction.

Magnetic sensors can be broadly categorized into different types such as Hall effect sensors, anisotropic magnetoresistive (AMR) sensors, giant magnetoresistive (GMR) sensors, tunnel magnetoresistive (TMR) sensors, and fluxgate sensors [45]. Among these, GMR and TMR sensors are widely used in high-sensitivity applications due to their excellent performance at nanoscale fields. Recent advances in materials like magnetic perovskites, multiferroics have enabled the development of highly sensitive and compact magnetic sensing devices. Manganite perovskite materials have been most studied for GMR and TMR sensor applications due to their high spin polarization, high transition temperature, and well-developed fabrication technology of thin manganite films particularly in multilayer structures [45]. Half metallic nature of $La_{1/2}Sr_{1/2}MnO_3$ (LSMO) contributes 95% spin polarization and high magnetoresistive effects [46]. GMR is observed in structures where ferromagnetic layers are separated by a thin non-magnetic layer. In LSMO-based spin valves, the GMR arises from the spin-dependent scattering of electrons as they travel through the layers with different magnetic orientations [47]. TMR is observed in magnetic tunnel junction [MTJ], where two ferromagnetic electrodes (like LSMO) are separated by a thin insulating layer. The TMR effect is due to the tunneling of spin-polarized electrons through the insulator, and its magnitude depends on the relative magnetization directions of the electrodes. Multilayer thin film heterostructures of $La_{1/2}Sr_{1/2}MnO_3/SrTiO_3$ represented 450% TMR at 4.5 K [47]. Apart from LSMO, double perovskite Sr_2FeMoO_6 (SFMO) represents high magnetoresistive effects for sensor applications. For spin filters and magnetoelectric sensors, $BiFeO_3$ has been the most studied material in the last decade due to room temperature magnetoelectric coupling.

12.16 Challenges and Future Outlook

Even though lead free magnetic perovskites offer promising alternative to their lead-based counterparts to address toxicity, several challenges hinder their widespread applications. The main concern remains to achieve magnetic ordering at room temperature because most of the lead-free perovskites have low magnetic transition temperature. Another challenge is to synthesise defect free hetero/nano structures with careful composition control to understand the electric and magnetic properties. However advanced theoretical modelling and computational calculations have opened new pathways for optimizing magnetic properties of these materials. With continued research, these materials are poised to paly a key role in development of high sensitivity magnetic sensors, spintronic devices with potential future applications.

12.17 Conclusion

In this chapter, various classes of lead-free magnetic perovskites have been comprehended to understand their magnetic properties and applications. The structural properties various classes of single and double perovskites have been discussed. The types magnetic interactions involved in lead free magnetic perovskites have been focused in detail to understand the origin of magnetism and magnetic properties. A brief discussion on the advanced synthesis and characterization techniques has also been presented. Lead free transition metal perovskites particularly $LaMnO_3$, its derivatives and $BiFeO_3$ have been focused due to their potential applications. Various classes of lead-free double perovskites and their applications have also been discussed in detail. A dedicated section has been focused on the applications and the perovskites used for this application to understand the future research aspects of lead-free magnetic perovskites.

References

1. S. Gosh, H. Shankar, P. Kar, Recent developments of lead-free halide double perovskites: a new superstar in the optoelectronic field. Mater. Adv. **3**(9), 3742 (2022). https://doi.org/10.1039/D2MA00071G
2. Y. Shimakawa, M. Mizumaki, Multiple magnetic interactions in a-site-ordered perovskite-structure oxides. J. Phys. Condens. Matter **26**(47), 473203 (2014). https://doi.org/10.1088/0953-8984/26/47/473203
3. K.W. Tan, D.T. Moore, et al., Thermally induced structural evolution and performance of mesoporous block copolymer-directed alumina perovskite solar cells. ACS Nano **8**(5), 4730 (2014). https://doi.org/10.1021/nn500526t
4. Z. Su, S. Sun, Y. Dai, X. Liu, J. Qiu, Femtosecond laser direct-writing of perovskite nanocrystals in glasses. J. Non Cryst. Solids X **18**, 100182 (2023). https://doi.org/10.1016/j.nocx.2023.100182
5. D. Zhou, T. Zhou, Y. Tian, X. Zhu, Y. Tu, Perovskite-based solar cells: materials, methods, and future perspectives. J. Nanomaterials **1**, 8148072 (2018). https://doi.org/10.1155/2018/8148072
6. M.W. Lufaso, P.M. Woodward, Jahn–Teller distortions, cation ordering and octahedral tilting in perovskites. Struct. Sci. **60**(1), 10 (2004). https://doi.org/10.1107/S0108768103026661
7. T. Jia, Z. Zeng, X. Zhang, P. Ohodnicki, B. Chorpening, G. Hackett, et al., The influence of oxygen vacancy on the electronic and optical properties of $ABO_{3-\delta}$ (A=La, Sr, B=Fe, Co) perovskites. Phys. Chem. Chem. Phys. **21**(36), 20454 (2019). https://doi.org/10.1039/C9CP03883C

8. M.M. Schmitt, Y. Zhang, A. Mercy, P. Ghosez, Electron-lattice interplay in $LaMnO_3$ from canonical Jahn-Teller distortion notations. Phys. Rev. B **101**(21), 214304 (2020). https://doi.org/10.1103/PhysRevB.101.214304
9. Y. Moritomo, S. Xu, A. Machida, T. Akimoto, E. Nishibori, M. Takata, M. Sakata, Electronic structure of double-perovskite transition-metal oxides. Phys. Rev. B **61**(12), R7827 (2000). https://doi.org/10.1103/PhysRevB.61.R7827
10. Q. Tang, Z. Xinhua, Half-metallic double perovskite oxides: recent developments and future perspectives. J. Mater. Chem. C **10**(41), 15301 (2022). https://doi.org/10.1039/D2TC03199J
11. O. Erten, O.N. Meetei, A. Mukherjee, M. Randeria, N. Trivedi, P. Woodward, Theory of half-metallic ferrimagnetism in double perovskites. Phys. Rev. Lett. **107**(25), 257201 (2011). https://doi.org/10.1103/PhysRevLett.107.257201
12. K. Ghosh, C.J. Lobb, R.L. Greene, S.G. Karabashev, D.A. Shulyatev, A.A. Arsenov, Y. Mukovskii, Critical phenomena in the double-exchange ferromagnet $La_{0.7}Sr_{0.3}MnO_3$. Phys. Rev. Lett. **81**(21), 4740 (1998). https://doi.org/10.1103/PhysRevLett.81.4740
13. P.W. Anderson, Antiferromagnetism. Theory of superexchange interaction. Phys. Rev. **79**(2), 350 (1950). https://doi.org/10.1103/PhysRev.79.350
14. R.E. Schaak, T.E. Mallouk, Perovskites by design: a toolbox of solid-state reactions. Chem. Mater. **14**(4), 1455 (2002). https://doi.org/10.1021/cm010689m
15. A. Saini, D. Singh, P. Kumar, K.K. Nagaraja, Advances in chemically tailored metal alkoxide as single-source precursors for engineered oxide nanostructures. Open Ceram. **100800** (2025). https://doi.org/10.1016/j.oceram.2025.100800
16. S.H. Feng, G.H. Li, Hydrothermal and solvothermal syntheses, in *Modern Inorganic Synthetic Chemistry*, (Elsevier, 2017), pp. 73–104. https://doi.org/10.1016/B978-0-444-63591-4.00004-5
17. F. Deganello, G. Marcì, G. Deganello, Citrate–nitrate auto-combustion synthesis of perovskite-type nanopowders: a systematic approach. J. Eur. Ceram. Soc. **29**(3), 439 (2009). https://doi.org/10.1016/j.jeurceramsoc.2008.06.012
18. J.A. Oke, T.C. Jen, Atomic layer deposition and other thin film deposition techniques: from principles to film properties. J. Mater. Res. Technol. **21**, 2481 (2022). https://doi.org/10.1016/j.jmrt.2022.10.064
19. V. Franco, B. Dodrill (eds.), *Magnetic Measurement Techniques for Materials Characterization* (Springer Nature, 2021). https://doi.org/10.1007/978-3-030-70443-8
20. J.S. Zhou, Structural distortions in rare-earth transition-metal oxide perovskites under high pressure. Phys. Rev. B **101**(22), 224104 (2020). https://doi.org/10.1103/PhysRevB.101.224104
21. V. Skumryev, F. Ott, J.M.D. Coey, et al., Weak ferromagnetism in $LaMnO_3$. Eur. Phys. J. B **11**, 401 (1999). https://doi.org/10.1007/BF03219176
22. R. Gupta, R.K. Kotnala, A review on current status and mechanisms of room-temperature magnetoelectric coupling in multiferroics for device applications. J. Mater. Sci. **57**(27), 12710 (2022). https://doi.org/10.1007/s10853-022-07377-4
23. J. Silva, A. Reyes, H. Esparza, et al., $BiFeO_3$: a review on synthesis, doping and crystal structure. Integr. Ferroelectr. **126**(1), 47 (2011). https://doi.org/10.1080/10584587.2011.574986
24. S. Belhamra, R. Masrour, A. Jabar, et al., A comparative study of the structural, electronic, magnetic properties and magnetocaloric effect of perovskite $LaRO_3$ (R=Mn, Cr and Fe). Polyhedron **193**, 114891 (2021). https://doi.org/10.1016/j.poly.2020.114891
25. A.T. Coşkun, Y.S. Ak, N. Güleç, et al., A comparative study of magnetic, and magnetocaloric properties of different transition metal-doped $La_{0.67}Sr_{0.33}AO_3$ (A: Mn, Co, Cr, and Fe) samples. J. Mater. Sci. Mater. Electron. **34**, 1257 (2023). https://doi.org/10.1007/s10854-023-10654-w
26. M. Nabi, S.S. Padelkar, J.J. Jasieniak, et al., Lead-free magnetic double perovskites for photovoltaic and photocatalysis applications. Phys. Rev. Appl. **21**(1), 014063 (2024). https://doi.org/10.1103/PhysRevApplied.21.014063
27. M. Irfan, G. Murtaza, N. Muhammad, et al., Experimental and theoretical studies of structural, electronic and magnetic properties of RE_2NiCrO_6 (RE=Ce, Pr and Nd) double perovskites. Physica E Low Dimensional Syst. Nanostruct. **148**, 115635 (2023). https://doi.org/10.1016/j.physe.2022.115635

28. J.P. Palakkal, T. Faske, M. Major, et al., Ferrimagnetism, exchange bias and spin-glass property of disordered La_2CrNiO_6. J. Magn. Magn. Mater. **508**, 166873 (2020). https://doi.org/10.1016/j.jmmm.2020.166873
29. R. Sivasamy, P. Venugopal, R. Espinoza-González, Structure, electronic structure, optical and magnetic studies of double perovskite Gd_2MnFeO_6 nanoparticles: first principle and experimental studies. Mater. Today Commun. **25**, 101603 (2020). https://doi.org/10.1016/j.mtcomm.2020.101603
30. K. Yi, Q. Tang, Z. Wu, et al., Structural, magnetic, and electrical transport properties of half-metallic double perovskite La_2CrNiO_6 oxides. J. Alloys Compd. **933**, 167742 (2023). https://doi.org/10.1016/j.jallcom.2022.167742
31. M. Kumar, A. Raj, A. Kumar, et al., Structural, electronic, magnetic and optical properties of double perovskite Nd_2CoMnO_6: first principle calculation. Optik **242**, 166764 (2021). https://doi.org/10.1016/j.ijleo.2021.166764
32. V.M. Gaikwad, M. Brahma, R. Borah, et al., Structural, optical and magnetic properties of Pr_2FeCrO_6 nanoparticles. J. Solid State Chem. **278**, 120903 (2019). https://doi.org/10.1016/j.jssc.2019.120903
33. M.D.I. Bhuyan, S. Das, M.A. Basith, Sol-gel synthesized double perovskite Gd_2FeCrO_6 nanoparticles: structural, magnetic and optical properties. J. Alloys Compd. **878**, 160389 (2021). https://doi.org/10.1016/j.jallcom.2021.160389
34. K. Aswathi, J.P. Palakkal, M. Raama Varma, Magnetization sign reversal, exchange bias, and Griffiths-like phase in orthorhombic perovskite Pr_2FeMnO_6. J. Magn. Magn. Mater. **476**, 45 (2019). https://doi.org/10.1016/j.jmmm.2018.12.044
35. S. Saini, N. Panwar, S. Kumar, Giant magnetocaloric properties of Gd-based double perovskite compounds in cryogenic temperature range. J. Magn. Magn. Mater. **614**, 172766 (2025). https://doi.org/10.1016/j.jmmm.2024.172766
36. T. Chakraborty, H. Nhalil, R. Yadav, et al., Magnetocaloric properties of R_2NiMnO_6 (R=Pr, Nd, Tb, Ho and Y) double perovskite family. J. Magn. Magn. Mater. **428**, 59 (2017). https://doi.org/10.1016/j.jmmm.2016.12.015
37. C. Gauvin-Ndiaye, A. Tremblay, R. Nourafkan, Electronic and magnetic properties of double-perovskites La_2MnRuO_6 and hole-doped La_2MnFeO_6 and their potential for magnetic refrigeration. Phys. Rev. B **99**, 125110 (2019). https://doi.org/10.1103/PhysRevB.99.125110
38. Z. Zhou, K. Zhang, Q. Leng, Tunneling magnetoresistance (TMR) materials and devices for magnetic sensors, in *Spintronics: Materials, Devices and Applications*, vol. 51 (2022). doi:https://doi.org/10.1002/9781119698968.ch3
39. M. Gobbi, E. Orgiu, The rise of organic magnetoresistance: materials and challenges. J. Mater. Chem. C **5**(23), 5572 (2017). https://doi.org/10.1039/C6TC04403D
40. K.I. Kobayashi, T. Kimura, H. Sawada, et al., Room-temperature magnetoresistance in an oxide material with an ordered double-perovskite structure. Nature **395**(6703), 677 (1998). https://doi.org/10.1038/27167
41. P.Z.Z. Nehan, O. Vitayaya, D.R. Munazat, et al., The magnetocaloric effect properties for potential applications of magnetic refrigerator technology: a review. Phys. Chem. Chem. Phys. **26**(20), 14476 (2024). https://doi.org/10.1039/D4CP01077A
42. C.R. Bahl, D. Velázquez, K.K. Nielsen, et al., High performance magnetocaloric perovskites for magnetic refrigeration. Appl. Phys. Lett. **100**(12) (2012). https://doi.org/10.1063/1.3695338
43. S.W. Cheong, M. Mostovoy, Multiferroics: a magnetic twist for ferroelectricity. Nat. Mater. **6**(1), 13 (2007). https://doi.org/10.1038/nmat1804
44. Y. Wang, J. Li, D. Viehland, Magnetoelectrics for magnetic sensor applications: status, challenges and perspectives. Mater. Today **17**(6), 269 (2014). https://doi.org/10.1016/j.mattod.2014.05.004
45. M. Bowen, M. Bibes, A. Barthélémy, et al., Nearly total spin polarization in $La_{2/3}Sr_{1/3}MnO_3$ from tunneling experiments. Appl. Phys. Lett. **82**(2), 233 (2003). https://doi.org/10.1063/1.1534619
46. C. Zhang, S. Ding, K. Qiao, et al., Large low-field magnetoresistance (LFMR) effect in free-standing $La_{0.7}Sr_{0.3}MnO_3$ films. ACS Appl. Mater. Interfaces **13**(24), 28442 (2021). https://doi.org/10.1021/acsami.1c03753

47. M. Viret, M. Drouet, J. Nassar, et al., Low-field colossal magnetoresistance in manganite tunnel spin valves. Europhys. Lett. **39**(5), 545 (1997). https://doi.org/10.1209/epl/i1997-00391-2

Chapter 13
Investigation of Lead-Free Halide Perovskites for Imaging Applications

Yukta, Qinqi Zhou, Sunardi Rahman, Vishnu Chauhan, and Maning Liu

Abstract Imaging technologies are very crucial in today's society, with critical importance in three major domains: security and risk detection, medical diagnosis, and industrial quality control. Imaging techniques enable the visualization, sensing, and detection of signals across the entire electromagnetic spectrum. Starting from early digital radiographs to complex hyperspectral or low-light techniques, the progress in imaging technologies is intrinsically linked to advancements in semiconductor detector architectures. Here, the system performance is largely governed by sensitivity, spatial resolution, and real-time acquisition efficiency. To date, conventional semiconductors have achieved substantial commercial success in this field; however, many roadblocks remain, including complex fabrication processes, higher manufacturing costs, toxicity of precursor materials, and restricted spectral coverage, which are primary limitations. Most recently, lead-free halide perovskites have arisen as a compelling, sustainable option for future semiconducting applications. The optical absorption of these materials can be systematically tuned from the ultraviolet region to the near-infrared through compositional engineering, and the elimination of toxic lead enhances their environmental compatibility. Furthermore, their solution processability, achieved through relatively lower-temperature synthesis methods, simplifies the fabrication process and reduces costs. The panchromatic absorption, superior ambient stability, and miraculous optoelectronic properties establish these materials as promising contenders for high-performance detection and imaging applications. In this chapter, we systematically explain the fundamental principles related to imaging, the mechanism underlying image generation, and the potential and

Yukta · Q. Zhou · S. Rahman · M. Liu (✉)
Centre for Analysis and Synthesis, Department of Chemistry, Lund University, 221 00 P.O. Box 124, Lund, Sweden
e-mail: maning.liu@chem.lu.se

Wallenberg Initiative Materials Science for Sustainability, Department of Chemistry, Lund University, 221 00 Lund, Sweden

NanoLund, Lund University, Professorsgatan 1, 223 63, Lund, Sweden

V. Chauhan
Department of Physics, Ramjas College, University of Delhi, New Delhi, India

V. Chauhan et al. (eds.), *Sustainable Lead-Free Perovskite Materials*, Topics in Applied Physics 157, https://doi.org/10.1007/978-3-032-24873-2_13

progress of lead-free perovskites in terms of imaging devices, with detection regimes spanning from X-rays to near-infrared wavelengths.

Keywords Imaging · Detectors · Lead-free perovskites · Optoelectronic properties · Sustainability · Sensitivity

13.1 Introduction

Imaging technologies constitute the foundation stone across contemporary medical, industrial, scientific, and security areas, enabling the mapping, visualization, and detection of various structures as well as dynamic processes spanning across a broad spectral range, from high-energy X-rays (~ 100 keV) to near infrared wavelengths (~ 0.7–1.2 eV) [1–3]. The advancement in semiconductor technologies, specifically in metal–oxide–semiconductor (MOS) devices for imaging sensors and silicon-based charge-coupled devices (CCDs), along with the development of advanced detector architectures, has primarily driven the progression of digital imaging [4, 5]. These technological advancements have led to major developments in the field of medical diagnosis, environmental monitoring, security and threat detection, industrial inspection, and low-light hyperspectral imaging [5]. In the late twentieth century, the first digital radiograph was taken; since then, significant progress has been made through advancements in spatial resolution, real-time data acquisition capabilities, and detector sensitivity [6]. The fundamental principle of radiation detection remains largely the same across the entire electromagnetic spectrum, comprising three main steps: radiation absorption and generation of charge carriers, the successive transport of these charge carriers, and finally, their efficient collection [7]. On the contrary, the choice of active detector materials depends strongly on the energy of the incident radiation. For example, semiconducting materials with a high effective atomic no. (Z_{eff}), strong radiation hardness, good X-ray absorption, and attenuation ability, as well as promising electrical properties such as a long mobility-lifetime ($\mu\tau$) product, are required for high-energy X-ray detection and imaging. In contrast, effective detection and imaging of NIR radiation requires materials with tunable and narrow band gaps, long charge carrier diffusion lengths, and high absorption coefficients. Therefore, identifying a single active semiconducting material that possesses all these characteristics across the entire range of the electromagnetic spectrum is very challenging, with the added constraints of low cost and environmental sustainability [7]. Consequently, the demand for higher sensitivity, faster response times, and high spectral and spatial resolution persistently drives the exploration for novel functional semiconducting materials for modern imaging technologies [8].

At present, conventional commercial semiconductor materials used for radiation detection are silicon (Si), amorphous selenium (α-Se), lead iodide (PbI_2), mercuric iodide (HgI_2), cadmium telluride (CdTe), and various III–V semiconducting compounds [9–12]. Despite the notable advantages of high sensitivity and excellent energy resolution, these conventional semiconductors are still constrained by various limitations. The limitations include high production costs associated with elevated temperature processing, suboptimal absorption in certain spectral wavelengths, incompatibility with flexible electronics, and their inherent toxicity in a few cases [13, 14]. In view of these existing challenges, lead-free perovskites appear to be a promising alternative that can function across the entire electromagnetic spectrum. These perovskites exhibit excellent optoelectronic properties, including strong panchromatic absorption, high absorption coefficients, long charge carrier diffusion lengths, high carrier mobilities, and inherent mechanical flexibility [15–17]. Substituting Pb^{2+} cations in halide perovskites with homovalent (Sn^{2+}, Ge^{2+}) or heterovalent (Sb^{3+}, Bi^{3+}) ions offers a viable pathway to mitigate toxicity while preserving the intrinsic advantages of the perovskite lattice. The replacement of water-soluble lead halides (a major source of environmental and health hazard) in perovskite structures enhances their sustainability as well [18]. Moreover, perovskite materials can be synthesized in almost all structural and electronic dimensionalities, ranging from single crystals to polycrystalline thin films, nanocrystals, and quantum dots, via low-temperature solution-processed fabrication routes, facilitating cost-effective device integration [19]. Additionally, recent advancements in dimensionality control, compositional engineering, and crystal growth methods have enabled substantial improvements in crystal quality, carrier transport properties, spectral tunability, defect formation and stability, and scalability of these lead-free perovskites [20, 21].

Conclusively, in this chapter, we will explain the basic principles of modern imaging and detection techniques, focusing on the image generation mechanism and processing. We will also elaborate on the crucial parameters that govern the performance matrices of these detectors, such as their sensitivity, resolution, response time, and operational stability. Subsequently, we will also focus on and highlight the unique structural and opto-electronic properties of the emerging lead-free perovskite materials, which render them promising active layer materials for next-generation imaging devices.

13.2 Detector Types, Operating Mechanisms, and Image Generation

Here within this chapter, following the sequence of increasing detection wavelengths, the detection and image generation systems are arranged from the X-ray region through to the NIR spectrum in the following sections.

13.2.1 X-Ray Detectors

X-ray detection and imaging are essential tools used in many fields, including non-destructive industrial testing, medical diagnostics, and security screening. This technique allows for accurate threat detection with high precision in case of security inspections [22–24]. In medicine, X-ray imaging has significantly advanced diagnostic methods through applications such as projection radiography, fluoroscopy, computed tomography (CT), and mammography [25]. All of these utilize the penetrating ability of high-energy photons to visualize internal structures without invasive procedures. For most medical purposes, hard X-rays in the 5–10 keV range are used, as they offer a good balance between penetration depth and image contrast [26]. Mainly, X-ray imaging is based on the principle of measuring the attenuated X-ray photons as they pass through an object, creating contrast based on differences in material density and composition. This principle supports both low-dose imaging for medical examinations and detailed inspections of non-living materials for quality control, structural assessment, and security purposes [27]. Commercially available X-ray detectors can be broadly categorized into **direct-conversion** and **indirect-conversion** systems, each employing distinct mechanisms for transducing incident X-ray photons into measurable electrical signals [28, 29]. In indirect-conversion detectors, high-energy X-rays are first absorbed by a scintillator material, which emits the visible photons, and these photons are subsequently detected by a photodiode, where they are converted into electrical signals that are amplified and digitized for image formation [30, 31]. In contrast, in the case of direct-conversion detectors, the X-ray energy is directly converted into electron–hole pairs and the resulting charge carriers are collected under an applied electric field, generating a current signal that is processed through a preamplifier, shaped, and analysed—often represented as a histogram using a shaping amplifier and multichannel analyser [32]. Although direct X-ray detectors offer higher spatial resolution and detection efficiency, indirect conversions remain largely used due to their cost-effectiveness, scalability for large area imaging applications, and stability. To date, commercial X-ray detectors have mostly relied on semiconducting materials such as α-Se, Si, CdTe, PbI_2, HgI_2, and cadmium zinc telluride (CdZnTe), among others [33–36]. Though these commercial semiconducting materials offer notable benefits, such as good energy resolution, integration compatibility with electronic readout systems, high spatial clarity in imaging, and high sensitivity, limitations still persist. High fabrication costs are due to elevated temperature processing, insufficient X-ray absorption linked with low Z_{eff}, low charge carrier mobility, and their inherent toxicity are a few prominent drawbacks related to these materials [37, 38]. Similarly, prominent traditional commercial scintillators such as bismuth germanate ($Bi_4Ge_3O_{12}$), thallium-doped cesium iodide (CsI:Tl), and lutetium yttrium oxyorthosilicate {$(Lu,Y)_2SiO_5$–(LYSO)} also have limitations, like time-consuming and complex high-temperature synthesis process, radioluminescence afterglow, non-tunable scintillation wavelengths, and low light yield [39–41]. Against this backdrop, halide perovskites stand out as a promising alternative active material for both direct and indirect X-ray detection. Their exceptional optoelectronic properties, including tunable bandgap energies, strong optical absorption, optimal carrier diffusion lengths, and low-temperature synthesis, address the fundamental performance-limiting factors of existing commercial detection and imaging technologies [42, 43]. However, despite the attractive optoelectronic properties, the intrinsic ambient instability of perovskites and the presence of water-soluble lead constituents are the major reasons impeding their adoption for commercial X-ray detectors [44, 45]. The strategic replacement of lead with less toxic, environmentally benign elements, combined with a robust encapsulation process, offers a viable path for enhancing material stability and scalability, thereby reducing health risks and facilitating their integration into commercial X-ray detection technologies [46, 47]. The upcoming sections provide a brief overview of the fundamental mechanisms of X-ray detection and imaging, as well as the detector performance parameters and matrices related to perovskites, making them suitable for imaging applications. Also, a concise overview of recent progress in lead-free perovskite X-ray detection devices is added simultaneously.

13.2.1.1 Direct and Indirect Conversion X-Ray Detectors

In the operation of a direct X-ray detector, incident X-ray photons are absorbed by the semiconducting layer, generating charge carriers that undergo trapping and re-emission, thereby creating the gain effect. These charge carriers are efficiently collected and further converted into signals by thin-film transistors (TFTs) or complementary metal–oxide–semiconductor (CMOS) arrays. TFT-driven flat panel detectors are commonly used in digital radiography and mammography, featuring pixel structures with storage capacitors and transistors that successively transfer charge to amplifiers for signal processing [33, 48]. As an alternative, CMOS arrays utilize a 4-transistor active pixel sensor (4T APS) architecture, which enables correlated double sampling to minimize noise. Voltage outputs are obtained during the readouts through the use of carefully timed switching and amplification. In the simplified configuration, direct X-ray detectors obtain images by positioning the object on a motorized translational stage and recording the X-ray-induced photocurrent as a function of spatial coordinates (x, y). In the case of indirect X-ray detectors, or scintillators, the detection mechanism predominantly involves three sequential processes. Initially, incident X-ray photons undergo absorption or scattering—primarily via the photoelectric effect, with Compton scattering dominating below ~ 1 MeV and the Auger process becoming significant above ~ 1.2 MeV—resulting in the generation of electron–hole pairs. Subsequently, these charge carriers migrate to luminescent centers within the material (timescale ~ 10^{-12} to 10^{-8} s). Finally, radiative recombination at these centers produces visible or near-visible photons, which are then guided onto a photodetector array where each pixel converts light into an electrical signal [32]. The difference in the working mechanism of both types of detectors is presented in Fig. 13.1.

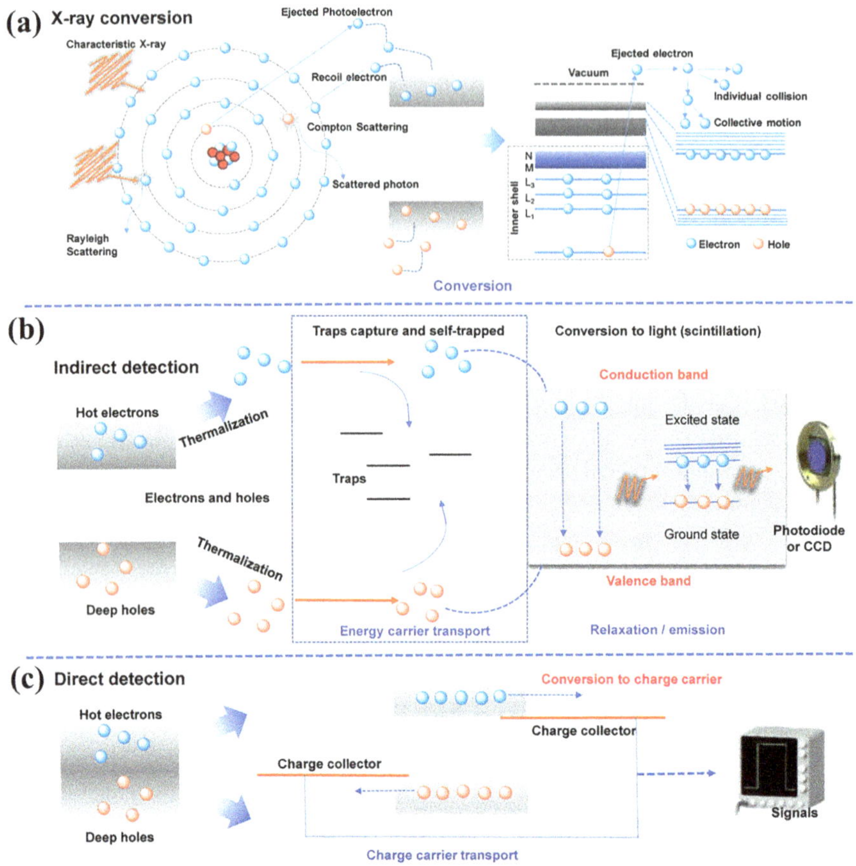

Fig. 13.1 **a** Schematics explaining different types of interaction between X-ray photons and active semiconductor layer, **b** process explaining detection mechanism in (**b**) Indirect X-ray detectors and **c** direct detectors [7] Copyright 2022. Reproduced with permission from Wiley

13.2.1.2 Key Concepts of Radiation Detecting Materials and Technologies

This section provides a detailed overview of the key properties (e.g., radiation attenuation ratio, mobility lifetime product, and ionization energy) of active materials relevant to X-ray detectors, alongside critical device-specific parameters including dark current, sensitivity, charge collection efficiency, limit of detection, and response time. The following discussion elaborates on the active semiconducting materials properties and performance parameters in concluding the overall efficiency and the functionality of the X-ray detector system with imaging applications.

(i) **Radiation attenuation ratio**

The incident X-ray radiation in semiconducting materials attenuates through various interaction mechanisms, such as the photoelectric effect, Rayleigh scattering, Compton scattering and electron–hole pair production. The extent to which a semiconductor can attenuate X-ray radiation is quantitatively described by the Beer–Lambert law:

$$I = I_o e^{\left[-\left(\frac{\mu}{\rho}\right)x\right]} \tag{13.1}$$

here I_o is the initial incident radiation intensity, I denotes the intensity of the transmitted (attenuated) radiation, ρ is the material's density, x is the penetration depth (or intensity) within the material, and μ signifies the linear attenuation coefficient. The linear attenuation coefficient μ is the crucial parameter here, as it describes the material's ability to absorb/scatter the incident radiation. A higher value of μ indicates the stronger attenuation of radiation by the material, which further implies that a comparatively thinner layer of material is sufficient to absorb the incident radiation. This coefficient largely depends on the material's physical properties and can be further approximated as:

$$\mu \propto \frac{\rho Z^n}{AE^m} \tag{13.2}$$

where, ρ is the density of the material, Z is the atomic number,A represents the atomic mass and E is the energy of the incident X-ray radiation. The exponents n and m depends on the dominant interaction mechanism (e.g., the photoelectric effect typically has $n \approx 4$ and $m \approx 3$).

Overall, materials with **higher atomic number *(Z)*** and **density** ($\boldsymbol{\rho}$), and optimized energy matching, exhibit **greater X-ray attenuation**, making them highly suitable for radiation detection and shielding applications [49].

(ii) **Ionization energy**

Ionization energy (W) is the energy required to generate an electron–hole pair in a semiconducting material. The ionization energy for a semiconducting material can be described empirically as:

$$W = 2E_g + 1.43\,\text{eV} \tag{13.3}$$

Here E_g is the bandgap of the material.

(iii) **Charge collection efficiency**

The charge collection efficiency (CCE) is defined as the ratio of charge carriers successfully collected at the electrodes to the total carriers generated within the active material after radiation exposure. CCE serves as a crucial metric for assessing how efficiently a material converts incident radiation into detectable electrical signals. The CCE is calculated upon the following expression:

$$CCE = \frac{\mu\tau V}{L^2}\left[1 - e^{\left(-\frac{L^2}{\mu\tau V}\right)}\right] \tag{13.4}$$

where u denotes the carrier mobility, τ represents the carrier lifetime, L is the thickness of active layer of material and V is the bias voltage. Higher the CCE value higher the material performance in detecting and converting incident radiation [50].

(iv) **Dark current**

The current–voltage relation for the single diode can be expressed by the following equation:

$$I = I_o\left[\exp\left(\frac{q(V - JR_s)}{nKT}\right) - 1\right] + \frac{V - JR_s}{R_{sh}} \tag{13.5}$$

Here n is the ideality factor, I_o. represents the reverse saturated current, R_s and R_{sh} are the series and shunt resistances, respectively. k is the Boltzmann constant, q is the electric charge and T is the temperature. Dark current refers to the reverse saturation current (I_o) that flows through a photodetector in the absence of incident light. The magnitude of dark current I_{dark} is inversely related to the resistivity of the semiconductor material, with higher resistivity leading to lower dark current. Minimizing dark current is crucial, as reduced dark current decreases electrical noise, thereby improving the signal-to-noise ratio (SNR) and enhancing the sensitivity of the detector for weak signal detection [51, 52].

(v) **Sensitivity**

Sensitivity (S) is a crucial parameter for the performance determination of radiation detectors. It defines the detector's capability to record radiation. A highly sensitive detector can detect even minimal radiation levels in shorter exposure times, thereby reducing the total radiation dose needed. High-sensitivity detectors are essential in X-ray imaging, where they can minimize radiation exposure times and help limit the overall radiation dosage to which the human body is exposed. The following equation describes the sensitivity of a radiation detector:

$$S = \frac{I_{radiation} - I_{dark}}{DA} \tag{13.6}$$

Here, D is the radiation dose rate, and A is the area of the detector and $I_{radiation}$ and I_{dark} is the output current with and without radiation exposure [53].

(vi) **Mobility-lifetime product**

The mobility-lifetime product ($\mu\tau$) of charge carriers in semiconducting materials is an important parameter for evaluating their charge transport properties. It provides information about the material capability to efficiently extract and collect the charge carriers generated deep within its bulk on radiation exposure. The $\mu\tau$ product can be easily calculated by the mathematical fitting of current–voltage (I–V) characteristic curves of the semiconducting material using the modified Hecht equation, which estimates the charge carrier drift and recombination behavior under an applied bias. Moreover, the $\mu\tau$ product could be further used to estimate the charge carrier diffusion lengths $\{L_D = (K_B T \mu\tau/e)^{1/2}\}$ Additionally, it conclusively connects the material's intrinsic properties to the detector performance and charge collection efficiency [54].

(vii) **Limit of detection**

The limit of detection (LOD) parameter for an X-ray detector is mainly calculated by analyzing its dynamic photocurrent response under very low X-ray dose rates. It is defined as the minimal detectable photocurrent signal, usually three times the noise current, that exceeds the noise signal by a statistically significant margin. This is often quantified using the SNR, which is expressed as $SNR = J_s/J_n$, where J_s is the signal current density generated by X-ray exposure and J_n represents the noise current density in the absence of radiation. A higher *SNR* reflects superior sensitivity and a lower detection limit, enabling accurate detection of weak X-ray signals [49].

(viii) **Response time or temporal resolution**

Response time is a key performance metric used to characterize the speed of a photodetector in reacting to incident radiation. It is defined as the duration required for the photocurrent to increase from 10 to 90% of its steady-state value (rising time) and subsequently decrease from 90 to 10% (decay or falling time) without radiation. Typically, the decay time exceeds the rise time, primarily due to the influence of trap states and the intrinsic crystalline quality of the material, which govern charge carrier recombination and collection dynamics [55, 56].

(ix) **Spatial resolution**

Spatial resolution is the ability of an X-ray imaging system to distinguish and display fine structural details by separating closely spaced anatomical features in the image. This parameter is generally measured in line pairs per millimetre (lp/mm) or evaluated using the modulation transfer function (MTF), which calculates the system's ability to preserve contrast at various spatial frequencies. More specifically, high spatial resolution is mainly important in applications as micro imaging and diagnostic radiology, where the accurate visualization of small structures is necessary [57].

(x) **Stability**

The long-term operational stability of an X-ray detector is crucial for sustaining detector performance under repeated or continuous radiation exposures. A radiation detector with high operational stability exhibits negligible variations in its output response over time, thereby maintaining constant image quality. This characteristic is significant in both industrial inspection and medical diagnostics, as even a minor deviation in the output signal could compromise diagnostic reliability and accuracy [58]. According to the commercial standard, perovskite solar cells should retain at least 80% of their initial efficiency after 9000 h in damp heat conditions (e.g., 85% relative humidity and a temperature of 85 °C), whereas perovskite-based X-ray detectors currently lack the necessary stability benchmarks and require significant advancements prior to their commercialization.

13.2.1.3 Advantages of Halide Perovskites for X-Ray Detection

The optoelectronic properties and key parameters that make metal halide perovskites superior for X-ray detection applications are discussed in this section. The bandgap (E_g) of the semiconducting material is one of the crucial parameters that influences its X-ray response. The Klein rule states that the ionization energy (W$\pm$), which is required to generate the electron–hole pairs in a semiconductor, is roughly proportional to (W$\pm \approx 3E_g$) its bandgap. Therefore, a narrower bandgap semiconductor requires less W$\pm$, and the number of charge carriers generated per absorbed X-ray photon is more, and can be quantified as $\Delta Q = eE_{X\text{-ray}}/W\pm$. The bandgap of halide perovskite can be tuned over a broad energy range (~ 1.5–3.1 eV) and well aligned in the optimal bandgap range (~ 2–3 eV) of traditional commercial X-ray detecting semiconductors [59]. In terms of the indirect X-ray detector or scintillators, emission wavelengths are fixed, and multicolor emission visualization from a single semiconductor is not possible. For halide perovskite semiconductors, compositional tuning {variation in halide anions (Cl^-, Br^-, I^-)} offers a pathway for multiple emission wavelengths spanning the entire visible spectrum in a single active scintillator material. Moreover, perovskites with narrower band gaps could attain exceptionally high photon conversion yields (approximately the order of 2.5×10^5 photons/MeV) as scintillation light yield is inversely correlated with the bandgap energy of the material [60].

Furthermore, the linear attenuation coefficient (μ) of a material, which dictates its X-ray attenuation capability, exhibits a strong dependence on the atomic number Z of its constituent elements and the energy E of the incident radiation, and can be evaluated as $\mu \propto Z^4/E^3$. This implies that the materials with constituent elements of higher atomic numbers are intrinsically more effective in X-ray attenuation. Therefore, metal halide perovskites, which incorporate the heavy metal elements like lead, exhibit comparatively high X-ray attenuation capability [42]. Intriguingly, in lead-free perovskites, the incorporation of bismuth (Bi-the heaviest stable element) can further increase X-ray attenuation efficiency, often surpassing that of lead-based counterparts. This poses Bi-based lead-free perovskites as one of the most promising candidates for environmentally benign high-performance X-ray detection. Moreover, perovskites exhibit defect tolerance and high charge carrier mobilities (of the order of 10^5–10^6 cm^2/V-sec), resulting in efficient charge collection, which further leads to accurate detection, faster response times, and is essential for high-resolution imaging [61]. The $\mu\tau$ product of perovskites even exceeds that of established commercialized X-ray detector materials, such as CdTe and CdZnTe. Markedly, perovskite exhibits remarkable structural versatility and could be synthesized in the form of single crystals (SCs), nanocrystals, and polycrystalline thin films using cost-effective and scalable, low-temperature solution-processed methods [62, 63]. SCs are most commonly synthesized using solid-state or solution-based methods, which are suitable for producing cost-efficient, large-area crystalline semiconducting materials. While polycrystalline thin films are prepared using the spin coating techniques, nanocrystals are generally synthesized using solution-based ligand-assisted reprecipitation (LARP) or hot injection methods [64]. These diverse synthesis routes enhance the scalability and economic viability of perovskite fabrication compared to high-temperature-based complex synthesis of III–V semiconductors used for traditional X-ray detectors, as the precursors used for perovskite synthesis are inexpensive and amenable to low-temperature processing [65]. Notably, all-inorganic lead-based perovskites demonstrate a material cost of approximately \$3.10 per gram, which is significantly lower than that of widely used scintillators such as lutetium yttrium orthosilicate (LYSO) at \$8.40/gram and cerium-doped yttrium aluminum perovskite (YAP:Ce) at around \$60/gram, underscoring their superior cost-effectiveness for large-scale deployment [48, 66].

13.2.1.4 Lead-Free Perovskite X-Ray Detectors and Scintillators

Direct X-Ray Detection and Imaging

The replacement of Pb^{2+} cations in halide perovskites with either homovalent (Sn^{2+}, Ge^{2+}) or heterovalent (Sb^{3+}, Bi^{3+}) ions represents a promising route to mitigate lead toxicity while retaining the intrinsic optoelectronic advantages of the perovskite structure. Among the group IV alternatives, Sn^{2+} is regarded as the most suitable Pb substitute for optoelectronic applications due to its similar electronic configuration (ns^2np^2), presence of a stereochemically active lone pair, and a nearly identical ionic radius [67, 68]. Compared to lead perovskites, tin perovskites exhibit a narrower bandgap and an enhanced absorption coefficient extending into the NIR region. Nevertheless, the lower electronegativity, reduced *s*-*p* orbital splitting, and higher atomic energy levels make Sn^{2+} highly susceptible to oxidation into Sn^{4+}, leading to increased instability under ambient atmospheric conditions [69]. Consequently, Sn-based devices often show performance degradation within hours to days, unless they are rigorously encapsulated. On the contrary, Sb^{3+} and Bi^{3+} are chemically robust and not prone to oxidation, thereby devices fabricated with these cations can sustain low leakage currents with higher operational lifetimes. Therefore, the heterovalent substitution of Pb^{2+} with Bi^{3+} or Sb^{3+} provides a promising route for developing stable, lead-free perovskite-based X-ray detectors. The beginning of direct-conversion perovskite X-ray detectors can be traced back to 2013, when Kanatzidis and co-workers successfully developed the melt-grown $CsPbBr_3$ perovskite SCs for X-ray detection [51]. Since then, lead-free perovskite systems have garnered significant attention (particularly with heterovalent substitution) to curb Pb^{2+} toxicity without compromising the favorable optoelectronic properties. Starting with $A_3B_2X_9$ types, a representative class of lead-free perovskite, $MA_3Bi_2I_9$, was first synthesized by Liu et al. They reported the first-ever fabrication of large (26 $\times$ 26 $\times$ 8 mm^3) zero-dimensional perovskite-inspired SCs using a precursor refinement strategy. This strategy effectively produces crystals with reduced intrinsic defect density, which is further translated into reduced ion migration and markedly enhanced environmental stability. These large area $MA_3Bi_2I_9$ SCs have shown exceptional x-ray detection performance with achieving a sensitivity of 1947 μC Gy_{air}^{-1} cm^{-2}-two orders of magnitude higher than the commercial α-Se detectors and a very low detection limit ~ 83 nGy_{air} s^{-1}. To assess spatial resolution, the authors employed detectors with a 1 × 1 mm^2 active area and a scanning step size of 0.5 mm to image two metallic keys. Figure 13.2a depicts the schematics of image generation using perovskite-based detector arrays. The system was capable of resolving fine structural features, including those concealed from the visible spectrum. As a striking example, a key partially embedded in black rubber was clearly visualized in the X-ray radiograph, with the metallic regions sharply distinguished from the surrounding polymer due to their pronounced differences in X-ray attenuation (Fig. 13.2b) [70]. They further investigated the similar crystal system again and changed the crystal growth technique to the low-temperature-gradient crystallization (LTGC) method. In this following work, a coplanar-structured detector was fabricated and used to image a variety of small objects with diverse geometries and thicknesses under X-ray illumination. By translating the target on an x–y scanning stage and recording the photocurrent generated by transmitted X-ray photons, two-dimensional (2D) radiographs were reconstructed using these detectors. To evaluate spatial imaging performance, the authors employed pixelated detectors with individual active areas of 0.6 × 0.6 mm^2. The resulting images of a metal gasket, nut, and wrench exhibited high contrast and clear edge definition, demonstrating the detector's ability to resolve fine structural details. Conclusively, the $MA_3Bi_2I_9$ single crystal have shown the advantageous material properties for X-ray detection, including a relatively high density (≈ 4.1 g cm^{-3}), substantial bulk resistivity (≈ 4.7 × 10^{10} Ω cm), and a large ion migration activation energy (0.83 eV), which collectively contribute to low dark current and stable operation under bias. The device performance metrics are equally notable demonstrating a minimum detectable dose rate of 31 nGy s^{-1}, sensitivity of 872 μC Gy_{air}^{-1} cm^{-2}, fast temporal response (~ 266 μs), spatial resolution of 4.22 lp mm^{-1} and prolonged operational stability in ambient environmental conditions [56]. Later studies have shown that completely inorganic $Cs_3Bi_2I_9$ SCs also achieved the high-performance matrices in direct X-ray detection with a sensitivity of 1652.3 μC Gy_{air}^{-1} cm^{-2} and detection limit around 131 nGy_{air} s^{-1} [71]. Building on this, Tao et al. improved the crystal quality of $Cs_3Bi_2Br_9$ SCs using the solid-state Bridgman crystal growth technique. The direct X-ray detectors based on these SCs exhibit a sensitivity of 1705 μC Gy_{air}^{-1} cm^{-2} and an ultralow detection limit of 0.58 nGy_{air} s^{-1}. The enhanced crystal quality and resultant detection performance achieved exceptional imaging performance under both soft (40 keV) and hard (120 keV) X-ray irradiation. However, in the soft X-ray mode, the detector captured the external silhouette of a screw embedded in pork due to low contrast and insufficient X-ray penetration. But at 120 keV (hard X-rays), it clearly resolved even the internal threads and structure of the screw due to deep X-ray penetration and effective attenuation contrast. This experiment precisely imitated a medical diagnosis scenario, as high X-rays detected and characterised the foreign embedded object within the soft tissue analogues [21]. Building on this, Tao et al. improved the crystal quality of $Cs_3Bi_2Br_9$ SCs using the solid-state Bridgman crystal growth technique. The direct X-ray detectors based on these SCs exhibit a sensitivity of 1705 μC Gy_{air}^{-1} cm^{-2} and an ultralow detection limit of 0.58 nGy_{air} s^{-1}. The enhanced crystal quality and resultant detection performance achieved exceptional imaging performance under both soft (40 keV) and hard (120 keV) X-ray irradiation. However, in the soft X-ray mode, the detector captured the external silhouette of a screw embedded in pork due to low contrast and insufficient X-ray penetration. But at 120 keV (hard X-rays), it clearly resolved even the internal threads and structure of the screw due to deep X-ray penetration and effective attenuation contrast. This experiment precisely imitated a medical diagnosis scenario, as high X-rays detected and characterised the foreign embedded object within the soft tissue analogues [72].

Later on, Zhang et al. transformed the 0D structural design of 329-type $MA_3Bi_2I_9$ perovskites by partially substituting iodide ions with bromide to form a 2D $MA_3Bi_2I_6Br_3$ configuration. This halide substitution enhanced the μτ product and slightly widened the bandgap by ~ 0.1 eV, thereby improving X-ray absorption efficiency. As a result, the material achieved a record-breaking X-ray detection sensitivity of 24,509 $\mu C\ Gy_{air}^{-1}\ cm^{-2}$, accompanied by an ultralow detection limit of 4.3 $nGy_{air}\ s^{-1}$. A linear detector array comprising 60 individually addressable electrodes was fabricated, yielding exceptional imaging uniformity and the highest reported spatial resolution for any lead-free perovskite X-ray detector to date—10.6 lp mm^{-1}. This high-resolution capability enabled the clear visualization of fine internal features, including concealed bone structures in a chicken leg, a miniature spring hidden within a ballpoint pen, and the internal circuitry of a test pen (Fig. 13.2d) [73].

Notably, beyond conventional lead-free perovskites, alternative structural classes have also been explored for X-ray detection, including $A_2B(I)B(III)X_6$ double perovskites and 2D hybrid layered perovskites in which organic amine molecules occupy the A-site. These materials deliver high-quality X-ray detection and imaging performance, while also enabling extensive compositional diversity. For example, Steele et al. employed $Cs_2AgBiBr_6$ SCs for direct detection and evaluated device performance at both room temperature and 77 K, revealing that cryogenic cooling extended the carrier lifetime from 700 ns (under ambient conditions) to 1500 ns (at 77 K). This enhancement in charge-transport characteristics contributed to a substantial improvement in X-ray sensitivity—from 316 $\mu C\ Gy_{air}^{-1}\ cm^{-2}$ to 988 $\mu C\ Gy_{air}^{-1}\ cm^{-2}$,, thereby significantly elevating the imaging quality [74]. Representative examples for 2D layered perovskites mainly include $(BA)_2AgBiBr_7$, $(4,4\text{-}DPP)_4AgBiI_8$, and $(F\text{-}PEA)_3BiI_6$, {here BA-*butyl ammonium,* 4,4-DPP-*4,4-difluoropiperidinium*, F-PEA-*fluoro-phenyl ethyl ammonium*} all of which have been successfully implemented in X-ray imaging devices [75–77]. Most recently, Xia et al. reported the synthesis of high-density post-perovskite phases (more popularly chalcogenides) $BiCuSCl_2$, $Bi_3In_2Se_7I$, and $BiAgSCl_2$ via a melt-cooling growth technique. Among these, $BiCuSCl_2$ exhibited a notable X-ray sensitivity of 4126 $\mu C\ Gy_{air}^{-1}\ cm^{-2}$ for 150 keV hard X-rays, combined with a low dark-current density of 3.2×10^{-10} A cm^{-2}. The method could also facilitate the fabrication of large-area detector arrays with effective imaging capability [78]. Despite these promising results, such unconventional perovskite systems still require significant optimization before their performance can rival that of established perovskite detector architectures.

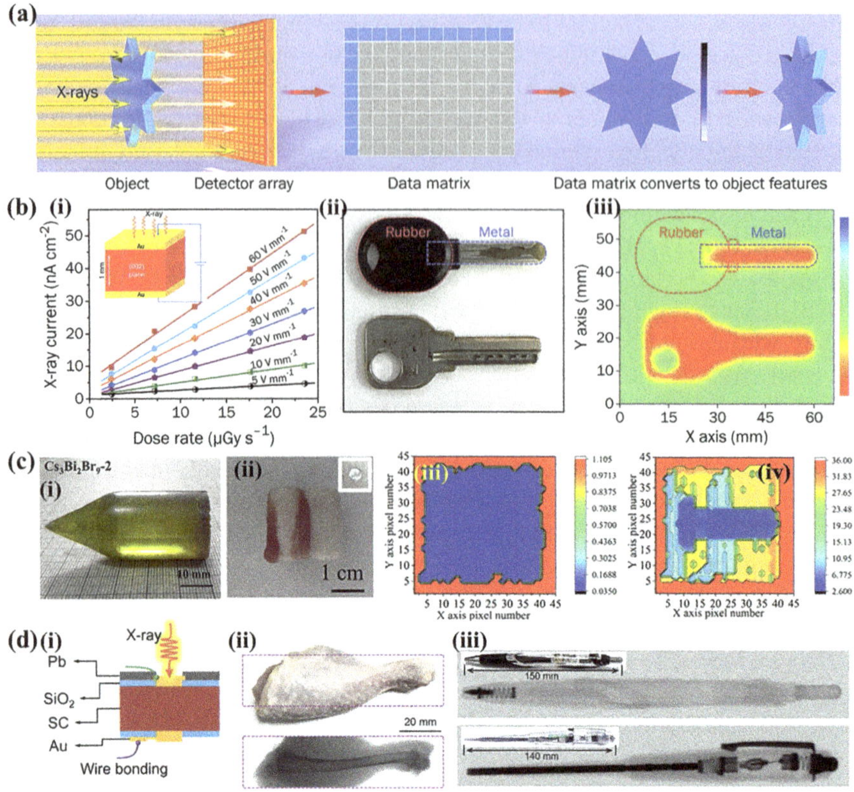

Fig. 13.2 **a** Generalized schematic depiction of the process employed in X-ray image acquisition [70] Copyright 2020. Reproduced with permission from Cell Press. **b** (i) X-ray photocurrent response of vertical $MA_3Bi_2I_9$ single-crystal detectors measured under varying X-ray dose rates, accompanied by (ii) photographic and (iii) radiographic images of a bare metal key and the same key partially encased in rubber [70] Copyright 2020. Reproduced with permission from Cell Press. **c** (i) Photograph of the as-grown $Cs_3Bi_2Br_9$ single crystal synthesized via the Bridgman technique, alongside an (ii) optical image of a screw embedded within a pork sample and its corresponding X-ray radiographs acquired using (iii) soft X-rays (~ 40 keV) and (iv) hard X-rays (~ 120 keV), where hard X-rays clearly resolve the embedded screw [21] Copyright 2022. Reproduced with permission from American Chemical Society. **d** (i) Schematic representation of the $MA_3Bi_2I_6Br_3$ linear array detector architecture, (ii) accompanied by an optical image of a chicken leg and the corresponding X-ray radiograph clearly revealing the embedded bone, (iii) Also shown is the X-ray image of a ball-point pen, captured using the same detector, depicting its intricate internal components, including the concealed spring [73] Copyright 2024. Reproduced with permission from Wiley

Indirect X-Ray Detection or Scintillation Detectors for Imaging Applications

Halide perovskites have emerged not only as efficient direct-conversion X-ray detectors but also as promising candidates for **indirect detection** in scintillation-based systems, owing to their highly tunable band gaps, near-unity photoluminescence quantum yields (PLQY), and photoluminescence (PL) lifetimes well-suited for rapid photon emission. Over the last decade, extensive research efforts have been made on developing high-performance perovskite scintillators, achieving detection limits down to 13 $nGy_{air}\,s^{-1}$, ultrafast response times of ~ 44.6 ns, and spatial resolutions comparable to commercial CsI:Tl flat-panel scintillators [60]. The field was initiated in 2016 with the first scintillation studies on $MAPbI_3$ and $MAPbBr_3$ SCs, followed by substantial progress in both lead-based and lead-free compositions. The following section highlights recent advances in lead-free perovskite scintillators for X-ray imaging applications [79, 80]. Tang et al. were among the first to explore lead-free perovskites for scintillation applications, employing lanthanide-based double perovskite systems, i.e., $Cs_2NaEuCl_6$ and $Cs_2NaTbCl_6$. Lanthanide perovskites are well known for their broad emission spectra and high PLQY, making them attractive for scintillator design. Under X-ray excitation, $Cs_2NaTbCl_6$ SCs exhibited an impressive light yield of 46,600 photons/MeV, surpassing that of most of many lead-based halide perovskites. However, their practical deployment in X-ray imaging remains limited [81]. Tang and coworkers utilized $Cs_3Cu_2I_5$ NCs (PLQY ~ 59%) for high resolution X-ray imaging. They reconstructed the projection images of an electrical chip and a 3D computed tomography image of a snail using $Cs_3Cu_2I_5$ NCs scintillator detector with a decay time of 38.94 ns, revealing both external morphology and intricate internal structures (Fig. 13.3a). High-resolution 3D reconstructions were obtained from 180 tomographic slices using a compressed sensing algorithm, which exhibited significantly low signal-to-noise ratios, and imaging artifacts [82]. In case of 0D $Cs_3Cu_2I_5$ NCs, few other reports are there, Lian et al. also developed a 0D $Cs_3Cu_2I_5$ NC-based scintillator that exhibited minimal self-trapped excitons (STEs) formation, thereby enabling enhanced radioluminescence free from self-absorption losses. Utilizing a drop-casted thin film of these NCs, they demonstrated high-quality X-ray imaging, which could successfully resolve intricate features of a circuit board and an enclosed metal spring within a ballpoint pen, as well as achieve a spatial resolution of 0.32 mm as shown in Fig. 13.3b [83]. Notably, 0D Sb^{3+} based organic–inorganic perovskites can benefit from the high atomic number of antimony ($Z = 51$), imparting stronger X-ray attenuation capabilities. The molecular stability, favorable electronic configuration, and strong electron–phonon coupling of $[SbX_n]^{n-3}$ clusters can enable these materials to achieve near-unity PLQYs, and their emission is typically dominated by broadband STEs recombination. Recently, Sun et al. fabricated the shape-engineered, optically transparent $(C_{20}H_{20}P)_2SbCl_5$ perovskite medium—(where $C_{20}H_{20}P^{+}$.. corresponds to ethyltriphenylphosphonium), it was synthesized using a low-temperature, solvent-free melt-quenching route, exhibiting intense orange emission and high optical clarity. This material demonstrated an exceptional spatial resolution of 30 lp mm^{-1} for transparent scintillation screenacross a thickness range of 700–1100 μm, significantly outperforming many conventional scintillators. By comparison, polycrystalline screens of the same composition with increasing thickness have shown the resolution degradation, emphasizing the advantage of a transparent medium in high-resolution X-ray imaging, as shown in Fig. 13.3c [84]. Tang et al. further fabricated one-dimensional (1D) Rb_2CuBr_3 perovskite SCs with a 98.6% PLQY and a high density of STEs, as evidenced by a substantial Stokes shift of ~ 85 nm. Finally, X-ray scintillation measurements carried out using these SCs, performed using a silicon photomultiplier, revealed a phenomenal light yield of 91,056 photons/MeV and an ultralow detection limit of 121.5 $nGy_{air}\ s^{-1}$. These performance parameters surpassed the benchmark of commercial LYSO scintillators by a substantial margin (Fig. 13.3c) [85]. Furthermore, Wang et al. synthesized lanthanide-based halide perovskites in the form of Cs_3LnCl_6 (Ln = Gd^{3+}, Y^{3+}, Lu^{3+}), incorporating Ce^{3+} dopants to achieve efficient blue–violet emission. In the case of Ce:Cs_3LuCl_6, the material exhibited a high PLQY of approximately 82%, along with a carrier lifetime of 34 ns. This composition delivered a remarkable scintillation light yield of 8120 photons/MeV and demonstrated an exceptionally low X-ray detection limit of 36.8 $nGy_{air}\ s^{-1}$. For scintillation applications, thin films were fabricated by dispersing 20% Ce:Cs_3LuCl_6 within a PMMA/toluene polymer matrix. Spatial resolution was evaluated through X-ray imaging of a standard resolution test pattern, revealing a resolvable feature range of approximately 8–10 lp mm^{-1} while yielding a measured resolution of 8.38 lp mm^{-1}. Additionally, digital radiographs of a microchip placed on the 20% Ce:Cs_3LuCl_6 scintillator film were captured under both ambient daylight and X-ray irradiation, confirming the material's capability for high-resolution imaging (Fig. 13.3e) [86]. Interestingly, 2D tin-based nanocrystals (NCs) have also emerged as a highly promising class within the domain of lead-free perovskite scintillators. Most recently, Birowosuto et al. developed a series of 2D Sn-based mixed-halide perovskites incorporating three distinct aromatic ammonium ligands—phenylmethylammonium (PMA), phenethylammonium (PEA), and phenylpropylammonium (PPA)—yielding the compositions $(PMA)_2SnBr_3I$, $(PEA)_2SnBr_3I$, and $(PPA)_2SnBr_3I$. Systematic ligand variation enabled precise modulation of interlayer spacing, optoelectronic characteristics, octahedral distortion, and charge-carrier dynamics. Among these, $(PMA)_2SnBr_3I$ exhibited the fastest PL decay (τ_{avg} ~ 1.1 ns) and the highest light yield, reaching around 10–15 photons/keV. Density functional theory (DFT) analysis attributed this superior scintillation performance to cation engineering, wherein shorter ligands and reduced octahedral distortion can diminish antibonding interactions between Sn 5 s and I 5p orbitals, thereby promoting ultrafast exciton recombination [87].

13.2.2 Photodetectors (PDs) and Imaging

13.2.2.1 Device Architecture and Mechanism

In imaging, PDs serve as the fundamental light-sensing elements that convert an optical scene into an electrical signal upon the photoelectric effect, enabling the capture, processing, and reproduction of images [88]. Based on the spatial configuration of the photoactive medium relative to the electrodes, PDs are typically categorized into vertical and lateral architectures [89]. Furthermore, PDs are generally classified as photodiodes, photoconductors, or phototransistors, as these device types embody fundamentally distinct structural designs and operational mechanisms [90] as illustrated in Fig. 13.4. Photodiodes operate on the principle of the p–n junction,

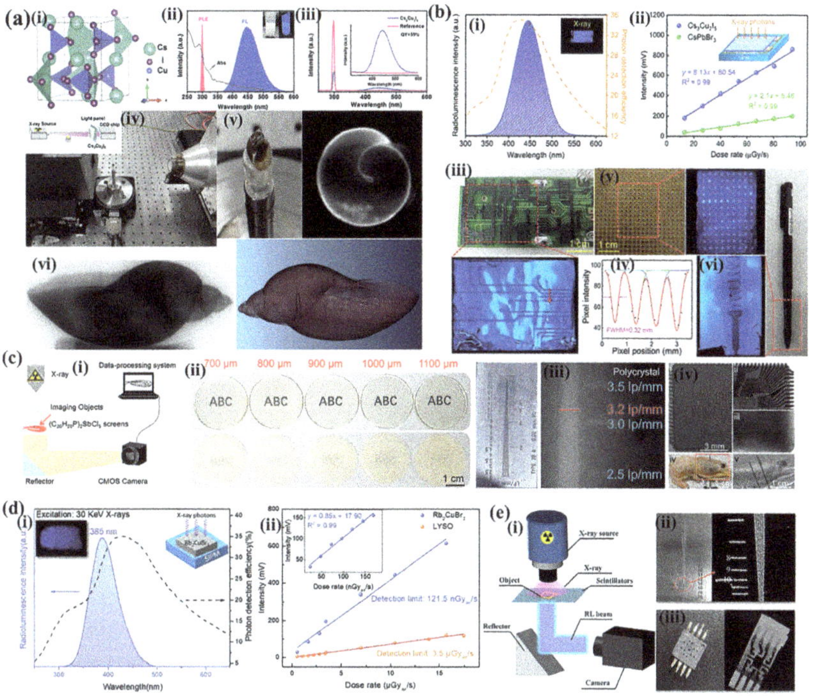

Fig. 13.3 **a** (i) Schematic representation of the crystal structure of $Cs_3Cu_2I_5$ NCs, (ii) PL spectrum measured under 300 nm excitation and photoluminescence excitation (PLE) spectrum recorded at emission wavelength of 445 nm, (iii) PLQY spectrum of $Cs_3Cu_2I_5$ nanocrystals measured using an integrating sphere, (iv) Schematic and image of the X-ray detection setup (consists of X-ray source tube and $Cs_3Cu_2I_5$ NCs based detector) used for imaging, (v) snail sample and its projection image, (vi) slice reconstructed and volume rendering image of the snail [82] Copyright 2021. Reproduced with permission from Wiley. **b** (i) radioluminescence (RL) spectrum $Cs_3Cu_2I_5$ NCs thin films under the irradiation of 50 keV X-rays, (ii) Comparison of X-ray photocurrent intensity of $Cs_3Cu_2I_5$ and $CsPbBr_3$ NCs-based detectors at various dose rates, (iii) photograph and relating X-ray image of the circuit board, (iv) Intensity profile point spread function fitted and analysed by Gaussian function, photographic and X-ray image of the (v) circuit board and (vi) ball point pen [83] Copyright 2020. Reproduced with permission from Wiley. **c** (i) image of the homemade X-ray imaging setup for $(C_{20}H_{20}P)_2SbCl_5$ detectors, (ii) photographic images of the transparent and polycrystalline screens of various thicknesses (700–1100 μm), (iii) results representing spatial resolutions (3.2 and 30 lp mm^{-1}) of polycrystalline and transparent screens, (iv) a clear and obscure X-ray image of the electronic chip based on the transparent and polycrystalline screen; high-clarity X-ray image of a shrimp containing two concealed needles [84] Copyright 2024. Reproduced with permission from American Chemical Society. **d** (i) Radioluminescence spectrum recorded under 30 keV X-ray excitation, along with the wavelength-dependent photon detection efficiency of the silicon photomultiplier (SiPM). The left inset presents a photograph of Rb_2CuBr_3 crystals exposed to X-ray irradiation, while the right inset illustrates a schematic of the experimental setup used to measure emission intensity with the SiPM, (ii) Comparison of emission intensity for Rb_2CuBr_3 and LYSO with various dose rates; the inset shows the detection limit curve of Rb_2CuBr_3 detectors (< 200nGyairs-1) [85] Copyright 2019. Reproduced with permission from Wiley. **e** (i) Schematic illustration of the X-ray imaging system—X-ray imaging was performed by positioning the target specimen above the scintillation screen and irradiating it using a Mini-X X-ray source within a custom-built imaging setup, (ii) photographic image and X-ray image of the resolution test plate, (iii) photographic image and X-ray image of the electronic chip (X-ray dose rate ~ $4.5mGy_{air}s^{-1}$) [86] Copyright 2024. Reproduced with permission from Elsevier

with the vertical device configuration being the most prevalent design in perovskite-based photodiodes. In such vertical PDs the reduced electrode spacing minimizes the carrier transit distance, thereby enhancing charge collection efficiency [88]. In the case of photoconductors, they usually have the lateral PD device structure, consisting of a photoactive layer paired with two parallel electrodes (Fig. 13.4e). The photogenerated carriers in the perovskite absorber are separated by a driving voltage and collected by two electrodes; thus, the applied voltage is indispensable for photoconductors, which are distinct from photodiode detectors that can directly convert light into electricity (photovoltage or photocurrent) without an external bias voltage [91]. Phototransistors usually have a three-terminal lateral device structure (i.e., source, drain, and gate), which is similar to the device structure of field-effect transistors (FETs), as shown in Fig. 13.4c and f. In comparison with photoconductors, phototransistors offer an additional degree of control by enabling modulation of the semiconductor channel conductance—either under illumination or in darkness—through adjustment of the gate voltage. This capability facilitates amplification of the photocurrent or suppression of the dark current [92].

Currently, commercial imaging systems are largely dominated by silicon and III–V semiconductor photodetectors, such as InGaAs and InP, due to their mature fabrication and reliable performance. However, their fixed band gaps, rigidity, and high production costs limit further progress. Lead-free halide perovskite photodetectors, with tunable band gaps, strong absorption, low or no toxicity, and low-cost solution processability, present a promising alternative for next-generation imaging technologies.

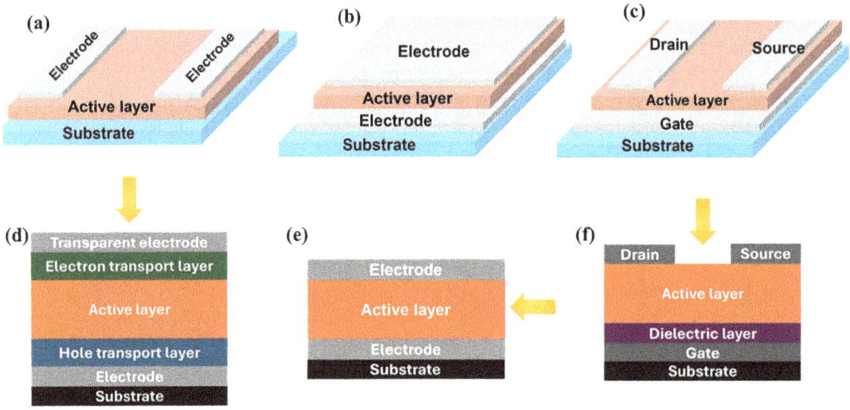

Fig. 13.4 Schematic structures of **a** a lateral PD, **b** a vertical PD with two parallel electrodes, and **c** a lateral PD with three metal electrodes (source, drain, and gate), **d** a photodiode with a vertical multilayer stack structure, **e** a two terminal photoconductor structure, and **f** a phototransistor with a bottom-gate and top-contact structure

13.2.2.2 Key Performance Parameters of PDs

In practice, image sensors are typically composed of arrays of PDs, with each PD functioning as an individual pixel. Consequently, the overall performance of an image sensor is intrinsically determined by the photoelectric characteristics of its constituent PDs [93]. To enable quantitative evaluation and facilitate direct comparison of reported perovskite PDs for imaging applications, the following section introduces several key performance parameters:

(i) **Responsivity**

Responsivity (R) quantifies the ability of a PD to convert incident radiation into an electrical output signal. It is formally defined as the ratio of the photocurrent generated by the device to the incident optical power at a given wavelength. It can be described as

$$R = \frac{I_p - I_d}{AP} \tag{13.7}$$

here, I_p and I_d represent photocurrent and dark current at the radiation wavelength λ, respectively. P is power density of the incident light, and A refers to the effective device area of the actual light-absorbing region of the device [88].

(ii) **Detectivity**

Detectivity (D*) is a key figure of merit that characterizes a PD's ability to discern weak optical signals in the presence of intrinsic and extrinsic noise. This parameter integrates both the device's sensitivity and its noise performance at a specific wavelength, providing a standardized metric for comparing detectors with different active areas and bandwidths. It can be calculated as

$$D^* = \frac{(S\Delta f)^{\frac{1}{2}}}{NEP} = \frac{R}{(2eI_n\Delta f)^{1/2}} \tag{13.8}$$

where Δf is the bandwidth, NEP is the noise equivalent power, e is the electron charge, and I_n is the root mean square value of the total noise current [94].

(iii) **External Quantum Efficiency**

External quantum efficiency (EQE) is defined as the ratio between the number of photogenerated charge carriers (electrons or holes) collected by the electrodes and the number of incident photons. This parameter reflects the photon-to-charge conversion efficiency of a PD, providing a direct measure of its light-harvesting and carrier-collection capabilities. It is described as follows:

$$EQE = \frac{Rhc}{e} \tag{13.9}$$

where e, h, and c represent the electron charge, the Planck constant, and the speed of light, respectively [95].

(iv) **On/Off Ratio**

The on/off ratio quantifies the contrast between a PD's performance under illumination (on state) and in darkness (off state). It serves as a metric for evaluating the device's capability to discriminate between light and dark conditions. This parameter is typically calculated using the following expression:

$$on/off\ ratio = \frac{I_p}{I_d} \tag{13.10}$$

here, I_p and I_d represent photocurrent and dark current at the radiation wavelength λ, respectively [96].

(v) **Signal to Noise Ratio (SNR)**

SNR is defined as the ratio of signal power to noise power, which can be described as follows:

$$SNR = \frac{I_{singal}}{I_{noise}} \tag{13.11}$$

in which I_{signal} is the signal current, I_{noise} is the noise induced current [88].

13.2.2.3 Advances in Lead-Free Perovskite PDs

In this section, we mainly focused on Sn-based and Bi^{3+} based lead free perovskites [97]. Sn halide perovskites with an optical bandgap below 1.4 eV, are promising for NIR-PDs that facilitate optical detection ranging from visible to NIR [98, 99]. Recently, Liu et al. deposited non-toxic quasi-single-crystalline $CH(NH_2)_2SnI_3$ ($FASnI_3$) perovskite thin films for fabricating a high-resolution NIR imaging PD array, which consists of 64 × 64 pixels on the thin-film transistors (TFT) backplanes (Fig. 13.5a) [100]. Owing to their high structural integrity and efficient NIR response, these imaging arrays have been demonstrated for hidden object recognition, as shown in Fig. 13.5b. This work represents the first reported instance of a perovskite-based imaging array by achieving such high functionality. Sn-based perovskite thin films could be conveniently deposited on both rigid and flexible substrates through low-temperature solution processes, enabling them to be integrated with flexible and wearable electronics [101]. However, Sn^{2+} is easily oxidized to Sn^{4+} even under trace amounts of moisture and oxygen (< 10 ppm), the fabrication of high-quality, dense active films with low defect density and effective encapsulation is thus crucial for enhancing the performance of Sn-based perovskite PDs. In recent years, additive engineering has been applied to improve Sn-based perovskite stability and film quality. For example, He et al. introduced phenyl–ethyl ammonium (PEA^+) additive in pristine $FASnI_3$, achieving $FA_{0.8}PEA_{0.2}SnI_3$ thin films with highly crystalline order and flexibility [102]. Fig. 13.5c, d shows the uniform coverage and smooth surface morphology of $FASnI_3$ films with 20% of PEA^+ substitutions. The PD array, fabricated from $FA_{0.8}PEA_{0.2}SnI_3$ thin film and comprising 64 × 64 pixels, is presented in Fig. 13.5e, with its corresponding circuit schematic depicted in Fig. 13.5f. In the PD array, the $FA_{0.8}PEA_{0.2}SnI_3$ thin film was deposited onto a polyethylene naphthalate (PEN) flexible substrate patterned with strip-shaped ITO electrodes (Fig. 13.5g). Figure 13.5h presents the dark current and photocurrent measurements of this flexible PD array, and the clear image of the letter "D" results, respectively, demonstrating the high-resolution imaging capability of $FA_{0.8}PEA_{0.2}SnI_3$ PDs. Therefore, Sn-based halide perovskites hold strong potential for high-resolution photodetector applications and flexible imaging arrays; however, their inherent instability necessitates effective mitigation strategies, with particular emphasis on optimizing film fabrication quality.

Benefiting from the variable oxidation states of Bi, Bi-based halide perovskites can adopt diverse crystal structures—as 0D, 1D, 2D, and 3D—corresponding to various stoichiometries such as $A_3Bi_2X_9$, ABi_aX_{3a+1}, and A_bBiX_{b+3}, as well as double perovskite A_2BBiX_6 (A = Cs, Ag, MA, FA, K, Rb, Cu; B = Ag^+, Na^+ and so on; X = I, Br, Cl) [103]. Among these, $Cs_2AgBiBr_6$ is recognized as one of the most promising candidates due to its excellent structural stability and high resistance to heat and moisture. However, the thin film fabrication of $Cs_2AgBiBr_6$ is relatively simple and solution-processable; however, achieving uniform film coverage, good pinhole-free film morphology, and long-range crystallinity remains challenging [18, 104]. Consequently, several film optimization strategies, including additive engineering and post-treatment methods, have been employed to achieve uniform, pinhole-free $Cs_2AgBiBr_6$ thin films. One prominent approach involves treating the perovskite film with isopropyl alcohol (IPA) and fabricating a heterojunction photodetector (PD) by depositing the perovskite layer onto an SnO_2 substrate. The SnO_2 layer promotes high-quality perovskite film growth, resulting in films with larger grains and fewer pinholes. For example, Li et al. employed SnO_2 as the growth template in $Cs_2AgBiBr_6$-based PDs and observed that post-treatment with isopropanol (IPA) significantly enhanced the photodetection performance of the $SnO_2/Cs_2AgBiBr_6$ devices [39]. Figure 13.6a presents the schematic design of the $SnO_2/Cs_2AgBiBr_6$ PD-based transmittance imaging system, along with imaging results (Fig. 13.6b) obtained under illumination intensities of 0.5 and 5 μWcm^{-2}, as well as the corresponding signal-to-noise ratio (SNR) responses. These results demonstrate that the optimized $SnO_2/Cs_2AgBiBr_6$ PD exhibits outstanding weak-light imaging capability with detectable light intensity of 42 nW/cm^2 [105]. Recently, Lv et al. developed advanced photodetectors using Bi^{3+}-based 2D $Cs_3Bi_2Br_9$ perovskites with an $A_3Bi_2X_9$ crystal structure. Their approach employed a dual-interface engineering strategy, incorporating carrier-selective charge transport layers—CuO_x and PCBM—on pristine $Cs_3Bi_2Br_9$ films to enhance device performance. The optimized $CuO_x/Cs_3Bi_2Br_9$/PCBM/BCP (CCPB) device demonstrated remarkable performance improvements, including a 9333-fold increase in response speed and a 503-fold enhancement in the on/off ratio in comparison to pristine $Cs_3Bi_2Br_9$/Ag

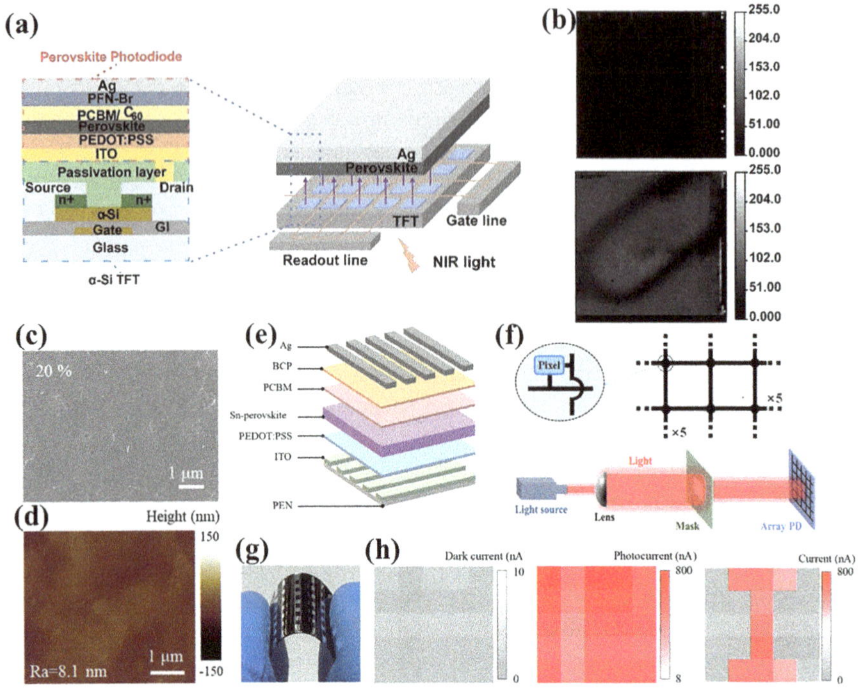

Fig. 13.5 **a** Schematic structure of the NIR PD array based on quasi-single-crystal $FASnI_3$ perovskite. **b** Image of paperclip inside an envelope captured by imaging array under visible (520 nm, top) and NIR (785 nm, bottom) illumination [100] Copyright 2024. Reproduced with permission from Wiley. **c** The SEM image and **d** corresponding AFM images of $FASnI_3$ perovskite film with 20% PEA^+ substitution. **e** The layer structure of the 5 × 5 array device based on $FA_{0.8}PEA_{0.2}SnI_3$ thin film. **f** The corresponding circuit diagram (top) and illustration (bottom) of the imaging test system for the 5 × 5 array device. **g** Picture of the PD array under a bending condition. **h** The dark current, photocurrent histograms, and imaging result for letter "D" (from left to right) [102] Copyright 2023. Reproduced with permission from Wiley

based PDs, and thereby achieved the specific detectivity of 2.95×10^{10} Jones (at 463 nm) and an ultrafast response time of 274 μs. Leveraging these advancements, the photodetectors were integrated into a fast Fourier transform single-detector imaging system, enabling the acquisition of high-quality 256 × 256 pixel images within just 130 s (detector schematics, mechanism, and image generation in Fig. 13.6d–f) [106]. Unlike conventional array-based detectors, this approach can achieve high-resolution imaging by employing a single detector that is combined with computational algorithms, which eventually paves a promising pathway for the next generation of imaging technologies.

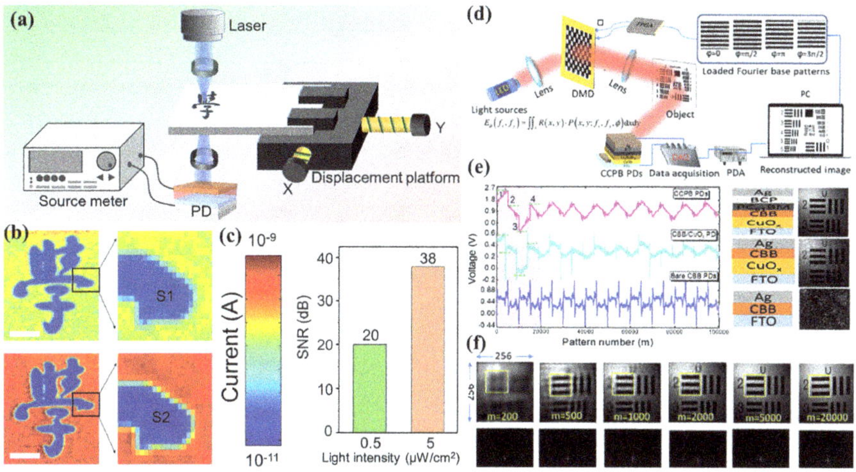

Fig. 13.6 **a** The schematic design of $SnO_2/Cs_2AgBiBr_6$ PD and **b** imaging results under a light intensity of 0.5 μW/cm^2 and 5 μW/cm^2, respectively (the scale bar is 0.4 cm), as well as **c** SNRs of the imaging results [105] Copyright 2022. Reproduced with permission from Elsevier. **d** Schematics for the image generation with CCPB PDs. **e** Representative photovoltage curves under the first four projection patterns, along with their corresponding reconstructed images. **f** Restored images with various projection numbers and sampling rates [106] Copyright 2024. Reproduced with permission from American Chemical Society

13.2.3 Perovskites for Fluorescence and NIR Bioimaging Applications

Perovskite nanocrystals (PNCs), including their quantum dots (QDs) form, have shown significant potential in bioimaging applications [107, 108]. They can be employed as efficient fluorescent probes for high-resolution visible light fluorescence imaging [109]. Fluorophores emitting within the visible spectral range, when combined with molecular specificity toward different cellular targets, are particularly valuable for resolving subcellular structures and dynamic processes, as most intrinsic cellular constituents lack inherent optical contrast. Additionally, perovskite emitting in NIR wavelengths (mostly ~ 700–1700 nm) allows deeper tissue penetration, thereby enabling advanced in vivo imaging [110]. In comparison to traditional fluorescence microscopy (utilizing visible light wavelengths), NIR imaging offers superior tissue penetration due to reduced photon scattering in biological media. Also, NIR imaging provides several (i) enhanced imaging contrast due to negligible absorption of NIR range photons by the biological tissues, (ii) comparatively high SNR due to effectively decreased tissue autofluorescence, and (iii) minimal photodamage in tissues due to light penetration, resulting in prolonged imaging durations without compromising tissue integrity [111, 112].

Due to their beneficial optoelectronic properties, perovskites are also promising candidates for bioimaging applications [113]. The optoelectronic properties of perovskites relevant for bioimaging include high PLQYs, long PL lifetimes, narrow emission bandwidths with low full-width at half maximum (FWHM), and low-intensity driven efficient excitation beneficial for live-cell imaging, exceptional two-photon absorption cross-sections, and strong photostability [114, 115]. However, the intrinsic instability and sensitivity of halide perovskites in aqueous biological environments continue to pose a major challenge for their use in bioimaging applications. For their use in biomedical imaging, PNCs must retain their stability in water and should be optically and structurally stable in the complex aqueous biological media. However, there are several different approaches, such as ligand modification, hydrophobic surface coatings, and foreign metal ion incorporation, that are used for improving the biocompatibility of PNCs, but are only partially successful and often lead to nanocrystal aggregation, which affects the size-dependent optical properties of PNCs and can reduce PLQY at the end [116, 117]. Hence, more precise solutions are required, particularly with a focus on addressing the high surface energies and inherent ionic nature of perovskites. The ultimate goal is to develop perovskite stabilization strategies that preserve and enhance their optimal photophysical properties. For instance, Vedi et al. synthesized lead-free $Cs_3Sb_2I_9$ nanocrystals (NCs) using the conventional colloidal hot-injection process and subsequently encapsulated the NCs with (3-aminopropyl) trimethoxysilane (APTES), yielding $Cs_3Sb_2I_9@SiO_2$ composites. These nanocomposites are further utilized for bioimaging studies using L929 mouse fibroblast cells for fluorescence phase contrast microscopy. This study enabled efficient cellular attachment and its clear visualization under blue-green wavelength excitation, establishing these probes as potential fluorescent biocompatible markers [118]. Furthermore, Huang et al. also synthesized lead-free $Rb_7Bi_3Cl_{16}$ NCs, which were further passivated with a silica shell using APTES Silica encapsulation significantly enhanced their aqueous dispersibility while improving the PLQYs from 28.43% to 37.03% (emission centred at 445 nm, FWHM $\approx$ 92 nm). The resulting $Rb_7Bi_3Cl_{16}$ NCs@SiO_2 composites exhibited excellent water stability, strong luminescence, and low cytotoxicity. These properties enabled their application in adenosine-5′-triphosphate disodium salt (ATP) sensing and bioimaging of HePG2 cancer cells using confocal fluorescence microscopy (Fig. 13.7a). The composites were clearly localized within tumor cells, demonstrating their promise as fluorescent probes for cancer imaging [119].

Particularly, NIR imaging can further leverage electromagnetic radiation ranging from 700–1700 nm, by involving NIR excitation or NIR emissions, or both. The well-known lead-free perovskites, $CsSnX_3$ (X = I, Br, or Cl), exhibit NIR emissions as described previously, with $CsSnI_3$ and $CsSnBr_3$ specifically showing emission peaks at around 940 nm and featuring a narrow FWHM of less than 50 nm [120]. However, broadband NIR emission is particularly desirable for bioimaging applications; the lead-free double perovskite $Cs_2Sn(IV)Cl_4Br_2$, when doped with Sb^{3+}, demonstrates a promising broadband emission ranging from 550–850 nm [121]. Nevertheless, the range still does not cover the entire desired NIR emission range. Lanthanide ions such as Er^{3+}, Yb^{3+}, and Nd^{3+} are particularly attractive for generating NIR emissions. Despite their intrinsically low absorption coefficients, these rare-earth ions can be effectively employed as dopants (Table 13.1). When combined with double

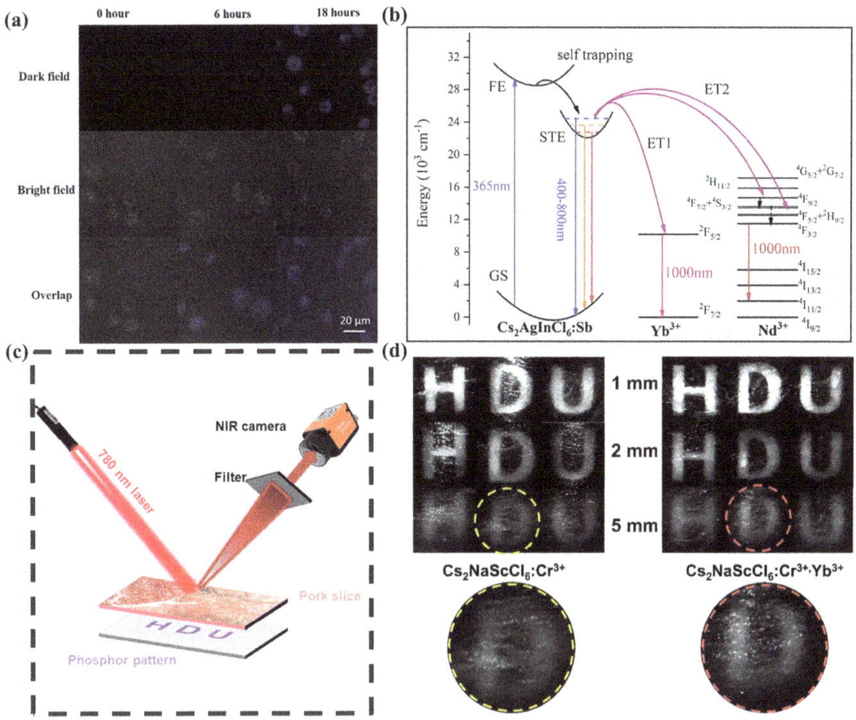

Fig. 13.7 **a** Fluorescence microscopic images of HePG2 cancer cells after different incubation time with $Rb_7Bi_3Cl_{16}$ NCs@SiO_2 composites [119] Copyright 2023. Reproduced with permission from Wiley. **b** Energy-level diagram of Yb^{3+}/Nd^{3+}-doped $Cs_2AgInCl_6$:Sb microcrystals, illustrating the suggested luminescence mechanism [123] Copyright 2022. Reproduced with permission from Elsevier. **c** Graphic showing instrumentation for the NIR bioimaging setup. **d** NIR camera images of the phosphor pattern "HDU" obtained when covered with pork slices of 1, 2, and 5 mm thickness. The highlighted circles identify that Yb^{3+} doped $Cs_2NaScCl_6$:Cr^{3+} composite perovskite film shows better contrast and a more defined pattern in comparison to the pristine perovskite film [124] Copyright 2025. Reproduced with permission from American Chemical Society

perovskites that possess high absorption coefficients, lanthanides serve as complementary emitters, thereby enhancing NIR emitting performance [122]. Meanwhile, the NIR emission could be enriched with co-doping strategies, which are generally used to garner multimodal emission or enhance the PLQYs. A similar mechanism has been observed in $Cs_2AgInCl_6$:Sb^{3+}–Nd^{3+}/Yb^{3+} where the presence of the rare earth metals in stable pristine double perovskite creates additional energy levels in the NIR emission range, i.e. Yb^{3+} show transition in $^2F_{5/2}$ to $^2F_{7/2}$ (Fig. 13.7b) [123]. Subsequent work by Zhong et al. demonstrated that co-doping $Cs_2NaScCl_6$:Yb^{3+} perovskites with Cr^{3+} significantly enhances the photoluminescence quantum yield (PLQY), increasing it from approximately 28.3% to 76.9%. This improvement is attributed to efficient energy transfer mechanisms between Cr^{3+} and Yb^{3+} ions, potentially facilitated by Cr^{3+} occupying defect states near the band edges of the host lattice, thereby altering its electronic band structure. Furthermore, a composite film comprising $Cs_2NaScCl_6$:Cr^{3+},Yb^{3+} powder dispersed in epoxy resin was fabricated to form an 'HDU' pattern, which remained undetectable under ambient lighting conditions when obscured by pork tissue. Under 780 nm laser excitation (200 mW), the letters became clearly visible to an NIR camera, even through a 5 mm slice of pork (Fig. 13.7c, d) [124]. Most recently, Zhang et al. demonstrated NIR bioimaging using Ce-doped double perovskite phosphors, i.e., $Cs_2NaBiCl_6$:Yb^{3+}/Cr^{3+}/RE^{3+} (RE = Er, Tm, Pr). In comparison to $Cs_2NaBiCl_6$:Yb^{3+}/RE^{3+} phosphors, codoping with Cr^{3+} ions significantly enhanced the NIR-II emission intensity of Er^{3+} at 1540 nm ($^4I_{13/2} \rightarrow {}^4I_{15/2}$ transition), achieving a 10.7-fold increase. These optimized phosphors were further employed for vascular imaging, where their strong NIR-II (1000–1700 nm) luminescence realized a high-resolution mapping of finger veins. The emission efficiently penetrated biological tissue, and the vascular structures were clearly visualized with an infrared camera, highlighting the material's potential for deep-tissue bioimaging applications [125]. Overall, lanthanide-doped lead-free perovskites significantly advance biomedical applications by endowing deep tissue penetration and high-contrast imaging, with Yb^{3+}-doped systems, in particular, offering highly suitable NIR emission for broad bioimaging applications. Alternatively, a new subclass of perovskites, called layered double perovskite, also presents the compositional engineering of $Cs_4CdSb_2Cl_{12}$ and $Cs_4MnSb_2Cl_{12}$ doped with Yb^{3+} and Er^{3+}. This kind of perovskite exhibits the multispectral emission of 983 nm from Yb^{3+} presence, 1541 nm from Er^{3+}, and ~ 650 nm for the Mn^{2+}, indicating a futuristic strategy for covering the entire visible range and NIR emission range, even from the X-ray range [124, 126–128]. These several recent examples illustrate the growing potential of PNCs in the bioimaging domain, though further optimization is needed to translate these materials into robust, biologically compatible imaging platforms.

13.3 Conclusions and Outlook

Conclusively, this chapter provided a comprehensive overview of the key parameters and mechanisms behind image generation across various detection technologies, particularly in X-ray detectors, PDs, and NIR detectors (limited to bioimaging applications). We reviewed the detailed image generation mechanism using various types of detectors and imaging wavelengths utilizing lead-free perovskite semiconductors. Particular emphasis was placed on elaborating the various beneficial optoelectronic properties of lead-free perovskites as active semiconducting materials. The discussion simplified how material composition, photophysical properties, interfacial design of the detector, and structural engineering influence the charge transport and detection efficiency across these devices. Collectively, these understandings establish the potential of lead-free perovskites in advancing sustainable, next-generation, and high-performance imaging technologies, although their

Table 13.1 Brief summary of the lead-free perovskite materials and their emission characteristic with the NIR emission range

Compounds	Excitation (nm)	NIR emission range (nm)	PLQY (%)	Forms
$CsSnX_3$ (X: I^-, Br^-, Cl^-)	445/532/632/730	475–940	–	NCs [120]
$Cs_2SnCl_4Br_2$: Sb^{3+}	410	550–850	–	Single crystals [121]
$Cs_2NaScCl_6$:Er^{3+}	365	808–1550	~ 28.3	Bulk [129]
$Cs_2NaScCl_6$:Cr^{3+}/Yb^{3+}	365	800–1300	76.9	Bulk [124]
$Cs_2AgInCl_6$:Sb^{3+}-Nd^{3+}/Yb^{3+}	365	975–1075	–	Bulk [127]
$Cs_2AgInCl_6$:Cr^{3+}-Er^{3+}	360/583/815	1010/1540	57.5	Bulk [130]
$Cs_2NaInCl_6$:Sb^{3+}/Yb^{3+}	318	896–1358	48.95	Single crystals [131]
$Cs_2AgBiCl_6$: Na^+	365	500–1100	51	Single crystals [132]
$Cs_2AgBiBr_6$: Yb^{3+}	360	950–1050	82.5	Film [133]
$Cs_4(Cd/Mn)Sb_2Cl_{12}$: Yb^{3+}/Er^{3+}	$\geq$ 400 nm	983 and 1541	< 1% (for $Cs_4CdSb_2Cl_{12}$:Yb^{3+}/Er^{3+}) and 8.5/18.4% (for $Cs_4MnSb_2Cl_{12}$:Yb^{3+}/Er^{3+})	NCs [128]

commercial integration and implementation still encounter several unresolved challenges. However, the toxicity issue of lead-based perovskites is partially resolved by replacing lead with other metal ions and developing environmentally benign lead-free perovskites, but their intrinsic instability towards atmospheric factors still remains a critical limitation. Their evident instability to environmental factors such as heat, moisture, and polar solvents (e.g., DMF, DMSO) obstructs their compatibility with traditional photolithography processes, which are used for the fabrication of CMOS and CCD imaging technologies. This incompatibility creates a significant barrier to incorporating the perovskite devices into high-resolution multipixel array architectures. However, encapsulation is a prominent strategy widely employed to mitigate the instability issue in perovskite devices, but its efficacy remains subpar. Alternatively, single-pixel imaging can be a promising alternative to circumvent these manufacturing constraints. Single-pixel devices offer simple fabrication, reduced design complexity, and improved performance compared to traditional silicon-based multipixel sensor architectures. Furthermore, single-pixel imaging has the potential to achieve high-quality image reconstruction beyond the sensitivity of silicon sensors, paving the way for advanced imaging technologies.

Nonetheless, both multipixel and single-pixel perovskite-based devices need high-quality multilayer architectures with optimal interfacial properties. One of the major fabrication challenges is to fabricate high-quality, uniform, large-area perovskite thin films or SCs, ensuring good active layer-electrode contact and better interfaces with charge transport layers. However, film optimization strategies, such as additive engineering, surface passivation, and interface engineering, are expected to overcome these limitations; yet, achieving a large area perovskite thin film with long-range crystallinity remains challenging. A recent breakthrough has occurred in the growth of single-crystalline perovskite thin films, which generally offers enhanced crystallinity and superior structural order, as studied by Bakar and co-workers [134]. Therefore, optimizing the synthesis process of perovskite films or crystals is crucial in order to minimize structural defects and disorders, as the uniformity of perovskite strongly influences its optoelectronic properties, and consequently, the performance of perovskite detectors and their imaging sensitivity. Moreover, interfacial matching between the electrodes and the perovskite also plays a critical role in the photoelectric gains, specifically in the case of perovskite X-ray detectors. Conclusively, mitigating these challenges by optimizing the perovskite synthesis conditions and interfacial engineering could significantly multiply the reliability and performance of existing perovskite-based imaging devices. Ultimately, these advancements could facilitate the commercial integration of perovskite-based detectors into next-generation imaging technologies.

References

1. F. Cao, X. Xu, D. Yu et al., Lead-free halide perovskite photodetectors spanning from near-infrared to X-ray range: a review. Nanophotonics **10**(8), 2221–2247 (2021). https://doi.org/10.1515/nanoph-2020-0632
2. A.P.H. Butler, N.G. Anderson, R. Tipples et al., Bio-medical X-ray imaging with spectroscopic pixel detectors. Nucl. Instrum. Methods Phys. Res., Sect. A **591**(1), 141–146 (2008). https://doi.org/10.1016/j.nima.2008.03.039
3. T.W. Rogers, N. Jaccard, E.J. Morton et al., Automated X-ray image analysis for cargo security: critical review and future promise. J. Xray Sci. Technol. **25**(1), 33–56 (2017). https://doi.org/10.3233/XST-160606
4. X. Wang, L. Ren, H. Zong et al., Perovskite single pixel imaging exceeding the visible towards X-ray and THz. J. Mater. Chem. C **12**(29), 10857–10873 (2024). https://doi.org/10.1039/d4tc02080d
5. H. Song, S. Sasada, T. Kadoya et al., Cross-correlation of confocal images for excised breast tissues of total mastectomy. IEEE Trans. Biomed. Eng. **71**(5), 1705–1716 (2024). https://doi.org/10.1109/TBME.2023.3348480
6. M.D. Murphey, J.L. Quale, N.L. Martin et al., Computed radiography in musculoskeletal imaging: state of the art. AJR Am. J. Roentgenol. **158**(1), 19–27 (1992). https://doi.org/10.2214/ajr.158.1.1727344

7. Y. Wu, J. Feng, Z. Yang et al., Halide perovskite: a promising candidate for next-generation X-ray detectors. Adv. Sci. **10**(1), 2205536 (2023). https://doi.org/10.1002/advs.202205536
8. Y. Namdev, J. Ghosh, M.A. Afroz et al., Efficient and highly stable X-ray detection and imaging using 2D $(BA)_2PbI_4$ perovskite single crystals. ACS Photonics **9**(11), 3529–3539 (2022). https://doi.org/10.1021/acsphotonics.2c00776
9. Z. Luo, J.G. Moch, S.S. Johnson, C. Chon Chen, A review on X-ray detection using nanomaterials. Curr. Nanosci. **13**(4), 364–372 (2017). https://doi.org/10.2174/1573413713666170329164615
10. R.H. Dyck, G.P. Weckler, Integrated arrays of silicon photodetectors for image sensing. IEEE Trans. Electron Devices **15**(4), 196–201 (1968). https://doi.org/10.1109/T-ED.1968.16166
11. Y. Eisen, A. Shor, I. Mardor, CdTe and CdZnTe X-ray and gamma-ray detectors for imaging systems. IEEE Trans. Nucl. Sci. **51**, 1191–1198 (2004). https://doi.org/10.1109/TNS.2004.829437
12. R.A. Street, S.E. Ready, K. Van Schuylenbergh et al., Comparison of PbI_2 and HgI_2 for direct detection active matrix X-ray image sensors. J. Appl. Phys. **91**(5), 3345–3355 (2002). https://doi.org/10.1063/1.1436298
13. G. Blasse, Scintillator materials. Chem. Mater. **6**(9), 1465–1475 (1994)
14. S. Li, X. Xie, J. Xiong et al., Review: perovskite X-ray detectors (1997–present). Crystals **12**(11), 1563 (2022). https://doi.org/10.3390/cryst12111563
15. Z. Shi, J. Guo, Y. Chen et al., Lead-free organic-inorganic hybrid perovskites for photovoltaic applications: recent advances and perspectives. Adv. Mater. **29**, 1605005 (2017). https://doi.org/10.1002/adma.201605005
16. F. Giustino, H.J. Snaith, Toward lead-free perovskite solar cells. ACS Energy Lett. **1**(6), 1233–1240 (2016). https://doi.org/10.1021/acsenergylett.6b00499
17. W. Ke, M.G. Kanatzidis, Prospects for low-toxicity lead-free perovskite solar cells. Nat. Commun. **10**(1), 965 (2019). https://doi.org/10.1038/s41467-019-08918-3
18. Z. Xiao, Z. Song, Y. Yan, From lead halide perovskites to lead-free metal halide perovskites and perovskite derivatives. Adv. Mater. **31**(47), 1803792 (2019). https://doi.org/10.1002/adma.201803792
19. M. Wang, W. Wang, B. Ma et al., Lead-free perovskite materials for solar cells. Nano-Micro Lett. **13**(1), 62 (2021). https://doi.org/10.1007/s40820-020-00578-z
20. X. Meng, Y. Wang, J. Lin et al., Surface-controlled oriented growth of $FASnI_3$ crystals for efficient lead-free perovskite solar cells. Joule **4**(4), 902–912 (2020). https://doi.org/10.1016/j.joule.2020.03.007
21. L. Xue, Y. Du, Y. Li, J. Zhao, X. Wang, C. Liu, D. C. Onwudiwe, B. S. Bae, M. Ertuğrul, Y. Zhu, W. Lei, Suppression of ion migration with a composite polymer film in perovskite single-crystal photodetectors for sensitive and stable x-ray imaging. Laser Photonics Rev., e01802 (2026). https://doi.org/10.1002/lpor.202501802
22. H. Wei, Y. Fang, P. Mulligan et al., Sensitive X-ray detectors made of methylammonium lead tribromide perovskite single crystals. Nat. Photonics **10**(5), 333–339 (2016). https://doi.org/10.1038/nphoton.2016.41
23. C. Connolly, X-ray systems for security and industrial inspection. Sens. Rev. **28**(3), 194–198 (2008). https://doi.org/10.1108/02602280810882535
24. M. Hoheisel, Review of medical imaging with emphasis on X-ray detectors. Nucl. Instrum. Methods Phys. Res. A **563**(1), 215–224 (2006). https://doi.org/10.1016/j.nima.2006.01.123
25. J.A. Rowlands, Current advances and future trends in X-ray digital detectors for medical applications. IEEE Trans. Instrum. Meas. **47**(6), 1415–1418 (2002). https://doi.org/10.1109/19.746705
26. R. Behling, F. Grüner, Diagnostic X-ray sources—present and future. Nucl. Instrum. Methods Phys. Res., Sect. A **878**, 50–57 (2018). https://doi.org/10.1016/j.nima.2017.05.034
27. M.H. Miah, M.U. Khandaker, M.A. Islam et al., Perovskite materials in X-ray detection and imaging: recent progress, challenges, and future prospects. RSC Adv. **14**(10), 6656–6698 (2024). https://doi.org/10.1039/d4ra00433g

28. Y. Zhou, J. Chen, O.M. Bakr et al., Metal halide perovskites for X-ray imaging scintillators and detectors. ACS Energy Lett. **6**(2), 739–768 (2021). https://doi.org/10.1021/acsenergylett.0c02430
29. E. Samei, M.J. Flynn, An experimental comparison of detector performance for direct and indirect digital radiography systems. Med. Phys. **30**(4), 608–622 (2003). https://doi.org/10.1118/1.1561285
30. O. Balitskii, M. Sytnyk, W. Heiss, Recent developments in halide perovskite nanocrystals for indirect X-ray detection. Adv. Mater. Technol. **9**(20), 2400150 (2024). https://doi.org/10.1002/admt.202400150
31. P.A. Douissard, A. Cecilia, X. Rochet et al., A versatile indirect detector design for hard X-ray microimaging. J. Instrum. **7**(09), P09016 (2012). https://doi.org/10.1088/1748-0221/7/09/P09016
32. A. Datta, Z. Zhong, S. Motakef, A new generation of direct X-ray detectors for medical and synchrotron imaging applications. Sci. Rep. **10**(1), 20097 (2020). https://doi.org/10.1038/s41598-020-76647-5
33. C. Szeles, CdZnTe and CdTe materials for X-ray and gamma ray radiation detector applications. physica status solidi (b) **241**(3), 783–790 (2004). https://doi.org/10.1002/pssb.200304296
34. M. Schieber, A. Zuck, H. Gilboa et al., Reviewing polycrystalline mercuric iodide x-ray detectors. IEEE Trans. Nucl. Sci. **53**(4), 2385–2391 (2006). https://doi.org/10.1109/TNS.2006.877043
35. M.R. Squillante, K.S. Shah, L. Moy, Stabilization of HgI_2 X-ray detectors. Nucl. Instrum. Methods Phys. Res., Sect. A **288**(1), 79–81 (1990). https://doi.org/10.1016/0168-9002(90)90467-K
36. M. Guerra, M. Manso, S. Longelin et al., Performance of three different Si X-ray detectors for portable XRF spectrometers in cultural heritage applications. J. Instrum. **7**(10), C10004 (2012). https://doi.org/10.1088/1748-0221/7/10/C10004
37. Y. Su, W. Ma, Y.M. Yang, Perovskite semiconductors for direct X-ray detection and imaging. J. Semicond. **41**(5), 051204 (2020). https://doi.org/10.1088/1674-4926/41/5/051204
38. X. Ou, X. Chen, X. Xu, L. Xie, X. Chen, Z. Hong, H. Yang et al., Recent development in x-ray imaging technology: future and challenges. Research (2021). https://doi.org/10.34133/2021/9892152
39. M. Nikl, A. Yoshikawa, Recent R&D trends in inorganic single-crystal scintillator materials for radiation detection. Adv. Opt. Mater. **3**(4), 463–481 (2015). https://doi.org/10.1002/adom.201400571
40. H. Feng, V. Jary, E. Mihokova, D. Ding, M. Nikl, G. Ren, R. Kucerkova et al., Temperature dependence of luminescence characteristics of $Lu_{2(1-x)}Y_{2x}SiO_5$:Ce^{3+} scintillator grown by the Czochralski method. J. Appl. Phys. **108**(3) (2010). https://doi.org/10.1063/1.3457856
41. B.K. Cha, J.H. Shin, J.H. Bae et al., Scintillation characteristics and imaging performance of CsI: Tl thin films for X-ray imaging applications. Nucl. Instrum. Methods Phys. Res., Sect. A **604**(1–2), 224–228 (2009). https://doi.org/10.1016/j.nima.2009.01.177
42. Q. Chen, J. Wu, X. Ou et al., All-inorganic perovskite nanocrystal scintillators. Nature **561**(7721), 88–93 (2018). https://doi.org/10.1038/s41586-018-0451-1
43. M. Roknuzzaman, C. Zhang, K. Ostrikov et al., Electronic and optical properties of lead-free hybrid double perovskites for photovoltaic and optoelectronic applications. Sci. Rep. **9**(1), 718 (2019). https://doi.org/10.1038/s41598-018-37132-2
44. M. Ren, X. Qian, Y. Chen et al., Potential lead toxicity and leakage issues on lead halide perovskite photovoltaics. J. Hazard. Mater. **426**, 127848 (2022). https://doi.org/10.1016/j.jhazmat.2021.127848
45. B. Conings, J. Drijkoningen, N. Gauquelin et al., Intrinsic thermal instability of methylammonium lead trihalide perovskite. Adv. Energy Mater. **5**(15), 1500477 (2015). https://doi.org/10.1002/aenm.201500477
46. J. Li, J. Duan, X. Yang et al., Review on recent progress of lead-free halide perovskites in optoelectronic applications. Nano Energy **80**, 105526 (2021). https://doi.org/10.1016/j.nanoen.2020.105526

47. S. Ma, G. Yuan, Y. Zhang et al., Development of encapsulation strategies towards the commercialization of perovskite solar cells. Energy Environ. Sci. **15**(1), 13–55 (2022). https://doi.org/10.1039/d1ee02882k
48. H. Wei, J. Huang, Halide lead perovskites for ionizing radiation detection. Nat. Commun. **10**(1), 1066 (2019). https://doi.org/10.1038/s41467-019-08981-w
49. M.H. Miah, M.U. Khandaker, M.A. Islam, M. Nur-E-Alam, H. Osman, M.H. Ullah, Perovskite materials in X-ray detection and imaging: recent progress, challenges, and future prospects. RSC Adv. **14**(10), 6656–6698 (2024). https://doi.org/10.1039/d4ra00433g
50. Š. Uxa, R. Grill, E. Belas, Evaluation of the mobility-lifetime product in CdTe and CdZnTe detectors by the transient-current technique. J. Appl. Phys. **114**(9) (2013). https://doi.org/10.1063/1.4819891
51. C.C. Stoumpos, C.D. Malliakas, J.A. Peters et al., Crystal growth of the perovskite semiconductor $CsPbBr_3$: a new material for high-energy radiation detection. Cryst. Growth Des. **13**(7), 2722–2727 (2013). https://doi.org/10.1021/cg400645t
52. J. Ghosh, P.J. Sellin, P.K. Giri, Recent advances in lead-free double perovskites for X-ray and photodetection. Nanotechnology **33**(31), 312001 (2022). https://doi.org/10.1088/1361-6528/ac6884
53. L. Basiricò, A. Ciavatti, B. Fraboni, Solution-Grown organic and perovskite X-ray detectors: a new paradigm for the direct detection of ionizing radiation. Adv. Mater. Technol. **6**(1), 2000475 (2021). https://doi.org/10.1002/admt.202000475
54. G.A. Elbaz, D.B. Straus, O.E. Semonin et al., Unbalanced hole and electron diffusion in lead bromide perovskites. Nano Lett. **17**(3), 1727–1732 (2017). https://doi.org/10.1021/acs.nanolett.6b05022
55. L. Li, H. Chen, Z. Fang, X. Meng et al., An electrically modulated single-color/dual-color imaging photodetector. Adv. Mater. **32**(24) (2020), 1907257. https://doi.org/10.1002/adma.201907257
56. Y. Liu, Y. Zhang, Z. Yang et al., Large lead-free perovskite single crystal for high-performance coplanar X-ray imaging applications. Adv. Opt. Mater. **8**(19), 2000814 (2020). https://doi.org/10.1002/adom.202000814
57. G. Hajdok, J.J. Battista, I.A. Cunningham, Fundamental X-ray interaction limits in diagnostic imaging detectors: spatial resolution. Med. Phys. **35**(7Part1), 3180–3193 (2008). https://doi.org/10.1118/1.2924219
58. J.W. Kingsley, S.J. Weston, D.G. Lidzey, Stability of X-ray detectors based on organic photovoltaic devices. IEEE J. Sel. Top. Quantum Electron. **16**(6), 1770–1775 (2010). https://doi.org/10.1109/JSTQE.2010.2048557
59. Z. Yang, A. Qin, H. Qin et al., Ion migration and dark current suppression in quasi-2D perovskite-based X-ray detectors. RSC Adv. **14**(27), 19124–19133 (2024). https://doi.org/10.1039/d4ra02606c
60. W. Pan, H. Wu, J. Luo et al., Cs2AgBiBr 6 single-crystal X-ray detectors with a low detection limit. Nat. Photonics **11**(11), 726–732 (2017). https://doi.org/10.1038/s41566-017-0012-4
61. S. Deumel, A. van Breemen, G. Gelinck et al., High-sensitivity high-resolution X-ray imaging with soft-sintered metal halide perovskites. Nat. Electron. **4**(9), 681–688 (2021). https://doi.org/10.1038/s41928-021-00644-3
62. Z. Yang, Y. Deng, X. Zhang et al., High-performance single-crystalline perovskite thin-film photodetector. Adv. Mater. **30**(8), 1704333 (2018). https://doi.org/10.1002/adma.201704333
63. A. Soultati, M. Tountas, K.K. Armadorou et al., Synthetic approaches for perovskite thin films and single-crystals. Energy Adv. **2**(8), 1075–1115 (2023). https://doi.org/10.1039/d3ya00098b
64. Yukta, S. Rahman, Q. Shi et al., M(III) site-driven structural engineering on lead-free layered double perovskite nanocrystals with enhanced photoelectrochemical activity. Small Struct., 2500179 (2025). https://doi.org/10.1002/sstr.202500179
65. F. Cao, D. Yu, W. Ma et al., Shining emitter in a stable host: design of halide perovskite scintillators for X-ray imaging from commercial concept. ACS Nano **14**(5), 5183–5193 (2019). https://doi.org/10.1021/acsnano.9b06114

66. H. Wei, D. DeSantis, W. Wei et al., Dopant compensation in alloyed $CH_3NH_3PbBr_{3-x}Cl_x$ perovskite single crystals for gamma-ray spectroscopy. Nat. Mater. **16**(8), 826–833 (2017). https://doi.org/10.1038/nmat4927
67. M. Awais, R.L. Kirsch, V. Yeddu et al., Tin halide perovskites going forward: frost diagrams offer hints. ACS Mater. Lett. **3**(3), 299–307 (2021). https://doi.org/10.1021/acsmaterialslett.0c00571
68. G. Nasti, A. Abate, Tin halide perovskite ($ASnX_3$) solar cells: a comprehensive guide toward the highest power conversion efficiency. Adv. Energy Mater. **10**(13), 1902467 (2020). https://doi.org/10.1002/aenm.201902467
69. T. Wu, D. Cui, X. Liu et al., Additive engineering toward high-performance tin perovskite solar cells. Solar RRl **5**(5), 2100034 (2021). https://doi.org/10.1002/solr.202100034
70. Y. Liu, Z. Xu, Z. Yang et al., Inch-size 0D-structured lead-free perovskite single crystals for highly sensitive stable X-ray imaging. Matter **3**(1), 180–196 (2020). https://doi.org/10.1016/j.matt.2020.04.017
71. X. Li, P. Zhang, Y. Hua et al., Ultralow detection limit and robust hard X-ray imaging detector based on inch-sized lead-free perovskite $Cs_3Bi_2Br_9$ single crystals. ACS Appl. Mater. Interfaces **14**(7), 9340–9351 (2022). https://doi.org/10.1021/acsami.1c24086
72. M. Chen, X. Dong, D. Chu et al., Interlayer-spacing engineering of lead-free perovskite single crystal for high-performance X-ray imaging. Adv. Mater. **35**(18), 2211977 (2023). https://doi.org/10.1002/adma.202211977
73. Y. Zhang, J. Hao, Z. Zhao et al., Lead-free perovskite single crystal linear array detector for high-resolution X-ray imaging. Adv. Mater. **36**(24), 2310831 (2024). https://doi.org/10.1002/adma.202310831
74. J.A. Steele, W. Pan, C. Martin et al., Photophysical pathways in highly sensitive $Cs_2AgBiBr_6$ double-perovskite single-crystal X-ray detectors. Adv. Mater. **30**(46), 1804450 (2018). https://doi.org/10.1002/adma.201804450
75. C.F. Wang, H. Li, M.G. Li et al., Centimeter-sized single crystals of two-dimensional hybrid iodide double perovskite $(4,4\text{-difluoropiperidinium})_4AgBiI_8$ for high-temperature ferroelectricity and efficient X-ray detection. Adv. Func. Mater. **31**(13), 2009457 (2021). https://doi.org/10.1002/adfm.202009457
76. M. Li, H. Li, W. Li et al., Oriented 2D perovskite wafers for anisotropic X-ray detection through a fast tableting strategy. Adv. Mater. **34**(8), 2108020 (2022). https://doi.org/10.1002/adma.202108020
77. Z. Xu, X. Liu, Y. Li et al., Exploring Lead-free hybrid double perovskite crystals of $(BA)_2CsAgBiBr_7$ with large mobility-lifetime product toward X-ray detection. Angew. Chem. Int. Ed. **58**(44), 15757–15761 (2019). https://doi.org/10.1002/ange.201909815
78. M. Xia, Y. Li, C. Li et al., High-density post-perovskite for ultra-sensitive hard X-ray detection. Angew. Chem. **137**(19), e202425448 (2025). https://doi.org/10.1002/anie.202425448
79. A. Xie, T.H. Nguyen, C. Hettiarachchi, M.E. Witkowski, W. Drozdowski, M.D. Birowosuto, H. Wang, C. Dang, Thermal quenching and dose studies of X-ray luminescence in single crystals of halide perovskites. J. Phys. Chem. C **122**(28), 16265–16273 (2018). https://doi.org/10.1021/acs.jpcc.8b03622
80. M.D. Birowosuto, D. Cortecchia, W. Drozdowski, K. Brylew, W. Lachmanski, A. Bruno, C. Soci, X-ray scintillation in lead halide perovskite crystals. Sci. Rep. **6**(1), 37254 (2016). https://doi.org/10.1038/srep37254
81. Q. Hu, Z. Deng, M. Hu et al., X-ray scintillation in lead-free double perovskite crystals. Sci. China Chem. **61**(12), 1581–1586 (2018). https://doi.org/10.1007/s11426-018-9308-2
82. J. Zhou, K. An, P. He et al., Solution-processed lead-free perovskite nanocrystal scintillators for high-resolution X-ray CT imaging. Adv. Opt. Mater. **9**(11), 2002144 (2021). https://doi.org/10.1002/adom.202002144
83. L. Lian, M. Zheng, W. Zhang et al., Efficient and reabsorption-free radioluminescence in Cs_3Cu2I_5 nanocrystals with self-trapped excitons. Adv. Sci. **7**(11), 2000195 (2020). https://doi.org/10.1002/advs.202000195

84. T. Feng, Z.A. Zhou, Y.N. An et al., Large-area transparent antimony-based perovskite glass for high-resolution X-ray imaging. ACS Nano **18**(26), 16715–16725 (2024). https://doi.org/10.1021/acsnano.4c01761
85. B. Yang, L. Yin, G. Niu et al., Lead-free halide Rb_2CuBr_3 as sensitive X-ray scintillator. Adv. Mater. **31**(44), 1904711 (2019). https://doi.org/10.1002/adma.201904711
86. X. Li, H. Cui, Y. Zhong et al., Lead-free Ce-doped perovskite scintillators with high figure of merit. J. Energy Chem. **99**, 74–82 (2024). https://doi.org/10.1016/j.jechem.2024.07.037
87. S. Hartati, T. Haposan, A.A. Afkauni et al., Tunable optical and scintillation properties of two-dimensional tin mixed-halide perovskites. J. Mater. Chem. C (2025). https://doi.org/10.1039/d5tc01768h
88. L. Li, S. Ye, J. Qu, F. Zhou, J. Song, G. Shen, Recent advances in perovskite photodetectors for image sensing. Small (2021). https://doi.org/10.1002/smll.202005606
89. D. Zheng, T. Pauporté, Advances in optical imaging and optical communications based on high-quality halide perovskite photodetectors. Adv. Func. Mater. **34**(11), 2311205 (2024). https://doi.org/10.1002/adfm.202311205
90. H. Lei, D. Hardy, F. Gao, Lead-free double perovskite $Cs_2AgBiBr_6$: fundamentals, applications, and perspectives. Adv. Funct. Mater. (2021). https://doi.org/10.1002/adfm.202105898
91. Y. Mo, X. Huang, Q. Hu, C. Hu, P.K. Chu, J. Li, Recent progress in perovskite photodetectors: from carrier dynamics to device structures and applications. Mater. Horiz. (2025). https://doi.org/10.1039/D5MH00032G
92. C. Xie, C.K. Liu, H.L. Loi et al., Perovskite-based phototransistors and hybrid photodetectors. Adv. Func. Mater. **30**(20), 1903907 (2020). https://doi.org/10.1002/adfm.201903907
93. R.D. Jansen-van Vuuren, A. Armin, A.K. Pandey, P.L. Burn, P. Meredith, Organic photodiodes: the future of full color detection and image sensing. Adv. Mater. **28**(24), 4766–4802 (2016). https://doi.org/10.1002/adma.201505405
94. X. Gong, M. Tong, Y. Xia et al., High-detectivity polymer photodetectors with spectral response from 300 to 1450 nm. Science **325**(5948), 1665–1667 (2009). https://doi.org/10.1126/science.1176706
95. G. Li, Y. Wang, L. Huang et al., Research progress of high-sensitivity perovskite photodetectors: a review of photodetectors: noise, structure, and materials. ACS Appl. Electron. Mater. **4**(4), 1485–1505 (2022). https://doi.org/10.1021/acsaelm.1c01349
96. J.P. Moy, Recent developments in X-ray imaging detectors. Nucl. Instrum. Methods Phys. Res., Sect. A **442**(1–3), 26–37 (2000)
97. Y. Zhang, Y. Ma, Y. Wang et al., Lead-free perovskite photodetectors: progress, challenges, and opportunities. Adv. Mater. **33**(26), 2006691 (2021). https://doi.org/10.1002/adma.202006691
98. Q. Tai, X. Guo, G. Tang et al., Antioxidant grain passivation for air-stable tin-based perovskite solar cells. Angew. Chem. Int. Ed. **58**(3), 806–810 (2019). https://doi.org/10.1002/ange.201811539
99. M. Pitaro, E.K. Tekelenburg, S. Shao et al., Tin halide perovskites: from fundamental properties to solar cells. Adv. Mater. **34**(1), 2105844 (2022). https://doi.org/10.1002/adma.202105844
100. T. Liu, L. Wang, Z. Yuan et al., Quasi-single-crystal tin halide perovskite films with high structural integrity for near-infrared imaging array enabling hidden object recognition. Angew. Chem. Int. Ed. **64**(6), e202418470 (2025). https://doi.org/10.1002/anie.202418470
101. K.R. Yun, T.J. Lee, S.K. Kim et al., Fast and highly sensitive photodetectors based on Pb-free Sn-based perovskite with additive engineering. Adv. Opt. Mater. **11**(1), 2201974 (2023). https://doi.org/10.1002/adom.202201974
102. M. He, Z. Xu, C. Zhao et al., Sn-based self-powered ultrafast perovskite photodetectors with highly crystalline order for flexible imaging applications. Adv. Func. Mater. **33**(24), 2300282 (2023). https://doi.org/10.1002/adfm.202300282
103. X. Zhang, X. Liu, B. Sun et al., Ultrafast, self-powered, and charge-transport-layer-free ultraviolet photodetectors based on sequentially vacuum-evaporated lead-free $Cs_2AgBiBr_6$ thin

films. ACS Appl. Mater. Interfaces **13**(30), 35949–35960 (2021). https://doi.org/10.1021/acsami.1c08613
104. Q. Fan, G.V. Biesold-McGee, J. Ma, Q. Xu, S. Pan, J. Peng, Z. Lin, Lead-free halide perovskite nanocrystals: crystal structures, synthesis, stabilities and optical properties. Angew. Chem. **132**(3), 1042–1059 (2020). https://doi.org/10.1002/ange.201904862
105. Z. Li, Y. Liu, J. He et al., $Cs_2AgBiBr_6$-based heterojunction photodetector for weak-light imaging application. Surf. Interfaces **29**, 101705 (2022). https://doi.org/10.1016/j.surfin.2021.101705
106. Y. Lv, S. Chen, J. Chen et al., Lead-free perovskite photodetectors for fast Fourier single-detector imaging. ACS Photonics **11**(9), 3621–3629 (2024). https://doi.org/10.1021/acsphotonics.4c00628
107. W. Song, D. Wang, J. Tian et al., Encapsulation of dual-passivated perovskite quantum dots for bio-imaging. Small **18**(42), 2204763 (2022). https://doi.org/10.1002/smll.202204763
108. K.K. Chan, D. Giovanni, H. He, T.C. Sum, K.T. Yong, Water-stable all-inorganic perovskite nanocrystals with nonlinear optical properties for targeted multiphoton bioimaging. ACS Appl. Nano Mater. **4**(9), 9022–9033 (2021). https://doi.org/10.1021/acsanm.1c01621
109. M.J. Tan, D. Ravichandran, H.L. Ang, E.W.Y. Ong, C.Q.X. Lim, G.M. Kam, Z.K. Tan et al., Magneto-fluorescent perovskite nanocomposites for directed cell motion and imaging. Adv. Healthc. Mater. **8**(23), 1900859 (2019). https://doi.org/10.1002/adhm.201900859
110. Y. Yang, Q. Jiang, F. Zhang, Nanocrystals for deep-tissue in vivo luminescence imaging in the near-infrared region. Chem. Rev. **124**(2), 554–628 (2023). https://doi.org/10.1021/acs.chemrev.3c00506
111. J. Zhang, Y. Zhu, Exploiting the photo-physical properties of metal halide perovskite nanocrystals for bioimaging. ChemBioChem **25**(5), e202300683 (2024). https://doi.org/10.1002/cbic.202300683
112. Y. Chen, S. Spinelli, Z. Pan, Deep-tissue NIR-II bioimaging performance of Si-based and InGaAs-based imaging devices using short-wave infrared persistent luminescence. J. Mater. Chem. C **12**(21), 7542–7551 (2024). https://doi.org/10.1039/d4tc01323a
113. J.S. Manser, J.A. Christians, P.V. Kamat, Intriguing optoelectronic properties of metal halide perovskites. Chem. Rev. **116**(21), 12956–13008 (2016). https://doi.org/10.1021/acs.chemrev.6b00136
114. N. Mondal, A. De, A. Samanta, Achieving near-unity photoluminescence efficiency for blue-violet-emitting perovskite nanocrystals. ACS Energy Lett. **4**(1), 32–39 (2018). https://doi.org/10.1021/acsenergylett.8b01909
115. J. Chen, W. Zhang, T. Pullerits, Two-photon absorption in halide perovskites and their applications. Mater. Horiz. **9**(9), 2255–2287 (2022). https://doi.org/10.1039/d1mh02074a
116. C. Liu, J. Sun, W.L. Tan et al., Alkali cation doping for improving the structural stability of 2D perovskite in 3D/2D PSCs. Nano Lett. **20**(2), 1240–1251 (2020). https://doi.org/10.1021/acs.nanolett.9b04661
117. N. Fiuza-Maneiro, K. Sun, I. Lopez-Fernandez et al., Ligand chemistry of inorganic lead halide perovskite nanocrystals. ACS Energy Lett. **8**(2), 1152–1191 (2023). https://doi.org/10.1021/acsenergylett.2c02363
118. S. Vedi, T. Dheivasigamani, G.S. Selvam et al., Growth optimization of single-phase novel colloidal perovskite $Cs_3Bi_2I_9$ nanocrystals and $Cs_3Bi_2I_9@SiO_2$ core–shell nanocomposites for bio-medical application. Biomater. Sci. **10**(20), 5956–5967 (2022). https://doi.org/10.1039/d2bm00773h
119. S. Huang, Z.C. Yu, J.L. Xie et al., Water-dispersable silica-supported nanocrystalline Pb-free perovskite $Rb_7Bi_3Cl_{16}@SiO_2$ for ATP-responsive bio-imaging. ChemistrySelect **8**(24), e202300176 (2023). https://doi.org/10.1002/slct.202300176
120. M. Han, J. Sun, M. Peng et al., Controllable growth of lead-free all-inorganic perovskite nanowire array with fast and stable near-infrared photodetection. J. Phys. Chem. C **123**(28), 17566–17573 (2019). https://doi.org/10.1021/acs.jpcc.9b03289
121. G. Xiong, L. Yuan, Y. Jin et al., Highly efficient and stable broadband near-infrared-emitting lead-free metal halide double perovskites. J. Mater. Chem. C **9**(38), 13474–13483 (2021). https://doi.org/10.1039/D1TC02867G

122. G. Zhang, P. Dang, L. Tian et al., Boosting energy transfer from self-trapped exciton to Er^{3+} through Sb^{3+} doping in $Cs_2Na(Lu/Er)Cl_6$ double perovskites. Adv. Opt. Mater. **11**(3), 2202369 (2023). https://doi.org/10.1002/adom.202202369
123. J. Liao, P. Zhang, X. Niu et al., Co-doping of stibium and rare earth (Nd, Yb) in lead-free double perovskite for efficient near-infrared emission. J. Alloy. Compd. **911**, 164946 (2022). https://doi.org/10.1016/j.jallcom.2022.164946
124. T. Zhang, X. Zhang, Q. Mao et al., NIR-I excitable and NIR-II emissive Cr, Yb, Co-doped $Cs_2NaScCl_6$ double perovskite halides for biodetection. ACS Appl. Electron. Mater. **7**(6), 2610–2618 (2025). https://doi.org/10.1021/acsaelm.5c00106
125. Y. Zhang, C. Sun, H. Tian et al., Enhancing near-infrared luminescence of $Cs_2NaBiCl_6$:Yb^{3+}/RE^{3+} (Er, Pr, Tm) double perovskite through codoping Cr^{3+} ions and bioimaging applications. Ceram. Int. **51**(2), 36660–36670 (2025). https://doi.org/10.1016/j.ceramint.2025.05.377
126. T. Wang, J. Sun, Z. Teng et al., Near-infrared emission perovskites for multifunctional bioimaging. Small Sci. **5**, 2500033 (2025). https://doi.org/10.1002/smsc.202500033
127. P. Saghy, A.M. Brown, C. Chu et al., Lanthanide double perovskite nanocrystals with emissions covering the UV-C to NIR spectral range. Adv. Opt. Mater. **11**(12), 2300277 (2023). https://doi.org/10.1002/ADOM.202300277
128. T. Cai, W. Shi, R. Wu et al., Lanthanide doping into all-inorganic heterometallic halide layered double perovskite nanocrystals for multimodal visible and near-infrared emission. J. Am. Chem. Soc. **146**(5), 3200–3209 (2024). https://doi.org/10.1021/jacs.3c11164
129. C. Zhao, Y. Gao, T. Song et al., An Er^{3+}-doped $Cs_2NaScCl_6$ lead-free double perovskite with efficient broadband visible to near-infrared emission and multimodal upconversion luminescence. J. Phys. Chem. Lett. **14**(40), 9011–9018 (2023). https://doi.org/10.1021/acs.jpclett.3c02321
130. W. Gan, L. Cao, S. Gu et al., Broad-band sensitization in Cr^{3+}–Er^{3+} co-doped $Cs_2AgInCl_6$ double perovskites with 1.5 μm near-infrared emission. Chem. Mater. **35**(14), 5291–5299 (2023). https://doi.org/10.1021/acs.chemmater.3c00446
131. X. Yun, H. Hu, H. Zhong et al., Enabling efficient near-infrared emission in lead-free double perovskite via a codoping strategy. Inorg. Chem. Front. **11**(18), 6146–6155 (2024). https://doi.org/10.1039/D4QI01518E
132. G. Zhang, D. Wang, B. Lou et al., Efficient broadband near-infrared emission from lead-free halide double perovskite single crystal. Angew. Chem. Int. Ed. **61**(33), e202207454 (2022). https://doi.org/10.1002/ANIE.202207454
133. M.N. Tran, I.J. Cleveland, J.R. Geniesse et al., High photoluminescence quantum yield near-infrared emission from a lead-free ytterbium-doped double perovskite. Mater. Horiz. **9**(8), 2191–2197 (2022). https://doi.org/10.1039/D2MH00483F
134. M.N. Lintangpradipto, H. Zhu, B. Shao et al., Single-crystal methylammonium-free perovskite solar cells with efficiencies exceeding 24% and high thermal stability. ACS Energy Lett. **8**(11), 4915–4922 (2023). https://doi.org/10.1021/acsenergylett.3c01935

Chapter 14
Lead-Free Perovskite Materials for Miniaturized Patch Antennas

Saarthak Kharbanda, Manish Kumar, Rajesh Kumar, Ashutosh Anand, Divya Deep Yadav, Tarun Kumar, and Vishnu Chauhan

Abstract The increasing demand for small and effective antennas in contemporary wireless communication, aerospace, and medicinal systems has prompted research into materials that can facilitate antenna shrinking without compromising performance. Although they are frequently utilized, conventional substrates like PTFE and FR4 provide little opportunity for additional size reduction. Because of their high dielectric constants, lead-based perovskites have long been thought to be successful; nevertheless, safer alternatives are required due to their toxicity and environmental issues. Lead-free perovskite materials, including $BaTiO_3$, BCZT, $Na_{0.5}Bi_{0.5}TiO_3$, Aurivillius phases, and spinel-perovskite composites, are the main topic of this chapter. We go into great length about their dielectric characteristics, production methods, and suitability for patch antenna substrates. Their stability, permittivity, and dielectric loss all of which have a direct impact on antenna efficiency and miniaturization are given particular consideration. All things considered, lead-free perovskites offer a high-performing, sustainable family of materials with great potential for use in next-generation antenna applications.

Keywords Antenna miniaturization · Lead-free perovskites · Barium titanate · BCZT · Sodium bismuth titanate · Aurivillius phases · Spinel · Composites · Dielectric

---

S. Kharbanda
Amity Institute of Nanotechnology, Amity University Haryana, Gurugram, India

M. Kumar
Experimental Research Laboratory, Department of Physics, ARSD College, University of Delhi, New Delhi, India

R. Kumar · A. Anand
University School of Basic and Applied Sciences, Guru Gobind Singh Indraprastha University, New Delhi, India

D. D. Yadav · T. Kumar · V. Chauhan (✉)
Department of Physics, Ramjas College, University of Delhi, New Delhi, India
e-mail: vishnuchauhan@ramjas.du.ac.in

V. Chauhan et al. (eds.), *Sustainable Lead-Free Perovskite Materials*, Topics in Applied Physics 157, https://doi.org/10.1007/978-3-032-24873-2_14

14.1 Introduction

Compact and high-performance antennas have become essential due to the quick spread of wireless communication technologies and Internet-of-things (IoT) devices [1–3]. Because of its low profile, ease of production, and compatibility with integrated circuits, patch antennas are one of the many types of antennas used in mobile, aeronautical, and healthcare applications. Antennas are crucial components of wireless communication networks because of their capacity to send and receive electromagnetic waves across great distances. Microstrip patch antennas (MPAs) are essential for modern wireless communication systems in order to solve new problems with transceiver system gain, bandwidth, and compactness [4]. The microstrip patch antenna (MPA), which has numerous advantages such as simple circuit integration, low profile, lightweight, robust, shape flexibility, polarization diversity, easy fabrication, and easy mounting on vehicles or missiles, had been the focus of a renowned study of antenna research. Polymer-based composites like polytetrafluoroethylene (PTFE) and FR4 epoxy are frequently used to make commercial antenna devices [5–8].

A critical challenge in designing such antennas lies in their miniaturization without compromising performance metrics such as bandwidth, efficiency, and return loss. One promising route to achieve miniaturization is the use of high-permittivity substrates, which allow a reduction in patch size by decreasing the guided wavelength. Ceramic-based materials offer more possibilities for optimizing the dielectric and magnetic properties of the substrate than conventional substrates. Microstrip antennas have been the subject of serious analysis since their discovery in the 1970s in order to accommodate the advancements in wireless communication technology, such as miniaturization and the attenuation of microwave signals. They are well known for their robust design, wide range of applications, high performance, and robust manufacturing. Therefore, searching for high dielectric and permeability values with low loss substrate materials is an achievable approach to reduce antenna size while preserving signal quality. Due to their many uses in microwave communication and energy storage, high dielectric materials have attracted a lot of attention. Traditionally, high-εr materials have been based on lead-containing perovskites such as $Pb(Zr,Ti)O_3$ and $PbTiO_3$ [9]. However, the hunt for lead-free substitutes has accelerated due to the serious risks that lead-based compounds bring to the environment and human health. Perovskite materials devoid of lead, especially those based on barium titanate ($BaTiO_3$) [10, 11], sodium bismuth titanate ($Na_{0.5}Bi_{0.5}TiO_3$) [12], and their solid solutions, have emerged as viable substitutes. These materials have favorable dielectric qualities, thermal stability, and environmental friendliness. The use of lead-free perovskite materials in miniature patch antennas is covered in this chapter along with its basic principles, material classes, characterisation methods, and antenna performance considerations [13].

14.1.1 Fundamentals of Microwave Dielectric Materials

Ceramic substrates provide the ideal combination of material properties for antenna applications. A low dielectric loss tangent and a strong dielectric constant guarantee effective antenna performance. Because of their exceptional thermal stability, they can also withstand high temperatures and a range of climatic conditions. Strong structural support and resistance to physical damage are provided by the inherent mechanical strength of ceramics. Their high electrical insulating properties further lessen unwanted signal interference and crosstalk between antenna parts. Notably, the consistent and predictable electrical conduct of ceramics throughout time leads to reliable and repeatable antenna performance. By varying the dielectric constant during material production, these substrates offer the flexibility to optimize antenna characteristics for specific applications. Ceramic substrates so significantly improve antenna performance, durability, and efficiency, particularly in high-frequency and compact designs. In contrast to conductors and semiconductors, electrical insulators are unable to block the flow of electronic current due to the huge band gaps in their electronic structure. Additionally, they don't have free charge carriers that can move macroscopically. Nevertheless, insulators also known as dielectric materials polarize when subjected to an external electric field due to the presence of tiny electrostatic dipoles made up of distinct positive and negative charge centers. Dielectrics can store energy because of this polarization; the degree of polarization and, thus, the amount of energy stored are measured by the material's permittivity (ε), also called the dielectric constant [14]. Relative permittivity (ε_r), dielectric loss tangent (tanδ), temperature coefficient of resonant frequency (τ_f), and quality factor (Q) are the four main characteristics of microwave dielectric materials. These factors affect the efficiency, thermal stability, and compactness of microwave parts like filters, resonators, and antennas.

- **Relative Permittivity (ε_r)**: The four primary properties of microwave dielectric materials are temperature coefficient of resonant frequency (τ_f), quality factor (Q), dielectric loss tangent (tanδ), and relative permittivity (ε_r). The efficiency, thermal stability, and compactness of microwave components such as filters, resonators, and antennas are all impacted by these characteristics. The four primary properties of microwave dielectric materials are temperature coefficient of resonant frequency (τ_f), quality factor (Q), dielectric loss tangent (tanδ), and relative permittivity (ε_r). The efficiency, thermal stability, and compactness of microwave components such as filters, resonators, and antennas are all impacted by these characteristics.
- **Dielectric Loss (tan δ)**: The performance of a miniaturized antenna is largely determined by dielectric loss, which is represented by tan δ. More electromagnetic energy is stored inside the substrate when high-permittivity materials are used to reduce the size of the antenna. Efficiency, gain, and overall antenna performance are decreased if the material has a high tan δ because a large amount of that energy is lost as heat rather than radiated. Because of the narrowly contained fields of tiny antennas, this effect is considerably more apparent. Because of this, having a

substrate with very low dielectric loss (preferably tan δ below 0.01 at the working frequency) is just as crucial for downsizing as having a high dielectric constant.

- **Quality Factor (Q_f)**: The quality factor (Q_f) of a dielectric material is a measure of how well it stores energy compared to how much it loses. To put it simply, a high Qf indicates that the material can support efficient radiation and has low losses, which is crucial for employing high-permittivity substrates to reduce the size of antennas. Any losses are more noticeable in smaller designs because the electromagnetic fields are tightly contained in the substrate. The antenna experiences poor efficiency, decreased gain, and a limited useful bandwidth if the Q_f is low. However, you can benefit from downsizing without significantly compromising performance when you use a high Q_f material. This is why researchers look for materials that combine high permittivity (for size reduction) with a high Q_f (for low loss and better efficiency).
- **Temperature Coefficient (τ_f)**: The extent that an antenna's operating frequency changes with temperature is indicated by the temperature coefficient of resonant frequency (τ_f). This phenomenon becomes particularly important in tiny antennas, which frequently use high-permittivity substrates, since even little variations in material characteristics can result in discernible detuning. In practical applications, a substrate with a large positive or negative τ_f will reduce stability and reliability by causing the antenna's frequency to drift with heating or cooling. Because they enable the antenna to perform consistently in a variety of settings, materials having a near-zero τ_f are therefore quite desired. Selecting a substrate with a stable τ_f guarantees that the advantages of compactness are not lost due to frequency instability in tiny devices where precision is already difficult. Understanding and optimizing these parameters are vital when selecting materials for high-frequency applications. The generalized frequency-dependent behavior of the real and imaginary components of a dielectric medium's permittivity is shown in Fig. 14.1. Permittivity magnitude and macroscopic polarization are the outcome of the combined effects of multiple physical events happening at various frequencies at the microscopic level.

Electronic polarization occurs when an electric field is applied at frequencies up to the UV/VIS band (~10^{15} Hz). This action creates an electric dipole moment and separates the centers of gravity of the two atoms by uprooting the negatively charged electron cloud that surrounds the positively charged nucleus of a neutral atom. When exposed to an alternating current (AC) field, the electron cloud oscillates around the nucleus at the same frequency as the applied field, much like in a damped spring-mass resonant system. When electromagnetic waves interact with atoms, or cations and anions within the lattice, at frequencies as high as the infrared (~10^{12}–10^{13} Hz), atomic or ionic polarization results. These particles have lower resonance frequencies than electrons because of their much bigger masses. This phenomenon happens when an applied field forces cations away from their equilibrium positions in the lattice toward the negative electrode and anions toward the positive electrode. Dipolar, or orientational, polarization occurs in solids and liquids with dipolar moments because the center of gravity of the molecule's negative and positive charges do not coincide.

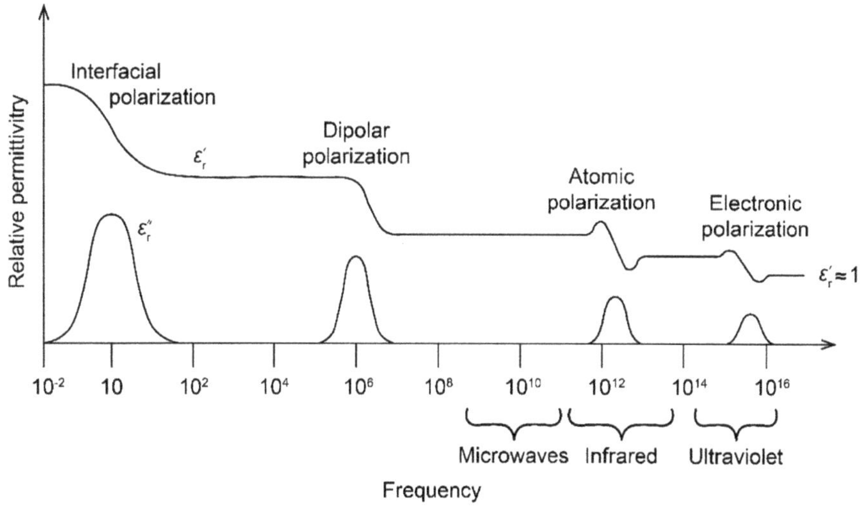

Fig. 14.1 This qualitative illustration demonstrates how the AC frequency usually affects the real and imaginary components of permittivity for a dielectric medium. Reprinted from [15] (open access)

Dipolar polarization shows a smoother fall in the real component of permittivity 'ε' in the microwave frequency range up to approximately 10^{10} Hz. At very low frequencies or when subjected to a direct current (DC) field, all polarization mechanisms, including interface or space charge polarization, contribute to the real fraction of permittivity 'ε', because the electric field varies slowly enough to allow other mobile charges to travel and accumulate at interfaces, regions of net positive or negative charge can emerge at low frequencies [16].

14.2 Lead-Free Perovskite Families for Microwave Applications

For microwave applications, several families of lead-free perovskites are being developed as eco-friendly alternatives to lead-based ceramics. Barium titanate ($BaTiO_3$) and its modified forms remain the most established, offering high permittivity and tunability, though they often need doping or composite approaches to reduce dielectric loss. Barium strontium titanate (BST) is widely used in tunable devices like phase shifters and filters, since strontium lowers the Curie point and makes its properties suitable for microwave frequencies. Potassium sodium niobate (KNN) is another promising family, valued for its good dielectric and piezoelectric properties, while $Na_{0.5}Bi_{0.5}TiO_3$ (NBT) and related systems are also explored despite their higher losses. More complex perovskite solutions, such as $Ba(Zr,Ti)O_3$ (BZT), $Ba(Ca,Ti)O_3$

(BCT), and $KTaO_3$ (KTO), are being engineered to achieve a balance of high permittivity, low dielectric loss, and stable temperature behavior. When combined, these families of lead-free perovskites provide a solid route toward small, adjustable, and effective microwave components without the negative environmental effects of lead-based materials.

14.2.1 Sn-Based Halide Perovskites

In view of its similar electrical structure to lead and reduced toxicity, tin halide perovskite (THP) is currently one of the most promising possibilities for an active layer in optoelectronic and electric devices like solar cells, photodiodes, and field-effect transistors (FETs). Recently, photovoltaics and FETs have demonstrated promising performances because of their advantageous electrical and optical properties [17, 18]. However, THP's broad use is significantly hampered by its easy oxidation from Sn^{2+} to Sn^{4+}, facile tin vacancy creation, uncontrolled film shape and crystallinity, and interface instability. Sn-based halide perovskites such as $CH_3NH_3SnI_3$ ($MASnI_3$) and $CsSnBr_3$ have garnered attention due to their structural similarity to lead halide perovskites. However, their use in microwave applications is limited by their intrinsic semiconducting nature and instability. Tin ions are prone to oxidation (Sn^{2+} to Sn^{4+}), leading to self-doping and increased carrier concentration, which in turn raises dielectric losses. Despite this, research continues on passivation techniques and structural engineering to improve their dielectric behavior [18].

14.2.2 Oxide Perovskites

Oxide-based perovskites are more thermally and chemically stable compared to halide counterparts. They are widely explored for their excellent dielectric properties:

- **$BaTiO_3$ (Barium Titanate)**: Numerous nanostructured materials are being investigated as a result of the present trend in electronic device miniaturization. Due to its superior dielectric and ferroelectric qualities, the ferroelectric material barium titanate ($BaTiO_3$) has attracted a lot of interest in the last ten years. Significant advancements in synthetic methods have resulted in the production of high-quality $BaTiO_3$ nanocrystals (NCs) with controlled crystal phases (such as cubic, tetragonal, and multi-phase) and well-defined morphologies (such as nanoparticles, nanorods, nanocubes, and nanowires). Research on the fascinating yet intricate dielectric characteristics of individual $BaTiO_3$ NCs and $BaTiO_3$/polymer nanocomposites is made possible by the capacity to create nanoscale $BaTiO_3$ with regulated properties. Exhibits high permittivity and tunability, especially in doped and modified forms. Figure 14.2 shows the XRD diffraction planes of thin films of Barium Strontium Titanate (BST) grown on Pt/Ti/SiO_2/Si substrate [20, 21].

- **BCZT**: Due to its remarkable qualities, such as its extremely high dielectric constant, low dielectric loss, and low temperature coefficient of dielectric constant, barium titanate (BTO) is a prototypical electroceramic that is extensively utilized in the capacitor industry. It displays three transitions and four polymorphs: (i) rhombohedral (R)-orthorhombic (O); (ii) ferro-ferroelectric phase transitions (O-tetragonal (T); and (iii) T-cubic (C), which denotes the ferroelectric to paraelectric transformation. Its electromechanical coupling factor is around 0.50, and its piezoelectric charge coefficient is approximately 190 pC/N. However, in comparison to PZT, it performs poorly as a piezoelectric material, which restricts its broader use as actuators and sensors. Several $BaTiO_3$-based solid solutions with various A- and B-site substituents (usually A = Ca, Sr, La; B double bond Nb, Ta, Zr) have been investigated in order to improve the piezoelectric capabilities. The MPB composition ($Ba_{0.85}Ca_{0.15}$) ($Zr_{0.10}Ti_{0.90}$)O_3 [henceforth referred to as BCZT] is of particular significance and has attracted a lot of research because of its extraordinarily high piezoelectric coefficient. It also demonstrates a high ε_r (~1500), low tan δ, and temperature stability [20, 22–26].
- **NBT ($Na_{0.5}Bi_{0.5}TiO_3$) based systems**: The majority of temperature-stable relaxors fall into one of two broad categories: (a) materials that achieve high upper temperature limits of stable εr (e.g., 300–500 °C) with mid value εr values of 1000–1200 but with a minimum working temperature (e.g., 50–100 °C) well above the required EIA standard value of −55 °C; or (b) materials that achieve stable εr over the target temperature range of −55 to 300 °C, but have a modest εr value of approximately 450–500, which is harmful to capacitor volumetric efficiency. [27–30]. Additionally, some materials show unexpected high dielectric loss tangents for at least part of the temperature range of stable εr. In light of this, it is imperative to discover novel and enhanced compositions that demonstrate low dielectric losses across the entire temperature range and attain a temperature range of stable and high relative permittivity from −55 °C to 300 °C. Because of its high peak εr values, $Na_{0.5}Bi_{0.5}TiO_3$ can be used as the parent relaxor material in compositional engineering to create a flattened ε_r-T response. These have a moderate permittivity and are frequently alloyed to improve dielectric performance with $BaTiO_3$ or $BiFeO_3$ [31–33].
- **Manganese Oxide-based perovskites**: Manganese oxide-based perovskites, generally written as $AMnO_3$ (where A can be La, Ca, Sr, Ba, and others), are a fascinating group of materials because they combine rich electrical and magnetic behaviors in a single structure. They are best known for their colossal magnetoresistance (CMR) effect, where their resistance changes sharply under a magnetic field, making them useful in sensors and spintronic devices. By carefully choosing or substituting the A-site ions, their properties can be tuned such as controlling the balance between Mn^{3+} and Mn^{4+} ions, which directly affects conductivity, magnetism, and dielectric response $La_{0.7}Sr_{0.3}Mn_{1+x}O_3$, a perovskite manganese

oxide, was produced by a solid-state process. For the stoichiometric composition ($x = 0$), a single-phase perovskite structure was produced. Nonetheless, secondary phases of Mn_2O_3 and Mn_3O_4 developed and expanded with an increase in Mn excess ($x > 0$), whereas La_2O_3 and Sr-rich phases appeared when Mn is inadequate ($x < 0$). Since the presence of Mn is the source of both charge transport and magnetism, the samples' electrical conductivity, permittivity, and permeability all indicated notable changes with x. A transmission line theory was used to compute RL maps using complex permittivity (ε', ε") and permeability (μ', μ") spectra. Due to significant EM wave reflection for samples with $x > 0$, which displayed relatively high values of ε' and ε, there was no appreciable absorption of electromagnetic (EM) waves in the examined frequency range ($0.1 \leq f < 18$ GHz). However, the Mn-deficient samples ($x = -0.2, -0.4$) demonstrated exceptional EM wave absorption characteristics in the GHz range. The Mn-deficient samples have favorable impedance matching properties because their ε' and ε" values are significantly lower. The sample with $x = -0.2$ in particular demonstrated the most amazing EM wave absorption capabilities, with an RL value of −56 dB at a frequency of 2.33 GHz and a thickness of 7.59 mm. It also achieved an RL value of −50 dB at 9.8 GHz and 2.90 mm thickness [34].

- **Aurivillius Phases**: There is currently a lot of interest in the ferroelectric characteristics of compounds of the Aurivillius type due to the possibility of using them to create ferroelectric nonvolatile memories (Fe-RAM). Over fifty years ago, Aurivillius made the discovery of this family of bismuth oxides. It can be explained by the regular stacking of $[A_{n-1}B_nO_{3n+1}]^{2-}$ perovskite blocks and $[M_2O_2]^{2+}$ slabs. Even though the perovskite blocks offer a lot of compositional freedom, Bi^{3+} cations nearly completely occupy the cation sites in the interleaved $[M_2O_2]$ slabs, resulting in $[Bi_2O_2]^{2+}$ slabs. Crystallographic studies have demonstrated the critical role of the A-site cation in the ferroelectric behavior of these materials, and a direct correlation between the average ionic radii of the A-site cation and the temperature of the ferroelectric–paraelectric transition (T_c) can be established. In addition to the polar displacements, cooperative tiltings of the BO_6 octahedra constitute another feature of the ferroelectric phases. In a recent work, we demonstrated that these structural characteristics are, in fact, shared by all Aurivillius ferroelectric phases and can be unifiedly characterized for members $n = 2$–∞ using the (3 + 1)D superspace group formalism. Layered structures such as $Bi_4Ti_3O_{12}$ show promise for their thermal stability and high ε_r [35, 36]. Aurivillius $Bi_5Ti_3FeO_{15}$ (BTF) multiferroic (MF) nanofibers (NFs) with a diameter of approximately 400 nm were developed using calcination and electrospinning. Benign micro-ferroelectricity and an effective microepiezoelectric coefficient of 35 pm/V are demonstrated by the BTF NFs. The behavior of the macro-ferroelectric and magnetoelectric couplings, which are smaller than those of the BTF ceramic or film due to the detrimental effects of filled air, was thoroughly examined for the first time using non-sintering and pressing. Up to 14 $mVcm^1Oe^1$ is the maximum magnetoelectric (ME) voltage coefficient. Weak ferromagnetism was also found at RT. Even though the sintering temperature was as low as 600 °C, MF BTF NFs as multiferroic materials were successfully produced. This offers fresh insights into the development and application of potential lead-free MF materials for cutting-edge gadgets. The inset of Fig. 14.3 displays the crystal structure schematic of the Aurivillius layered perovskite $Bi_{m+1}Ti_3Fe_{m-3}O_{3m+3}$ Aurivillius phase (m = 4), which also shows the magnetization (M) vs magnetic field (H) curves of the BTF NFs at room temperature [37].

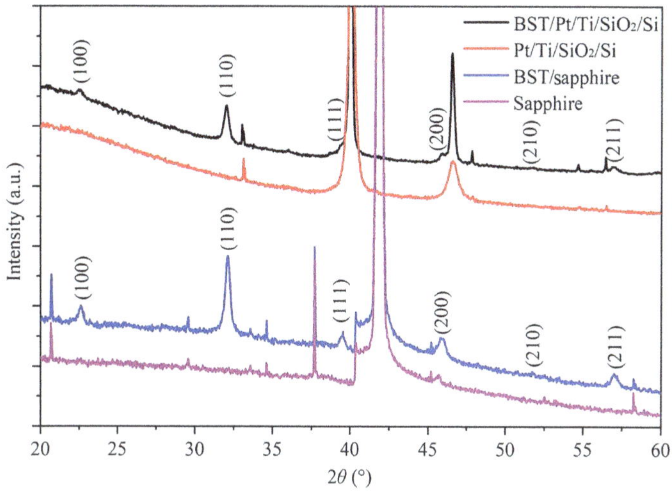

Fig. 14.2 BST thin-film XRD patterns developed on sapphire or $Pt/Ti/SiO_2/Si$ substrates. The primary perovskite phase reflections are indexed. Reprinted from [19], open access

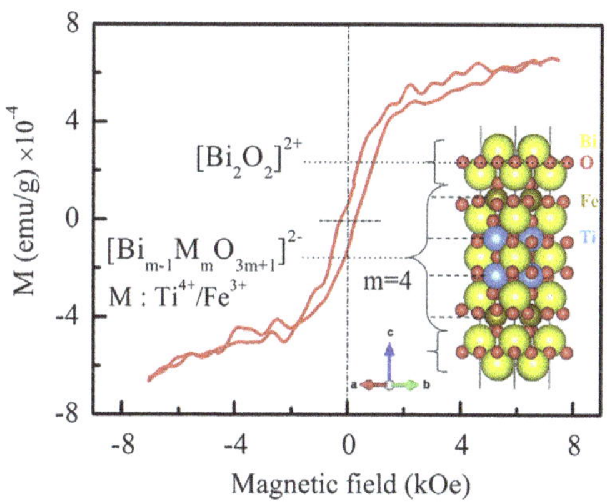

Fig. 14.3 The inset displays the crystal structure schematic of the Aurivillius layered perovskite $Bi_{m+1}Ti_3Fe_{m-3}O_{3m+3}$ Aurivillius phase (m = 4) together with the room temperature magnetization (M) versus magnetic field (H) curves of the BTF NFs. Reprinted with permission from Elsevier, Lic No. 6117500248546 [37]

14.2.3 Microwave-Optimized Spinel-Perovskite Hybrids

Over the past ten years, multiferroics with simultaneous magnetic and electric ordering have attracted a lot of attention due to a special phenomena called magneto-electric (ME) coupling [38]. This linkage ensures its noble applications in multi-state memory systems, tunable microwave devices, magnetoelectric transformers, and magnetic field sensors. The ME coupling can manifest directly in single-phase multiferroics or indirectly through strain in two-phase composite materials. The intrinsic materials known as single phase multiferroics show weak ME coupling at temperatures significantly lower than room temperature (RT). The rhombohedrally deformed perovskite structure of $BiFeO_3$ (BFO), a rare single phase multiferroic compound, is an excellent ferroelectric above room temperature ($T_N = 370$ °C & TC = 830 °C) and a G-type antiferromagnetic [12, 39]. BFO is an amazing material for technical applications because of its intriguing properties. BFO stands out as a top contender among multiferroics due to its antiferromagnetic properties and high remanent polarization value (~90μC/cm^2). Phase stability, high leakage current, and weak magnetization are some of the inherent issues with BFO that prevent its useful practical application. The connection between the magnetic and ferroelectric orders is weak in pure BFO because their origins are linked to distinct ions [40–42]. By coupling a combination with high magnetic properties with a good ferroelectric material, the challenges of merging magnetic and electrical ordering in a single phase have been avoided. Because of the strain coupling caused by the lattice mismatch at the interface, spinel-perovskite composites have been the subject of intense research among the wide variety of multiferroic composites. The perovskite in these heterostructured composites acts as a piezoelectric material, while the spinel phase acts as a magnetostrictive. These hybrids are particularly relevant for microwave antenna substrates operating at X- and Ku-bands [5, 43].

14.3 Dielectric Characterization and Suitability for Patch Antennas

Accurate dielectric characterization is essential for antenna substrate design. The key characterization methods include, impedance analyzer measurements, split-post dielectric resonator (SPDR) for high-frequency (5–20 GHz) low-loss materials and microstrip ring resonators that allow characterization in the exact geometry used in antennas. Several simple and economical techniques for creating and analyzing $Ba_{0.6}Sr_{0.4}TiO_3$ (BST) thin films at microwave frequencies were presented. When the applied field is in the out-of-plane direction, the produced BST thin films here show homogeneous structures and excellent tunability (η) throughout a broad frequency and temperature range. The large concentration of polar nanoclusters is responsible for the remarkable tunability. A new non-destructive near-field scanning microwave microscopy (NSMM) method was used to directly visualize their reaction to the applied direct current (DC) field. Figure 14.4 represents the dielectric studies of Barium-strontium titanates (BST) [19]. The $Sr_{2-x}Bi_xTi_{2-x}Fe_xO_{67}$ perovskites have been made into microcrystalline powders using the liquid-mix process. The dielectric response appears to be associated with ferroelectric behavior of the relaxor type. Values for tangent loss, conductivity, and capacitance show how common intergrain conductive contribution is in the electrical response. The intergrain contribution dominates the impedance response, which is typified by an equivalent circuit of one or more parallel resistance-capacitance (RC) components. There was relaxor ferroelectric behavior in the permittivity as a function of temperature. The semiconductive character of the compounds is consistent with electrical conductivity values that generally follow a typical Arrhenius pattern. Temperature-dependent variations in the concentration of oxygen vacancies can be somewhat linked to the values of the observed activation energies, which range from 0.5 to 1.3 eV [44].

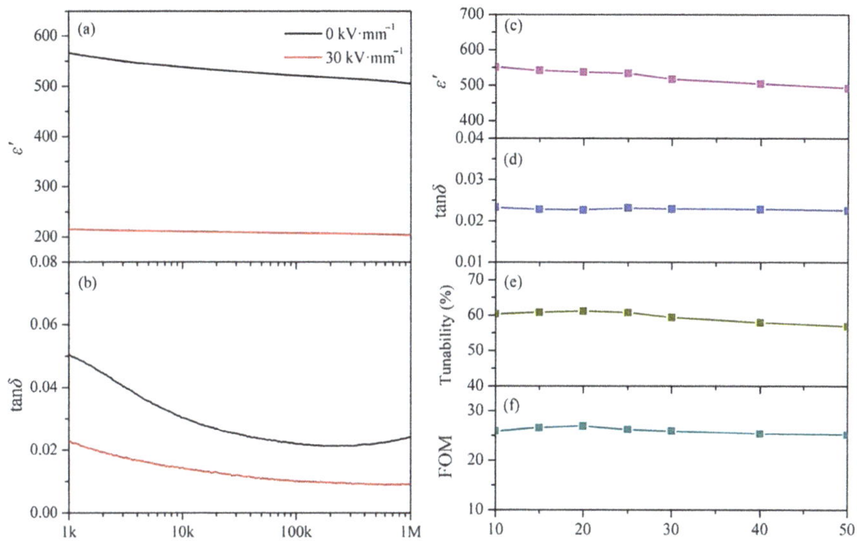

Fig. 14.4 BST thin film growth on a $Pt/Ti/SiO_2/Si$ substrate from 1 kHz to 1 MHz at 25 °C with **a** ε' and **b** tan δ without bias and with an applied DC electric field of 30 $kV{\cdot}mm^{-1}$ (DC bias of 15 V). BST thin film growth on a $Pt/Ti/SiO_2/Si$ substrate without bias from 10 to 50 °C at 100 kHz is shown by **c** ε' and **d** tan δ. **e** BST thin-film tunability and **f** FOMs grown on $Pt/Ti/SiO_2/Si$ substrate at 100 kHz with an electric field of 30 $kV{\cdot}mm^{-1}$ and temperatures ranging from 10 to 50 °C. Reprinted from [19], open access

14.4 Microstrip Patch Antenna Substrate Thickness Variation

Variation of substrate thickness plays a crucial role in the performance of microstrip patch antennas based on lead-free perovskite substrates such as $BaTiO_3$, BST, or KNN [45]. A thinner substrate combined with high dielectric permittivity (ε_r) enables strong miniaturization by reducing patch size, but it also narrows bandwidth and increases dielectric losses if the material has a high loss tangent. Increasing substrate thickness broadens impedance bandwidth and improves radiation efficiency, yet excessive thickness (typically $h/\lambda_0 > 0.05$) excites surface waves, leading to pattern distortion and gain reduction. Since lead-free perovskites often exhibit very high ε_r and moderate to high $\tan\delta$, composite approaches or thin-film integration are recommended to balance miniaturization with efficiency. Thus, careful optimization of thickness relative to wavelength, along with accurate high-frequency dielectric characterization, is essential to exploit the advantages of lead-free perovskite substrates for miniaturized and efficient antenna designs. The dual-band rectangular strip patch antenna application prototype on a $Bi_{0.5}Na_{0.5}TiO_3$ based perovskite substrate material is parametrically improved by substrate thickness change in order to validate the suggested design. Figure 14.5a shows the antenna's return loss, which is observed to be resonating in the X-band region's 8.5–9GHz and 10–10.5GHz ranges. The observed antenna in Fig. 14.5c, on the other hand, produces a slight shift towards higher frequency from 10.34GHz to 12.2GHz (t = 1.02 mm); 10.34–12GHz (t = 1 mm); and for t = 0.9 mm, the simulated result shows two bands, whereas the measured antenna shows one band at about 11.5–12 GHz. At about 9 and 10.5 GHz, the antenna resonates with the greatest return loss, which is estimated for varying thickness. The simulated antenna is constructed on a $Bi_{0.5}Na_{0.5}TiO_3$ (BNT) ceramic plate for t = 0.9 mm, t = 1 mm, t = 1.02 mm, and t = 1.04 mm, as shown in Fig. 14.5f. This is verified by analyzing the voltage standing wave ratio (VSWR), which is less than 2 in the same range. This indicates that the antenna and the transmission line or RF source have a reasonably good impedance match, and that power is being reflected back with these specific resonant frequencies. Better power transfer efficiency and less signal loss are indicated by a lower VSWR value (<2) [46].

The performance of the antenna is demonstrated by the simulated peak gain of the proposed layout. In Fig. 14.6a, the gain increases in the frequency range of 8–10.5 GHz as the thickness increases from 0.7 to 1.04 mm. The gain of the microstrip patch antenna is expected to grow as the substrate thickness increases. The reason is fringing fields close to the patch's margins, which get stronger as the antenna's effective area increases and the thickness of the antenna substrate increases. The capacity of an antenna to concentrate energy in a particular direction during transmission or to more effectively accept energy from a particular direction during reception is known as directivity. An antenna with a directivity of 1 (or 0 dB) and nearly zero directionality broadcasts equally in all directions. The directivity of the planned antenna for different thicknesses in the X-band area is shown in Fig. 14.6b. Directivity is seen to rise with substrate thickness until it reaches wavelength parity. Another crucial

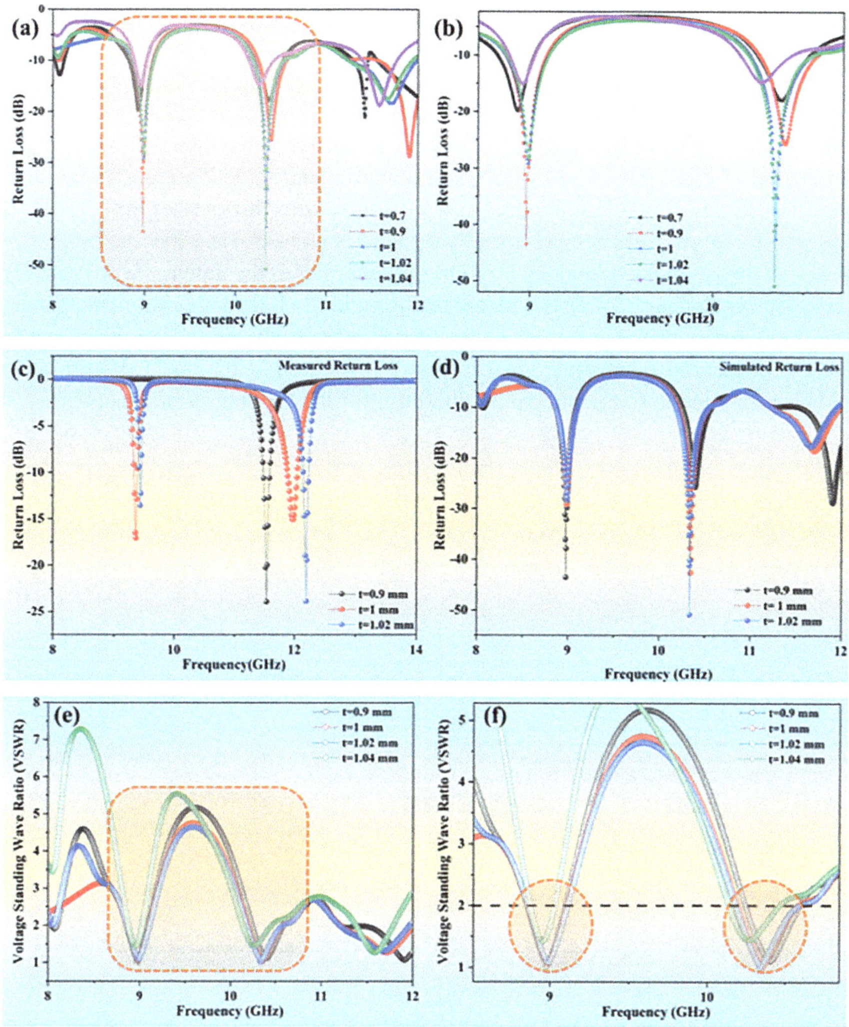

Fig. 14.5 **a** Return loss (in dB) as a function of frequency (in GHz) with different substrate thicknesses **b** return loss (in dB) as a function of frequency (in GHz) for resonances at 9–9.5 GHz and 10–10.5 GHz at t = 0.7 mm, t = 0.9 mm, t = 1 mm, t = 1.02 mm, and t = 1.04 mm **c** the constructed $Bi_{0.5}Na_{0.5}TiO_3$ based ceramic antenna's observed return loss as a function of frequency **d** the $Bi_{0.5}Na_{0.5}TiO_3$ based ceramic antenna's simulated return loss as a function of frequency **e** the voltage standing wave ratio of the simulated antenna versus frequency **f** the higher voltage standing wave ratio of the simulated antenna as a function of frequency. Reprinted with permission from Elsevier, Lic No. 6117100752599 [46].

metric that illustrates how well an antenna transmits and receives radio frequency signals is radiation efficiency, which is calculated as the ratio of total power radiated by the antenna to total input power received from the generator. The input power is efficiently radiated into free space using an antenna with a high radiation efficiency. Internal antenna losses such metal conduction, dielectric, and magnetic losses are the main cause of input power loss when radiation efficiency is low. Figure 14.6c illustrates the efficiency in relation to frequency using various thicknesses for 0.7, 0.9, 1, 1.02, and 1.04 mm. An antenna is usually designed to transmit at a specific frequency known as the resonant frequency (f_r). On the other hand, the antenna will typically focus on this resonant frequency and radiate throughout a little wider frequency range. This frequency range, referred to as the impedance bandwidth (BW), is shown in Fig. 14.6d. However, substrate thicknesses of 0.7, 0.9, 1, 1.02, and 1.04 mm were used to calculate the fractional bandwidth ratio for the top four iterations [46].

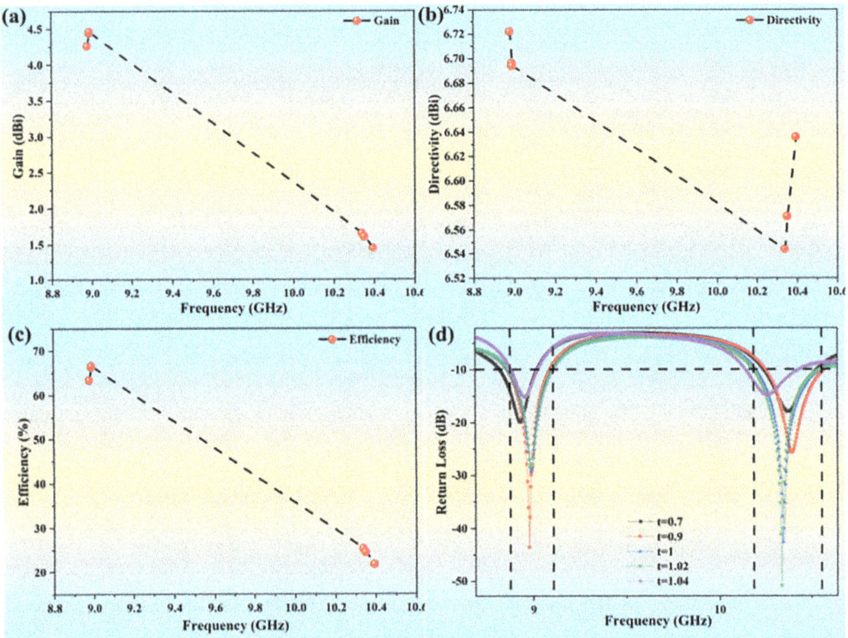

Fig. 14.6 The proposed simulated antenna is plotted along the following axes: gain to frequency, directionality to frequency, efficiency to frequency, and reflection loss to frequency. Reprinted with permission from Elsevier, Lic No. 6117100752599 [46]

14.5 Antenna Miniaturization Mechanisms Using High Dielectric Materials

The search for dielectric ceramics that can operate at temperatures beyond 200 °C much higher than the present limit of high volumetric efficiency capacitor materials is reviewed in light of recent developments. The characteristics of compositional systems based on lead-free relaxor dielectrics with mixed cation site occupancy on the perovskite lattice are summarized and compared. Engineering features resulting from greater dielectric peak broadening and changes to peak temperatures might lead to a plateau in the relative permittivity temperature response (ε_r-T). As a result, εr has a consistency of at least ±15% throughout a wide temperature range. This article groups materials having expanded upper temperature limits of 300, 400, and even 500 °C based on the parent component of the solid solution, such as $BaTiO_3$ and $Na_{0.5}Bi_{0.5}TiO_3$ [31, 32]. The resonant frequency (f_r) of a patch antenna is inversely proportional to the square root of the effective dielectric constant (ε_{eff}) of the substrate. Additionally, it has been shown that alloying titanates and materials with lower melting temperatures improves the ferroelectric and dielectric properties of lead-free perovskite oxide thin films. $BaTiO_3$ (or $SrTiO_3$) with a 25% addition of $BiFeO_3$ has much higher crystalline perfection because of the lower melting point of $BiFeO_3$, which improves growth kinetics. The maximum dielectric peak temperature of $BaTiO_3$ is raised by roughly 200 °C, while leakage currents are reduced by up to a factor of 100. Up to 300 °C, the loss tangent falls by a factor greater than 14 at room temperature. From room temperature to 500 °C, the dielectric constant is greater than 1000 and the dielectric breakdown strength is nearly three times higher (> 2200 kV cm^{-1}). Furthermore, the dielectric constant changes by just around 9% between room temperature and 330 °C, compared to roughly 110% for $BaTiO_3$. The maximum polarization (P_{max}) is double that of $BaTiO_3$ at 125.3 μC cm^{-2} [20, 47, 48]. The film matches Pb-based ferroelectric films with large energy storage densities of >52 J cm^{-3} at 2050 kV cm^{-1}. Applications in energy storage and high temperature (up to 300 °C) capacitors, as well as broader use in other electronic and energy technologies, would benefit from the significantly enhanced performance [49]. Different performance parameters for different samples are discussed in Table 14.1 (Figs. 14.7 and 14.8).

Therefore, by increasing ε_r, the physical dimensions of the patch can be significantly reduced. For example, a patch resonating at 10 GHz on an ε_r = 2.2 (RT

Table 14.1 Dielectric properties of lead-free perovskites for antenna miniaturization

Composition	ε_r	$tan\delta$	f_0 (GHz)	Tunability	Process	Ref.
$(Ba_{0.6}Sr_{0.4})\ TiO_3$	554	0.06	1	50% @300 kV/cm	Sputtering	[50]
$(Ba_{0.6}Sr_{0.4})\ TiO_3$	68	0.07	1	54% @500 kV/cm	Sol-gel	[19]
$(Ba_{0.6}Sr_{0.4})\ TiO_3$	470	0.15	1	71%	PLD	[51]
$BaTi_{0.7}Zr_{0.3}O_3$	–	0.05	3	78% @550 kV/cm	Sputtering	[49]

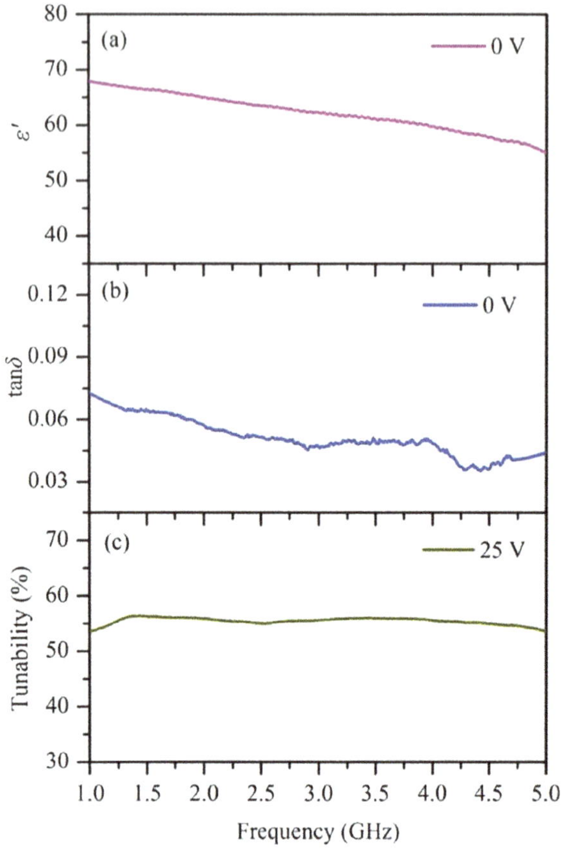

Fig. 14.7 The ε′ and tan δ of the BST thin film that was produced and developed on the $Pt/Ti/SiO_2/Si$ substrate in the 1–5 GHz microwave frequency range without a DC bias applied. Reprinted from [19], open access

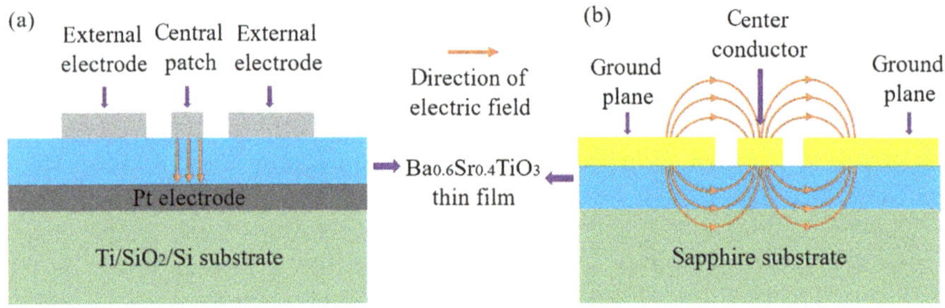

Fig. 14.8 Electric field distributions in (a) the parallel plate electrode's out-of-plane direction on the $BST/Pt/Ti/SiO_2/Si$ sample and (b) the CPW transmission line's in-plane direction on the BST/sapphire sample. Adapted from [19], open access

Duroid) substrate may be ~10 mm × 10 mm. The same design on $\varepsilon_r = 25$ substrate can reduce size to ~4 mm × 4 mm. However, high ε_r comes with challenges such as, narrow bandwidth, increased surface wave losses. Increased Q factor, sensitivity to fabrication tolerances Balancing these factors is crucial for successful antenna miniaturization [52] (Shlivinski et al. 1997).

14.6 Conclusions

The development of lead-free perovskite materials has become essential to achieving high-performance, sustainable substrates for small patch antennas. Despite their capacity to provide high dielectric constants and tunability, conventional lead-based systems are no longer appropriate because of their toxicity and environmental risks. On the other hand, lead-free substitutes like $BaTiO_3$, BCZT, and $Na_{0.5}Bi_{0.5}TiO_3$ provide similar dielectric performance while guaranteeing environmental safety. Patch antennas may be made smaller without significantly sacrificing efficiency or bandwidth thanks to their capacity to combine high permittivity with low dielectric loss. Unique benefits are offered by various material classes. Because of their strong ferroelectric behavior and high permittivity, oxide perovskites like $BaTiO_3$ are still the most researched. For reliable functioning across a broad temperature range, solid-solution systems like BCZT provide exceptional thermal stability and low losses. While layered Aurivillius phases expand the design area by fusing ferroelectricity and structural stability, NBT-based relaxor systems provide temperature stability and mild permittivity. Furthermore, hybrid spinel perovskite composites offer a chance to take advantage of magnetoelectric coupling, which makes them appealing for both miniaturization and tunable, multipurpose microwave devices.

The chapter has also emphasized the significance of dielectric characterization methods, such as microstrip resonators, split-post dielectric resonators, and impedance analysis, which enable precise assessment of material performance under circumstances pertinent to antenna design. These techniques are essential for linking microstructural characteristics with electromagnetic behavior and for optimizing substrate selection.

References

1. C.A. Balanis, Antenna theory: A review. Proc. IEEE **80**(1), 7–23 (1992). https://doi.org/10.1109/5.119564
2. R.C. Hansen, M. Burke, Antennas with magneto-dielectrics. Microw. Opt. Technol. Lett. **26**(2), 75–78 (2000). https://doi.org/10.1002/1098-2760(20000720)26:2<75::AID-MOP3>3.0.CO;2-W
3. H. How, P. Rainville, F. Harackiewicz, C. Vittoria, Radiation frequencies of ferrite patch antennas. Electron. Lett. **28**(15), 1405–1406 (1992). https://doi.org/10.1049/el:19920893

4. S.D. Ram Prasath, S. Balaji, M.S. Khanna, S. Raju, V. Abhaikumar, Miniaturization of patch antennas using magneto-dielectric materials. Indian Antenna Week: A Workshop Adv. Antenna Technol. **2010**, 1–4 (2010). https://doi.org/10.1109/AAT.2010.5545951
5. K. Carver, J. Mink, Microstrip antenna technology. IEEE Trans. Antennas Propag. **29**(1), 2–24 (1981). https://doi.org/10.1109/TAP.1981.1142523
6. F. Ferrero, C. Luxey, G. Jacquemod, R. Staraj, Dual-band circularly polarized microstrip antenna for satellite applications. IEEE Antennas Wirel. Propag. Lett. **4**, 13–15 (2005). https://doi.org/10.1109/LAWP.2004.841622
7. R. Munson, Conformal microstrip antennas and microstrip phased arrays. IEEE Trans. Antennas Propag. **22**(1), 74–78 (1974). https://doi.org/10.1109/TAP.1974.1140723
8. D.M. Pozar, V. Sanchez, Magnetic tuning of a microstrip antenna on a ferrite substrate. Electron. Lett. **24**(12), 729–731 (1988). https://doi.org/10.1049/el:19880491
9. A.B. Schäufele, K. Heinz Härdtl, Ferroelastic properties of Lead Zirconate Titanate ceramics. J. Am. Ceram. Soc. **79**(10), 2637–2640 (1996). https://doi.org/10.1111/j.1151-2916.1996.tb09027.x
10. H.F. Kay, Preparation and properties of crystals of barium titanate, BaTiO3. Acta Crystallogr. **1**(5), 229–237 (1948). https://doi.org/10.1107/S0365110X4800065X
11. C. Wong, Y.Y. Teng, J. Ashok, P.L.H. Varaprasad, Barium Titanate (BaTiO3), in *Handbook of Optical Constants of Solids*, (Elsevier, 1997), pp. 789–803. https://doi.org/10.1016/B978-012544415-6.50080-7
12. V.A. Isupov, Ferroelectric $Na_{0.5}Bi_{0.5}TiO_3$ and $K_{0.5}Bi_{0.5}TiO_3$ perovskites and their solid solutions. Ferroelectrics **315**(1), 123–147 (2005). https://doi.org/10.1080/001501990910276
13. Y. Yang, H. Zhang, J. Li, Y. Rao, G. Wang, G. Gan, Bi^{3+} doping-adjusted microstructure, magnetic, and dielectric properties of nickel zinc ferrite ceramics for high frequency LTCC antennas. Ceram. Int. **46**(16), 25697–25704 (2020b). https://doi.org/10.1016/j.ceramint.2020.07.046
14. S. Kharbanda, P. Thakur, F. Wan, A. Sharma, Y. Slimani, A. Thakur, Investigation of magnetic and electrochemical properties of NiBixFe2-xO4 ferrite nanoparticles for enhanced functional applications. J. Magn. Magn. Mater. **596**, 171968 (2024). https://doi.org/10.1016/j.jmmm.2024.171968
15. Y. Yamada, Dielectric properties of textile materials: Analytical approximations and experimental measurements—A review. Textiles **2**(1), 50–80 (2022). https://doi.org/10.3390/textiles2010004
16. A. Saini, A. Thakur, P. Thakur, Matching permeability and permittivity of Ni0.5Zn0.3Co0.2In0.1Fe1.9O4 ferrite for substrate of large bandwidth miniaturized antenna. J. Mater. Sci. Mater. Electron. **27**(3), 2816–2823 (2016). https://doi.org/10.1007/s10854-015-4095-8
17. T. Leijtens, R. Prasanna, A. Gold-Parker, M.F. Toney, M.D. McGehee, Mechanism of tin oxidation and stabilization by Lead substitution in tin halide perovskites. ACS Energy Lett. **2**(9), 2159–2165 (2017). https://doi.org/10.1021/acsenergylett.7b00636
18. W. Yang, F. Igbari, Y. Lou, Z. Wang, L. Liao, Tin halide perovskites: Progress and challenges. Adv. Energy Mater. **10**(13) (2020a). https://doi.org/10.1002/aenm.201902584
19. H. Ruan, T.G. Saunders, H. Giddens, H. Zhang, A.A. Ihalage, J.F. Kolb, M. Blunt, S. Haq, H. Yan, Y. Hao, Microwave characterization of two $Ba_{0.6}Sr_{0.4}TiO_3$ dielectric thin films with out-of-plane and in-plane electrode structures. J. Adv. Ceram. **12**(8), 1521–1532 (2023) https://doi.org/10.26599/JAC.2023.9220769
20. B. Jiang, J. Iocozzia, L. Zhao, H. Zhang, Y.-W. Harn, Y. Chen, Z. Lin, Barium titanate at the nanoscale: Controlled synthesis and dielectric and ferroelectric properties. Chem. Soc. Rev. **48**(4), 1194–1228 (2019). https://doi.org/10.1039/C8CS00583D
21. Y. Luo, I. Szafraniak, N.D. Zakharov, V. Nagarajan, M. Steinhart, R.B. Wehrspohn, J.H. Wendorff, R. Ramesh, M. Alexe, Nanoshell tubes of ferroelectric lead zirconate titanate and barium titanate. Appl. Phys. Lett. **83**(3), 440–442 (2003). https://doi.org/10.1063/1.1592013
22. I. Coondoo, N. Panwar, D. Alikin, I. Bdikin, S.S. Islam, A. Turygin, V.Y. Shur, A.L. Kholkin, A comparative study of structural and electrical properties in lead-free BCZT ceramics: Influence

of the synthesis method. Acta Mater. **155**, 331–342 (2018). https://doi.org/10.1016/j.actamat.2018.05.029
23. X. Ji, C. Wang, S. Li, S. Zhang, R. Tu, Q. Shen, J. Shi, L. Zhang, Structural and electrical properties of BCZT ceramics synthesized by sol–gel process. J. Mater. Sci. Mater. Electron. **29**(9), 7592–7599 (2018a). https://doi.org/10.1007/s10854-018-8751-7
24. B. Tiwari, T. Babu, R.N.P. Choudhary, Piezoelectric lead zirconate titanate as an energy material: A review study. Mater. Today Proc. **43**, 407–412 (2021). https://doi.org/10.1016/j.matpr.2020.11.692
25. D. Viehland, J.-F. Li, X. Dai, Z. Xu, Structural and property studies of high Zr-content lead zirconate titanate. J. Phys. Chem. Solids **57**(10), 1545–1554 (1996). https://doi.org/10.1016/0022-3697(96)00025-X
26. T. Yamada, T. Ueda, T. Kitayama, Piezoelectricity of a high-content lead zirconate titanate/polymer composite. J. Appl. Phys. **53**(6), 4328–4332 (1982). https://doi.org/10.1063/1.331211
27. J. Gomah-Pettry, Sodium-bismuth titanate based lead-free ferroelectric materials. J. Eur. Ceram. Soc. **24**(6), 1165–1169 (2004). https://doi.org/10.1016/S0955-2219(03)00473-4
28. A.R. Paterson, H. Nagata, X. Tan, J.E. Daniels, M. Hinterstein, R. Ranjan, P.B. Groszewicz, W. Jo, J.L. Jones, Relaxor-ferroelectric transitions: Sodium bismuth titanate derivatives. MRS Bull. **43**(8), 600–606 (2018). https://doi.org/10.1557/mrs.2018.156
29. N. Rhimi, N. Dhahri, M. Khelifi, E.K. Hlil, J. Dhahri, Structural, morphological, optical and dielectric properties of sodium bismuth titanate ceramics. Inorg. Chem. Commun. **146**, 110119 (2022). https://doi.org/10.1016/j.inoche.2022.110119
30. F. Yang, M. Li, L. Li, P. Wu, E. Pradal-Velázquez, D.C. Sinclair, Defect chemistry and electrical properties of sodium bismuth titanate perovskite. J. Mater. Chem. A **6**(13), 5243–5254 (2018). https://doi.org/10.1039/C7TA09245H
31. K.M. Kuruvila, D. Dhayanithi, N.V. Giridharan, Investigation of structural and enhanced piezoelectric properties in low lead content Na0.5Bi0.5TiO3-BaTiO3-PbTiO3 system. J. Alloys Compd. **897**, 163195 (2022). https://doi.org/10.1016/j.jallcom.2021.163195
32. A. Zeb, S. Jan, F. Ullah Bamiduro, D.A. Hall, S.J. Milne, Temperature-stable dielectric ceramics based on Na0.5Bi0.5TiO3. J. Eur. Ceram. Soc. **38**(4), 1548–1555 (2018). https://doi.org/10.1016/j.jeurceramsoc.2017.12.032
33. X. Zhou, G. Xue, H. Luo, C.R. Bowen, D. Zhang, Phase structure and properties of sodium bismuth titanate lead-free piezoelectric ceramics. Prog. Mater. Sci. **122**, 100836 (2021). https://doi.org/10.1016/j.pmatsci.2021.100836
34. M.-G. Kang, Y.-M. Kang, Effect of Mn content on the structure, electromagnetic properties, and electromagnetic wave absorption characteristics in La0.7Sr0.3Mn1+O3. Ceram. Int. **49**(24), 40043–40050 (2023). https://doi.org/10.1016/j.ceramint.2023.09.334
35. C.H. Hervoches, A. Snedden, R. Riggs, S.H. Kilcoyne, P. Manuel, P. Lightfoot, Structural behavior of the four-layer Aurivillius-phase ferroelectrics SrBi4Ti4O15 and Bi5Ti3FeO15. J. Solid State Chem. **164**(2), 280–291 (2002). https://doi.org/10.1006/jssc.2001.9473
36. J. Tellier, P. Boullay, M. Manier, D. Mercurio, A comparative study of the Aurivillius phase ferroelectrics CaBi4Ti4O15 and BaBi4Ti4O15. J. Solid State Chem. **177**(6), 1829–1837 (2004). https://doi.org/10.1016/j.jssc.2004.01.008
37. Y. Zhao, H. Fan, G. Liu, Z. Liu, X. Ren, Ferroelectric, piezoelectric properties and magnetoelectric coupling behavior in aurivillius Bi5Ti3FeO15 multiferroic nanofibers by electrospinning. J. Alloys Compd. **675**, 441–447 (2016). https://doi.org/10.1016/j.jallcom.2016.03.017
38. L. Li, R. Ubic, X.M. Chen, On the τ measurement of microwave dielectric ceramics. Mater. Res. Bull. **178**, 112920 (2024). https://doi.org/10.1016/j.materresbull.2024.112920
39. B. Sattibabu, T.D. Rao, A.K. Bhatnagar, V.S. Murthy, J.A. Chelvane, S. Rayprol, Structural and magnetic properties of bi substituted nickel ferrite. Mater. Today Proc. **39**(xxxx), 1482–1486 (2020). https://doi.org/10.1016/j.matpr.2020.05.371
40. G. Catalan, J.F. Scott, Physics and applications of bismuth ferrite. Adv. Mater. **21**(24), 2463–2485 (2009). https://doi.org/10.1002/adma.200802849
41. S. Chaturvedi, I. Sarkar, M.M. Shirolkar, U.S. Jeng, Y.Q. Yeh, R. Rajendra, N. Ballav, S. Kulkarni, Probing bismuth ferrite nanoparticles by hard x-ray photoemission: Anomalous

occurrence of metallic bismuth. Appl. Phys. Lett. **105**(10) (2014). https://doi.org/10.1063/1.4895672
42. I. Sosnowska, T.P. Neumaier, E. Steichele, Spiral magnetic ordering in bismuth ferrite. J. Phys. C Solid State Phys. **15**(23), 4835–4846 (1982). https://doi.org/10.1088/0022-3719/15/23/020
43. K. Borah, N. Bhattacharyya, Magneto-dielectric material with nano ferrite inclusion for microstrip antennas: Dielectric characterization. IEEE Trans. Dielectr. Electr. Insul. **17**(6), 1676–1681 (2010). https://doi.org/10.1109/TDEI.2010.5658216
44. Y. Saad, I. Álvarez-Serrano, M.L. López, M. Hidouri, Structural and dielectric characterization of new lead-free perovskites in the (SrTiO3)–(BiFeO3) system. Ceram. Int. **42**(7), 8962–8973 (2016). https://doi.org/10.1016/j.ceramint.2016.02.028
45. R. Gaur, N. Sharma, S. Kharbanda, S. Singh, Enhanced ferroelectric and electrocaloric properties in CuO-modified lead-free (Na0.5K0.5)NbO3 ceramics for solid-state cooling application. Mater. Sci. Eng. B **261**, 114767 (2020). https://doi.org/10.1016/j.mseb.2020.114767
46. R.R. Samal, D. Das, A. Priyadarsini Dikhsit, K. Parashar, S.K.S. Parashar, Development of new lead free ceramic substrate for rectangular patch antenna application in X-band communication. Mater. Today Commun. **40**, 109436 (2024). https://doi.org/10.1016/j.mtcomm.2024.109436
47. S. Cho, C. Yun, Y.S. Kim, H. Wang, J. Jian, W. Zhang, J. Huang, X. Wang, H. Wang, J.L. MacManus-Driscoll, Strongly enhanced dielectric and energy storage properties in lead-free perovskite titanate thin films by alloying. Nano Energy **45**, 398–406 (2018). https://doi.org/10.1016/j.nanoen.2018.01.003
48. A. Zeb, S.J. Milne, High temperature dielectric ceramics: A review of temperature-stable high-permittivity perovskites. J. Mater. Sci. Mater. Electron. **26**(12), 9243–9255 (2015). https://doi.org/10.1007/s10854-015-3707-7
49. A. Tumarkin, E. Sapego, A. Gagarin, A. Karamov, High Tunable BaTixZr1-xO3 films on dielectric substrate for microwave applications. Molecules **27**(18), 6086 (2022). https://doi.org/10.3390/molecules27186086
50. L. Yang, F. Ponchel, G. Wang, D. Rémiens, J.-F. Légier, D. Chateigner, X. Dong, Microwave properties of epitaxial (111)-oriented Ba0.6Sr0.4TiO3 thin films on Al2O3(0001) up to 40 GHz. Appl. Phys. Lett. **97**(16) (2010). https://doi.org/10.1063/1.3478015
51. T.S. Akhil Raman, B. Arun, C. Shivakumar, A. Joseph, K.C. James Raju, Influence of oxygen pressure during deposition on the microwave dielectric tunability of Ba0.5Sr0.5TiO3 thin films in PLD process. Appl. Surf. Sci. **669**, 160477 (2024). https://doi.org/10.1016/j.apsusc.2024.160477
52. G.-M. Yang, X. Xing, A. Daigle, M. Liu, O. Obi, J.W. Wang, K. Naishadham, N.X. Sun, Electronically Tunable miniaturized antennas on Magnetoelectric substrates with enhanced performance. IEEE Trans. Magn. **44**(11), 3091–3094 (2008). https://doi.org/10.1109/TMAG.2008.2003062

Chapter 15
Potential Design, Challenges, and Future Prospects of Lead-Free Perovskite Solar Cells

Vinay Pratap Verma, Vishnu Chauhan, Upkar Kumar Verma, and Sandeep Kumar Pundir

Abstract Perovskite solar cells (PSCs) are one of the most popular and got attention of the photovoltaic researchers of the twenty-first century. It is well characterized by their high-power conversion efficiency (PCE), adjustable bandgaps, and cost-effective production methodology. But, heavy use of toxic lead (Pb)-bearing in the active absorber layer has create considerable alarm over their environmental and toxicological impacts, thus initiation to transit from lead to lead-free substitutes. This chapter provides in-depth investigation of design principles of lead-free PSCs, with their syntheses process, performance trends, challenges, and prospects. Promising material like tin (Sn), germanium (Ge), bismuth (Bi), antimony (Sb), silver (Ag), copper (Cu), double perovskites, and chalcogenides are evaluated in terms of structural robustness, electronic quality, scalability, and environmental friendliness. The results show that Sn-based perovskites surpass 15% efficiency, but affected by oxidation hence accelerate degradation. Bi- and Sb-double perovskite films provide strong long-term stability, but efficiency is low due to large bandgaps; while chalcogenides exhibit high robustness in extreme conditions, but processing technology are complex. We cover syntheses such as solution deposition, vapor deposition, and environmentally friendly syntheses, noting them closely in terms of scalability and integrability in a device layer. The chapter further covers key market debut constraints, that is, intrinsic material unitability, efficiency plateaus, scalability issues, toxicities, and non-existence of standard recycling protocols. Also, prospects such as alloy engineering, 2D/3D hybrid geometries, computer-aided materials discovery,

V. P. Verma · S. K. Pundir (✉)
Department of Physics, University of Lucknow, Lucknow, Uttar Pradesh, India
e-mail: Pundir_s@lkouniv.ac.in

V. P. Verma
e-mail: vinaypratapverma@gmail.com

V. Chauhan
Department of Physics, Ramjas college, University of Delhi, New Delhi, India

U. K. Verma (✉)
Department of Physics, B. S. N. V. P. G. College, Lucknow, Uttar Pradesh, India
e-mail: upkarthegreat@gmail.com

V. Chauhan et al. (eds.), *Sustainable Lead-Free Perovskite Materials*, Topics in Applied Physics 157, https://doi.org/10.1007/978-3-032-24873-2_15

tandem schemes, and environmentally friendly syntheses reveal identified avenues to address such constraints. The chapter provides a comprehensive analysis potential for commercialization of lead-free PSCs replace Pb-containing PSCs. Also, their great potential for commercialization, such as indoor solar cells, flexible electronics, and usage in space or desert regions are discussed. Finally, the development of lead-free perovskite solar cells with using emerging technologies to achieve higher efficiency, good stability, and environment-friendly counterparts.

Keywords: Perovskite solar cells · Lead-free photovoltaics · Toxicity of lead · Sustainable photovoltaics

15.1 Introduction

Perovskite solar cells (PSCs) have emerged as the most promising photovoltaic technologies in recent times with their extraordinary power conversion efficiencies, tunable structure, and low-cost preparation possibilities. Conventional lead-based perovskite solar cells provide outstanding optoelectronic performance; but presence of lead's poses toxicity creating severe environmental and public health concerns [1–4]. Also, Lead pollution can negatively impact soil and waters' quality and imposes hazards to ecosystems and human beings equally [5, 6]. Therefore, it attract the intention to a vigorous research activity, to develop lead-free perovskite solar cells that will be retaining efficiency like conventional photovoltaics that ensuring greatly diminished toxicity and improved sustainability [7, 8].

Lead-free perovskites can be classified into several categories depending on their constituent elements [9, 10]. Among of these categories, tin-based perovskites and double perovskites stand out prominently. For instance, tin (Sn)-based perovskites have demonstrated highly promising power conversion efficiencies with some efficiencies higher than 17% [11] as shown in Fig. 15.1, through replacement of highly toxic lead with less toxic elements. Furthermore, halide double perovskites incorporating elements like bismuth (Bi), antimony (Sb), germanium (Ge), and combinations thereof represent a promising route to environmentally friendly solar technology [12, 13]. These new discoveries drive largely due to a compelling necessity to eliminate or control health risks attendant upon lead extraction, processing, and ultimate device disposal.

Requirement for sustainability stands at the core within renewable energy technology development [14]. With mounting social concerns regarding environmental decline, substituting lead-based materials in solar cells not only guarantees humankind's protection but also stimulates broader public acceptability of this novel technology [15, 16]. This chapter addresses design impediments still limiting lead-free perovskite solar cells' efficacy and future directions towards stable, efficient, and economical lead-free photovoltaic devices.

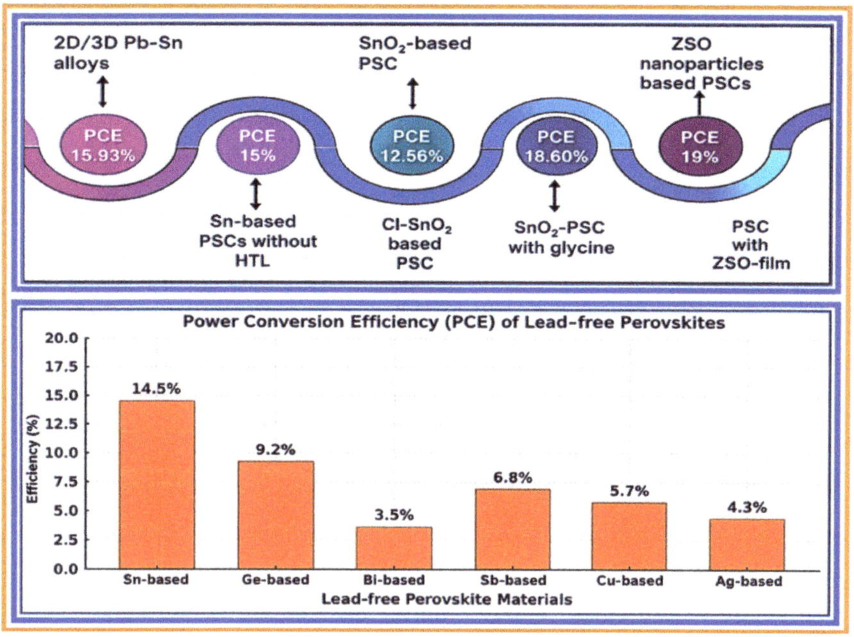

Fig. 15.1 Shematics figure shows efficiencies of different lead-free perovskite

15.1.1 Background of Lead-Free Perovskite Design

The typical PSC utilizes organic–inorganic halide perovskites of the generic formula ABX_3 [17], wherein A is a monovalent cation (e.g., methylammonium (MA^+), formamidinium (FA^+), or cesium (Cs^+), B is often a divalent metal cation like Pb^{2+}, and X is a halide anion (Cl^-, Br^-, or I^-). Lead (Pb)-based perovskites like methylammonium lead iodide ($MAPbI_3$) have achieved certified efficiency levels of over 26% in just a few years, competing with conventional crystalline silicon solar cells [18–20]. The natural Pb toxicity, however, together with instability towards environmental stressors, has given cause for alarm about large-scale implementation and even commercialization [21, 22]. Figure 15.2 illustrate Sn-Based perovskite with their benefits, application, and issues.

These issues have prompted lead-free perovskite studies with a goal of matching high functionality with environmental-friendliness. Replacing Pb^{2+} with other cations like Sn^{2+}, Ge^{2+}, Bi^{3+}, Sb^{3+}, or Ag^+, or with double perovskite structures like $Cs_2AgBiBr_6$ [23–26], gives a route to non-toxic, environmentally friendly substitutes [27] as shown in Figure 15.2. While tantalizing, such systems suffer from built-in problems such as fast oxidation of Sn^{2+} and Ge^{2+}, large bandgaps of Bi- and Sb-containing systems, and large-area device scalability difficulties [28, 29].

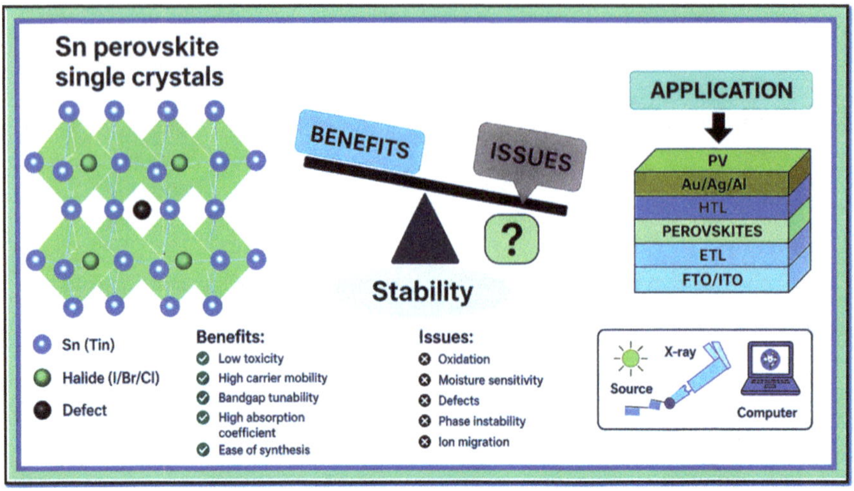

Fig. 15.2 Schematic figure illustrate Sn-based perovskite with their benefits and issues

15.1.1.1 Crystal Structure and Tolerance Factor

Stability and optoelectronic properties of perovskite are controlled by their crystal structure [30]. The optoelectronic properties of perovskite and their stability are defined by The Goldschmidt tolerance factor (T) [31–33] given as in eq. 15.1:

$$\mathrm{T} = \frac{r_A + r_X}{\sqrt{2}\,(r_B + r_X)} \tag{15.1}$$

where r_A, r_B, and r_x are A-site cation, B-site cation, and halide anion ionic radii, respectively, acts like a structural stability prediction parameter. For structural stability of a perovskite lattice, T tends to be between 0.8 and 1.0. Swapping Pb^{2+} for Sn^{2+} or Ge^{2+} tends to give tolerance factors close to that of $MAPBI_3$ and therefore maintains a perovskite structure [34]. However, substitutions with Bi^{3+} and Sb^{3+} often result in lattice distorting, such that compensating structural changes such as double perovskite design are necessary [35]. Figure 15.3 shows the general structure of Perovskite solar cell.

15.1.1.2 Electronic Band Structure and Bandgap Engineering

For best single-junction solar cell efficiency, it has been proposed that an ideal bandgap would be Nearly to1.34 eV by the Shockley-Queasier limit [36, 37]. Pb-perovskites like $MAPBI_3$, has band gap 1.55 eV [38] get close to this number. Sn-perovskites have smaller bandgaps 1.2 to1.4 eV [39, 40], having strong absorption

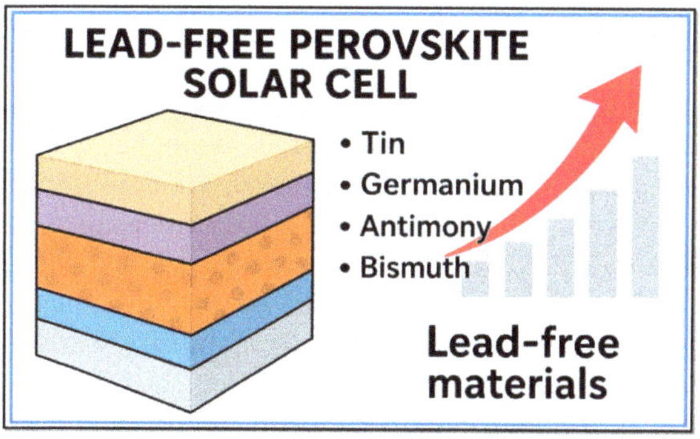

Fig. 15.3 Schematic general structure of perovskite solar cell

but with larger losses due to recombination because of defects [41]. Bi- and Sb-perovskite double perovskites have large bandgaps greater than 2 eV, restricting their solar absorption ability but being compatible to work as upper layers in tandem cells [42].

Bandgap tuning of absorbers through compositional alloying (i.e., alloying of halides, doping with alkali metals) is a major strategy for fine-tuning absorption efficiency with stability improvement [43].

15.1.1.3 Defect Tolerance and Carrier Dynamics

Lead halide perovskites are defect tolerant point defects prefer to create shallow states near band edges rather than deep trap states that would extinguish carrier lifetimes [44]. Few lead-free perovskites are defect tolerant, however. For example, Sn^{2+} oxidized to Sn^{4+} in moisture, creating deep trap states that degrades device efficiency. It must be addressed with surface passivation, doping, and encapsulation methods [45].

15.1.2 Recent Ongoing Research on Lead-Free Perovskite Solar Cells

The serious environmental and health issues with lead-based perovskites have driven active attention to researcher for appropriate substitutes. Most work has dealt with replacing Pb with safer elements such as tin (Sn) and bismuth (Bi) [46]. Tin-based perovskites, for instance, have been investigated intensively in prospect since tin can yield perovskite structures like Pb's. But operation of Sn-based PSCs is hindered due to rapid oxidation of Sn (II) to Sn (IV) and accompanying defects that adversely affect their optoelectronic performance [47].

Bismuth perovskites represent an alternative route to designing lead-free solar cells. Bismuth (Bi) was made attractive mostly because of its relatively low toxicity and capacity to synthesize compounds that possess perovskite-related structures. Initial studies on Bi-based materials involving perovskites indicate a promising route to green photovoltaics [48]. However, a critical problem accompanies bismuth-based perovskites, will naturally form low-dimensional materials whose inherent opto-electronic properties will be poorer compared to their three-dimensional lead-based material system equivalents [49]. With finite-dimensional material systems, exciton binding effects and charge carrier mobility constraints often lead to reduced device performance [50]. Owing to this fact, scientists have been compelled to explore further alterations. One such promising substitute is synthesizing Bi-based elpasolites that have represented a better alternative to conventional Bi-based perovskites owing to structural modifications potentially capable of abolishing limitations accompanying low dimensionality [51].

In summary, while promising alternatives like Sn- and Bi-based perovskites have been proposed to replace the toxicity due to lead, such a switch poses serious challenges. Present research in this area continues to address numerous issues: oxidation (for Sn), low dimensionality inherent to some Bi-based perovskites, and any resultant decrease in optoelectronic functionalities [52]. This section presents an in-depth description of lead's most promising lead-free alternatives, including associated technical difficulties in achieving parity to conventional lead-based devices.

15.2 Material Composition and Properties

Lead-free perovskite solar cells have numerous substances to replace lead with preserving the inherent advantages of the perovskite crystal structure. In conventional perovskite compositions with the ABX_3 formula, A is most likely to be replaced by organic cations like methylammonium or by formamidinium, B by a metal cation, while X is replaced by halide anions [53]. In lead-free perovskites, tin-based perovskites have received significant attention due to their similar electronic configurations like lead-containing ones. The tin (Sn^{2+}) cation has an electronic configuration that supports efficient light absorption and produces a band gap that can approximate the optimum value of about 1.4 eV, predicted by the Shockley–Queasier limit [54].

In addition to these advantageous, lead-free perovskites face numerous challenges. For example, the swift crystallization of films based on tin results in subpar film quality and results to nonradiative recombination. The instability of Sn^{2+}, which readily oxidizes to Sn^{4+}, introduces additional defects that diminish photovoltaic performance over time. Fluctuations in lattice structure and with variation in dimensionality reduced exciton binding energy, results to reduced carrier mobility rate [55]. Collectively, these elements contribute to a diminished power conversion efficiency in comparison to cutting-edge lead-based devices, despite some recent progress where certified power conversion efficiency values for Sn-based solar cells have achieved

Table 15.1 Comparison of key parameters of lead-free and lead-based perovskite solar cells

Parameter	Lead-based perovskites	Lead-free perovskites (e.g., Sn-based)
Toxicity	High	Low
Band gap	Approx. 1.5 eV	1.3 to 1.4 eV
Power conversion efficiency (PCE)	25.2% or higher	Approx.13.24%
Exciton binding energy	Low	Lower
Stability	Moderate (with encapsulation)	Poor (due to Sn oxidation and crystallization)

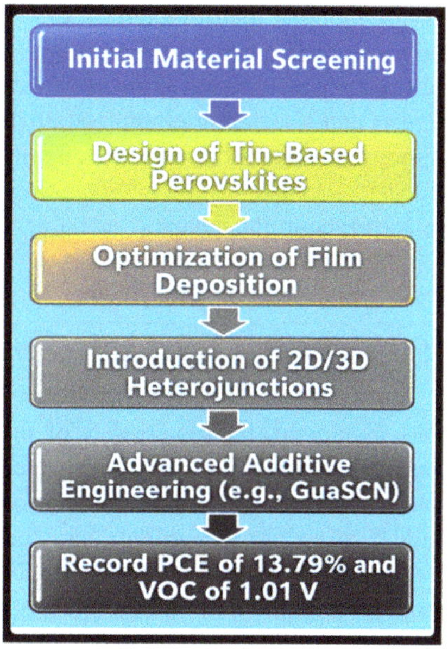

Fig. 15.4 Schematic of process involved during preparation in lead-based perovskite solar cells

13.24% [56]. The key parameters of lead-free perovskite solar cells and perovskite solar cells is surmised in Table 15.1 and process involved is shown Fig. 15.4.

15.3 Optimization of Device

Interfacial engineering is a key route to circumvent the weaknesses of lead-free perovskite solar cells, especially defect minimization, band energy level control, and efficient charge transport [57]. In contrast to bulk properties, alone, device performance is affected considerably by interfacial properties between the charge transport layers (CTLs) and perovskite absorber layers. Recent reports on Sn-containing perovskite solar cells elucidate that optimizing interfaces by advanced engineering can result in trap defect minimization, hence, VOC and FF improvement and nonradiative recombination loss suppression [58].

15.3.1 Passivation of Trap Defect

Defects located at the interfaces can serve as recombination centers for photogenerated charge carriers, thereby considerably diminishing the overall efficiency of the device [59]. Effective passivation approaches that utilize materials aimed at chemically stabilizing the surface have shown enhancements in carrier lifetime and light-harvesting efficiency [60]. For example, the integration of one-dimensional or two-dimensional perovskites at the interface has been advantageous in passivating surface defects and alleviating rapid crystallization issues related to Sn-based films. Using appropriate material synthesis techniques and controlling technique in the processing reduce defect-induced recombination, result to improved V_{OC} and enhanced overall performance of the device.

15.3.2 Energy Level Alignment

The alignment of energy levels of the perovskite absorber layer and the charge transport layers (CTLs) is essential criterion for effective separation and extraction of charges. The non-alignment of energy layers of these layers may lead to band bending, and generating energy barriers that hinder the easy transport of charge carriers at the device interfaces [61]. Numerous techniques have been used to optimize the alignment of energy levels at interfaces, such as chemical doping and the application of additive engineering. By proper adjusting the interfacial energy levels, it is reducing carrier recombination losses and improve the overall performance of photovoltaic systems. The open circuit voltage (V_{OC}), a parameter characterized by the disparity between the material's band gap (Eg) is a critical metric for evaluating device performance [62].

15.3.2.1 Incorporation of Low-Dimensional Perovskites

The use of low-dimensional perovskite at the interfaces can enhance stability. Such architectures exhibit increased bond strengths and reduced ion migration, that are critical aspects for lasting device functionality. The employment of two-dimensional perovskite layers in combination with three-dimensional absorbers layer has increase device stability, owing to its ability to establish a robust shield that prevents environmental degradation, especially moisture-induced degradation, to a large extent [63].

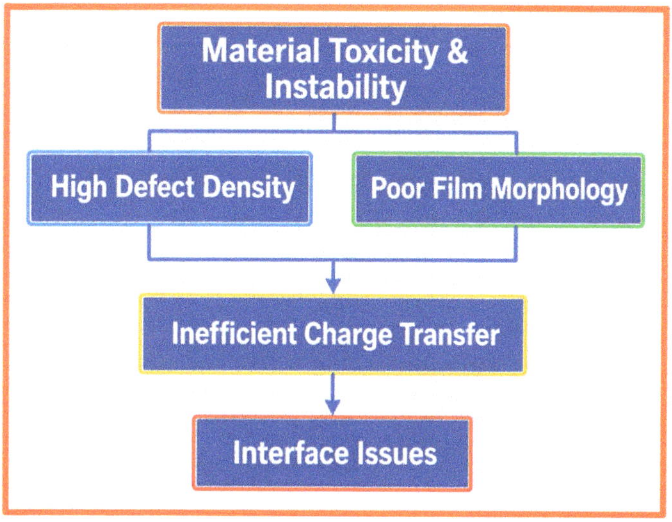

Figs. 15.5 Schematically challenges in lead-free perovskite solar cel. l

The key steps involved with interfacial engineering for lead-free perovskite solar cells are illustrated in Fig. 15.5.

15.4 Designing Challenges in Lead-Free Perovskite Solar Cells

Transitioning to lead-free perovskite solar cells has numerous challenges across a variety of axes that range from material inherent properties to device architecture and wider environmental and economic aspects [64]. We perform an in-depth analysis of these challenges in the discussion below.

15.4.1 Material-Level Challenges

Material Selection is crucial point in device designing due its different issue is reduced. Lead-free different absorber material is shown in Fig. 15.6 with increasing efficiency. Depth analysis of material level challenges is discussed in sub-sections.

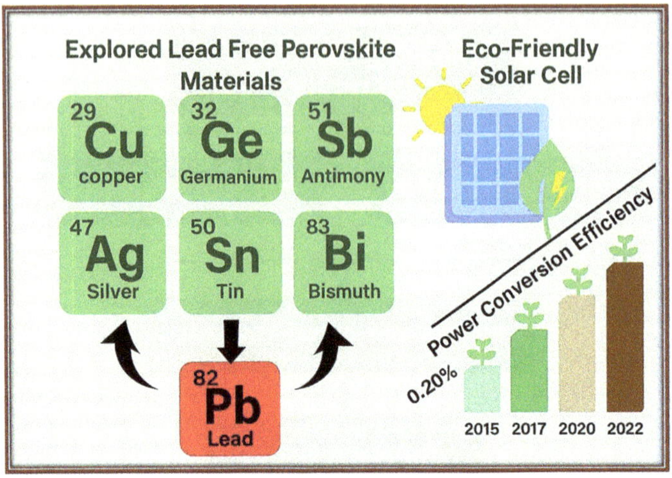

fig. 15.6 Schematic representation of lead-free absorber materials in pervoskite solar cell

15.4.1.1 Device Stability Issues

One of the crucial challenges in lead-free perovskite development is stability of the device. Unlike their lead-based counterparts, several lead-free materials especially tin-based materials are prone to quick degradation in response to ambient exposures. For example, tin-based perovskites can be oxidized from Sn^{2+} into Sn^{4+} in the exposure to air and moisture, results to degradation of device performance over time [65]. Additionally, structural stability in halide double perovskites could be impacted negatively due to environmental exposures such as temperature fluctuations, humidity levels, and prolonged illumination. Methods designed to boost stability such as encapsulation protection and passivation or compositional engineering existent today entail a lot of studies; however, long-term performance assurance still presents a large challenge [66].

From an engineering viewpoint, making perovskite layers long-lasting requires suitable technique and control over the synthesis process. The irregular Changes in crystallization conditions can introduced defects in the layer and irregular film morphologies results to degradation of device performance. Additionally, thermal cycle and mechanical flexing stress generally used in flexible or bendable devices further reinforce to search and design material having higher mechanical strength [67]. Recent studies emphasize that deeper research is needed on degradation issues systematically and employing in situ characterization methods can explores the essential information to solve the issue in the designing of better resilient perovskite materials.

15.4.1.2 Efficiency and Bandgap Tuning

Lead-based PSCs have realized impressive power conversion efficiencies (PCEs) above 25% in a vacuum setup, whereas lead-free substitutes usually fare behind. A prime limiting factor is a comparatively weaker optical absorption coefficient and inferior bandgap matching to the solar spectrum [68]. Tin-based perovskites have a smaller bandgap which have better infrared absorption. However, such enhancements in light harvesting are negated by fast degradation of device due to its structural instability and oxidation [69]. On the other hand, some double perovskites possess shorter charge carrier diffusion lengths and poor absorption properties compared to lead-based systems which results to lower efficiency [70].

Bandgap engineering is a path to reduce these efficiency roadblocks [38]. Material researchers are focused to study the effects of compositional tuning and doping engineering to get desirable bandgap that will suitably capture a larger portion of the solar spectrum while maintaining high charge separation efficiencies [71]. It is still a difficult task to achieve a trade-off between absorption of radiation, collection of charge, and material stability that provides a better counterpart of Lead.

15.4.1.3 Charge Carrier Dynamics

Solar energy conversion efficiencies in perovskite solar cells are greatly are the function of mobility of charge carriers and their lifetime after photogeneration. Charge carrier mobility in lead-free perovskites is observed and reported to be below the threshold. Thus, Charge carrier recombination are increased such that diffusion lengths become shorter [72]. Lower mobility limits charge extraction as well as non-radiative recombination losses is also increased that affect device efficiency.

Defect reduction technique with improved crystal quality and surface finishing methods may provide some better result. For example, improved film structure and increased material purity will results to increased charge carrier lifetimes and resolve the performance gap versus lead-based perovskite cells [73]. However, charge carrier dynamics is a very crucial that require further comprehension and optimization on a holistic level to get fruitful result.

15.4.1.4 Scalability and Manufacturing Issues

It's difficult task to obtain reproducible material quality in the fabricating process having large-area lead-free perovskite films. It is generally a condition where a lot of stoichiometry and uniformity variation of films is observed, and hence device performance is affected [74]. Reproducibility issues occur primarily due to lead-free materials' highly sensitive compositional space. Wider compositional landscape demands tighter control on process conditions, substrate preparation protocol, and precursor ratios to obtain uniformity in crystalline growth [75].

Processing methods like solution processing, vapor deposition, and printing need further optimization to yield large-area defect-free films reproducibly in lead-free formulations. Preparations of strong scalable fabrication techniques holding optimized device performance remain a focal research interest since the sector aims to move out of lab prototypes to market products [76].

15.4.2 Device-Level Problems

Device-related problems in lead-free perovskite solar cells (PSCs) are severe bottlenecks in their large-scale commercialization. While removing lead reduces toxicity-linked problems, most of the lead-free candidates (including Sn, Ge, Bi, Sb, and Cu-based perovskites) are plagued by several stability, efficiency, and fabrication issues at the device level. The predominant device-level problems in lead-free PSCs are instability due to oxidation (Sn/Ge), defect-mediated recombination, poor energy band alignment, interface instability, ionic migration, and comparative lower efficiency than in Pb-based perovskites [77]. These are resolved by material engineering advancements (doping, compositional tuning), interface engineering, and encapsulation techniques. The Detailed Device-Level Challenges is described below:

15.4.2.1 Interface Engineering and Charge Transport

Besides difficulties rooted in the active perovskite layer, the perovskite/adjacent layers interface (e.g., electron or hole transport layers) holds central position in ensuring device performance overall. For lead-free perovskite solar cells, charge extraction is impaired at a suboptimal level due to inefficiencies in interface engineering. Interfacial energy level alignment between perovskite absorber and charge transporting layers determines optimal suppression of recombination losses and enabling efficient carrier extraction [78].

Interface engineering with customized buffer layers, surface manipulations, and using material which have better alignment of energy gap between layers, will revels to remarkable success in optimizing interfacial charge transport [79]. Characterization tools such as atomic force microscopy (AFM) and Kelvin probe force microscopy (KPFM) are not valuable tools in mapping and optimizing electronic features within these interfaces [80]. Notwithstanding this progress, a challenge is always universally present, and forced to find better solution applicable at interface that can uniformly improve device performance across a variety of lead-free perovskite devices.

15.4.2.2 Mechanisms of Degradation in Device Operation

In addition to material and interfacial concerns, device degradation processes are also major hinder the consistent functioning of lead-free PSCs. Exposure of these PSCs with moisture, oxygen, and higher temperatures initiates degradation of device that degrade both the perovskite layer and the interfaces [81]. Ion migration is the major issue introduced by oxidation in perovskite materials, is responsible for creating harmful defect states and ultimate device failure. With help of protective encapsulation methods and adding stabilizers to buffer these problems of degradation. However, the highly complex nature of degradation have interactions between chemicals, physical forces, and environment, a system-level and integrated approach is needed to overcome it [82]. Systematic long-term studies with deeper understanding on reliability involving accelerated aging effect and actual cause of degradation in environmental exposers, need to be performed research thoroughly to mitigate these issues.

15.4.3 Environmental Challenges

Recent research requires a bifocal strategy one focus on emphasizes the development of advanced manufacturing technique and life-cycle analyses (LCAs) of lead-free perovskite solar cells (PSCs) for mitigating environmental impacts and, the other focuses on the chemical and technical challenges involved in the development of lead-free technologies [83].

15.4.3.1 Lifecycle Assessment and Environment Impacts

The major issue behind the replacing lead-based perovskite solar cells is the high environmental hazard introduced by lead toxicity. Lead is well known for neurotoxin activity and habitat that have water-polluting potential and hence creating a danger for both human and ecosystem. The whole lifecycle assessment (LCA) that based at the environmental impact of solar cells from material extraction to end-of-life waste is crucial to find the actual sustainable lead-free substitutes [84].

A full LCA of lead-free perovskite solar cells would entail production energy use comparisons, greenhouse emissions, and hazardous waste generation potential at device end-of-life [85]. Initial LCA studies have indicated that lead replacement can potentially greatly reduce environmental pollution potentials if lead-free solar cells are adequately recycled and disposed. More quantitative studies need to be done to comparatively evaluate lead-free and lead-based perovskite solar cell full lifecycle impacts [86]. Some LCA studies have also been conducted for the additional purpose of not only comparisons of the energy and CO_2 footprint of the perovskite solar cell alongside established technologies but also for the determination of potential hotspots in the production, operating, and disposal segments. Detailed LCAs, for instance, have indicated that even though the monolithic version of the perovskite module may have a low energy payback time (EPBT), the corresponding environmental impacts in terms of electricity production units can turn out larger under discrete degradation scenarios. Reports have indicated that even as the production process entails relatively low rates of lead use, the toxicity of the element can yet induce marine eutrophication and system-level impacts [87]. This highlights the necessity for better system performance ratios and lifetimes for the device in a bid to lower the overall system environmental footprint of the PSC.

15.4.3.2 Toxicity and Environmental Impact

The major environmental issue originates from the toxicity of Sn and Ge are less toxic than Pb but still pose hazards upon oxidation. Sb have low toxicity concerns and requires strict waste management [88]. When the degradation of the perovskite solar cells or improper waste disposal releases lead compounds in the environment, they can end up in water and soil and pose a risk for the damage of the ecosystems and human health [89]. It was indicated by research that the mechanisms of degradation for lead-free perovskites are not yet clear and this complicates the long-term environmental impacts. In addition, while life cycle analyses (LCAs) for photovoltaic perovskites have in general indicated the environmental competitiveness of the technologies in comparison to the classic silicon photovoltaic cells, the very specific role for lead toxicity is a source of concern [90].

15.4.3.3 Degradation Products and Ecosystem Risk

The other environmental issue pertains to the degradation products of and lead-free perovskite solar cells. Experimental experience indicates degradation triggered by heat, moisture, or mechanical force potentially degrade the Sn atoms or associated substances into the environment [91]. The products have the potential for gradual build-up and alter the regional ecology and cause food chain bioaccumulation. One scientist expressed concern about the long-term heavy damage potentially inflicted by perovskite technologies should degradation mechanisms not be properly controlled or understood. Consequently, the environmental hazard posed by these, highlights the need for the development of better, degradation-proof formulation or the use of effective recycling and disposal practices [92].

15.4.4 Economics of Perovskite Solar Cell Manufacturing

Economic viability constitutes a principal parameter primarily defining commercial viability of any new solar photovoltaic technology. Techno-economic evaluation (TEA) is useful tool to study whether lead-free perovskite solar cells can rival better result against existing photovoltaic technology with respect to both efficiency and cost parameters which is inherent within mass production and large deployment [93]. To study TEA for lead-free PSCs must have address various factors:

15.4.4.1 Production Costs and Costs of Raw Materials

Even lead-free materials use non-poisonous elements that are abundant on Earth (e.g., tin, bismuth, antimony), purity and processability can cause costs escalation if not efficiently optimized. From an economic point of view, solar cells using perovskites have considerable cost benefits relative to traditional silicon-based systems [94]. For instance, studies revealed the low-cost production possibility for perovskite modules at a value of $31.7 for the square meter down to an MSP value of about $0.41 Wp at base-case production in the United States. Such economic estimates stem from bottom-up cost modeling considering the material cost and the energy consumption during the production process in addition to labor costs [95].

15.4.4.2 Production Scalability

Challenges in production affect yield and reproducibility directly and thus influence overall production cost per watt [96]. Scalability of the production line has also been a core research emphasis, wherein the pilot lines are conceived in a way to support up to 100 MW annual production volumes.

15.4.4.3 Device Lifeline and Maintenance

Long-term stability is a major factor which effect levelized cost-of-energy (LCOE). Extended lifetimes reduce maintenance and replacement expenses during a system's lifetime. End-of-life (EOL) management strategies is another sustainable deployment pillar of perovskites [97]. Component-level safe recycling, component-level recovery, and component-level environmentally friendly disposal facilitated by a circular economy mode. It will be needed to avoid residual chemical impacts that have harmful effects after end of service life. Integration of prudent EOL practices in design and production cycles will be helpful to realize overall environmental benefits of lead-free perovskite solar cells [98].

15.4.4.4 Advanced Processing Methods

Recent Advancement in fabrication techniques decreases production costs with improved quality of perovskite film. One of the significant examples like rapid spray plasma processing (RSPP) technique that is useful in the fabrication of perovskite film at unprecedented scales and reducing fabrication costs day by day. In this technique spraying a perovskite solution of precursors onto a substrate and simultaneously exposing the film to the plasma to hasten crystallization. In RSPP technique post-fabrication annealing steps are not required, hence cutting the fabrication cost with less energy requirements and fabrication time [99]. Further, it is underscoring at large-scale industrial fabrication, which plays a vital role to cut the levelized cost of energy (LCOE).

15.4.4.5 World-Wide Production Cost Variations

Production cost is also affected in different geographic region. Materials expenses (e.g., encapsulants and glasses) depends on availability of raw materials in that geographic region. The different site locations have different intensity of the solar radiation at ambient levels can make a difference in the MSP, the LCOE, and the energy payback times (EPBT) [100]. For example, it is observed that in locations of high brightness like Spain and Egypt the EPBT can be as low as 0.6–1.1 years, yet for locations with lower irradiation is high.

15.4.4.6 Uncertainties and Sensitivity Issues

Due to the relatively young nature of PSC technology, and different modifications in it, economic models by necessity include uncertainties. Sensitivity studies with Monte Carlo simulations have been utilized in estimating the effect of various parameters e.g., module longevity, material price variations, and degradation rates are important economic indicators such as MSP and LCOE [101]. These studies explore important process parameters, which need to be optimized to make PSC cost-competitive with mature technologies like crystalline silicon photovoltaics. Nevertheless, uncertainties in scaling up the successes in a lab setting, but continuous and more studies are needed to confirming these economic models at commercial scales.

15.4.5 *Device Architecture Challenges*

The device architecture challenges faced by lead-free perovskite solar cells technologies. The discussion focuses on major concerns related to interface engineering, charge transport mechanism, and encapsulation technologies by which device efficacy and long-term reliability are affected [102]. Recent advancements with depth analysis of intrinsic and extrinsic challenges are helpful to improve the lead-free PSC technologies in future. Lead-free perovskite solar cells, particularly consisting of tin (Sn), have a high level of interest due to their low toxicity and harmless nature. Despite of extensive research activity, the lead-free materials continuously faces numerous challenges such as oxidation processes (e.g., the transformation from Sn^{2+} to Sn^{4+}), unstable lattice arrangements, and lower efficiencies with respect to their lead-based counterparts [103]. The substitution of lead-based perovskite by tin-based perovskite solar cells requires a delicate balance to maintain desirable optoelectronic properties. Addressing novel challenges in terms of stability and device architecture, involves the judicious selection and optimization of the charge transport layers, electrode materials, interfacial design, and encapsulation procedures [104].

15.4.5.1 Charge Transport Layer Challenges

The charge transport layers (CTLs) have major advantages in perovskite solar cell (PSC) devices. Since the extraction and transport of the photogenerated charge carrier is influenced by it and the non-suitable CTL layer is serving as selective barrier layers [105]. In the case of lead-free PSC architectures, the selection of proper electron transport layers (ETLs) and hole transport layers (HTLs) is important to achieve optimum efficiency and stability.

Electron Transport Layers (ETL)

The ETL in PSC cells is also implicated in the extraction of photogenerated electrons from the perovskite absorber and injecting them in the electrode [106]. Considering the lead-free cells, several challenges are involved in the selection and optimization of the ETL materials.

Aligning Bands: For the best extraction of electrons, the lowest unoccupied molecular orbital (LUMO) of the ETL would ideally be lower than the conduction band of the perovskite absorber. ETL with unaligned perovskite absorber layer can cause high energy loss and non-optimized device performance [107].

Material Selection Challenges: Material selection is a crucial point since it affects the device performance. For example, Titanium dioxide (TiO_2) is widely used since it is inexpensive and has facile processing procedures. However, the inherent oxygen vacancies and photocatalytic nature, particularly under the influence of ultraviolet (UV) radiation, may results to oxidation of the Sn^{2+} in tin-based perovskites and hence speedups material degradation [108].

Effect of Doping: The ETL material's resistivity, thickness, and dopant concentration affects the device performance and life-span so, it must be performed using suitable technique and under the precise control. A non-optimal ETL can slow down the mobility of electrons and create higher series resistance and therefore lower the overall power conversion efficiency (PCE) [109].

Hole Transport Layers (HTLs)

HTLs are optimized for photogenerated hole extraction and transport and suppression of electrons during the reduction to reduce the loss related to recombination [110]. In lead-free PSCs, the following limitations of the HTLs exist:

Energy Level Alignment: The valence band of the perovskite must also be above the highest occupied molecular orbital (HOMO) level of the HTL for effective hole extraction. Poor alignment can result in major Voc losses with reported losses of 0.4–0.5 eV due to such misalignments [111].

Material Challenges: Organic HTLs such as Spiro-OMeTAD, PTAA, and PEDOT: PSS have been widely employed, yet they usually carry limitations such as low long-term stability and the necessity of using dopants, which may themselves be hygroscopic [112]. Alternative inorganic materials such as NiO have been shown here and elsewhere to enhance device efficiency as well as enhance device stability.

Interface Barriers and Recombination: Impurities and trap states at the interface of the perovskite/HTL result in non-radiative recombination [113]. Surface passivation and surface treatment have been indicated to fix these defects but the choice of the HTL continues to play a significant role in the performance of the device.

15.4.5.2 Device Structure and Configuration Challenges

The device work of the lead-free PSCs is not solely dictated by the HTL/ ETLs. The device structure, i.e., ordering and layer structure, electrode nature, and global interface contact are also equally responsible. PSCs are typically fabricated in two main forms: the traditional (n-i-p) structure and the inverted.

In the n-i-p architecture, the electron selective layer is initially deposited onto the transparent conductive substrate, then the perovskite absorber and finally the hole selective layer [114]. While very common, this structure can experience the limitations inherent in TiO_2-based ETLs, for example oxygen vacancies which can oxidize the Sn^{2+} and hence destroy the active layer.

The inverted version p-i-n Architecture casts the HTL/ETL straight onto the substrate without any intermediate layer, then the perovskite and finally the top ETL [115]. The p-i-n structure possesses a few benefits for tin PSCs. Significantly, it reduces the necessity for the use of conducting metal oxides that can degrade the stability, and the pathway for the collection of holes is more suitable. The architecture consistently shows better performance and reduced hysteresis in the case of Sn-based devices.

15.4.5.3 Influence of Electrode

Structural material for the top electrode is also a highly significant parameter in the device's performance. The top electrode materials not only should have high electrical conductivity but also should chemically complement the subsequent layers. Traditional PSC design has commonly utilized silver, yet migration and chemical instability under ambient exposures have led many to pursue substitutions [116]. New literature shows the application of the use of the copper electrode for improved device stabilization and for reduced degradation upon illumination and use in ambient air.

Interfacial transition is also a crucial point such as the contact of the electrode and the hole transport layer (HTL) may initiate interfacial reactions which may increase recombination of the charge or initiate undesirable doping reactions. To reduce such challenges, the employment of the right barrier layers in the form of a thin BCP (2,9-dimethyl-4,7-diphenyl-1,10-phenanthroline) or any other hole-blocking materials is employed [117].

15.4.5.4 Role Interface Engineering Techniques

Optimizing interface engineering is very important in terms of not only boosting the efficiency but also improving the stability of lead-free PSCs. High-quality interface not just provides effective charge transport but also impacts the reduction in the creation of trap states adverse for the device's functioning [118].

Passivation and 2D/3D Heterostructures are one of the highly effective interface-quality improvement strategies is the implementation of low-dimension (2D) materials or the creation of 2D/3D heterostructures superimposed on the 2D or 3D perovskite absorber. Ultra-thin passivation layers in the range of bulky organic cations or 2D perovskites are dispersed onto the bulk absorber surface in a way which passivates defects and reduces surface recombination. The passivation also achieves the extra benefit of stopping moisture infiltration and providing better energy alignment along the CTLs. Hybrid materials which unite the high light capture and charge transport capabilities of 3D perovskites and the improved stability of 2D materials have been encouraging. As an example, a 2D phase which exclusively encapsulates the 3D grains can act not only as a protective covering but also as an effective interface for the transport of charges and hence enhance the performance and the stability upon ambient conditions [119]. Also**,** Suitable energy band alignment at the CTL/perovskite absorber interface is required for high-efficiency charge extraction and loss reduction due to recombination. The conduction band of the ETL should closely match the conduction band of the perovskite and similarly the HOMO energy level of the HTL should align with the valence band of the perovskite. Such mismatches would cause energy bottlenecks in rising series resistance or loss in Voc [120].

Interface engineering involves the use of self-assembled monolayers (SAMs) or dopants having abilities to encompass the alteration of the interface's local electronic structure. The layers are for the purpose of enhancing energy level offsets and diminishing the concentration of the interface trap states existing.

15.4.5.5 Role of Encapsulation for Environmental Preservation

The encapsulation of PSC devices is a best method for shielding form moisture, oxygen, temperature variations, and solar radiation [121]. It is especially necessary for lead-free PSCs as they are typically more prone to degradation by the environment in comparison with the corresponding Pb-based PSCs. Effective encapsulation overcomes the materials possessing barrier and maintain the device's electrical and

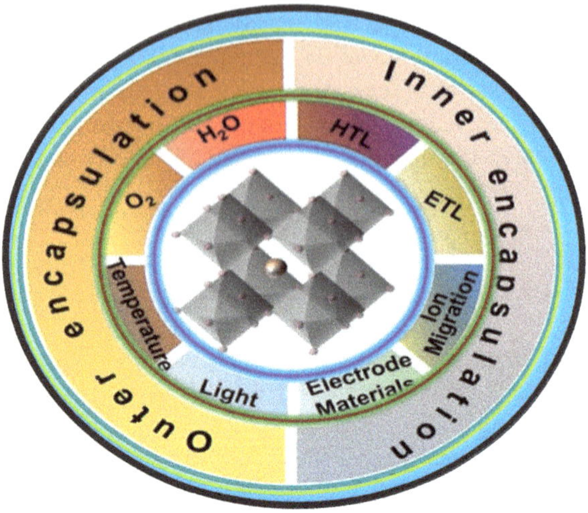

Fig. 15.7 Different encapsulation methodology

optical integrity. The major performance characteristics of the encapsulants have High OTR and WVTR values are important in preventing the entry of oxygen and moisture. These are important parameters in promoting the long-term stability of PSC devices. Also, for flexible devices, the encapsulate should also be mechanically strong and yet preserve the barrier performance. Multilayer stack architectures, typically consisting of inorganic and organic layers in combination, have successfully balanced the target barrier performance and suppressed rigidity [122]. In addition to this, the encapsulant should also be inert in the presence of the base perovskite layer. Incompatible materials may react with the perovskite and promote degradation instead of suppressing it.

Encapsulation Methodologies

Some encapsulation techniques have been employed for PSCs:

Glass Encapsulation: Rigid devices can be encapsulated in a glass encapsulate by applying sealants that create an effective barrier for moisture and oxygen. This method does not work for flexible devices [123]. Figure 15.7 shows different Encapsulation methodology.

Multilayer Barrier Films: For flexible PSCs, the encapsulants having the best performance so far reported are multilayered assemblies consisting of alternating inorganic and organic layers. Inorganic layers (e.g., Al_2O_3 by atomic layer deposition) have very good barrier capability and together with flexible organic layers create a good encapsulation holding down pinhole defects [122].

Pinhole Decoupling Method: Their approach seeks to reduce the multi-layer alignment of defects in such a way as to increase the overall barrier effectiveness [124].

15.4.5.6 Intrinsic Stability and Efficiency Restrictions

Besides the architectural challenges extrinsic to the active materials, a variety of intrinsic factors affect the efficiency and the stability of lead-free PSCs. These factors are usually inherent in the source material, the perovskite material, and the interface interactions known as Ion Migration, Hysteresis, and Trap.

Ion Migration is One major issue faced in perovskite solar cells is ion migration in the lattice structure of the perovskite [125]. Ionic migration is associated with the presence of vacancies and interstitial defects and facilitates ionic conduction upon the application of an extrinsic electric field or thermal stress. The phenomenon can lead to a change in the localized electric fields and cause hysteresis in the current-voltage (JV) behavior and consequently device degradation with time.

Trap States are Defect-induced trap states act as recombination centers for photogenerated carriers. At the interface between the perovskite and CTLs, these trap states reduce carrier lifetimes due to recombination of photo generated charge carriers, leading to lower power conversion efficiencies. Surface passivation strategies, as discussed earlier, are critical in mitigating these loss mechanisms [113].

15.4.5.7 Processing Method and Solvent Optimization Influence

The processing parameter also largely influences the thin-film quality of the perovskites. In the lead-free PSC and even more in the case of the Sn perovskite PSCs, introducing the solvent and employing the additives largely dictate the crystallization and the thin-film quality. Traditional solvents, including dimethyl sulfoxide (DMSO), have been shown to oxidize Sn^{2+}, resulting in the early deterioration of the perovskite solution before the fabrication of devices [126]. Now researchers are focused to search an alternative solvent or mixed solvent methodologies that mitigate or eradicate oxidation and improve film uniformity. Various additives like SnF_2 and $SnCl_2$ have been employed to control the oxidation of the Sn into Sn^{2+} and enhancing the quality of the film and that increases device reliability. These additives act chemically by sequestering the oxidized parts and physically by improving more homogeneous perovskite nucleation. Scanning condition—related parameters in the process of device characterization, such as the direction and scan rate also affects the effectiveness of device as seen hysteresis effect in the J-V curves. Advanced processing procedures with engineered material should be able to attain repeatable and optimal device performance.

15.5 Future Prospects of Lead-Free Perovskite Solar Cells

The future of lead-free perovskite solar cells (LF-PSCs) is filled with exceptional promise as global research efforts move toward replacing toxic lead with environmentally benign elements. Although lead halide perovskites have achieved record power conversion efficiencies (PCEs) exceeding 26%, their toxicity and instability limit large-scale commercialization [127]. Therefore, the focus has shifted to developing efficient, stable, and sustainable lead-free alternatives incorporating metals such as tin (Sn), germanium (Ge), bismuth (Bi), antimony (Sb), and copper (Cu), which possess suitable electronic and structural properties [128]. In future, major progress is expected through rational material design, compositional engineering, and interface optimization. Tin-based perovskites such as $CH_3NH_3SnI_3$ and $HC(NH_2)_2SnI_3$ have attracted attention due to their ideal bandgaps and high charge-carrier mobilities. However, the easy oxidation of Sn^{2+} to Sn^{4+} under ambient conditions leads to instability and rapid performance degradation [129]. Researchers are now exploring doping with halides, the introduction of reducing agents (e.g., SnF_2 or hydrazine), and encapsulation techniques to suppress oxidation and prolong device lifetimes [130]. In addition to Sn-based systems, double perovskites ($A_2B'B''X_6$) and vacancy-ordered halide perovskites (A_2BX_6) such as $Cs_2AgBiBr_6$ and Cs_2SnI_6 have gained attention for their inherent air stability and low toxicity [131]. The exploration of low-dimensional perovskites has further opened new frontiers for achieving stability and tunable optoelectronic properties. Two-dimensional (2D) Ruddlesden–Popper and Dion–Jacobson perovskites, formed by incorporating large organic spacer cations, can effectively suppress ion migration and moisture degradation, thereby extending device longevity [132]. These structures not only enhance chemical robustness but also facilitate flexible and wearable device integration. Interface and defect passivation strategies are likely to play a decisive role in determining the next phase of LF-PSC development. The introduction of self-assembled monolayers, surface ligands, and molecular interlayers has been demonstrated to minimize non-radiative recombination losses and improve open-circuit voltage and fill factor [133]. Surface passivation using halide additives, Lewis bases, and polymeric encapsulants also contributes to stabilizing the perovskite layer against oxygen and humidity exposure [134]. From a manufacturing perspective, future device fabrication will emphasize scalable, low-temperature, and environmentally friendly processes. The development of roll-to-roll compatible coating techniques such as blade coating, slot-die printing, and inkjet deposition promises to make large-area perovskite modules feasible [135]. Moreover, replacing toxic solvents like dimethylformamide (DMF) with greener alternatives and employing carbon or conducting polymer electrodes will promote cleaner and more sustainable production [136]. The improvement of long-term device stability remains a key challenge. Advances in encapsulation technology, such as atomic-layer-deposited barrier layers, cross-linked polymers, and multilayer films, have shown potential for protecting perovskite absorbers from moisture and thermal degradation [137]. When combined with intrinsic compositional and interfacial stability, these approaches could extend the operational lifetime of LF-PSCs to more than 20 years, meeting industrial benchmarks. Another exciting avenue for LF-PSCs is their integration into tandem and flexible architectures. Their tunable bandgaps enable efficient pairing with silicon, CIGS, or organic cells to form high-efficiency tandem devices [138]. In addition, their mechanical flexibility and lightweight nature make them suitable for emerging technologies such as building-integrated photovoltaics (BIPV), wearable power systems, and self-powered Internet-of-Things (IoT) devices. Finally, the future of LF-PSC innovation will be driven by data-centric design through computational modeling and machine learning. Predictive algorithms, combined with density functional theory (DFT), are increasingly being employed to screen potential lead-free compositions, evaluate defect formation energies, and predict electronic band structures [139]. We can transform wearable sensors, screens, and energy harvesting devices by integrating lead-free perovskites with eco-friendly, effective, and lightweight solutions. Cost-effectiveness and scalability are also crucial factors for the use of lead-free perovskites in practice. The assisted approaches will accelerate the identification of promising candidates and reduce experimental bottlenecks. In summary, the outlook for lead-free perovskite solar cells is exceptionally significant. The convergence of material innovation, advanced fabrication techniques, and sustainable design will pave the way for their commercialization. With continued progress in stability, efficiency, and environmental safety, LF-PSCs are poised to play a transformative role in the future of renewable energy, driving the global transition toward clean and sustainable photovoltaic technologies.

References

1. A.U.I. Shah, E.L. Meyer, Perovskite-based solar cells in photovoltaics for commercial scalability: Current progress, challenges, mitigations and future prospectus. Sol. Energy **286**, 113172 (2025). https://doi.org/10.1016/j.solener.2024.113172
2. L. Zhang, L. Mei, K. Wang, Y. Lv, S. Zhang, Y. Lian, et al., Advances in the application of perovskite materials. Nano-Micro Lett. **15**(1), 177 (2023). https://doi.org/10.1007/s40820-023-01140-3
3. S. Liu, V.P. Biju, Y. Qi, W. Chen, Z. Liu, Recent progress in the development of high-efficiency inverted perovskite solar cells. NPG Asia Mater. **15**(1), 27 (2023). https://doi.org/10.1038/s41427-023-00474-z
4. A.E. Magdalin, P.D. Nixon, E. Jayaseelan, M. Sivakumar, S.K.N. Devi, M.S.P. Subathra, et al., Development of lead-free perovskite solar cells: Opportunities, challenges, and future technologies. Results Eng. **20**, 101438 (2023). https://doi.org/10.1016/j.rineng.2023.101438
5. T. Münzel, O. Hahad, A. Daiber, P.J. Landrigan, Soil and water pollution and human health: What should cardiologists worry about? Cardiovasc. Res. **119**(2), 440–449 (2023). https://doi.org/10.1093/cvr/cvac082
6. S. Collin, A. Baskar, D.M. Geevarghese, M.N.V.S. Ali, P. Bahubali, R. Choudhary, et al., Bioaccumulation of lead (Pb) and its effects in plants: A review. J. Hazard. Mater. Lett. **3**, 100064 (2022). https://doi.org/10.1016/j.hazl.2022.100064

7. I. Ahmed, K. Prakash, S.M. Mobin, Lead-free perovskites for solar cell applications: Recent progress, ongoing challenges, and strategic approaches. Chem. Commun. **61**(37), 6691–6721 (2025). https://doi.org/10.1039/D4CC06835A
8. K. Ma, X. Li, F. Yang, H. Liu, Lead leakage of Pb-based perovskite solar cells. Coatings **13**(6), 1009 (2023)
9. V. Chauhan, D. Tripathi, P. Singh, A. Sharma, M.K. Khanna, R. Kumar, et al., Prospects for lead free perovskite for photovoltaic applications and biological impacts: Challenges and opportunities. Inorg. Chem. Commun. **157**, 111421 (2023). https://doi.org/10.1016/j.inoche.2023.111421
10. A. Ayyaz, S. Saidi, N. Dawas Alkhaldi, G. Murtaza, N. Sfina, Q. Mahmood, Lead-free double perovskites Rb_2TlSbX_6 (X=cl, Br, and I) as an emerging aspirant for solar cells and Green energy applications. Sol. Energy **279**, 112844 (2024). https://doi.org/10.1016/j.solener.2024.112844
11. K.J. Savill, A.M. Ulatowski, L.M. Herz, Optoelectronic Properties of Tin–Lead Halide Perovskites. ACS Energy Lett. **6**(7), 2413–2426 (2021). https://doi.org/10.1021/acsenergylett.1c00776
12. N. Neelu, N. Pandey, S. Chakrabarti, Synthesis, structural and optical properties of lead free $Cs_2CuBiCl_6$: A potential & promising eco-friendly double perovskite for solar cell applications. Opt. Mater. **143**, 114250 (2023). https://doi.org/10.1016/j.optmat.2023.114250
13. T. Wu, X. Liu, X. Luo, X. Lin, D. Cui, Y. Wang, et al., Lead-free tin perovskite solar cells. Joule **5**(4), 863–886 (2021). https://doi.org/10.1016/j.joule.2021.03.001
14. A.G. Olabi, K. Elsaid, K. Obaideen, M.A. Abdelkareem, H. Rezk, T. Wilberforce, et al., Renewable energy systems: Comparisons, challenges and barriers, sustainability indicators, and the contribution to UN sustainable development goals. Int. J. Thermofluids **20**, 100498 (2023). https://doi.org/10.1016/j.ijft.2023.100498
15. G. Schileo, G. Grancini, Lead or no lead? Availability, toxicity, sustainability and environmental impact of lead-free perovskite solar cells. J. Mater. Chem. C **9**(1), 67–76 (2021). https://doi.org/10.1039/D0TC04552G
16. C.E. Torrence, C.S. Libby, W. Nie, J.S. Stein, Environmental and health risks of perovskite solar modules: Case for better test standards and risk mitigation solutions. iScience **26**(1), 105807 (2023). https://doi.org/10.1016/j.isci.2022.105807
17. Z. Yi, N.H. Ladi, X. Shai, H. Li, Y. Shen, M. Wang, Will organic–inorganic hybrid halide lead perovskites be eliminated from optoelectronic applications? Nanoscale Adv. **1**(4), 1276–1289 (2019). https://doi.org/10.1039/C8NA00416A
18. A. Srivastava, J.A. Satrughna, A. Kanwade, M. Tiwari, S. Yadav, K. Bala, P. Shirage, Lead metal halide perovskite solar cells: Fabrication, advancement strategies, alternatives, and future perspectives. Mater. Today Commun. **35**, 105686 (2023). https://doi.org/10.1016/j.mtcomm.2023.105686
19. J. Qian, B. Xu, W. Tian, A comprehensive theoretical study of halide perovskites ABX_3. Org. Electron. **37**, 61–73 (2016). https://doi.org/10.1016/j.orgel.2016.05.046
20. M. Hichour, D. Rached, M. Rabah, S. Benalia, R. Khenata, F. Semari, Structural and elastic properties of antiperovskites $XNBa_3$ (X=as, Sb) under pressure effect. Phys. B Condens. Matter **404**(21), 4034–4038 (2009). https://doi.org/10.1016/j.physb.2009.07.154
21. A. Kumar, A. Kumar, M.M.S.C. Pinto, A.K. Chaturvedi, A.A. Shabnam, G. Subrahmanyam, et al., Lead toxicity: Health hazards, influence on food chain, and sustainable remediation approaches. Int. J. Environ. Res. Public Health **17**(7) (2020). https://doi.org/10.3390/ijerph17072179
22. J.-P. Jenny, O. Anneville, F. Arnaud, Y. Baulaz, D. Bouffard, I. Domaizon, et al., Scientists' warning to humanity: Rapid degradation of the world's large lakes. J. Great Lakes Res. **46**(4), 686–702 (2020). https://doi.org/10.1016/j.jglr.2020.05.006
23. F. Cao, L. Li, Progress of Lead-free halide perovskites: From material synthesis to photodetector application. Adv. Funct. Mater. **31**, 2008275 (2020). https://doi.org/10.1002/adfm.202008275

24. A. Shalan, S. Kazim, S. Ahmad, Lead-free perovskites: Metals substitution towards environmentally benign solar cell fabrication. ChemSusChem **12** (2019). https://doi.org/10.1002/cssc.201901296
25. W. Ke, M.G. Kanatzidis, Prospects for low-toxicity lead-free perovskite solar cells. Nat. Commun. **10**(1), 965 (2019). https://doi.org/10.1038/s41467-019-08918-3
26. E. Meyer, D. Mutukwa, N. Zingwe, R. Taziwa, Lead-free halide double perovskites: A review of the structural, optical, and stability properties as well as their viability to replace Lead halide perovskites. Meta **8**(9), 667 (2018)
27. A. Singh, S. Najman, A. Mohapatra, Y.-J. Lu, H. Chintam, C.-W. Pao, et al., Modulating performance and stability of inorganic Lead-free perovskite solar cells via Lewis-pair mediation. ACS Appl. Mater. & Interfaces (2020). https://doi.org/10.1021/acsami.0c06971
28. J.-K. Chen, B.-B. Zhang, Q. Liu, N. Shirahata, O. Mohammed, O. Bakr, H.-T. Sun, Advances and challenges in tin halide perovskite nanocrystals. ACS Materials Letters **3**, 1541–1557 (2021). https://doi.org/10.1021/acsmaterialslett.1c00444
29. F. Liu, J. Jiang, T. Toyoda, M.A. Kamarudin, S. Hayase, R. Wang, et al., Ultra-halide-rich synthesis of stable pure tin-based halide perovskite quantum dots: Implications for photovoltaics. ACS Appl. Nano Mater. **4** (2021). https://doi.org/10.1021/acsanm.1c00324
30. A. Kumar, K. Balasubramaniam, J. Kangsabanik, Vikram, A. Alam, Crystal structure, stability and optoelectronic properties of the organic-inorganic wide bandgap perovskite $CH_3NH_3BaI_3$: Candidate for transparent conductor applications. Phys. Rev. B **94** (2016). https://doi.org/10.1103/PhysRevB.94.180105
31. C. Bartel, C. Sutton, B. Goldsmith, R. Ouyang, L. Ghiringhelli, M. Scheffler, New tolerance factor to predict the stability of perovskite oxides and halides. Sci. Adv. **5**, eaav0693 (2019). https://doi.org/10.1126/sciadv.aav0693
32. J. Islah, E. Darkaoui, A. Abbassi, S. Taj, B. Manaut, H. Ez-Zahraouy, Stability, photovoltaic, and optoelectronic properties of eco-friendly $CsGeCL_{3-n}F_n$ halide perovskite for solar cells applications: Insights from DFT calculations. Eur. Phys. J. B **98**(3), 51 (2025). https://doi.org/10.1140/epjb/s10051-025-00895-5
33. J.W. Turnley, S. Agarwal, R. Agrawal, Rethinking tolerance factor analysis for chalcogenide perovskites. Mater. Horiz. **11**(19), 4802–4808 (2024). https://doi.org/10.1039/D4MH00689E
34. Z. Li, M. Yang, J.-S. Park, S.-H. Wei, J. Berry, K. Zhu, Stabilizing perovskite structures by tuning tolerance factor: Formation of Formamidinium and Cesium Lead iodide solid-state alloys. Chem. Mater. **28** (2015). https://doi.org/10.1021/acs.chemmater.5b04107
35. M. Liu, C.-K. Duan, P. Tanner, C.-G. Ma, M. Yin, Rationalizing the photoluminescence of Bi^{3+} and Sb^{3+} in double perovskite halide crystals. J. Phys. Chem. C **125** (2021). https://doi.org/10.1021/acs.jpcc.1c09069
36. F.H. Alharbi, S. Kais, Theoretical limits of photovoltaics efficiency and possible improvements by intuitive approaches learned from photosynthesis and quantum coherence. Renew. Sust. Energ. Rev. **43**, 1073–1089 (2015). https://doi.org/10.1016/j.rser.2014.11.101
37. Y. Xu, T. Gong, J.N. Munday, The generalized Shockley-Queisser limit for nanostructured solar cells. Sci. Rep. **5**(1), 13536 (2015). https://doi.org/10.1038/srep13536
38. M.H. Miah, M.U. Khandaker, M.B. Rahman, M. Nur-E-Alam, M.A. Islam, Band gap tuning of perovskite solar cells for enhancing the efficiency and stability: Issues and prospects. RSC Adv. **14**(23), 15876–15906 (2024). https://doi.org/10.1039/d4ra01640h
39. M. Hu, M. Chen, P. Guo, H. Zhou, J. Deng, Y. Yao, et al., Sub-1.4eV bandgap inorganic perovskite solar cells with long-term stability. Nat. Commun. **11**(1), 151 (2020). https://doi.org/10.1038/s41467-019-13908-6
40. S. Tahir, S. Mushtaq, J. McQueen, J. Iqbal, R.S. Almufarij, R.S. Alqurashi, et al., Computational insights into wide bandgap lead free perovskite solar cells for silicon based tandem configurations. Sci. Rep. **15**(1), 20020 (2025). https://doi.org/10.1038/s41598-025-04637-6
41. B. Zhang, Z. Zeng, H. Dong, W. Gao, C. Ran, Recent advances in tin halide perovskite solar cells: A critical review. J. Mater. Chem. A (2025). https://doi.org/10.1039/D5TA04568A
42. T. Ibn-Mohammed, S.C.L. Koh, I.M. Reaney, A. Acquaye, G. Schileo, K.B. Mustapha, R. Greenough, Perovskite solar cells: An integrated hybrid lifecycle assessment and review in

comparison with other photovoltaic technologies. Renew. Sust. Energ. Rev. **80**, 1321–1344 (2017). https://doi.org/10.1016/j.rser.2017.05.095
43. L. Malavasi, Band gap tuning through cation and halide alloying in Mechanochemical synthesized $Cs_3(Sb1\text{-}xBix)_2Br_9$ and $Cs_3Sb_2(I1\text{-}xBrx)_9$ solid solutions. (2023). https://doi.org/10.26434/chemrxiv-2023-62vzx
44. C. Wang, Z. Xie, Y. Wang, Y. Ding, M.K.H. Leung, Y.H. Ng, Defects of metal halide perovskites in photocatalytic energy conversion: Friend or foe? Adv Sci (Weinh) **11**(29), e2402471 (2024). https://doi.org/10.1002/advs.202402471
45. T. Mahmoudia, W.-Y. Rhob, M.H. Kohan, Y. Im, D.S. Mathur, Y.-B. Hahn, Suppression of Sn2+/Sn4+ oxidation in tin-based perovskite solar cells with graphene-tin quantum dots composites in active layer. Nano Energy **90**, 106495 (2021). https://doi.org/10.1016/j.nanoen.2021.106495
46. T. Seyisi, B.G. Fouda-Mbanga, J.I. Mnyango, Y.B. Nthwane, B. Nyoni, S. Mhlanga, et al., Major challenges for commercialization of perovskite solar cells: A critical review. Energy Rep. **13**, 1400–1415 (2025). https://doi.org/10.1016/j.egyr.2025.01.019
47. K.M. Anoop, D. Devadiga, S. Sunitha, T.N. Ahipa, Advancements in preventing Sn oxidation in tin-based perovskite solar cells: A Review. physica status solidi (a) **222** (2024). https://doi.org/10.1002/pssa.202400569
48. G.U.A. Shaikh, M.A. Ibrahim, M.S. Sadek, W. Zulhafizhazuan, S.M. Sultan, S.M. Hafiz, K. Sobayel, Synthesis of lead-free cesium bismuth iodide perovskite ink for solar cell using flow reactor. Sol. Energy **286**, 113150 (2025). https://doi.org/10.1016/j.solener.2024.113150
49. Z. Jin, Z. Zhang, J. Xiu, H. Song, T. Gatti, Z. He, A critical review on bismuth and antimony halide based perovskites and their derivatives for photovoltaic applications: Recent advances and challenges. J. Mater. Chem. A **8**(32), 16166–16188 (2020). https://doi.org/10.1039/D0TA05433J
50. B.P. Carwithen, T.R. Hopper, Z. Ge, N. Mondal, T. Wang, R. Mazlumian, et al., Confinement and exciton binding energy effects on hot carrier cooling in Lead halide perovskite nanomaterials. ACS Nano **17**(7), 6638–6648 (2023). https://doi.org/10.1021/acsnano.2c12373
51. M.S. Shadabroo, H. Abdizadeh, M.R. Golobostanfard, Elpasolite structures based on A_2AgBiX_6 (A: MA, Cs, X: I, Br): Application in double perovskite solar cells. Mater. Sci. Semicond. Process. **125**, 105639 (2021). https://doi.org/10.1016/j.mssp.2020.105639
52. G.G. Njema, J.K. Kibet, A review of chalcogenide-based perovskites as the next novel materials: Solar cell and optoelectronic applications, catalysis and future perspectives. Next Nanotechnol. **7**, 100102 (2025). https://doi.org/10.1016/j.nxnano.2024.100102
53. I. Jeon, K. Kim, E. Jokar, M. Park, H.-W. Lee, E.W.-G. Diau, Environmentally compatible Lead-free perovskite solar cells and their potential as light harvesters in energy storage systems. Nano **11**(8), 2066 (2021)
54. E. Aktas, N. Rajamanickam, J. Pascual, S. Hu, M.H. Aldamasy, D. Di Girolamo, et al., Challenges and strategies toward long-term stability of lead-free tin-based perovskite solar cells. Commun. Mater. **3**(1), 104 (2022). https://doi.org/10.1038/s43246-022-00327-2
55. A. Kama, S. Tirosh, A. Itzhak, M. Ejgenberg, D. Cahen, New Pb-free stable Sn–Ge solid solution halide perovskites fabricated by spray deposition. ACS Appl. Energy Mater. **5**(3), 3638–3646 (2022). https://doi.org/10.1021/acsaem.1c04115
56. Y. Li, W. Sun, W. Yan, S. Ye, H. Rao, H. Peng, et al., 50% Sn-based planar perovskite solar cell with power conversion efficiency up to 13.6%. Adv. Energy Mater. **6** (2016). https://doi.org/10.1002/aenm.201601353
57. X. Shi, K. Xu, Y. He, Z. Peng, X. Meng, F. Wan, et al., Strategies for enhancing energy-level matching in perovskite solar cells: An energy flow perspective. Nano-Micro Lett. **17**(1), 313 (2025). https://doi.org/10.1007/s40820-025-01815-z
58. S. Palei, G. Murali, C.-H. Kim, I. In, S.-Y. Lee, S.-J. Park, A review on Interface engineering of MXenes for perovskite solar cells. Nano-Micro Lett. **15**(1), 123 (2023). https://doi.org/10.1007/s40820-023-01083-9

59. L.M. Benhafsa, Y. Azzaz, N. Benseddik, N. Moulay, D. Bensaid, K. Bencherif, Optoelectronic and photovoltaic properties of Lead-free Ba_2AgBiX_6 (X = I, Br and cl) double perovskites: As a sustainable solar cell study based on DFT and SCAPS-1D. J. Inorg. Organomet. Polym. Mater. (2025). https://doi.org/10.1007/s10904-025-04014-w
60. Y.S. Lee, J. Jae Do, J.W. Jung, A comparative study of surface passivation of p-i-n perovskite solar cells by phenethylammonium iodide and 4-fluorophenethylammonium iodide for efficient and practical perovskite solar cells with long-term reliability. J. Alloys Compd. **988**, 174060 (2024). https://doi.org/10.1016/j.jallcom.2024.174060
61. M. Cheng, C. Zuo, Y. Wu, Z.A. Li, B. Xu, Y. Hua, L. Ding, Charge-transport layer engineering in perovskite solar cells. Science Bulletin. (2020). https://doi.org/10.1016/j.scib.2020.04.021
62. J. Qiu, X. Mei, M. Zhang, G. Wang, S. Zou, L. Wen, et al., Dipolar chemical bridge induced $CsPbI_3$ perovskite solar cells with 21.86% efficiency. Angew. Chem. (Int. Ed. Engl.) **63**, e202401751 (2024). https://doi.org/10.1002/anie.202401751
63. M. Rani, M.M. Khan, A. Numan, M. Khalid, S.M. Abbas, M. Iqbal, M.A. Mansoor, Breaking barriers: Addressing challenges in perovskite solar cell development. J. Alloys Compd. **1010**, 177648 (2025). https://doi.org/10.1016/j.jallcom.2024.177648
64. C. Yang, W. Hu, J. Liu, C. Han, Q. Gao, A. Mei, et al., Achievements, challenges, and future prospects for industrialization of perovskite solar cells. Light Sci. Appl. **13**(1), 227 (2024). https://doi.org/10.1038/s41377-024-01461-x
65. Y. Bai, Y. Meng, M. Yang, R. Tian, J. Wang, B. Jiao, et al., Lattice stabilization and strain homogenization in Sn-Pb bottom subcells enable stable all-perovskite tandems solar cells. Nat. Commun. **16**(1), 7344 (2025). https://doi.org/10.1038/s41467-025-62661-6
66. A.J. Olasoji, J.K. Park, H.J. Lee, Y. Song, D.S. Lee, S.H. Im, Metal halide perovskites: A platform for next-generation multifunctional devices. Adv. Ind. Eng. Chem. **1**(1), 12 (2025). https://doi.org/10.1007/s44405-025-00011-2
67. Z. Zhang, W. Kim, M.J. Ko, Y. Li, Perovskite single-crystal thin films: Preparation, surface engineering, and application. Nano Converg. **10**(1), 23 (2023). https://doi.org/10.1186/s40580-023-00373-7
68. S. Bhattarai, Highly efficient lead-free perovskite solar cells design through SCAPS-1D simulations. Indian J. Phys. **2023**, 1–8 (2023). https://doi.org/10.1007/s12648-023-02977-5
69. P. Li, X. Cao, J. Li, B. Jiao, X. Hou, F. Hao, et al., Ligand engineering in tin-based perovskite solar cells. Nano-Micro Lett. **15**(1), 167 (2023). https://doi.org/10.1007/s40820-023-01143-0
70. M.H. Miah, M.U. Khandaker, M.J. Hossen, E.A. Noor, I. Jahan, M. Shahinuzzaman, et al., Lead-free alternatives and toxicity mitigation strategies for sustainable perovskite solar cells: A critical review. Mater. Adv. **6**(9), 2718–2752 (2025). https://doi.org/10.1039/D5MA00010F
71. N.S. Peighambardoust, E. Sadeghi, U. Aydemir, Lead halide perovskite quantum dots for photovoltaics and Photocatalysis: A review. ACS Appl. Nano Mater. **5**(10), 14092–14132 (2022). https://doi.org/10.1021/acsanm.2c02787
72. M. Gantumur, M. Shahiduzzaman, M.I. Hossain, M. Akhtaruzzaman, M. Nakano, M. Karakawa, et al., Revolutionizing light capture: A comprehensive review of back-contact perovskite solar cell architectures. Mater. Today (2025). https://doi.org/10.1016/j.mattod.2025.08.017
73. L. Bai, L. Liu, M. Esquivel, B.L. Tardy, S. Huan, X. Niu, et al., Nanochitin: Chemistry, structure, assembly, and applications. Chem. Rev. **122**(13), 11604–11674 (2022). https://doi.org/10.1021/acs.chemrev.2c00125
74. A. Soultati, M. Tountas, K.K. Armadorou, A.R. Yusoff, b. M., Vasilopoulou, M., & Nazeeruddin, M. K., Synthetic approaches for perovskite thin films and single-crystals. Ener. Adv. **2**(8), 1075–1115 (2023). https://doi.org/10.1039/d3ya00098b
75. Alba-Rubio, A., Schweitzer, N., Gounder, R., Rioux, R., Bhan, A., Flaherty, D., ... Meyer, R. (2023). Addressing Rigor and Reproducibility in Thermal, Heterogeneous Catalysis
76. S.O. Jeje, P. Mpofu, N. Malatji, L.R. Kanyane, M.B. Shongwe, Multi-principal element alloy coatings: A review of deposition techniques, applications, and future prospects. Met. Mater. Int. (2025). https://doi.org/10.1007/s12540-025-01943-6

77. M. Wang, W. Wang, B. Ma, W. Shen, L. Liu, K. Cao, et al., Lead-free perovskite materials for solar cells. Nano-Micro Lett. **13**(1), 62 (2021). https://doi.org/10.1007/s40820-020-00578-z
78. A. Yusuf, R. Abubakar Phd, A. Abubakar, I. Mohammed, Progress on electron transport layers for perovskite solar cells. Niger. J. Phys. **32**, 81–90 (2024). https://doi.org/10.62292/njp.v32i4.2023.156
79. Y. Sun, C. Li, D. Liu, F. Zhang, J. Xue, Q. Zheng, Surface and interfacial engineering for multifunctional Nanocarbon materials. ACS Nano **19**(2), 1944–1980 (2025). https://doi.org/10.1021/acsnano.4c14128
80. F. Schulz, J. Ritala, O. Krejčí, A.P. Seitsonen, A.S. Foster, P. Liljeroth, Elemental identification by combining atomic force microscopy and kelvin probe force microscopy. ACS Nano **12**(6), 5274–5283 (2018). https://doi.org/10.1021/acsnano.7b08997
81. H.J. Kim, G.S. Han, H.S. Jung, Managing the lifecycle of perovskite solar cells: Addressing stability and environmental concerns from utilization to end-of-life. eScience **4**(2), 100243 (2024). https://doi.org/10.1016/j.esci.2024.100243
82. A. Henao-Ardila, M.X. Quintanilla-Carvajal, F.L. Moreno, Emulsification and stabilisation technologies used for the inclusion of lipophilic functional ingredients in food systems. Heliyon **10**(11), e32150 (2024). https://doi.org/10.1016/j.heliyon.2024.e32150
83. J.M. Lee, H.J. Kim, H.S. Jung, Life cycle assessment of lead recycling processes in perovskite solar cells. Chem. Commun. **61**(9), 1850–1853 (2025). https://doi.org/10.1039/D4CC05121A
84. K. Ganesan, C. Valderrama, Anticipatory life cycle analysis framework for sustainable management of end-of-life crystalline silicon photovoltaic panels. Energy **245**, 123207 (2022). https://doi.org/10.1016/j.energy.2022.123207
85. N. Singh, O.A. Ogunseitan, Disentangling the worldwide web of e-waste and climate change co-benefits. Circ. Econ. **1**(2), 100011 (2022). https://doi.org/10.1016/j.cec.2022.100011
86. Clegg, C., Mei, J., Fuensanta, A., Ibn-Mohammed, T., & Pecunia, V. (2024). Evaluating Lead-based vs. Lead-Free Perovskites for Environmentally Sustainable Indoor Photovoltaics
87. W. Akram, X. Li, S. Ahmed, Z. Ouyang, G. Li, A review of life cycle assessment and sustainability analysis of perovskite/Si tandem solar cells. RSC Sustainability **3**(1), 21–36 (2025). https://doi.org/10.1039/D4SU00431K
88. G. Ondrasek, J. Shepherd, S. Rathod, R. Dharavath, M.I. Rashid, M. Brtnicky, et al., Metal contamination—A global environmental issue: Sources, implications & advances in mitigation. RSC Adv. **15**(5), 3904–3927 (2025). https://doi.org/10.1039/d4ra04639k
89. M.L. Rencheck, C. Libby, A. Montgomery, J.S. Stein, Managing potential environmental and human health risks of lead halide perovskite photovoltaic modules. Sol. Energy **269**, 112337 (2024). https://doi.org/10.1016/j.solener.2024.112337
90. S. Gerbinet, S. Belboom, A. Léonard, Life cycle analysis (LCA) of photovoltaic panels: A review. Renew. Sust. Energ. Rev. **38**, 747–753 (2014). https://doi.org/10.1016/j.rser.2014.07.043
91. D. Zhang, D. Li, Y. Hu, A. Mei, H. Han, Degradation pathways in perovskite solar cells and how to meet international standards. Commun. Mater. **3**(1), 58 (2022). https://doi.org/10.1038/s43246-022-00281-z
92. P. Dissanayake, K.M. Yeom, B. Sarkar, D. Alessi, D. Hou, J. Rinklebe, et al., Environmental impact of metal halide perovskite solar cells and potential mitigation strategies: A critical review. Environ. Res. **219**, 115066 (2022). https://doi.org/10.1016/j.envres.2022.115066
93. J. Zhang, N. Chang, C. Fagerholm, M. Qiu, L. Shuai, R. Egan, C. Yuan, Techno-economic and environmental sustainability of industrial-scale productions of perovskite solar cells. Renew. Sust. Energ. Rev. **158**, 112146 (2022). https://doi.org/10.1016/j.rser.2022.112146
94. P. Čulík, K. Brooks, C. Momblona, M. Adams, S. Kinge, F. Maréchal, et al., Design and cost analysis of 100 MW perovskite solar panel manufacturing process in different locations. ACS Energy Letters **7**(9), 3039–3044 (2022). https://doi.org/10.1021/acsenergylett.2c01728
95. F. Field, R. Kirchain, R. Roth, Process cost modeling: Strategic engineering and economic evaluation of materials technologies. JOM **59**, 21–32 (2007). https://doi.org/10.1007/s11837-007-0126-0

96. Y. Liu, Z. Zhang, T. Wu, W. Xiang, Z. Qin, X. Shen, et al., Cost effectivities analysis of perovskite solar cells: Will it outperform crystalline silicon ones? Nano-Micro Lett. **17**(1), 219 (2025). https://doi.org/10.1007/s40820-025-01744-x
97. V. Larini, M. Degani, S. Cavalli, G. Grancini, Sustainable decommissioning of perovskite solar cells: From waste to resources. Chem. Soc. Rev. **54**(15), 7252–7270 (2025). https://doi.org/10.1039/D5CS00359H
98. E. Iacovidou, C.A. Velis, P. Purnell, O. Zwirner, A. Brown, J. Hahladakis, et al., Metrics for optimising the multi-dimensional value of resources recovered from waste in a circular economy: A critical review. J. Clean. Prod. **166**, 910–938 (2017). https://doi.org/10.1016/j.jclepro.2017.07.100
99. M. Hovish, N. Rolston, F. Hilt, K. Brüning, C. Tassone, R. Dauskardt, Crystallization kinetics of rapid spray plasma processed multiple cation perovskites in open air. J. Mater. Chem. A **8** (2019). https://doi.org/10.1039/C9TA07980G
100. N. Alam, T.K. Tulka, M.M. Hossain, Materials, stability, and environmental aspects of perovskite solar cell: A topical review. E-Prime—Adv. Electr. Eng. Electron. Energy **13**, 101093 (2025). https://doi.org/10.1016/j.prime.2025.101093
101. W. Akram, X. Li, S. Ahmed, Z. Ouyang, G. Li, A review of life cycle assessment and sustainability analysis of perovskite/Si tandem solar cells. RSC Sustain. **3**(1), 21–36 (2024). https://doi.org/10.1039/d4su00431k
102. R. Wang, J. Wang, S. Tan, Y. Duan, Z.-K. Wang, Y. Yang, Opportunities and challenges of Lead-free perovskite optoelectronic devices. Trends Chem. **1**(4), 368–379 (2019). https://doi.org/10.1016/j.trechm.2019.04.004
103. D.M. Ikram, R. Malik, R. Raees, M. Imran, F. Wang, S. Ali, et al., Recent advancements and future insight of lead-free non-toxic perovskite solar cells for sustainable and clean energy production: A review. Sustain Energy Technol Assess **53**, 102433 (2022). https://doi.org/10.1016/j.seta.2022.102433
104. L. Serrano-Luján, N. Espinosa, T. Trofod, J. Abad, A. Urbina, F. Krebs, Tin- and Lead-based perovskite solar cells under scrutiny: An environmental perspective. Adv. Energy Mater. (2015). https://doi.org/10.1002/aenm.201501119
105. Z. Khan, M.I. Khan, M. Ismail, M. Farooq, Harnessing Sr_3SbI_3 perovskites with optimized charge transport layers for high-efficiency solar energy conversion. Results Eng. **27**, 106954 (2025). https://doi.org/10.1016/j.rineng.2025.106954
106. A. Aftab, R. Murshed, Recent progress in electron transport layers for enhancing the performance of tin-based perovskite solar cells. Sol. Energy **300**, 113866 (2025). https://doi.org/10.1016/j.solener.2025.113866
107. M. Aliaghayee, An overview of the numerical modeling, simulation, and optimization methods: Toward the design of highly efficient planer perovskite solar cells. Discov. Electron. **1**(1), 35 (2024). https://doi.org/10.1007/s44291-024-00026-x
108. C. Anucha, I. Altin, E. Bacaksız, V. Stathopoulos, Titanium dioxide (TiO_2)-based Photocatalyst materials activity enhancement for contaminants of emerging concern (CECs) degradation: In the light of modification strategies. Chem. Eng. J. Adv. **10**, 100262 (2022). https://doi.org/10.1016/j.ceja.2022.100262
109. M. Elhadi, M.T. Tazwar, A.S. Zishan, M.A. Faruque, H.Z. Anonto, M.M.A. Eid, et al., Optimization and comparative analysis of electron transport layers for high-performance InGeCl3-based perovskite solar cells. ChemistrySelect **10**(26), e01364 (2025). https://doi.org/10.1002/slct.202501364
110. Wang, Q., Shao, S., Xu, B., Duim, H., Dong, J., Adjokatse, S. et al. (2020). Impact of the Hole Transport Layer on the Charge Extraction of Ruddlesden–Popper Perovskite Solar Cells. ACS Appl. Mater. Interfaces, 12(26), 29505–29512. doi:https://doi.org/10.1021/acsami.0c05290
111. R. Belisle, P. Jain, R. Prasanna, T. Leijtens, M. McGehee, The minimal effect of hole-transport material ionization potential on the open-circuit voltage of perovskite solar cells. ACS Energy Lett. **1** (2016). https://doi.org/10.1021/acsenergylett.6b00270
112. S.N. Afraj, A. Velusamy, S. Sharma, J. Jacob, J.-P. Correa-Baena, M.-C. Chen, A. Facchetti, Recent progress in molecularly tailored organic hole transporting materials for highly efficient

perovskite solar cells. J. Mater. Chem. C **13**(30), 15234–15289 (2025). https://doi.org/10.1039/D5TC01870F

113. T.S. Sherkar, C. Momblona, L. Gil-Escrig, J. Ávila, M. Sessolo, H.J. Bolink, L.J.A. Koster, Recombination in perovskite solar cells: Significance of grain boundaries, Interface traps, and defect ions. ACS Energy Lett. **2**(5), 1214–1222 (2017). https://doi.org/10.1021/acsenergylett.7b00236
114. D.S. Utomo, L.M. Svirskaite, A. Prasetio, V. Malinauskiene, P. Dally, E. Aydin, et al., Nonfullerene self-assembled monolayers as electron-selective contacts for n-i-p perovskite solar cells. ACS Energy Lett. **9**(4), 1682–1692 (2024). https://doi.org/10.1021/acsenergylett.4c00306
115. A. Agresti, F. Di Giacomo, S. Pescetelli, A. Di Carlo, Scalable deposition techniques for large-area perovskite photovoltaic technology: A multi-perspective review. Nano Energy **122**, 109317 (2024). https://doi.org/10.1016/j.nanoen.2024.109317
116. S. Svanström, T.J. Jacobsson, G. Boschloo, E.M.J. Johansson, H. Rensmo, U.B. Cappel, Degradation mechanism of silver metal deposited on Lead halide perovskites. ACS Appl. Mater. Interfaces **12**(6), 7212–7221 (2020). https://doi.org/10.1021/acsami.9b20315
117. Y. Lin, Y. Fang, J. Zhao, Y. Shao, S.J. Stuard, M.M. Nahid, et al., Unveiling the operation mechanism of layered perovskite solar cells. Nat. Commun. **10**(1), 1008 (2019). https://doi.org/10.1038/s41467-019-08958-9
118. I. Grill, M. Aygüler, T. Bein, P. Docampo, N. Hartmann, M. Handloser, A. Hartschuh, Charge transport limitations in perovskite solar cells: The effect of charge extraction layers. ACS Appl. Mater. Interfaces **9** (2017). https://doi.org/10.1021/acsami.7b09567
119. T.M. Koh, V. Shanmugam, X. Guo, S.S. Lim, O. Filonik, E.M. Herzig, et al., Enhancing moisture tolerance in efficient hybrid 3D/2D perovskite photovoltaics. J. Mater. Chem. A **6**(5), 2122–2128 (2018). https://doi.org/10.1039/C7TA09657G
120. S. Malhotra, L. Gupta, H. Nandan, M.K.A. Mohammed, M.K. Hossain, J. Madan, et al., Tailoring the hole transport layer and understanding the impact of Sn oxidation for different mixed halide perovskite active layers: On a quest for the perfect match. Energy Fuel **38**(4), 3417–3427 (2024). https://doi.org/10.1021/acs.energyfuels.3c04392
121. P. Wu, S. Wang, X. Li, F. Zhang, Beyond efficiency fever: Preventing lead leakage for perovskite solar cells. Matter **5**(4), 1137–1161 (2022). https://doi.org/10.1016/j.matt.2022.02.012
122. M. Sang, K. Kim, J. Shin, K.J. Yu, Ultra-thin flexible encapsulating materials for soft bio-integrated electronics. Adv. Sci. (Weinh) **9**(30), e2202980 (2022). https://doi.org/10.1002/advs.202202980
123. K. Aitola, G. Gava Sonai, M. Markkanen, J. Jaqueline Kaschuk, X. Hou, K. Miettunen, P.D. Lund, Encapsulation of commercial and emerging solar cells with focus on perovskite solar cells. Sol. Energy **237**, 264–283 (2022). https://doi.org/10.1016/j.solener.2022.03.060
124. A.S.S. Bilal, M.M. Bilal, M.F. Zia, S. Feroz, M.N. Ullah, M.A. Khan, et al., Dual interfacial modification of hematite electron transport layer for efficient and stable perovskite solar cells. Results Eng. **25**, 103954 (2025). https://doi.org/10.1016/j.rineng.2025.103954
125. E. Bi, Z. Song, C. Li, Z. Wu, Y. Yan, Mitigating ion migration in perovskite solar cells. Trends Chem. **3**(7), 575–588 (2021). https://doi.org/10.1016/j.trechm.2021.04.004
126. X. Huang, B. Wu, N. Zheng, Optimizing solvent chemistry for high-quality halide perovskite films. Acc. Mater. Res. **6**(1), 40–51 (2025). https://doi.org/10.1021/accountsmr.4c00148
127. M.A. Green, E.D. Dunlop, J. Hohl-Ebinger, M. Yoshita, N. Kopidakis, X. Hao, Solar cell efficiency tables (version 60). Prog. Photovolt. Res. Appl. **30**(5), 421–437 (2022)
128. E.H. Jung et al., Progress and challenges of lead-free halide perovskite solar cells. Adv. Energy Mater. **11**(12), 2003878 (2021)
129. N.K. Noel et al., Lead-free organic–inorganic tin halide perovskites for photovoltaic applications. Energy Environ. Sci. **7**(9), 3061–3068 (2014)
130. W. Liao et al., Enhanced stability in tin-based perovskite solar cells via additive engineering. J. Mater. Chem. A **11**(14), 7271–7283 (2023)

131. G. Volonakis et al., Lead-free halide double perovskites via heterovalent substitution of lead with bismuth and silver. J. Phys. Chem. Lett. **8**(4), 772–778 (2017)
132. G. Grancini, M.K. Nazeeruddin, Dimensional tailoring of hybrid perovskites for photovoltaics. Nat. Rev. Mater. **4**(1), 4–22 (2019)
133. Y. Xu et al., Interface engineering for highly stable and efficient lead-free perovskite solar cells. Nano Energy **99**, 107405 (2022)
134. Y. Zhou et al., Surface passivation of perovskite solar cells for improved performance and stability. ACS Energy Lett. **5**(3), 871–878 (2020)
135. Y. Zhang et al., Scalable fabrication of lead-free perovskite solar modules by slot-die coating. Adv. Sci. **8**(23), 2102021 (2021)
136. D. Li et al., Green solvent processing of perovskite films for eco-friendly photovoltaics. J. Power Sources **574**, 233032 (2023)
137. T.-B. Song et al., Encapsulation of perovskite solar cells for long-term stability. Adv. Energy Mater. **10**(5), 1903243 (2020)
138. M. Jung et al., Flexible and tandem lead-free perovskite solar cells: Current status and outlook. Nat. Energy **7**(5), 381–391 (2022)
139. J. Kim et al., Machine learning and computational design of lead-free perovskites for solar energy conversion. Mater. Today **47**, 79–91 (2021)

Chapter 16
Finishing Remarks on Sustainable Lead Free Perovskite Materials: Unique Potential of Applications in New Generation Devices

Manish Kumar, Khushboo Gupta, Prachi Chaudhary, Vivek Mani Tripathi, Rohit Kumar Singh Gautam, Manvandra Kumar Singh, Deepika Dwivedi, and Vishnu Chauhan

Abstract Lead-free perovskite materials have emerged as sustainable alternatives to lead-based systems, fueled by advancements in nanoscience and growing environmental concerns. This book provides a concise and comprehensive description of the synthesis, properties, and device performance of leading lead-free perovskites, with emphasis on improving efficiency, stability, and eco-friendliness. Recent advances in the field of material design and nanoscale engineering are included in the context of applications in photovoltaics, optoelectronics, energy storage, sensing, catalysis and hydrogen production. This book is a source of valuable information on the subject

M. Kumar (✉) · P. Chaudhary
Experimental Research Laboratory, Department of Physics, ARSD College, University of Delhi, New Delhi, India
e-mail: mkumar2@arsd.du.ac.in

K. Gupta
Department of Physics, School of Sciences, IILM University, Greater Noida, Uttar Pradesh, India

V. M. Tripathi
Department of Mechanical Engineering, Uttar Pradesh Textile Technology Institute, Kanpur, India

R. K. S. Gautam
Department of Mechanical Engineering, Teerthanker Mahaveer University, Moradabad 244001, India

M. K. Singh
Department of Mechanical Engineering, Graphic Era (Deemed to be University), Dehradun, Uttarakhand, India

D. Dwivedi
Dr. B.R. Ambedkar Center for Biomedical Research (ACBR), University of Delhi, New Delhi, India

V. Chauhan (✉)
Department of Physics, Ramjas College, University of Delhi, New Delhi, India
e-mail: vishnuchauhan@ramjas.du.ac.in

V. Chauhan et al. (eds.), *Sustainable Lead-Free Perovskite Materials*, Topics in Applied Physics 157, https://doi.org/10.1007/978-3-032-24873-2_16

of lead-free perovskites since it outlines the existing developments and outlook of the topic as good candidates in the next generation of technologies.

Keywords Perovskite solar cell · Lead-free perovskites · Photovoltaics · Future outlook

16.1 Introduction

With the advent of nanoscience and nanotechnology, lead free sustainable perovskites show a diversity of effects in varieties of materials and devices. The modification of lead free perovskites properties and their applications mainly for the reduction of environment toxicity for efficiency of various devices has been central focus in the global trend for their use in industrial, medical, and engineering fields. Thus, researchers are working in the field of to lead free perovskites exploration their utilizations to increase the efficiency of the mainly photovoltaics and optoelectronic devices. The purpose of this book is to provide an advanced level of understanding of the synthesis, properties and efficiency of emerging lead free perovskite materials and devices. The technological needs of modern society cannot be envisioned without modifications of the existing materials and discoveries of new materials. Lead free perovskites presents new paradigms in this area by exploring the nanomaterials for environment protection, research and innovation. Moreover, the behaviour of lead containing perovskites in environmental and industrial applications is crucial for many purposes, such as optimum efficiency for solar cells, LEDs, gas sensors, catalysis, energy storage, thin film transistors, piezoelectric devices (actuators and sensors), water splitting (hydrogen production) etc. Metal halide perovskites have emerged as one of the most transformative classes of materials in modern materials science, owing to their exceptional optoelectronic properties, low-cost processing, and structural versatility. Since the first report of perovskite-based solar cells in 2009, their power conversion efficiency has increased dramatically, surpassing 26%, rivaling well-established silicon technologies [1–3]. Beyond photovoltaics, perovskites have demonstrated remarkable potential in light-emitting diodes (LEDs), photodetectors, lasers, radiation detectors, and memory devices, positioning them as cornerstone materials for next-generation optoelectronic and electronic applications [4–6]. Regardless of these successes, there are serious environmental and health concerns regarding the common use of traditional lead-based perovskites, which is usually represented as ABX_3 (with A being monovalent cation, B being Pb^{2+} and X being a halide ion). Lead poisoning is extremely dangerous in material production, device fabrication, operational degradation, and disposal at the end of life-cycle, creating a regulatory and sustainability problem to commercialization [7–9]. These issues have precipitated intense worldwide studies on the development of lead-free perovskite compounds that still have the desirable characteristics of their lead-based counterparts, but ones that do not pose any environmental risks.

Sustainability has become a central guiding principle in advanced materials research, emphasizing environmental safety, resource abundance, energy efficiency, and long-term stability. In this context, lead-free perovskites represent a crucial step toward eco-friendly optoelectronic technologies. The substitution of Pb^{2+} with non-toxic or less toxic metal cations such as Sn^{2+}, Ge^{2+}, Bi^{2+}, Sb^{2+}, Cu^{2+}, Ag^{2+}, and In^{2+} has enabled the design of a diverse family of perovskite-inspired materials with tunable structural and electronic properties [10–12].

Among these alternatives, tin-based perovskites (e.g., $FASnI_3$ and $CsSnI_3$) have attracted considerable attention due to their favorable bandgaps, high carrier mobilities, and strong optical absorption. However, their susceptibility to oxidation from Sn^{2+} to Sn^{4+} leads to instability and self-doping effects, limiting device performance [13, 14]. Germanium-based perovskites, while less toxic, suffer from similar oxidation and stability issues. On the other hand, bismuth- and antimony-based perovskites offer superior chemical stability and reduced toxicity, albeit often at the expense of indirect bandgaps and lower charge transport efficiency [15–17]. To overcome these challenges, researchers have explored double perovskites ($A_2B'B''X_6$), vacancy-ordered perovskites, and low-dimensional perovskite derivatives, which provide enhanced structural stability and compositional flexibility. Notably, silver–bismuth-based double perovskites such as $Cs_2AgBiBr_6$ have demonstrated improved moisture stability and promising optoelectronic performance, making them strong candidates for sustainable device applications [18–20].

Lead-free perovskites have a number of inherent strengths that render them appealing in the next-generation devices. They have high absorption coefficients, tolerance to defects, long carrier diffusion lengths and can be fabricated by solution-process due to their low temperature and ability to work on flexible and lightweight substrates [21–23]. Furthermore, compositional engineering and dimensional control enables accurate adjustment of bandgaps in the visible and near-infrared range enabling multifunctional device integration. Notably, numerous lead-free perovskites have been shown to be thermally, chemically and photostable than lead-based systems especially in all-inorganic and half-breed perovskites. These characteristics are essential to long-term operational stability in actual applications in outdoor photovoltaics, wearable electronics, and sensors in harsh environments [24, 25]. Moreover, lead-free compositions pose less of a threat to the environment, which is in line with international standards like RoHS and REACH and helps them go through the adoption process faster.

Lead-free perovskites have a wide range of advanced technologies, owing to their versatility. Lead-free perovskite solar cells have shown consistent increases in efficiency, stability, and scalability in photovoltaics, and tin- and bismuth-based systems have a specific potential in tandem and indoor energy-harvesting [26–28]. Lead-free perovskites in light-emitting devices have narrow emission lines, high color purity, and tuned photoluminescence and can be used in display and lighting applications [29, 30]. In other fields, lead-free perovskites are becoming popular in photocatalysis, radiation detection, thermoelectrics and memory devices, where the defect tolerance, and ionic migration characteristics can be used advantageously [31–33]. Their future application in flexible electronics, wearable devices and Internet-of-Things (IoT) platforms further highlight their enabling material aspect in future sustainable technologies. This concluding perspective emphasizes that while lead-free perovskite materials may not yet fully match the peak performance of lead-based counterparts, their environmental compatibility, structural diversity, and application versatility provide compelling advantages. Continued progress in defect passivation, interface engineering, compositional optimization, and scalable fabrication techniques is expected to bridge existing performance gaps. Ultimately, sustainable lead-free perovskites represent not merely substitutes for lead-based materials, but a distinct and evolving class of functional materials poised to drive innovation in next-generation devices.

Altogether, this book provides readers with the comprehensive understanding of synthesis, properties, performance/efficiency of the device and extensive literature review. The authors of the chapters to be included in this book will provide a comprehensive examination of lead free perovskites synthesis and characterizations and various applications. Connecting to the above, the book also reports the recent advancements in development of new perovskites, a promising technique for creating better materials suitable for various applications. Lastly, the book highlights new materials showcasing promising technologies and enhanced properties, unlocking potential possibilities in various domains.

16.2 Summary of the Work

Lead-free perovskite materials are sustainable alternatives to conventional lead-based perovskites, addressing critical environmental, health, and regulatory concerns while striving to maintain high functional performance. The content is systematically organized to guide readers from fundamental concepts to advanced applications, thereby presenting a holistic picture of the current research landscape and future prospects of lead-free perovskites across diverse technological domains. Figure 16.1 shows the various next generation device applications of perovskite materials.

Chapter 1 in this book establishes the conceptual and scientific base of the book by presenting the structural chemistry, electrical behavior, chemical stability and environmental benefits of lead free perovskites. These materials are grouped together in the chapter according not only to chemical composition (tin-, bismuth-, and antimony-based and hybrid organic metal-inorganic systems) but also to structure dimensionality (zero dimension, one dimensional, two dimensional, dimensional or three-dimensional). This chapter compares lead-free systems and traditional lead-based perovskites critically and identifies the inherent material differences that determine the performance of the device. It also highlights critical challenges like lower efficiency and stability that should be conquered thus formulating the scientific questions that are answered in the following chapters.

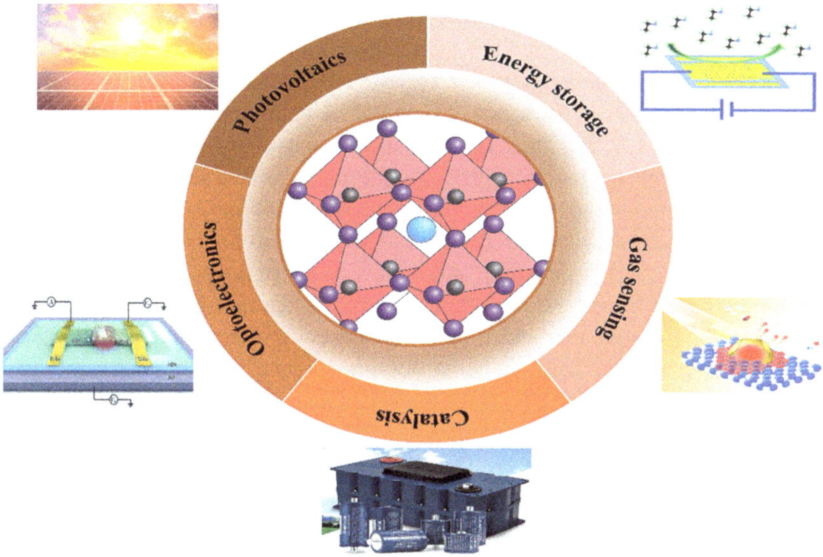

Fig. 16.1 Schematics of next-generation perovskite based devices

Based on these basics, Chap. 2 further elaborates on the synthesis approach, crystal structure and optoelectronic characteristics of perovskites that are lead-free. Sophisticated routes of fabrication such as solution processing, hot-injection and solid-state methods are discussed with respect to thin films and nanocrystals. There is a specific focus on Sn-, Ge-, Bi-, Sb-based systems and double perovskites and a focus on their enhanced environmental friendliness and structural stability. The chapter is a critical analysis on the unsolved issues including oxidation, lack of charge transport and performance differences compared to lead-based counterparts, and gives the direction of future research. Further, Chap. 3 focuses on the development of lead-free perovskite solar cells as environmentally sustainable alternatives to lead-based photovoltaic technologies. The chapter highlights the growing interest in these materials driven by concerns over lead toxicity and the need for eco-friendly energy solutions. Lead-free perovskites are discussed in the context of circular economy principles, particularly their potential for safer recycling and reduced hazardous waste generation. Although their power conversion efficiencies currently lag behind lead-based counterparts, the chapter emphasizes the rapid growth of research efforts aimed at overcoming performance limitations. A comprehensive literature review is presented, covering material systems, device architectures, and efficiency trends. In addition, a bibliometric analysis based on 1894 Scopus-indexed publications provides quantitative insights into research progress, emerging themes, and existing gaps, offering valuable guidance for future investigations in lead-free photovoltaic technologies.

Expanding the optoelectronic scope, Chap. 4 provides a detailed review of the progress achieved in lead-free perovskite-based optoelectronic devices, including light-emitting diodes, photodetectors, and emerging neuromorphic memristors. The chapter critically evaluates the suitability of alternative cations such as Sn, Ge, Bi, and Cu as replacements for toxic Pb^{2+}, with particular emphasis on their electronic band structures, optical absorption and emission characteristics, carrier transport behavior, and electrical performance. Device-level demonstrations are discussed to highlight advances in brightness, responsivity, switching behavior, and operational stability. Despite encouraging progress, the chapter identifies key performance bottlenecks related to charge recombination, material instability, and defect-induced losses. Finally, it outlines future strategies such as compositional engineering, interface optimization, defect passivation, and dimensional control to bridge the performance gap with lead-based systems, underscoring the central role of material innovation in enabling sustainable and high-performance optoelectronic technologies. In Chap. 5, the concept of environmental monitoring and sensing is covered, and the lead-free perovskites to detect gases are discussed. The benefits of perovskites based on nanoparticles as the lead-free material are discussed in the chapter, such as high surface area and sensitivity. The sensing mechanisms that include surface adsorption, ion exchange and trap modulation are examined in detail and issues of long term stability and selectivity under real world operating conditions. Chapter 6 is about the potential of the photocatalysis and renewable energy conversion by the use of lead-free perovskites. This chapter is a systematic discourse on synthetic paths, crystallographic classification, and property relation to structure that applies in photocatalytic applications. The importance is put on the knowledge of defect chemistry, separation of charges and stability and is critically concerned with scalability and processing constraints. The chapter gives the road map to create efficient and long lasting lead-free perovskite photocatalysts.

Energy storage technologies are comprehensively reviewed in Chap. 7, which examines lead-free perovskites as electrode materials for supercapacitors, rechargeable batteries, and hybrid energy storage systems. The chapter correlates crystal structure and compositional tunability with electrochemical performance metrics such as capacity, rate capability, and cycling stability. It also highlights future research directions to optimize these materials for high-performance, sustainable energy storage devices.

In Chap. 8, the book addresses lead-free perovskite thin-film transistors (TFTs), a critical component of next-generation electronics. The chapter discusses device architectures, charge transport mechanisms, fabrication techniques, and performance parameters, underscoring the potential of lead-free perovskites for displays and low-power electronics. Challenges related to mobility, stability, and large-area processing are critically examined.

Chapter 9 is devoted to lead-free perovskite piezoelectric materials and gives a specific examination to the major systems, which include KNN, BNT-BT, and BZT-BCT. There is the discussion of advanced processing techniques and domain engineering as well as phase boundary effects in terms of improved electromechanical performance. Significant practical uses in sensors, actuators, energy harvesters, and biomedical devices are also brought out in this chapter. Chapter 10 also explains the possibility of lead-free perovskites in hydrogen production and is an overview of photochemical, electrochemical, and photoelectrochemical water-splitting reactions.

The doping strategy, defect engineering, and surface modification are discussed as one of the methods to enhance efficiency and durability, making lead-free perovskites one of the promising candidates to be used in the creation of a sustainable hydrogen manufacturer. Chapter 11 deals with emerging computing paradigms and addresses lead-free perovskite-based memristors and neuromorphic devices. The chapter covers the materials choice, the device architecture, the fabrication methods, and the working principles of these materials, which have the potential to reduce power consumption, create flexible, scalable neuromorphic systems. Chapter 12 covers the transformation of waste heat into electricity with the use of lead-free perovskites, and makes a review of the developments in the thermoelectric performance. The essential parameters that affect the figure of merit (zT) are examined, and the perspectives of enhancing the thermoelectric efficiency by designing materials and nanostructuring are described. Chapter 13 focuses on the development of lead-free perovskites for magnetic applications, driven by the demand for environmentally sustainable and multifunctional materials. Substitution of toxic Pb^{2+} with magnetically active ions such as Bi^{3+}, Mn^{2+}, and Fe^{3+} enables diverse magnetic behaviors, including ferro-, antiferro-, and multiferroicity. Oxide-based and La-based double perovskites exhibit robust magnetic ordering and magnetoelectric coupling, supporting applications in spintronics and data storage. Magnetically doped halide perovskites are also explored for magneto-optical and quantum-related functionalities. Despite challenges related to low Curie temperatures and stability, continued compositional and synthetic optimization highlights the promise of lead-free perovskites for sustainable magnetic technologies. In Chap. 14, the book investigates lead-free perovskites for imaging and detection technologies, spanning applications from X-ray to near-infrared detection. The chapter emphasizes tunable optical properties, low-temperature processing, and improved environmental compatibility, positioning these materials as viable alternatives to conventional semiconductors. The application of lead-free perovskites in antenna technology is discussed in Chap. 15, where high-dielectric materials such as $BaTiO_3$- and BCZT-based perovskites are evaluated for antenna miniaturization and performance enhancement. Processing strategies and dielectric stability are emphasized in relation to next-generation wireless systems.

Lastly, Chap. 16 offers a detailed and futuristic evaluation of lead-free perovskite solar cells, including the material systems, synthesis pathways, efficiency development, commercialization issues and future opportunities. This chapter combines the content of prior chapters and emphasizes how the realization of lead-free perovskite photovoltaics is not solely an environmental priority but also a progress towards the realization of robust, scalable and high-performance solar technologies. The highlights of all the chapter of the current book are illustrated in Table 16.1. To summarize, the book provides an interdisciplinary and coherent outlook on lead-free perovskites, encompassing both fundamental science and device engineering and sustainability. The systematic approach to the design and optimization of materials and application-oriented issues makes the work a source of indisputable information to the researcher, engineer, and policymaker who wants to focus on the development of perovskite-based technologies that are environmentally friendly.

Table 16.1 Major findings and highlights of the current book

Chapters	Highlights
Chapter 1	• Lead-free perovskites provide eco-friendly alternatives to Pb-based materials. • Composition and dimensionality strongly affect performance. • Stability and efficiency remain key challenges.
Chapter 2	• Sn-, Ge-, bi-, Sb-, and double perovskites show tunable optoelectronic properties. • Advanced synthesis improves film quality and stability. • Oxidation and charge transport limit performance.
Chapter 3	• Lead-free PSCs offer sustainable photovoltaics. • Efficiency lags behind Pb-based PSCs. • Interface and material engineering are crucial.
Chapter 4	• Lead-free perovskites enable LEDs, photodetectors, and neuromorphic devices with reduced toxicity. • Sn, Ge, bi, and cu substitutions provide suitable optical and electrical properties. • Device stability and performance optimization are critical for commercialization.
Chapter 5	• Lead-free perovskites are promising gas-sensing materials. • Nanostructures enhance sensitivity. • Stability limits practical deployment.
Chapter 6	• Lead-free perovskites exhibit tunable optoelectronic properties for photocatalysis. • Structure–defect–property correlations govern catalytic performance. Stability, scalability, and processing control are essential for practical deployment.
Chapter 7	• Lead-free perovskites show good ion transport. • Structural tunability enhances storage performance. • Electrode optimization is required.
Chapter 8	• Pb-free perovskites support sustainable TFTs. • Charge transport controls device performance. • Large-area stability remains challenging.
Chapter 9	• Lead-free perovskites replace toxic PZT. • Phase engineering improves piezoelectric response. • Suitable for sensors and energy harvesting.
Chapter 10	• Lead-free perovskites enable water splitting. • Doping and defect engineering improve efficiency. • Long-term stability is critical.
Chapter 11	• Lead-free perovskites enable low-power memristors. • Multiple systems show stable switching. • Scalability and endurance need improvement.
Chapter 12	• Lead-free perovskites demonstrate competitive thermoelectric performance ($zT \approx 0.8–1.0$). • Seebeck coefficient, electrical conductivity, and thermal conductivity govern efficiency. • Nanostructuring and compositional tuning are key to future improvements.
Chapter 13	• Lead-free perovskites exhibit ferro-, antiferro-, and multiferroic behavior. • Compositional engineering enhances magnetic ordering and magnetoelectric coupling. • Applications include spintronics, sensors, and data storage technologies.

(continued)

Table 16.1 (continued)

Chapters	Highlights
Chapter 14	• Lead-free perovskites enable tunable, low-toxicity imaging materials. • Solution-processable fabrication lowers cost and enhances scalability. • Applications span X-ray, visible, and near-infrared detection regimes.
Chapter 15	• Lead-free perovskites provide high dielectric constants for antenna miniaturization. • Materials such as $BaTiO_3$ and BCZT offer low dielectric loss and stability. • These materials enable compact, efficient next-generation wireless antennas.
Chapter 16	• Sn-based PSCs exceed 15% efficiency, while bi/Sb systems offer superior stability. • Scalable, eco-friendly synthesis routes are critical for commercialization. • Alloy engineering, tandem structures, and AI-assisted design will drive future progress.

16.3 Concluding Remarks

The book is a complete summary of the lead-free perovskite materials as sustainable alternatives to lead-based systems, which is in response to critical environmental and health issues, yet does not loses its functionality. Blending the basic ideas with recent developments in the field of synthesis, characterization and engineering of devices, the book draws attention to the high speed of development and the increased versatility of lead-free perovskites in diverse applications, such as energy conversion and storage, optoelectronics, sensing, and multifunctional devices. Despite the problems of efficiency, stability, and scalability, the overall knowledge offered in the chapters suggests the unmistakable opportunities of improving the material and developing technologies. Overall, this work consolidates current knowledge and offers a forward-looking perspective, positioning lead-free perovskites as promising materials for enabling sustainable, high-performance next-generation technologies.

References

1. A. Kojima et al., J. Am. Chem. Soc. **131**, 6050–6051 (2009)
2. NREL Best Research-Cell Efficiency Chart, 2024
3. M.A. Green et al., Prog. Photovolt. **30**, 3–12 (2022)
4. B.R. Sutherland, E.H. Sargent, Nat. Photonics **10**, 295–302 (2016)
5. Z.K. Tan et al., Nat. Nanotechnol. **9**, 687–692 (2014)
6. L. Dou et al., Science **349**, 1518–1521 (2015)
7. A. Babayigit et al., Energy Environ. Sci. **9**, 2452–2466 (2016)
8. W. Ke, M.G. Kanatzidis, Nat. Commun. **10**, 965 (2019)
9. D.P. McMeekin et al., Science **351**, 151–155 (2016)
10. A.H. Slavney et al., J. Am. Chem. Soc. **138**, 2138–2141 (2016)
11. A.M. Ganose et al., Chem. Mater. **30**, 3827–3835 (2018)
12. G. Volonakis et al., J. Phys. Chem. Lett. **7**, 1254–1259 (2016)
13. N.K. Noel et al., Energy Environ. Sci. **7**, 3061–3068 (2014)

14. F. Hao et al., Nat. Photonics **8**, 489–494 (2014)
15. B.W. Park et al., Adv. Mater. **27**, 6806–6813 (2015)
16. A.J. Lehner et al., Chem. Mater. **27**, 7137–7148 (2015)
17. M.R. Filip, F. Giustino, Phys. Rev. B **94**, 140302 (2016)
18. E.T. McClure et al., Chem. Mater. **28**, 1348–1354 (2016)
19. G. Volonakis et al., J. Phys. Chem. Lett. **8**, 772–778 (2017)
20. W. Pan et al., Nat. Photonics **11**, 726–732 (2017)
21. S.D. Stranks, H.J. Snaith, Nat. Nanotechnol. **10**, 391–402 (2015)
22. L.M. Herz, Annu. Rev. Phys. Chem. **67**, 65–89 (2016)
23. M. Grätzel, Nat. Mater. **13**, 838–842 (2014)
24. X.G. Zhao et al., J. Am. Chem. Soc. **139**, 6718–6725 (2017)
25. Z. Yang et al., Adv. Funct. Mater. **30**, 1908765 (2020)
26. W. Ke et al., Adv. Mater. **30**, 1803230 (2018)
27. R. Wang et al., Energy Environ. Sci. **12**, 2054–2079 (2019)
28. M.C. Jung et al., Adv. Energy Mater. **11**, 2003060 (2021)
29. Y. Cao et al., Nature **562**, 249–253 (2018)
30. G. Li et al., Adv. Opt. Mater. **8**, 1902089 (2020)
31. P.V. Kamat, J. Phys. Chem. Lett. **10**, 447–458 (2019)
32. H. Wei et al., Nat. Photonics **11**, 315–321 (2017)
33. J. Yin et al., Nat. Rev. Mater. **6**, 977–999 (2021)

GPSR Compliance
The European Union's (EU) General Product Safety Regulation (GPSR) is a set of rules that requires consumer products to be safe and our obligations to ensure this.

If you have any concerns about our products, you can contact us on

ProductSafety@springernature.com

In case Publisher is established outside the EU, the EU authorized representative is:

Springer Nature Customer Service Center GmbH
Europaplatz 3
69115 Heidelberg, Germany

www.ingramcontent.com/pod-product-compliance
Ingram Content Group UK Ltd.
Pitfield, Milton Keynes, MK11 3LW, UK
UKHW021614260926
475829UK00010B/15
9783032248725